ADVANCES IN CHEMICAL PHYSICS

VOLUME 141

ADVANCES IN
CHEMICAL PHYSICS

VOLUME 141

Series Editor

STUART A. RICE

Department of Chemistry
and
The James Franck Institute
The University of Chicago
Chicago, Illinois

 **WILEY**

A JOHN WILEY & SONS, INC., PUBLICATION

Library of Congress Catalog Number: 58-9935

ISBN: 978-0470-41713-3

Printed in the United States of America

10 9 8 7 6 5 4 3 2 1

CONTRIBUTORS TO VOLUME 141

BIMAN BAGCHI, Solid State and Structural Chemistry Unit, Indian Institute of Science, Bangalore 560 012, India

GAVIN A. BUXTON, Department of Chemistry, University of Durham, DH1 3LE, United Kingdom

DWAIPAYAN CHAKRABARTI, Department of Chemistry, University of Cambridge, Cambridge CB2 1EW, United Kingdom

NIGEL CLARKE, Department of Chemistry, University of Durham, DH1 3LE, United Kingdom

DERRICK S. F. CROTHERS, Queen's University Belfast, Belfast, Northern Ireland BT7 1NN

R. A. W. DRYFE, School of Chemistry, University of Manchester, Manchester M13 9PL, United Kingdom

JOHN H. D. ELAND, Physical and Theoretical Chemistry Laboratory, University of Oxford, Oxford, OX1 3QZ, United Kingdom

VOLKER ENGEL, Universität Würzburg, Institut für Physikalische Chemie and Röntgen Research Centre for Complex Material Systems, Am Hubland, 97074 Würzburg, Germany

VLADIMIR I. GAIDUK, Institute of Radio Engineering and Electronics, Russian Academy of Sciences, Fryazino, 141120 Moscow Region, Russia

CHRISTOPH MEIER, Laboratoire Collisions, Agrégats et Reactivité, IRSAMC, Université Paul Sabatier, 31062 Toulouse, France

SANG-HEE SHIM, Department of Chemistry, University of Wisconsin—Madison, Madison, Wisconsin 53706 USA

DAVID B. STRASFELD, Department of Chemistry, University of Wisconsin—Madison, Madison, Wisconsin 53706 USA

DAVID J. TANNOR, Department of Chemical Physics, Weizmann Institute of Science, 76100 Rehovot, Israel

MARTIN T. ZANNI, Department of Chemistry, University of Wisconsin—Madison, Madison, Wisconsin 53706 USA

INTRODUCTION

Few of us can any longer keep up with the flood of scientific literature, even in specialized subfields. Any attempt to do more and be broadly educated with respect to a large domain of science has the appearance of tilting at windmills. Yet the synthesis of ideas drawn from different subjects into new, powerful, general concepts is as valuable as ever, and the desire to remain educated persists in all scientists. This series, *Advances in Chemical Physics*, is devoted to helping the reader obtain general information about a wide variety of topics in chemical physics, a field that we interpret very broadly. Our intent is to have experts present comprehensive analyses of subjects of interest and to encourage the expression of individual points of view. We hope that this approach to the presentation of an overview of a subject will both stimulate new research and serve as a personalized learning text for beginners in a field.

STUART A. RICE

CONTENTS

NEW ADVANCES IN MID-IR PULSE SHAPING AND ITS APPLICATION TO 2D IR SPECTROSCOPY AND GROUND-STATE COHERENT CONTROL

DAVID B. STRASFELD, SANG-HEE SHIM, AND MARTIN T. ZANNI

Department of Chemistry, University of Wisconsin—Madison, Madison, Wisconsin 53706

CONTENTS

I. INTRODUCTION

Science often advances only as quickly as technology allows. This trend is especially evident in the field of molecular spectroscopy, where the science of studying molecular structures, dynamics, and chemical reactions has progressed in step with the technology of manipulating electromagnetic fields. For example, nuclear magnetic resonance (NMR) spectroscopy, which is now used to determine the structures of proteins, was made possible by radio-frequency technology developed in the 1950s. Likewise, laser-guided control of excited-state chemistry became possible in the 1990s with technology for temporally shaping visible and

Advances in Chemical Physics, Volume 141, edited by Stuart A. Rice

near-infrared (near-IR) femtosecond laser pulses [1–3]. In this review chapter, we highlight a new generation of shaping technology—femtosecond mid-infrared (mid-IR) pulse shaping—and review some of the novel spectroscopies and experiments that can now be implemented. With this new shaper, ground-state vibrational motions can be coherently controlled [4] and sophisticated new two-dimensional infrared (2D IR) spectroscopies implemented [5]. Experiments will be reviewed that incorporate mid-IR pulse shaping to expand our knowledge of vibrationally excited molecules in condensed phases, give us control over vibrational motions, and allow us to probe intricate molecular motions such as those that occur during protein folding.

Our desire for pulse shaping in the mid-IR stems from the impact such technology could have on multidimensional IR spectroscopies and coherent control of ground-state vibrations, among other possibilities. In the absence of a mid-IR pulse shaper, the only shaped pulses that are straightforward to generate in the mid-IR are linearly chirped with either material dispersion or pairs of gratings. In fact, the first experiments with linearly chirped pulses were performed in 1998 by Heilweil and co-workers [6]. They showed that it was possible to improve vibrational excitation and even invert vibrational populations with linearly chirped picosecond pulses in a process that has become known as ladder climbing. Ladder climbing is now possible using broad-bandwidth femtosecond pulses so that much higher vibrational states can be accessed. It is often routine to see excitations reach $\upsilon = 6$ or 7 in metal–carbonyl systems even without chirped pulses. While impulsive interactions that rely on transform limited electric fields can excite numerous vibrational quanta, achieving control of specific vibrational motions within complex potential landscapes requires equally complex laser pulses.

An interesting consequence of the slow development in mid-IR pulse-shaping technology is that the theory for manipulating molecules lies far ahead of experiments [1, 2, 7]. There are a number of theoretical research articles simulating possible experiments that might be done with shaped mid-IR light. These include articles using shaped pulses to enhance specific features in 2D IR spectra [7], controlling vibrational excitation on ground molecular states [8, 9], controlling chemical reactions like proton transfer [10], condensing Bose–Einstein condensates in gases [11], and using vibrations for quantum computing [12, 13]. Complicated pulse shapes have also been suggested for enhanced ladder climbing, such as those by Meier and Heitz, who predicted that pulses with sophisticated time and frequency profiles can improve vibrational excitation and attain at least partial quantum selectivity [8]. Many of these control experiments rely on aspects of abiabatic rapid passage and STIRAP (stimulated Raman adiabatic passage) methodologies [1, 14–16]. Theory could only be compared to experiment in the case of ladder climbing, as arbitrarily shaped mid-IR pulses have only recently become available. Thus, while a few experiments have been performed with shaped mid-IR pulses, theory suggests that the field of mid-IR-based coherent control is in its infancy.

Femtosecond pulse shaping in the visible and near-IR regions of the spectrum has been possible for more than a decade [17]. Indeed, there are numerous methods available for shaping visible pulses. The most widely adopted method (Fig. 1) disperses the femtosecond pulses using a grating so that the pulse spectrum is spread over a finite spatial distance. With the spectrum spatially resolved, the amplitudes and phases in the pulse can be manipulated in the Fourier plane using a programmable modulator and then transformed back into the time-domain using a second grating. The final pulse shape is simply the inverse Fourier transform of the mask, convoluted with the pulse spectrum. Thus, waveforms can be synthesized with a level of complexity only constrained by the bandwidth of the pulse and the specifications of the modulator, including its frequency resolution, phase resolution, and absorption profile.

The major problem with extending these shaping methods into the mid-IR is the availability of a suitable modulator. There are many different types of phase and amplitude modulators that work in or near the visible, including liquid crystal modulators (LCM), TeO_2 and fused silica acoustooptic modulators

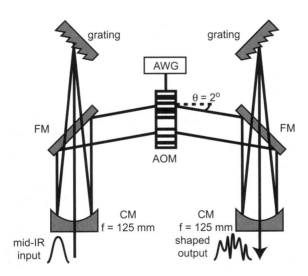

Figure 1. Experimental setup of the pulse shaper: AOM, Ge acousto-optic modulator; AWG, arbitrary waveform generator; CM, cylindrical mirror; grating, 200-g/mm ruled grating. Pulses are dispersed by the grating, focused at a CM, shaped at the AOM, collimated at a CM, and recombined at a grating.

(AOM), and deformable mirrors, but none are suitable for the mid-IR: Liquid crystal modulators transmit in the visible and near-IR, TeO_2 and fused silica acoustooptic modulators operate in the visible and ultraviolet regions of the spectrum, respectively, and deformable mirrors only allow full-phase control for wavelengths <900 nm. Even though these modulators do not currently operate directly in the mid-IR, they can be used to indirectly create modulated mid-IR pulses by shaping in the near-IR and then transferring that shape to the mid-IR via difference frequency mixing [18–21]. Parametric transfer makes a large-wavelength regime available for phase and amplitude tailoring. However, difference frequency mixing scales nonlinearly with electric field, and thus the intensity of mid-IR light generated in this manner depends on the desired pulse shape. Furthermore, the frequency resolution is suboptimal, because the near-IR pulses are convoluted through the finite thickness of the mixing crystal. As a result, indirect shaping methods have limited utility in IR spectroscopy and coherent control. It is better to shape directly in the mid-IR so that the resolution and efficiency is dictated only by the design of the pulse shaper.

In the past two years, we have developed a pulse shaper that operates directly on mid-IR light. Our design is largely based on the work of Warren Warren [22, 23], who developed pulse shaping for visible frequencies using a TeO_2 acoustooptic modulator. Our shaper uses a germanium acoustooptic modulator that permits shaping between about 2 and 18 μm with fidelity comparable or better than most visible masks. By working directly in the mid-IR rather than by difference frequency mixing, it is possible to create intense and accurately shaped mid-IR pulses. With this shaper, many of the theoretically proposed experiments are now possible. In what follows, we first review some simple experiments demonstrating that ground-state vibrational motion can be controlled with even very simply shaped pulses. We then illustrate how our mid-IR shaper can be used to simplify and automate 2D IR data collection for improved insight into vibrational couplings and molecular structures. We think that this new technology will permit rapid advances in both controlling and understanding ground-state chemistry.

II. MID-IR PULSE-SHAPING TECHNOLOGY

The germanium pulse shaper works by modulating the intensity and phase of the frequency spectrum of each mid-IR pulse. The broader the frequency spectrum, the more intricately shaped the pulses can be. We generate mid-IR pulses with >500-nm bandwidth using a modified optical parametric amplifier (OPA) pumped by a short-pulse Ti:Sapphire regenerative amplifier (800 nm, <50 fs). The optics and nonlinear crystals in the OPA have been optimized for large bandwidth and high mid-IR intensity (typically 4.5 μJ at 6 μm when pumped with 800 μJ of 800 nm) [24]. These pulses serve as the shaper input.

A schematic of the pulse shaper is shown in Fig. 1. The pulse shaper functions through the use of a pair of diffraction gratings (200 grooves/mm) and a pair of cylindrical mirrors ($f = 125$ mm). In this geometry the gratings, cylindrical mirrors and AOM are all separated by the mirror focal length with the Ge AOM placed in the center of the Fourier plane, as is typical of the 4-f geometry. The shaper design incorporates gratings placed in quasi-Littrow configuration and tilted vertically, allowing for easy adjustment to multiple light sources, such as HeNe or mid-IR. Before inserting the Ge AOM, the 4-f geometry of the cylindrical mirrors and gratings was initially set using three parallel propagating HeNe beams diffracted from the first grating in 7th, 8th, and 9th order. The Ge AOM, when placed in the Fourier plane, deflects the mid-IR at a Bragg angle of $\sim 2°$ with amplitude and phase according to the acoustic wave passing through the crystal. Since the HeNe beam does not transmit through Ge, we compensate for the $2°$ angular deviation using the folding mirrors (FM, Fig. 1) immediately before and after the Ge AOM. To create a desired mid-IR pulse, the properly shaped acoustic wave is coupled into an RF amplifier using a 300-Msample/s arbitrary waveform generator. The amplified waveform drives a piezoelectric bonded to the Ge crystal, creating an acoustic wave across the crystal aperture. Since the acoustic wave velocity is negligible relative to the mid-IR pulses traversing through the Ge AOM, this wave acts like a static modulated grating. Thus, by shaping the acoustic wave, the desired mid-IR frequencies are deflected with amplitude and phase specified by that acoustic wave. As a result, the phase and intensity of the shaped mid-IR pulses are set by the phase and intensity of the acoustic wave.

The frequency resolution of the shaper is dictated by the product of the time aperture Δt and the usable RF bandwidth, Δf. The acoustic wave propagates along the length of the AOM at 5.5 mm/μs, the speed of sound in Ge. Given a crystal aperture of 5.5 cm $\times$ 1 cm, the time aperture of the AOM is $\Delta t = 10$ μs. The AOM was designed to operate at a 75-MHz center frequency and has a bandwidth of $\Delta f = 50$ MHz. Thus, for our design, $\Delta t \cdot \Delta f = 500$, which indicates a maximum of 500 equivalent resolvable elements across the crystal aperture.

III. PERFORMANCE OF Ge AOM SHAPER

Figure 2 illustrates the phase and amplitude resolution of our device. Shown in this figure is a spectrum collected for a shaped pulse generated by a series of 3-ns acoustic waves separated by 665 ns, shown in Fig. 2b. Each wave deflects a particular frequency in the mid-IR pulse spectrum to create a comb of frequencies. The resulting peaks each have an fwhm of 5 nm and are separated by 63 nm, demonstrating the high resolution and contrast ratio of the pulse shaper. Considering that the 3-ns duration of each acoustic pulse is smaller than

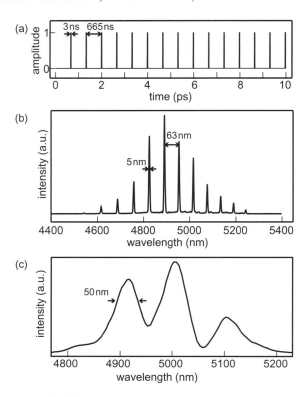

Figure 2. (a) Periodic RF signal used to modulate the frequency spectrum. (b) Spectrum resulting from the modulated RF signal in (a). (c) Pulse shaped by two π-jumps in an effort to create a comb spectrum using parametric transfer.

the design resolution, we determine an actual resolution of 190 equivalent pixels across the entire aperture, after adjustment for monochromator resolution. The ideal resolution is not reached because under the focusing conditions of our 4-f geometry, we have a spot size of 250 μm at the Ge crystal, which is roughly twice the size of a resolvable frequency element. Tighter focusing at the crystal by either expanding the mid-IR beam or using shorter-focal-length cylindrical mirrors will improve the resolution. Nonetheless, the intensity and phase of each of these resolvable elements can be adjusted to create sophisticated shaped pulses. For example, the resolution dictates the maximum time delay, τ, by which two pulses in a pulse pair can be separated, where $\tau = 2\pi/\omega$. Thus, with 190 resolvable elements, we can generate a pair of pulses separated by a time delay as large as 13 ps. With 500-pixel resolution, a delay of 35 ps can be achieved. Given that most condensed-phase vibrational lifetimes are <10 ps, our configuration is adequate for properly resolving spectral features without obfuscation from the ringing that arises with temporal truncation.

The frequency comb spectrum also serves to calibrate the position of the Ge AOM aperture to the wavelength. To accurately shape in the time domain, a second calibration is also necessary to compensate for the chirp in the input mid-IR pulses emanating from the OPA (which are not usually compressed) as well as a linear chirp created by the Bragg deflection angle of the AOM. To account for these chirps, we use an autocorrelator with a 0.5-mm-thick type I AgGaS$_2$ ($\theta = 33°$) doubling crystal and vary the second- and third-order dispersion coefficients (ϕ_2 and ϕ_3) to maximize the second harmonic signal. Neglecting acoustic nonlinearities, the desired temporal shape is simply a Fourier transform of the product of the acoustic profile and frequency spectrum, adjusted by the two calibrations.

IV. COMPARISON TO PARAMETRIC TRANSFER

For comparison, we also show in Fig. 2c the comb spectrum generated not with our Ge AOM shaper, but by parametric transfer from the near-IR. Before developing our Ge AOM shaper, we first began creating shaped mid-IR pulses following the methods developed by other groups using difference frequency mixing to shift a shaped near-IR pulse into the mid-IR [18–21]. Following their methods, we used a 4-f geometry and a phase controllable liquid crystal modulator to shape the signal beam in an optical parametric amplifier, which was then difference frequency mixed with the idler beam stretched to 1 ps. The spectrum in Fig. 2c was created by phase modulating the signal beam in an effort to place a series of evenly spaced dips in the mid-IR spectrum. At best, we were able to create a frequency profile consisting of three distinct peaks, each approximately 50 nm in width. Clearly, the frequency resolution attainable from parametric transfer is much poorer than by shaping directly in the mid-IR, although we do note that other groups have reported better resolution using indirect transfer [18, 20]. Another drawback of parametric transfer is that the intensity of the tailored mid-IR depends strongly on its pulse shape, because the transfer is a nonlinear optical process. As a result, much more intense pulses can be made when shaping directly in the mid-IR.

V. COHERENT CONTROL OF GROUND-STATE VIBRATIONAL DYNAMICS

Advances in the theory and implementation of femtosecond pulse shaping have spurred a revolution in the coherent control of atomic and molecular processes. By creating complex vibrational and/or electronic wavefunctions with tailored pulses, it is now possible to guide many chemical and physical processes such as photodissociation, ionization, electron transfer, fluorescence and vibrational excitation [1, 3]. Most studies so far have focused on the excited electronic states of systems because the wavelength ranges accessible to existing pulse shapers only span the visible and near-IR spectral regions. This limited wavelength range is unfortunate, because simulations suggest that it should be possible to control chemical reactions using ground-state vibrations [10]. Among these simulations are studies concluding that tailored pulse shapes can control intra- and intermolecular proton transfer [10, 25–28], that electron transfer can be controlled using vibrational excitation [29], and that mid-IR excitation can direct isomerization [30, 31]. Thus, extending the techniques developed for visible pulse shaping into the mid-IR promises to open a new class of experiments geared toward understanding and optimizing ground state control.

The prior lack of a suitable mid-IR pulse shaper hindered scientific progress toward ground-state control. Without a shaper that works directly in the mid-IR, experimental control of ground-state vibrational excitation is limited to using either stimulated Raman excitation with shaped Stokes and anti-Stokes pulses [32] or vibrational up-pumping with mid-IR pulses that are linearly chirped with material dispersion or a grating pair [6, 33–35]. While useful, neither of these methods provides the intensity or flexibility in pulse shape necessary for arbitrary generation of ground-state wavepackets. Taking advantage of the capabilities afforded us by a mid-IR pulse shaper, we set about optimizing vibrational ladder climbing in the ground state.

Vibrational ladder climbing is one of the most fundamental coherent control strategies. Most experimental schemes proposed for chemical control, whether it is proton transfer, electron transfer, or isomerization, require that vibrational energy be selectively deposited into one or more eigenstates with programmable phase and amplitude. Ladder climbing with linearly chirped pulses is the most intuitive way of accomplishing this task. The vibrational selection rules of harmonic potentials limit vibrational excitation to one quantum at a time. Therefore, at least near the minima of potential energy surfaces, vibrational excitation requires a driving field that is initially on resonance with the fundamental frequency and chirped to match the anharmonicity of the potential, thereby sequentially exciting the system through a series of vibratonal states analogous to climbing the rungs on a ladder. Linearly chirped mid-IR pulses can be easily generated with materials or a grating pair, and the effect of linear chirp on ladder climbing efficiency has been studied [6, 33]. Experiments on $W(CO)_6$, carbonmonoxy myoglobin (Mb-CO), and other compounds have found that a negative chirp improves vibrational excitation [6, 33–35]. If enough energy is deposited, photodissociation occurs, at least in the gas phase [34, 35].

One might expect that linearly chirped pulses are sufficient to ladder climb Morse potentials, because the slope of the chirp can be adjusted to match the anharmonic spacing of the Morse potential (the frequency difference between vibrational levels υ and $\upsilon + 1$ is constant because Morse potentials only have a single anharmonic term in their expansion). Following this line of thought, a more sophisticated pulse would only be necessary for potentials with more complex curvature, such as at high excitation energies where potentials usually deviate from Morse behavior or on multidimensional surfaces where the curvature depends on couplings to other modes. However, recent simulations by Meier and Heitz show that complex pulse shapes dramatically increase ladder climbing efficiency even on Morse potentials [8]. Shown in Fig. 3a is the Wigner diagram of a shaped pulse, optimized to maximize the vibrational excitation of the carbon monoxide stretch bound to the heme of a myoglobin protein (Mb). The Wigner

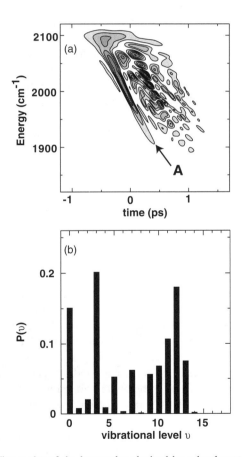

Figure 3. (a) Wigner plot of the laser pulse obtained by a local control algorithm used to optimize higher-level vibrational populations in carbon monoxyhemoglobin. (b) The vibrational distribution created by the laser pulse depicted in (a). (These figures are reproduced with permission of C. Meier and M.-C. Heitz.)

diagram plots the frequency of the shaped pulse as a function of time, revealing the pulse's complex shape. The slanted feature, labeled A, has a linearly chirped intensity that is the traditional requirement for ladder climbing, but intertwined in time and frequency with A are a series of features with complex shapes. Shown in Fig. 3b is the vibrational distribution that would be created by the shaped pulse. What is remarkable about this distribution is that it is peaked at high vibrational levels ($\upsilon = 13$), it is inverted (there is more population in vibrational states $\upsilon > 8$ than in $\upsilon < 8$), and it has some states that are preferentially populated over their neighbors (such as $\upsilon = 3$). These three capabilities are the necessary ingredients for successful ground-state coherent control: high vibrational excitation to surmount chemical barriers, population inversion to maximize product yield, and population selectivity to control the deposition of energy. Simulations with linearly chirped pulses show none of these effects; they have a vibrational distribution that is peaked at $\upsilon = 1$ and monotonically decreases with quantum number (not shown). Moreover, these simulations were carried out under the experimental conditions recently used to study ladder climbing of myoglobin-CO with linearly chirped mid-IR pulses (2.2 µJ with a 40-µm focal spot) [33]. With our pulse shaper, we achieve comparable pulse intensities and, additionally, are capable of controlling the phase and amplitude of our pulses in seeking to control ladder-climbing processes.

Simulations such as those by Meier and Heitz [8] suggest that mid-IR shaping technology is currently the limiting factor in ground-state coherent control. We now have the experimental capabilities to test complex waveforms and to begin testing the limits of ladder climbing. Besides the complexity of waveforms, there will still be hurdles to extremely high vibrational excitation, set by the dynamics of the molecules themselves. These hurdles include (1) population relaxation times that will redistribute the deposited vibrational energy even while the pulse is still interacting with the system, (2) solvent interactions that will dephase the coherences, (3) degenerate modes that may hinder excitation on multidimensional potentials, (4) correlations between fluctuations of energy levels that may limit the number of molecules efficiently excited in an ensemble, and (5) the broad absorption bands of condensed phase systems that could reduce mode selectivity. These factors will need to be characterized and understood to optimize ground-state coherent control. For this reason, the experiments presented below are aimed at both optimizing ground-state coherent control as well as understanding how vibrational dynamics limit coherent control.

To date, our work has focused mostly on the selective vibrational excitation of $W(CO)_6$ in hexane. $W(CO)_6$ exhibits three degenerate T_{1u} antisymmetric carbonyl stretches whose oscillation strengths are enhanced by the metal center. $W(CO)_6$ is a good model system for ladder climbing due to its high signal strengths and the fact that the anharmonicity of the T_{1u} mode is larger than the linewidths of the sequence bands. Thus, individual $\upsilon, \upsilon + 1$ bands are resolved to greater than $\upsilon = 8$. Shown in Fig. 4a is the pump-probe spectrum measured for a transform limited pump pulse (55-fs duration). The spectrum consists of a negative feature at 1983 cm^{-1} followed by a series of positive and progressively weaker peaks at lower frequencies. This spectrum is typical of femtosecond pump-probe studies. The negative feature corresponds to a bleach of the ground state and $\upsilon = 1$ to 0 stimulated emission. The positive peaks are sequence bands arising from $\upsilon = 1$–2, 2–3, 3–4, and so on, absorptions, the first three appearing at 1970, 1952, and 1933 cm^{-1}. Peak spacings increase due to the character of the normal mode and match previously reported data.

The absorption bands in Fig. 4a correspond to the differences in populations between successive vibrational levels. To interpret the pump-probe spectra, one needs to consider that the probe pulse interrogates the vibrational populations through both absorption $(\upsilon \rightarrow \upsilon + 1)$ and stimulated emission $(\upsilon + 1 \rightarrow \upsilon)$. Absorption causes a positive peak in the pump-probe spectrum, whereas stimulated emission peaks are negative. Thus, a positive peak signifies that $\upsilon + 1$ has a smaller population than υ. Conversely, a negative peak appears when $\upsilon + 1$ has a larger population than υ (population inversion). Therefore, the decreasing intensities of the positive peaks in Fig. 4a indicate that the pump pulse excites vibrational levels up to $\upsilon \sim 6$, which is typical for transform limited femtosecond pulses. The $\upsilon = 0$–1 transition is always negative because the spectrum is a difference between the probe spectrum with and without the pump. It is always negative so long as population is transferred out of $\upsilon = 0$.

Each spectrum in Fig. 4a can be collected simultaneously using an IR array detector. Thus, the relative populations of all vibrational levels accessible within the bandwidth of our ~ 50-fs pulses can be monitored at the same time with every laser shot. Since the data are generated very rapidly, this configuration allows for coherent control strategies to be implemented using a genetic learning algorithm and a feedback loop. Using the difference in targeted peak amplitudes as the fitness parameter in our optimization studies, we were able to create bleaches (and, consequently, population inversions) at the $\upsilon = 1$–2 and $\upsilon = 2$–3 absorptions as evidenced in Fig. 4a. Peak ratios were improved by 2.2–5.8 times the transform limited pulse.

We can also improve overall ladder climbing efficiency by targeting higher-level absorptions, such as $\upsilon = 7$–8. By using the difference between the $\upsilon = 7$–8 peak height at 1824 cm^{-1} and an adjacent trough at 1835 cm^{-1} as our fitness parameter, population in $\upsilon = 8$ could be maximized separate from processes leading to population in other vibrational levels. Figure 4b offers an expanded view of the absorption bands produced by a transform limited pulse as compared to those produced by a pulse optimized to enhance ladder climbing. The result of optimization enhances not only the population of $\upsilon = 8$, but also the population of vibrational levels lying lower in the ground-state potential. It is

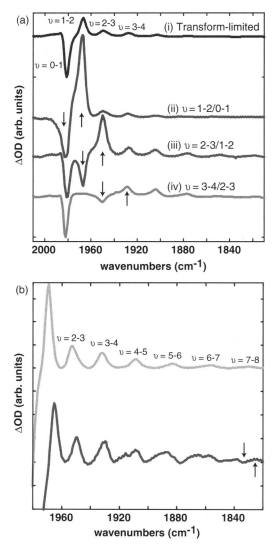

Figure 4. (a) A series of pump probe spectra for $W(CO)_6$ in which (i) a transform limited pulse was implemented, (ii) a pulse optimizing $\upsilon = 1–2$ versus $\upsilon = 0–1$ was implemented, (iii) a pulse optimizing $\upsilon = 2–3$ versus $\upsilon = 1–2$ was implemented, and (iv) a pulse optimizing the ratio of $\upsilon = 3–4$ versus $\upsilon = 2–3$ was implemented. (b) Comparison of (top) the absorptive features in a transform limited pump probe spectrum and (bottom) the features in a spectrum in which the $\upsilon = 7–8$ feature was optimized relative to an adjacent minima. Arrows indicate the features in the pump-probe spectra that were optimized relative to one another, where an up arrow signifies maximization and a down arrow signifies minimization.

also interesting to note the optimized pulse appears to create a hole in the $\upsilon = 4$–5, $\upsilon = 5$–6, $\upsilon = 6$–7, and $\upsilon = 7$–8 absorption bands. It has been established that the $\upsilon = 0$–1 transition is homogeneously broadened, which is not affected by hole burning. From our results it appears that hole burning is possible at $\upsilon = 4$–5 and above, suggesting that inhomogeneous broadening is beginning to contribute. Although this needs to be explored further before reaching a definitive conclusion, such an effect would signify that the nonharmonic character of the potential is beginning to contribute to the frequency fluctuations induced by the environment.

To better understand how the pulses to which the genetic learning algorithm converged induce population transfer over the duration of the pulse/sample interaction, we systematically explored the population transfer induced by each part of the pump through a series of experiments using truncated pulses. The phase and intensity mask used to generate an optimized pulse is represented mathematically by the following expression:

$$E_{opt}(\omega) = E_{in}(\omega)H_{opt}(\omega)\exp\lfloor i\phi_{opt}(\omega)\rfloor \tag{1}$$

where the input pulse, $E_{in}(\omega)$, is modulated according to the optimized phase and intensity masks, ϕ_{opt} and H_{opt}, respectively. The variable space spanned by ϕ_{opt} and H_{opt} is searched using a genetic algorithm and a feedback loop to generate the final electric field $E_{opt}(\omega)$ that maximizes a desired process. To understand how $E_{opt}(\omega)$ reaches the desired output, we use a mask to truncate the optimized pulse in the time domain using the following equation:

$$M_{trun}(\omega) = \frac{2\pi E_{in}(\omega)}{\int\limits_{-\infty}^{\infty}\left[\int\limits_{-\infty}^{\infty}E_{opt}(\omega)e^{-i\omega t}d\omega\right]_{=0 \text{ for } t>a}e^{i\omega t}dt} \tag{2}$$

where a (in units of time) sets the point at which the pulse is truncated in time. Figure 5 shows an optimized pulse truncated in steps of 0.5 ps and plotted as a running Fourier transform measured using cross-correlation with an unshaped mid-IR pulse. By measuring the probe spectrum for each of these truncated pulses, we can measure the vibrational populations induced by each segment of the shaped excitation pulse to better understand how it achieves its desired goal. The pulse featured in Fig. 5 was created to maximize the intensity of the $\upsilon = 1$–2 feature against the bleach.

By stepping a over the duration of the pulse and collecting a pump-probe spectrum for each time value of a, the vibrational population transfer over the course of the pulse is assessed. To quantitatively extract relative vibrational populations from these pump-probe spectra, we fit peak intensities assuming a harmonic scaling law for the transition dipoles [e.g., $(n + 1)|\mu_{n,n+1}|^2 = (n + 2)$

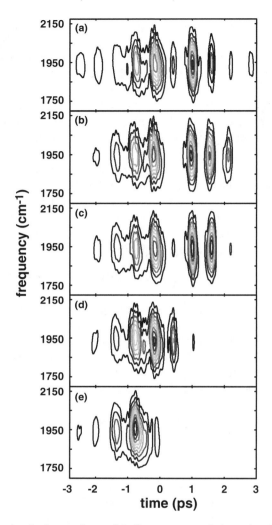

Figure 5. Running fourier transforms of the linear cross-correlations taken for a series of laser pulses systematically truncated in the time domain. The entire pulse (a) is truncated in 500-fs intervals so that approximately half the pulse remains in panel (e). The center of a transform limited pulse was used to define the center ($t = 0$) of the optimized pulse.

$|\mu_{n+1,n+2}|^2$]. Figure 6 details how the peak amplitudes and vibrational populations change as a pulse optimized to create a bleach at the $\upsilon = 1-2$ absorption interacts with the sample. The $\upsilon = 2-3/\upsilon = 1-2$ optimized pump pulse is composed of a series of pulses with tight spacing (~ 250 fs) and alternating intensities. This pulse train improves the ratio by increasing the $\upsilon = 2-3$ absorption peak intensity at the expense of the fundamental $\upsilon = 0-1$

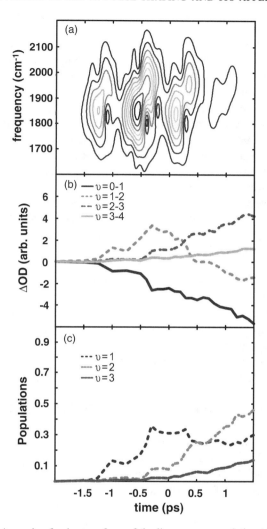

Figure 6. (a) A running fourier transform of the linear cross-correlation taken for the Laser pulse generated to optimize the $\upsilon = 2$–$3/1$–2 feature ratio. Contour levels are shown in 10% spacings from minimum to maximum intensity. (b) Amplitudes of the first four peaks are plotted as a function of the pulse truncation. The portion of the pulse remaining after truncation increases from left to right, so that $+1.5$ ps corresponds to a full pump pulse. (c) Relative populations of the first three vibrational levels extracted from fits.

transition, thereby moving the population into the preferred excited vibrational level. At $t = 0.5$ ps, the pump begins to depopulate the $\upsilon = 1$ vibrational level in order to transfer the population to the $\upsilon = 2$ vibrational level, creating an inverted population. It is interesting that this population decrease occurs even while the fundamental transition continues to bleach, indicating that the population is still flowing out of $\upsilon = 0$. Thus, at the start of excitation the pulse moves population out of the ground state and into the excited states, while at the end of the pulse all the vibrational levels are coupled so that population can continue to flow out of the ground state while depopulating intermediate levels to achieve population inversion.

An ability to control vibrational populations portends an ability to control the myriad chemical processes that take place in the ground state. For systems in which a ground-state potential barrier is present, such as with isomerization, proton transfer, or electron transfer, creating a wavepacket with amplitude maximized to overcome such a barrier would provide selective access to desired chemical processes. Where shaped Raman pulses have been used previously for selective ground-state excitation [32, 36], shaped mid-IR pulses offer a complementary approach to selective ground-state excitation. Furthermore, since mid-IR spectroscopy is performed using pump-probe absorption geometry, the nature of the pulse excitation can be explored by systematically truncating the excitation pulse and monitoring population transfer. Such an approach could be used to optimize a pulse not by looking at just the final products as is usually done and as we have done here, but by optimizing intermediate populations on the way to the final desired state—for example, by shaping in the time domain. Such an approach may enhance and accelerate optimization, since future time evolution will be based on past performance, thereby narrowing the search space and guiding the trajectory.

VI.   AUTOMATED 2D IR SPECTROSCOPY

In the coherent control experiments above, a desired process (such as a population) is optimized using a feedback loop to experimentally modify the shape of the mid-IR pulses. Rather than arriving at an unknown pulse shape, another useful feature of pulse shaping is that desired shapes can be programmed to high accuracy. This has especially powerful implications for nonlinear spectroscopies whose goal is to manipulate vibrational motions in order to probe samples with high frequency and fast time resolution. Nonlinear spectroscopies use carefully crafted pulse sequences to measure quantities such as vibrational frequencies, combination band frequencies, or IR linewidths. These quantities are useful in structural studies because they are related to molecular conformations and environments. Two- and higher-dimensional IR spectroscopies are becoming important tools for probing the fast structural dynamics of chemical and biological systems [7, 27, 37–41]. This section of the chapter reviews our work on developing mid-IR pulse shaping as a means for generating sequences of pulses to measure 2D IR spectra. Traditional ways of collecting 2D IR spectra are hard to implement, cause distorted peak shapes, and result in poor time resolution and/or phase problems. The capabilities of 2D IR spectroscopy for elucidating time-evolving structures is enhanced by a programmable mid-IR pulse shaper that greatly improves the ease, speed, and accuracy of data collection.

2D IR spectroscopy is analogous in many ways to 2D NMR spectroscopy, except that it is the dynamics and couplings of vibrational dipoles, rather than spin dipoles, that are probed. As a result, the technique has a unique combination of structural sensitivity and fast time resolution (fs/ps) that makes this technique and its variants [37, 42–44] especially adept at monitoring dynamics of evolving structures or the kinetics of chemical reactions [45–48]. Although it is a powerful technique, implementing 2D IR spectroscopy is technically challenging. In a typical 2D IR spectrometer, each pulse has its own optical path composed of several mirrors to route the beam to the sample and a mechanical delay stage that controls the time delay by changing the physical length of the optical path. All of the pulses have the same frequencies and shapes unless additional optics are added such as an etalon to narrow a pulse bandwidth [31, 49] or a second mixing crystal to generate pulse sequences with two center frequencies [50, 51]. Thus, implementing even the simplest pulse sequence is a tremendous amount of work. This is in sharp contrast to 2D NMR spectroscopy, where NMR pulse sequences are easily programmed with precisely set frequencies, time delays, phases, and intensities. As a result, it is straightforward in NMR to change the pulse sequences or add additional pulses in order to create the best spectrum for measuring the desired information. Mid-IR pulse shaping takes 2D IR spectroscopy a step closer toward automating it like 2D NMR spectroscopy, because we can now programmably generate 2D IR pulse sequences with controlled phases, delays, and shapes [43, 52]. In this review, we first discuss traditional 2D IR methods and then replicate these experiments using our shaper. We then show how we can improve upon their accuracy, speed, and ease of implementation with several alternative pulse sequences optimized to maximize the structural content of the spectra.

Hamm, Lim, and Hochstrasser collected the first 2D IR spectrum in 1998 using a hole-burning approach [31]. They monitored the changes in the intensity of the protein carbonyl stretch band (amide I band) with a broadband femtosecond pulse while using an etalon to scan a narrow-band mid-IR pulse across the protein absorption spectrum. Changes in probe intensity indicate coupled vibrational modes, which appear as cross-peaks when plotted in a 2D representation. An etalon consists of two partially reflective mirrors, so that a pulse entering on one end bounces back and forth between the two partial reflectors, leaking a portion of the pulse intensity with each bounce. The frequency is narrowed because the phases of the leaked portions interfere with one another, creating a Lorentzian shaped profile in the frequency domain and an exponential decay in the time domain. The reflectivity of the mirrors determines the bandwidth, which is usually set to the homogeneous linewidth of the sample.

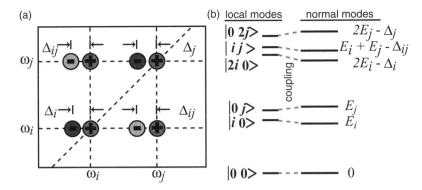

Figure 7. (a) A schematic representing a typical 2D-IR spectrum for a coupled two-vibrator system. On-diagonal peaks are separated by the diagonal anharmonicity, while the off-diagonal anharmonicity separates off-diagonal peaks. Positive peaks arise from fundamental transitions. Negative peaks arise due to overtone transitions. (b) The energy-level diagram for such a two-vibrator system. The energy levels on the left are shown without coupling, while those on the right are shifted due to coupling.

A schematic of a typical 2D IR spectrum is shown in Fig. 7 for a coupled two-vibrator system whose energy level diagram is also shown [42, 53, 54]. We call the two local modes $|i>$ and $|j>$, the coupling of which gives rise to the normal modes observed in the spectrum, which are the symmetric and antisymmetric stretch normal modes. Likewise, the overtone and combination band of the local modes also couple to create delocalized normal mode eigenstates. In the hole-burning approach, a 2D IR spectrum is generated by scanning the pump frequency across the fundamental transitions while the probe monitors transitions to the overtone and combination band normal mode eigenstates, transitions back to the ground state, and transitions to the ground-state bleach. Probe interactions that only interrogate the fundamentals and ground state result in four positive peaks that form a square in the 2D IR spectrum. Transitions to the overtone and combination bands produce four negative peaks. The frequency difference between the negative and positive peaks are usually small, so that the "diagonal" and "cross-peaks" appear in pairs. The diagonal peaks are separated by the diagonal anharmonicity, Δ_i and Δ_j, of the system while the cross-peaks are separated by the off-diagonal anharmonicity, Δ_{ij}. By measuring these anharmonicities, the couplings between the local modes can be deduced. This is a very useful quantity, because it is a strong indicator of structure. Besides being useful for measuring the coupling, the peaks also have shapes that reflect the vibrational dynamics of the system and the relative frequency fluctuations between the modes.

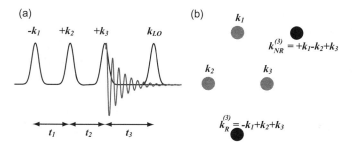

Figure 8. (a) A pulse train typically used to generate 2D-IR spectra. (b) Phase-matching geometries to generate both rephrasing, R, and non-rephasing, NR, signals.

The hole-burning technique uses a frequency domain approach to collecting 2D IR spectra. In 2000, Hochstrasser and co-workers implemented a pulsed version of 2D IR spectroscopy that is analogous in many ways to modern pulsed 2D NMR spectroscopy [42, 55]. Rather than scanning in the frequency domain, the pulsed version uses a sequence of femtosecond mid-IR pulses to impulsively excite the sample (Fig. 8). The resulting photon echo (or free induction decay) is measured by heterodyning the emitted electric field with a fourth pulse, called a local oscillator, and the signal is Fourier-transformed along two time dimensions to give the 2D IR spectrum (t_1 and t_3 in Fig. 8). Each pulse in the sequence is typically generated from a series of beamsplitters that reflect a portion of an intense mid-IR pulse generated from one or two optical parametric amplifiers. Each pulse traverses its own optical path so that the relative delays of the pulses can be controlled with translation stages or wedged optics before impinging on the sample. The directions of the beams are described by their wavevectors, k_n, so while the pulse sequence drawn in Fig. 8a looks collinear, in practice the three excitation beams arrive at the sample from three different directions. This beam geometry results in interference of their electric fields so that the sample emits in unique spatial directions, dictated by the wavematching geometry. A commonly used wave geometry is shown in Fig. 8b, where the three excitation beams form an equilateral triangle. In this geometry, the signal is often measured in the $-k_1 + k_2 + k_3$ direction, because a rephasing signal (a photon echo) is emitted in this direction when the pulses arrive at the sample in that order. When the timing of the pulses are ordered $+k_2 - k_1 + k_3$, then a free induction decay is emitted in this direction, which is one of several non-rephasing signals. Most often, both the rephasing and non-rephasing signals are measured and added together to remove phase twist from the spectra, which optimizes the frequency resolution [56]. Besides these two signals, other wavematching directions and/or pulse delays can be monitored for additional information [43]. For instance, the direction $-2k_1 + 2k_2 + k_3$ has been measured in order to generate a fifth-order pulse sequence and collect 3D IR spectra [43, 57, 58]. In the hole-burning approach to 2D IR spectroscopy, the phase-matching geometry is much simpler since only two beams are used to collect the 2D IR spectra (the frequency-narrowed pump and the femtosecond probe). As a result, both the non-rephasing and rephasing signals are generated in the direction of the probe, which serves as both an excitation pulse as well as the local oscillator. Since both signals are generated simultaneously, this beam geometry automatically removes phase twist, resulting in high-resolution absorptive spectra.

Both versions of 2D IR spectroscopy are now being used in many research laboratories across the world. As the field has evolved, the strengths and weaknesses of each approach have become apparent; the hole-burning approach is simpler to implement since only two beams, not four, are required, portions of the 2D IR spectra can be monitored by tuning the hole-burning pulse without having to scan and Fourier transform the entire signal, and the spectra automatically remove phase twist. However, the pulsed version has better time resolution because only femtosecond pulses are used; it is also more versatile for measuring particular molecular responses via more sophisticated phase matching and pulse sequences.

With mid-IR pulse shaping, alternative methods for collecting 2D IR spectra that overcome many of the weaknesses of the two original methods while adding many new capabilities are now available. Femtosecond mid-IR pulses that enter the shaper have the phases and amplitudes of their spectrum tailored so that they exit with the desired shape. In this manner, one pulse can be turned into a train of pulses with computer-controllable characteristics like is done in NMR. Thus, the original hole-burning method for collecting 2D IR spectra implemented by mimicking the pulse shape created by an etalon [59] or a sequence of femtosecond pulses can be created for the pulsed method. However, these sequences can now be improved. For instance, the spectrum of the narrowed pulses in the hole-burning method no longer need to be Lorentzian-shaped. They can be Gaussian-shaped instead (or any other shape), which leads to improved frequency resolution since Lorentzians have long wings that can interfere and overlap. Alternatively, the pulsed version can now be implemented using collinear excitation pulses so that the rephasing and non-rephasing spectra are emitted together like they are in the hole-burning method. As a result, phase twist can be automatically removed while retaining high time resolution and removing many of the challenges of spatially aligning four mid-IR pulses. As we demonstrate below, our collinear method retains the best aspects of both techniques: (a) properly phased absorptive spectra and (b) ultrafast time resolution with improved collection efficiency and phase stability.

Shown in Fig. 9 are four series of 2D IR spectra, using four different pulse sequences, collected for $W(CO)_6$ solvated in hexane. Each series is generated

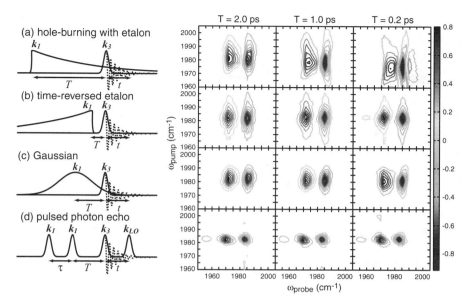

Figure 9. Pulse sequences for various 2D-IR methods. Leftmost column: (a) The traditional pulse sequence used for hole burning with an etalon. Pulse sequences in which the pump pulse is a time-reversed etalon (b), a Gaussian (c), and a pulse pair (d). 2D-IR spectra of W(CO)$_6$ for each of the pump pulse shapes appear to the right of their corresponding pulse train. Spectra were taken at $T = 2.0$, 1.0 and 0.2 ps. Fifteen contours were drawn from minimum to maximum of normalized intensity.

using a different pump shape. The first three methods utilize a hole-burning approach, while the last one is based on a pulsed method. W(CO)$_6$ has three degenerate modes at 1983 cm^{-1} that are created by antisymmetric stretches of the metal carbonyl groups, but for the purposes of this chapter the molecule responds like a single, uncoupled vibrator. As a result, the 2D IR spectra shown below consist of only a pair of diagonal peaks with no cross-peaks. The first series of 2D IR spectra shown in Fig. 9a mimic the spectra that would be generated had an etalon been used to narrow the pump spectrum. An etalon generates a Lorentzian-like spectrum according to

$$I(\omega) = \left| \frac{1 - R}{1 - Re^{+i(2\pi\Delta\omega)}} \right|^2 \tag{3}$$

where R is the reflection coefficient of the mirrors in the etalon and Δ is the path difference between each proceeding reflection [59]. To generate the 2D IR spectra in the first row of Fig. 9, we set the bandwidth to 4.6 cm^{-1}, scanned ω, and plotted the change in absorption of the probe with and without the pump. 2D

IR spectra were collected at three different pump-probe delay times, $T = 2000$, 1000, and 200 fs. At $T = 2000$ fs, a negative and positive peak pair is observed (blue and red in Fig. 9, respectively). The negative peak appears along the diagonal with $\omega_{pump} = \omega_{probe} = 1983$ cm^{-1}, which is the fundamental frequency of W(CO)$_6$. The positive peak is shifted by $\Delta\omega_{probe} = 13.3$ cm^{-1}, which is the anharmonicity of the W(CO)$_6$ antisymmetric stretch [60]. The fwhm of the peaks are 13.4 cm^{-1} along ω_{pump} and 5.3 cm^{-1} along ω_{probe} (16.5 and 7.5 cm^{-1}, respectively, for the first overtone). This spectrum has many desirable characteristics. The peaks have absorptive lineshapes because both the rephasing and non-rephasing spectra are measured simultaneously and the spectra are properly phased. However, the peaks are broader than the intrinsic W(CO)$_6$ linewidth, especially along ω_{pump}, and the spectra exhibit aberrations that become especially grievous for $T = 200$ fs. With regard to the widths, the width along ω_{probe} is a convolution of the intrinsic linewidth with the monochromator resolution (2.2 cm^{-1}) whereas the width along ω_{pump} involves a convolution with the pump bandwidth (4.6 cm^{-1}). The monochromator resolution can be improved with a different grating, but narrowing the pump bandwidth leads to larger aberrations because the aberrations are caused by the exponential tail of the pump pulse overlapping in time with the probe pulse. For the pump bandwidth of 4.6 cm^{-1} used here, the exponential tail has a time constant of 1.2 ps. Thus, the spectra become distorted for $T = 1000$ and 200 fs because the probe overlaps with the exponential tail of the pump. Distortion can be avoided by collecting spectra with T delays longer than the exponential decay, but the signal strength decreases proportionately because of population relaxation. Thus, many applications are best done with T delays as short as possible.

An etalon can only control the frequency and bandwidth of narrowed pulse, but our shaper can create any pulse shape within the limits of its resolution. Shown in the second row of Fig. 9 are three 2D IR spectra collected using a "time-reversed" etalon, generated by changing the sign of the exponential in Eq. (1). The pump pulse frequency distribution is identical for the normal versus the time-reversed etalon, but the shape of the pulse is now reversed in the time domain (Fig. 9b). Thus, the pump and probe pulses do not overlap until much shorter time delays, resulting in smaller peak distortions. Because T can be smaller without causing distortions, the signal strength is also stronger for a time-reversed etalon pulse shape. The improvement in signal strength is not very dramatic for W(CO)$_6$ because the vibrational lifetime is 140 ps, but for the amide I band of a peptide or protein where the vibrational lifetime is $\sim$1400 fs, a three-time enhancement in signal strength would occur for $T = 200$ versus 1800 fs.

As mentioned above, a drawback with the etalon method is that the pulses have Lorentzian-shaped frequency profiles. A better profile would be Gaussian, since Gaussians tail to zero much more quickly than Lorentzians with the same full width at half-maximum. With our shaper, we also collected

2D IR spectra using a Gaussian shaped pump pulse, shown in the third row of Fig. 9. For these spectra we set the fwhm of the Gaussian spectrum close to the etalon-style pump (fwhm $= 7.0\,cm^{-1}$) and the phase to zero so that the pulse is transform-limited with 2.2 ps fwhm in the time domain. The resulting spectra are similar in some respects to the time-reversed etalon. The peaks have the correct phases and little distortion even at $T = 200\,fs$. The main difference is that the peak shapes along ω_{probe} now have Gaussian rather than Lorentzian profiles and thus better resolution. The reason for the change in profile is because the measured spectrum is a convolution of the natural system response with the pump-pulse (see Discussion). Since the natural linewidth of $W(CO)_6$ ($4.4\,cm^{-1}$) is much narrower than the fwhm of the pump pulses, the measured spectrum closely resembles the shape of the pump. Gaussian shaped pump pulses are a better choice than Lorentzians because Gaussians decay to baseline more quickly. A comparison of the peak profiles along ω_{pump} are shown in Fig. 10.

The fourth method that we report more closely resembles the pulsed 2D IR method by Hochstrasser and co-workers [42, 53, 55] except that the experiment is performed in a pump-probe beam geometry. In this method, the pulse shaper is used to create two transform-limited Gaussian pulses. Rather than scan the frequencies like was done above, the 2D IR spectrum is instead generated by collecting the signal as a function of the time delay between the two pulses, τ, and Fourier transforming it to give the ω_{pump} axis. To mimic the way that pulse delays are typically generated using translation stages, the phases of the two pump pulses inside their respective envelopes were held constant ($\varphi_1 = \varphi_2 = 0$). The resulting

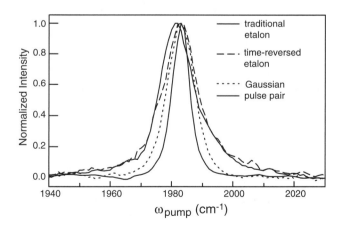

Figure 10. Slices of 2D IR spectra of $W(CO)_6$ in hexane for $T = 2.0\,ps$ in Fig. 9 at $\omega_{pump} = 1983\,cm^{-1}$. Pump pulses are shaped as a traditional etalon (solid line), a time-reversed etalon (dashed line), a Gaussian (dotted line), and a pulse pair (heavy line).

spectra are shown in the fourth row of Fig. 9 for $\tau = 0$ to 10,000 fs in 14-fs steps. Like the first two methods, the spectra are automatically phased and have absorptive lineshapes, but now the narrowest $W(CO)_6$ pump linewidth is measured and there is no discernable distortion in the peaks even at $T = 200$ fs. Thus, we are able to generate the ideal 2D IR spectrum in a straightforward manner. The small peak that appears at $\omega_{probe} = 1956 \, \text{cm}^{-1}$ is caused by a slight amount of overpumping. Overpumping is uncommon in the hole-burning approach because the picosecond pulses have low peak powers and narrower bandwidths. As is explained in more detail below, this third method retains the phasing and absorptive features of the pump-probe spectra because both rephasing and non-rephasing signal are collected simultaneously and the time zeros of the pulse delays are perfectly set, resulting in a properly phased spectrum. Moreover, femtosecond pump pulses allow linewidths closest to the intrinsic resolution to be measured. We also point out that the peak shapes are symmetrical about ω_{pump} and that there are no spurious ghost images, indicating that the shaper time resolution is sufficiently accurate. Asymmetric shapes and ghost images are common problems that occur when 2D IR spectra are collected using translation stages to increment the time delays [58, 61].

In Fig. 11, we compare the 2D IR spectra for a coupled oscillator system that exhibits both cross-peaks and diagonal peaks. The sample is a nickel metal dicarbonyl, and the spectra are collected using the Gaussian and pulse-pair pump methods. Nearly identical 2D IR spectra are measured with these two methods, illustrating that both techniques contain the same information about the molecular couplings and linewidths. To obtain linewidths with the hole-burning method that are comparable to the pulsed method, the Gaussian linewidth was set to $7 \, \text{cm}^{-1}$, which is much smaller than the homogeneous linewidth of $20 \, \text{cm}^{-1}$.

Besides its improved accuracy and ease of implementation, 2D IR spectroscopy via pulse shaping has several other advantages. First, data collection is much more rapid. In a traditional setup, most of the data collection time is spent moving the translation stages to increment the time delays or the etalon to adjust the wavelength. With our mid-IR pulse shaper, a new pulse sequence can be generated at laser repetition rates up to 100 kHz. Thus, the frequency or time delay can be incremented between every laser shot so that no time is wasted. For a typical peptide sample with a lifetime of 1.5 ps, an entire 2D IR spectrum is measured every 0.3 s with the pulsed method. Another advantage is that phase cycling can now be implemented into the pulse sequences. Phase cycling is often used in NMR spectroscopy to isolate specific signals. In 2D IR spectroscopy, phase cycling can be used to isolate the rephasing and non-rephasing signals; it can also be used to shift the frequency into the rotating frame for faster data collection; and it can be used to average background scatter which is a common problem in infrared spectroscopies.

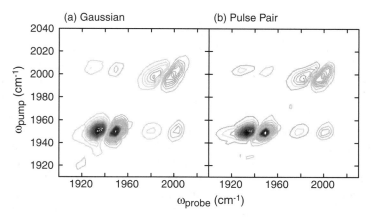

Figure 11. 2D IR spectra of bis(triphenylphosphine)dicarbonylnickel in THF:hexane (1:3) with pump pulses shaped as a Gaussian (a) and a pulse pair (b).

Perhaps the biggest advantage of mid-IR pulse shaping with regard to collecting 2D IR spectroscopy is the ease with which spectra can now be collected. Pulse sequences can now be generated by simple programming rather than time-consuming optical rearrangements. Furthermore, many uncertainties are removed. 2D IR spectra collected in a traditional four-wave mixing geometry require that the phase of the data be calibrated. This step is necessary because the absolute phase of the spectra is not known unless the time delays between the pulses can be set with absolute certainty. In a four-wave mixing geometry where each pulse follows a different optical path, it is not possible to set the time delays perfectly. Instead, the 2D IR spectra are phased by projecting them onto a pump-probe absorption spectrum [62, 63]. A pump-probe spectrum is perfectly phased because the pump pulse is responsible for the first two pulse interactions (e.g., $t_1 = 0$) and the probe acts as the third excitation beam and as the local oscillator (e.g., $t_3 = 0$). As a result, there is no uncertainty in the timing of the pulses. In the methods described here, the pulse delays are also perfectly certain, because the shaper perfectly sets the t_1 delay and the t_3 delay is set by the probe. Thus, the spectra are properly phased without any post-data corrections. This eliminates perhaps the most difficult challenge and least accurate step associated with traditional pulsed methods for collecting 2D IR spectra. Furthermore, since the first two excitation beams are collinear as is the probe beam with the signal, the pulse sequence is also perfectly phase-stable (at least between these two pairs of pulses). As a result, the time delays do not drift with time, a problem in the traditional method where time drift leads to distortions in the spectra [58]. Thus, extremely accurate and highly resolved 2D-IR spectra can now be collected with our pulse shaper, making many new experiments possible.

VII. SUMMARY

Pulse shaping directly in the mid-IR opens up many new experiments in both coherent control and multidimensional IR spectroscopy. This technological advance is very recent, and so only the most rudimentary uses have been implemented so far [4, 5]. In fact, coherent control and multidimensional spectroscopies might be combined to strongly complement one another. In coherent control methods using feedback, the bottleneck is often understanding how the process is optimized by the pulses. If 2D IR spectroscopy were used in conjunction with coherent control experiments (and truncated pulse sequences), the vibrational pathways selected by the control path could be monitored, lineshapes could be monitored to understand how solvent dephasing alters selectivity, and energy transfer could be mapped to understand how inter-molecular relaxation inhibits the excitation process. Understanding these processes will likely lead to improved control strategies. Coherent control strategies might also be used to enhance 2D IR spectra [7]. In NMR spectroscopy, radio-frequency control over nuclear spins allows for precise control over pulse sequences. Similar methodology might be possible in multidimensional spectroscopy with pulse shaping. For instance, cross-peaks might be enhanced using shaped pulses that preferentially excite the combination bands. With pulse shaping, implementing coherent control and 2D IR spectroscopy may be straightforward. At least in principle, combining coherent control and 2D IR spectroscopy is just a matter of computer programming, whereby the shaped pulse is incorporated into the 2D pulse sequence, or the 2D pulse sequence just follows a shaped pulse. Whether or not these experiments are possible, it is clear that even with the rudimentary experiments performed so far, mid-IR pulse shaping is an insightful new tool to apply toward understanding and manipulating the vibrational modes of molecules.

Acknowledgments

The authors gratefully acknowledge financial support from the National Science Foundation (No. CHE-0350518), the Beckman Foundation, and the David-Lucile Packard Foundation.

References

1. R. J. Gordon and S. A. Rice, *Annu. Rev. Phys. Chem.* **48**, 601 (1997).

2. R. S. Judson and H. Rabitz, *Phys. Rev. Lett.* **68**, 1500 (1992).

3. V. V. Lozovoy and M. Dantus, *ChemPhysChem* **6**, 1970 (2005).

4. D. B. Strasfeld, S.-H. Shim, and M. T. Zanni, *Phys. Rev. Lett.* **99** (2007).

5. S. H. Shim, D. B. Strasfeld, Y. L. Ling, and M. T. Zanni, *Proc. Natl. Acad. Sci. USA* **104**, 14197 (2007).

6. V. D. Kleiman, S. M. Arrivo, J. S. Melinger, and E. J. Heilweil, *Chem. Phys.* **233**, 207 (1998).

7. D. Abramavicius and S. Mukamel, *J. Chem. Phys.* **120**, 8373 (2004).

8. C. Meier and M. C. Heitz, *J. Chem. Phys.* **123** (2005).

9. C. Gollub, B. M. R. Korff, K. L. Kompa, and R. de Vivie-Riedle, *Phys. Chem. Chem. Phys.* **9**, 369 (2007).

10. N. Doslic, K. Sundermann, L. Gonzalez, O. Mo, et al., *Phys. Chem. Chem. Phys.* **1**, 1249 (1999).

11. Q. Ren, G. G. Balint-Kurti, F. R. Manby, M. Artamonov, et al., *J. Chem. Phys.* **125** (2006).

12. B. M. R. Korff, U. Troppmann, K. L. Kompa, and R. de Vivie-Riedle, *J. Chem. Phys.* **123** (2005).

13. C. M. Tesch and R. de Vivie-Riedle, *Phys. Rev. Lett.* **89** (2002).

14. N. V. Vitanov, T. Halfmann, B. W. Shore, and K. Bergmann, *Annu. Rev. Phys. Chem.* **52**, 763 (2001).

15. Q. Shi and E. Geva, *J. Chem. Phys.* **119**, 11773 (2003).

16. M. Demirplak and S. A. Rice, *J. Chem. Phys.* **116**, 8028 (2002).

17. A. M. Weiner, D. E. Leaird, G. P. Wiederrecht, and K. A. Nelson, *Science* **247**, 1317 (1990).

18. H. S. Tan, E. Schreiber, and W. S. Warren, *Opt. Lett.* **27**, 439 (2002).

19. H. S. Tan and W. S. Warren, *Opt. Exp.* **11**, 1021 (2003).

20. T. Witte, K. L. Kompa, and M. Motzkus, *Appl. Phys. B: Lasers Opt.* **76**, 467 (2003).

21. T. Witte, D. Zeidler, D. Proch, K. L. Kompa, et al., *Opt. Lett.* **27**, 131 (2002).

22. M. A. Dugan, J. X. Tull, and W. S. Warren, *J. Opt. Soc. Am. B* **14**, 2348 (1997).

23. C. W. Hillegas, J. X. Tull, D. Goswami, D. Strickland, et al., *Opt. Lett.* **19**, 737 (1994).

24. S. H. Shim, D. B. Strasfeld, and M. T. Zanni, *Opt. Exp.* **14**, 13120 (2006).

25. K. Ando and J. T. Hynes, *J. Phys. Chem. A* **103**, 10398 (1999).

26. Y. Ohta, T. Bando, T. Yoshimoto, K. Nishi, et al., *J. Phys. Chem. A* **105**, 8031 (2001).

27. E. Geva, *J. Chem. Phys.* **116**, 1629 (2002).

28. M. Petkovic and O. Kuhn, Chem. Phys. **304**, 91 (2004).

29. S. S. Skourtis, D. H. Waldeck, and D. N. Beratan, *J. Phys. Chem. B* **108**, 15511 (2004).

30. G. K. Paramonov and P. Saalfrank, *Chem. Phys. Lett.* **301**, 509 (1999).

31. R. Schanz, V. Bolan, and P. Hamm, *J. Chem. Phys.* **122** (2005).

32. D. Zeidler, S. Frey, W. Wohlleben, M. Motzkus, et al., *J. Chem. Phys.* **116**, 5231 (2002).

33. C. Ventalon, J. M. Fraser, M. H. Vos, A. Alexandrou, et al., *Proc. Natl. Acad. Sci. USA* **101**, 13216 (2004).

34. L. Windhorn, T. Witte, J. S. Yeston, D. Proch, et al., *Chem. Phys. Lett.* **357**, 85 (2002).

35. T. Witte, T. Hornung, L. Windhorn, D. Proch, et al., *J. Chem. Phys.* **118**, 2021 (2003).

36. A. M. Weiner, D. E. Leaird, G. P. Wiederrecht, and K. A. Nelson, *J. Opt. Soc. Am. B* **8**, 1264 (1991).

37. M. H. Cho, *PhysChemComm*, **40** (2002).

38. A. T. Krummel and M. T. Zanni, *J. Phys. Chem. B* **110**, 13991 (2006).

39. P. Mukherjee, I. Kass, I. Arkin, and M. T. Zanni, *Proc. Natl. Acad. Sci. USA* **103**, 3528 (2006).

40. H. S. Chung, Z. Ganim, K. C. Jones, and A. Tokmakoff, *Proc. Nat. Acad. Sci. USA* **104**, 14237 (2007).

41. J. A. Ihalainen, J. Bredenbeck, R. Pfister, J. Helbing, et al., *Proc. Natl. Acad. Sci. USA* **104**, 5383 (2007).

42. W. M. Zhang, V. Chernyak, and S. Mukamel, *J. Chem. Phys.* **110**, 5011 (1999).

43. S. H. Shim, D. B. Strasfeld, E. C. Fulmer, and M. T. Zanni, *Opt. Lett.* **31**, 838 (2006).

44. E. C. Fulmer, F. Ding, P. Mukherjee, and M. T. Zanni, *Phys. Rev. Lett.* **94** (2005).

45. H. S. Chung, M. Khalil, A. W. Smith, Z. Ganim, et al., *Proc. Natl. Acad. Sci. USA* **102**, 612 (2005).

46. Y. S. Kim and R. M. Hochstrasser, *Proc. Natl. Acad. Sci. USA* **102**, 11185 (2005).

47. J. R. Zheng, K. Kwak, J. Asbury, X. Chen, et al., *Science* **309**, 1338 (2005).

48. C. Kolano, J. Helbing, M. Kozinski, W. Sander, et al., *Nature* **444**, 469 (2006).

49. V. Cervetto, J. Helbing, J. Bredenbeck, and P. Hamm, *J. Chem. Phys.* **121**, 5935 (2004).

50. I. V. Rubtsov, J. P. Wang, and R. M. Hochstrasser, *J. Phys. Chem. A* **107**, 3384 (2003).

51. K. Heyne, N. Huse, E. T. J. Nibbering, and T. Elsaesser, *Chem. Phys. Lett.* **369**, 591 (2003).

52. S. H. Shim, D. B. Strasfeld, and M. T. Zanni, *Opt. Express* **14**, 13120 (2006).

53. M. T. Zanni, S. Gnanakaran, J. Stenger, and R. M. Hochstrasser, *J. Phys. Chem. B* **105**, 6520 (2001).

54. O. Golonzka, M. Khalil, N. Demirdoven, and A. Tokmakoff, *Phys. Rev. Lett.* **86**, 2154 (2001).

55. M. C. Asplund, M. T. Zanni, and R. M. Hochstrasser, *Proc. Natl. Acad. Sci. USA* **97**, 8219 (2000).

56. M. Khalil, N. Demirdoven, and A. Tokmakoff, *Phys. Rev. Lett.* **90**, 047401 (2003).

57. F. Ding, E. C. Fulmer, and M. T. Zanni, *J. Chem. Phys.* **123** (2005).

58. F. Ding, P. Mukherjee, and M. T. Zanni, *Opt. Lett.* **31**, 2918 (2006).

59. G. Hernandez, *Fabry–Perot Interferometers*, Cambridge University Press, New York, 1986.

60. T. Witte, J. S. Yeston, M. Motzkus, E. J. Heilweil, et al., *Chem. Phys. Lett.* **392**, 156 (2004).

61. V. Volkov, R. Schanz, and P. Hamm, *Opt. Lett.* **30**, 2010 (2005).

62. T. Brixner, I. V. Stiopkin, and G. R. Fleming, *Opt. Lett.* **29**, 884 (2004).

63. J. B. Asbury, T. Steinel, and M. D. Fayer, *J. Lumin.* **107**, 271 (2004).

LOCAL CONTROL THEORY: RECENT APPLICATIONS TO ENERGY AND PARTICLE TRANSFER PROCESSES IN MOLECULES

VOLKER ENGEL

Universität Würzburg, Institut für Physikalische Chemie and Röntgen Research Centre for Complex Material Systems, Am Hubland, 97074 Würzburg, Germany

CHRISTOPH MEIER

Laboratoire Collisions, Agrégats et Reactivité, IRSAMC, Université Paul Sabatier, 31062 Toulouse, France

DAVID J. TANNOR

Department of Chemical Physics, Weizmann Institute of Science, 76100 Rehovot, Israel

CONTENTS

Advances in Chemical Physics, Volume 141, edited by Stuart A. Rice
Copyright © 2009 John Wiley & Sons, Inc.

I. INTRODUCTION

The invention of schemes to induce and observe dynamical processes in molecules employing ultrashort laser pulses has founded research areas with the names Femtochemistry and Femtobiology [1–8], culminating with the Nobel Prize awarded to Ahmed Zewail for his achievements in femtosecond spectroscopy [9,10]. More recently, the technology of pulse shaping [11,12] has opened up the field of laser control of chemical reactions, until then populated only by theorists [13–17]. This has been an active area of research in recent years [18–23], and textbooks [24–26] as well as many reviews [27–40] have been published on the topic.

From a theoretical point of view, different control schemes have been introduced. As a general strategy for laser control, the interaction of a system with an external field $W(t)$ is tailored such that a transition from an initial state $|\psi_i\rangle$ at a time t_i to a final (or target) state $|\psi_f\rangle$ at a time t_f is induced. In achieving this goal, one might distinguish global and local control schemes. In the first one, the control fields are constructed employing information on the entire dynamics from time t_i to time t_f, whereas in a local control scheme, the field is determined instantaneously, taking the system's response into account. For a review on early work on local optimization, see Ref. 41.

In this chapter we review our recent applications of a local control scheme to various problems in chemical physics. The approach we follow is called *Local Control Theory (LCT)*. The idea first appeared in the formulation of Optimal Control Theory introduced by Kosloff, Rice, and Tannor [42] and has been extensively developed since then [41,43–51]. The local control concept was used to model laser cooling or heating of molecular vibrational motion on the electronic ground state, in which an excited state is used by changing the phase of the radiation field [44, 50] (throughout this review, heating and cooling are not used in the thermodynamic sense, but in the sense of adding or taking away internal energy from the molecule). Furthermore, this approach has been used for a generalization of stimulated Raman adiabatic passage (STIRAP) [52] to an N-level quantum system [48]. In all these applications, the heating/cooling or population transfer were achieved by locally designing external fields under certain constraints like, for example, the locking of population in specific electronic states.

The basic idea of the LCT approach is to choose a control field in order to ensure an increase or decrease of the expectation value of an observable—that is, impose a condition on the sign of its time derivative. An extension to this approach is to specify not only the sign, but also the temporal evolution, of the time derivative [53]. Further developments include the treatment of multiple objectives leading to the possibility to force the system dynamics along a predescribed path through Hilbert space [54–57].

A different theory of local control has been derived from the viewpoint of global optimization, applied to finite time intervals [58–60]. This approach can also be applied within a classical context, and local control fields from classical dynamics have been used in quantum problems [61]. In parallel, Rabitz and co-workers developed a method termed "tracking control," in which Ehrenfest's equations [26] for an observable is used to derive an explicit expression for the electric field that forces the system dynamics to reproduce a predefined temporal evolution of the control observable [62, 63]. In its original form, however, this method can lead to singularities in the fields, a problem circumvented by several extensions to this basic idea [64–68]. Within the context of ground-state vibration, a procedure similar to tracking control has been proposed in Ref. 69. In addition to the examples already mentioned, the different local control schemes have found many applications in molecular physics, like population control [55], wavepacket control [53, 54, 56], control within a dissipative environment [59, 70], and selective vibrational excitation or dissociation [64, 71]. Further examples include isomerization control [58, 60, 72], control of predissociation [73], or enantiomer control [74, 75].

As far as the general formulation is concerned, it is interesting to note that there exist similarities between monotonically convergent algorithms for optimal control and local control methods. This relationship has only recently been elucidated by Salomon and Turinici [76].

There is an additional important point to be mentioned, which is that the control fields constructed within the LCT scheme can be interpreted in a straight-forward way. This is not always possible for fields derived from other control theories. In the examples presented below, this aspect is of central importance.

The structure of this review is as follows: After presenting the theoretical foundation and some introductory examples, LCT is first applied to photo-dissociation processes (Section III) and a selective mode excitation in Section IV. In the latter examples, the field is constructed such that it dissipates energy into the system under investigation; that is, *heating* is taking place. The opposite process, where energy is taken out of the system, is discussed in Section V *(cooling)*, which investigates photoassociation processes. Following this line of thought, the next step is to combine heating and cooling fields to achieve, for example, a particle transfer between local minima of potential energy surfaces. Examples for such processes are presented in Section VI. The excitation of electronic degrees of freedom is treated in Section VII, where it is demonstrated that signatures of electronic and vibrational motion described within the Born–Oppenheimer approximation determine the control fields. The phenomenon of *optical paralysis* is illustrated in Section VIII before (in Section IX) several aspects of electronic and vibrational excitation are reinvestigated. The former examples are all treated without taking rotational degrees of freedom into account. Therefore, the role of orientation is discussed separately in Section X. Finally, we discuss the problem of a coupled electronic and nuclear motion and present a study of a control beyond the Born–Oppenheimer adiabatic approximation Section (XI). The chapter is concluded with some final remarks in Section XII.

II. LOCAL CONTROL THEORY

A. General Formulation

We study systems with a Hamiltonian of the form

$$H(t) = T(P) + V(R) + W(R,t) = H_0 + W(R,t) \tag{1}$$

Here R and P are sets of coordinate and momentum operators, respectively. The unperturbed system is described by the Hamiltonian H_0, which is the sum of the kinetic energy $(T(P))$ and the potential energy $(V(R))$ operators. The time-dependent perturbation is denoted as $W(R,t)$. In the examples presented below, the latter is an electric dipole interaction with an external field $E(t)$:

$$W(R,t) = -\mu(R)E(t) \tag{2}$$

where $\mu(R)$ is the projection of the dipole operator on the polarization vector of the external field. The time-dependent Schrödinger equation for the state vector $|\psi(t)\rangle$ is expressed as

$$i\hbar\frac{\partial}{\partial t}|\psi(t)\rangle = H(t)|\psi(t)\rangle \tag{3}$$

One now considers an operator A that is assumed to be independent of time, and the purpose is to control the temporal change of its expectation value. Employing the dynamical equation, Eq. (3), the rate is evaluated as

$$\frac{d\langle A\rangle_t}{dt} = \frac{d}{dt}\langle\psi(t)|A|\psi(t)\rangle = \frac{i}{\hbar}\langle\psi(t)|[H_0,A]|\psi(t)\rangle + \frac{i}{\hbar}\langle\psi(t)|[W,A]|\psi(t)\rangle \quad (4)$$

where $[X,Y] = XY - YX$ denotes the commutator of the operators X and Y. It is important to note that the second term in the latter equation contains the interaction $W(R,t)$ and thus the electric field $E(t)$. This hints at the possibility, if the operators W and A do not commute, to influence the temporal change of the expectation value of A by a properly chosen external field. This is what local control theory is all about, and it could not be expressed in a simpler form.

B. Quantum Versus Classical Mechanics

It will be demonstrated below that, within the LCT approach, many aspects of the combined particle/field dynamics can be understood in the classical limit. The temporal change of a quantity A (not explicitly depending on time) is given by

$$\frac{dA}{dt} = \frac{\partial A}{\partial R}\frac{\partial R}{\partial t} + \frac{\partial A}{\partial P}\frac{\partial P}{\partial t} = \{A,H_0\} + \{A,W\} \quad (5)$$

where we have used Hamilton's equation of motion for the canonical coordinates R and momenta P,

$$\frac{dR}{dt} = \frac{\partial H}{\partial P}, \frac{dP}{dt} = -\frac{\partial H}{\partial R} \quad (6)$$

and employed the definition of the Poisson brackets:

$$\{F,G\} = \frac{\partial F}{\partial R}\frac{\partial G}{\partial P} - \frac{\partial F}{\partial P}\frac{\partial G}{\partial R} \quad (7)$$

The classical expression, Eq. (5), corresponds to the quantum mechanical one of Eq. (4). In fact, starting from the classical expression and, as suggested by Dirac [77], replacing the Poisson bracket of two classical quantities F and G by the commutator of the respective operators as

$$\{F,G\} \leftrightarrow \frac{[F,G]}{i\hbar} \quad (8)$$

directly establishes the connection between the classical and quantum expressions. As is to be expected from Ehrenfest's theorems [26], a classical and quantum treatment of a problem will yield similar results if the time-dependent wavepacket $\psi(R,t) = \langle R|\psi(t)\rangle$ is sufficiently localized if compared to the system's extension [78], for numerical studies on this connection within the context of coherent control, see, e.g., Refs. 14 and 79–87.

C. Analytical Examples

To become accustomed with the ideas of local control theory, we give two simple analytical examples in what follows.

1. Deceleration of a Free Particle

First, the one-dimensional motion of a free classical particle with mass m is treated. Thus, the unperturbed motion is described by the (classical) Hamiltonian

$$H_0(P) = \frac{P^2}{2m} \tag{9}$$

The objective is to stop the motion of the particle by an external field $E(t)$ which interacts via a linear (dipole) coupling as

$$W(R,t) = -R(t)E(t) \tag{10}$$

It is then sufficient to impose the condition that the energy of the system decreases at any instant of time. Thus, the observable to be regarded is the Hamiltonian H_0, i.e., $A = H_0(P)$. Then, the first term in Eq. (5) vanishes and one finds

$$\frac{dA}{dt} = \frac{dH_0}{dt} = \{H_0(P), W(R,t)\} = \frac{P(t)}{m}E(t) = u(t)E(t) \tag{11}$$

where $u(t)$ is the particle velocity. In order to ensure this energy rate to be negative, a simple choice of the field is to take it proportional to the velocity:

$$E(t) = -\lambda u(t) \tag{12}$$

with λ being a positive constant. This corresponds to a force acting on the particle, which is of the form

$$F(t) = -\lambda u(t) \tag{13}$$

Here we notice that local control theory, with the objective to stop a freely moving particle, directly leads to Stokes' law describing friction by a force being proportional to the velocity. In the absence of other forces, this leads to an exponential decrease of the velocity to zero [88].

2. Fermi's Golden Rule

Next, we treat a quantum mechanical two-level system with states $|0\rangle$ and $|1\rangle$ of energies E_0 and E_1, respectively. Thus, the system Hamiltonian can be written as

$$H_0 = |0\rangle E_0 \langle 0| + |1\rangle E_1 \langle 1| \tag{14}$$

Transitions from the initially populated state $|0\rangle$ to state $|1\rangle$ are discussed, which are induced by an interaction as

$$W(t) = -|1\rangle \mu_{10} (\lambda e^{-i\omega t}) \langle 0| \tag{15}$$

where $\mu_{10} = \langle 1|\mu|0\rangle$ denotes the dipole matrix element and only the term corresponding to absorption is kept [26]. The field depends on the frequency ω and the field strength λ, which are kept fixed in what follows. The state vector then is of the form

$$|\psi(t)\rangle = c_0(t)|0\rangle + c_1(t)|1\rangle \tag{16}$$

Within first-order perturbation theory the time-dependent coefficients are given as [26]

$$c_0(t) = e^{-iE_0 t/\hbar}$$

$$c_1(t) = \lambda \frac{i}{\hbar} \int_0^t dt' e^{-iE_1(t-t')/\hbar} e^{-i\omega t'} \mu_{10} e^{-iE_0 t'/\hbar} \tag{17}$$

where the initial condition is $c_0(0) = 1$. We now formulate the control objective by demanding that the field (here depending only on the frequency ω) be determined such that the population in the upper state $|1\rangle$ increases mono-tonically. This can be formulated within the framework of LCT, by defining the operator $A = |1\rangle\langle 1|$ so that, employing Eq. (4), we obtain

$$\frac{d\langle A\rangle_t}{dt} = \frac{d}{dt} \langle \psi(t)|1\rangle\langle 1|\psi(t)\rangle = -\frac{i}{\hbar} E(t) c_1^*(t) \mu_{10} c_0(t)$$

$$= \frac{\lambda^2}{\hbar^2} |\mu_{10}|^2 \int_0^t dt' e^{-i(\omega - (E_1 - E_0)/\hbar)(t-t')} \tag{18}$$

This expression will, depending on the value of the time t, assume positive and negative values. In order to assure that a monotonically increasing excited-state population is prepared, the simplest choice for the parameter ω is $\omega = (E_1 - E_0)/\hbar$, which is the resonance condition obtained from Fermi's Golden rule expression and ensures that the rate increases linearly with time [89]. To conclude, the present example shows that if the field oscillations ω are adjusted to match the system dynamics, characterized by $(E_1 - E_0)/\hbar$, an efficient population transfer is induced. Thus, the condition that, upon excitation, the population in a final state increases monotonically leads (employing the usual approximations) directly to Fermi's Golden rule expression for the transition rate.

III. PHOTODISSOCIATION

The photodissociation of small molecules has been studied extensively over the years [90]. The simplest fragmentation process is that of a bond breaking in a diatomic molecule in a single bound electronic state. This is the first problem that will be treated (numerically) to illustrate the power and simplicity of LCT (Section III.A). There exist, however, competing processes with those one wants to control. One example is a predissociation where fragments, due to nonadiabatic couplings, formed in an exit channel other than that of the target. This is discussed in Section III.B. Another complication that influences the efficiency of a control process is the coupling of different intramolecular modes. Here, the question of a possible selective mode excitation followed by a selective bond breaking arises. An example of how LCT deals with such problems is presented in Section III.C.

A. Direct Photodissociation

As a first numerical example for the application of LCT to molecular problems, we regard the photodissociation of a diatomic molecule in a single electronic state [91]. The model illustrating the key features is that of the excited-state fragmentation of NaI. In Fig. 1, we show the potentials $V_0(R)$ and $V_1(R)$ of two electronic states that exhibit an avoided crossing [92]. This behavior is found in many alkali halides and, due to nonadiabatic coupling, these molecules undergo predissociation upon UV excitation [93]. The characteristic decay dynamics of NaI was monitored in real time on the femtosecond timescale by the Zewail group [94, 95]. First control experiments on the same molecule were performed in the same lab [96]; for related studies, see Refs. 73 and 97–102.

For now, we study molecules prepared by a femtosecond excitation (field $E_1(t)$ in Fig. 1) from the ground state $|0\rangle$ to the excited state $|1\rangle$ and neglect the nonadiabatic coupling—that is, artificially close the predissociation channel; for a more extended treatment see Sections III.B and IX. The excitation process prepares a wavepacket $\psi_1(R, t)$ performing a vibrational motion which, if no external perturbations are present, will persist forever. In fact, even in the case of predissociation, quasi-bound vibrational motion can be observed for a long period [103–105]. The single-state system Hamiltonian then is $H_{0,1} = T(P) + V_1(R)$, where $V_1(R)$ is the upper potential curve shown in Fig. 1.

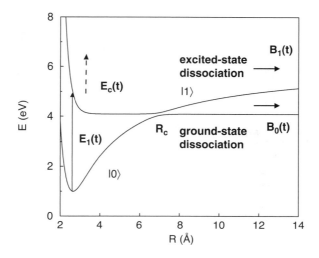

Figure 1. Potential energy curves of the NaI molecule. Two excitation processes are indicated, where a first field $E_1(t)$ induces a $|1\rangle \leftarrow |0\rangle$ electronic transition, and a second field $E_2(t)$ triggers an excited-state dissociation with yield $B_1(t)$. Due to the nonadiabatic coupling in the region around R_c, predissociation can also occur, leading to ground-state atomic fragments (yield $B_0(t)$).

Let us now formulate an objective, namely, that a control field $E(t)$ (field $E_c(t)$ in Fig. 1) is to be determined which causes a fragmentation. This scenario is sketched in Fig. 2. Intuitively it is clear that, in order to break the bond, energy has to be pumped into the system. In the formalism of LCT, this can be realized if the system's energy (represented by the Hamiltonian $H_{0,1}$) increases steadily as a function of time. To specify the interaction with the external field, we have to take the special features of the NaI molecule into consideration: Due to the ionic character of state $|1\rangle$ at distances larger than the point where the nonadiabatic coupling is present (marked as R_c in Fig. 1), the dipole moment is linear [106], so that the dipole interaction is $W(R, t) = -RE(t)$. Then, via Eq. (4) and for $A = H_{0,1}$, we find

$$\frac{d\langle A\rangle_t}{dt} = \frac{d\langle H_{0,1}\rangle_t}{dt} = \frac{i}{\hbar}\langle[-RE(t), H_{0,1}]\rangle_t = \frac{\langle P\rangle_t}{m}E(t) \qquad (19)$$

This is the quantum mechanical analog of the classical expression Eq. (11), with the canonical momentum replaced by the expectation value of the momentum operator $\langle P\rangle_t$. It is then clear that the choice

$$E(t) = \lambda\frac{\langle P\rangle_t}{m} \qquad (20)$$

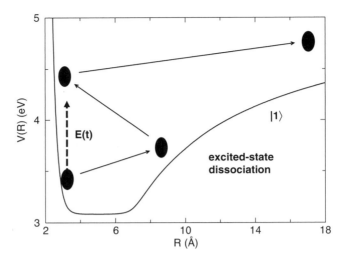

Figure 2. Scenario of a field induced photodissociation: upon energy absorption from an external field which follows the mean momentum of the system, the bond is broken. This is comparable to the dynamics of a classical particle which is driven out of the potential well.

will, for a positive parameter λ, lead to an energy increase in the system—that is, a heating. On the other hand, if λ is chosen to be negative, this leads to a cooling (see below). In passing, we note that the assumption of a linear dipole moment applies to the NaI molecule but, in general, is not valid. Then, the expression for the field (in the case of heating and cooling) contains a commutator between the kinetic energy operator and the dipole moment and thus is more complicated (see Section X.A). This often is accompanied by a loss in efficiency as was shown, for example, in the case of photoassociation [107].

We illustrate this field-induced photodissociation process in Fig. 3. The left-hand panels of the figure contain the coordinate expectation values (solid lines) for two cases that differ in the value of the field strength parameter λ. The lower panel corresponds to the lower field strength ($\lambda = 5 \times 10^{-6}$ a.u.), whereas the results in the upper panel derive from a stronger field ($\lambda = 6 \times 10^{-6}$ a.u.). Also shown is the bound-state population, calculated from the norm of the wavefunction for values of the bond length R smaller than 34.5 Å (dashed lines). It is seen that a bound-state motion with increasing amplitude takes place. For the weaker field, three complete vibrational periods occur until the energy in the system is sufficiently high so that parts of the wavepacket enter the asymptotic region. At that time (~ 9 ps), the bound-state population drops to about 50%. For the stronger field the same behavior is found, but the dissociation sets in at an earlier time and occurs with a higher efficiency. The difference in efficiency can

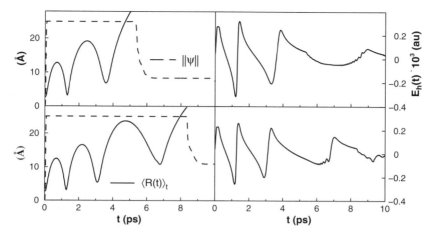

Figure 3. Photodissociation of the NaI molecule. Left-hand panels: Time-dependence of the coordinate expectation values (solid lines) for a weaker (lower panel) and stronger (upper panel) field. The bound-state population in each case is also shown (dashed lines). The latter exhibit a sudden decrease if photodissociation sets in. The driving LCT fields are displayed in the right-hand panels.

be related to the wavepacket dispersion, which diminishes the efficiency of the laser control. The right-hand panels of Fig. 3 display the LCT fields, which track the vibrational motion. Note that as soon as the wavepacket has bifurcated into a bound-state and continuum part (and thus is no longer strongly localized), the field exhibits irregular features.

What we have found is that a control field adapted to fit the instantaneous momentum expectation value leads to an effective photodissociation. The quantum treatment of the forced dissociation has its analogy in the classical treatment of an in phase-driven oscillator [108]. As is to be expected classically, the curves in Fig. 3 document that, because of the longer vibrational period at higher internal energy, the field exhibits a down-chirp—that is, a decrease of the carrier frequency as a function of time. Quantum mechanically, this corresponds to the phenomenon of *ladder climbing* [109, 110]: In order to achieve efficient dissociation, the momentary field frequency has to be adapted to the vibrational level spacing (see the more extended discussion included in Section IV). With increasing energy and due to anharmonicities in the potential curve, the level spacing decreases so that the transition frequency becomes smaller and the field is to be adapted accordingly. This illustrates that the fields derived from LCT can be interpreted clearly in terms of the properties of the perturbed system. Further examples for this connection are given below.

B. Predissociation Dynamics

A restricted model of the NaI molecule, where predissociation is excluded, was discussed in the last subsection. In what follows, we investigate the possibility of controlling the excited-state dissociation in the presence of the predissociation channel [111] (see Fig. 1). Again, the first step is a femtosecond excitation from the ground state, preparing an excited-state vibrational wavepacket.

The unperturbed molecular Hamiltonian in a diabatic representation (for a discussion on this representation, see Refs. 26 and 112–114 and references therein) is

$$H_0 = \sum_{n=0}^{1}\{|n\rangle(T(P) + V_n(R))\langle n|\} + |0\rangle V_c(R)\langle 1| + |1\rangle V_c(R)\langle 0| \qquad (21)$$

where $V_0(R)$, $V_1(R)$ are the diabatic potentials, and $V_c(R)$ is a potential coupling element. The adiabatic potentials displayed in Fig. 1 are obtained by diagonalization of the potential energy matrix; for details of the NaI potentials we refer to Ref. 92 and references therein.

First, we treat the heating of the system. That is, we obtain the field from Eq. (20), where the momentum expectation value is calculated in the excited state and a positive value of the strength parameter λ is chosen. Results for the values of $\lambda = 1 \times 10^{-6}$ a.u. and $\lambda = 4 \times 10^{-6}$ are displayed in Fig. 4. If no field is present, the predissociation proceeds unperturbed, producing ground state atoms Na and I. The respective yield $(B_0(t))$ is shown in the upper panel of the figure as a dashed line. Initially, a stepwise increase is seen which occurs at times the vibrational wavepacket passes the region of the nonadiabatic coupling [94, 115]. Here, the excited-state dissociation yield $(B_1(t))$ is zero. For the weaker LCT field $(s = 1)$, still, no dissociation is observed. On the other hand, the predissociation yield is larger than in the field free case. This is in accord with Landau–Zener theory [26, 116, 117], which predicts an increasing transition probability (in the adiabatic picture) with increasing energy; for an extensive discussion, see Ref. 111.

That the LCT field indeed dissipates energy into the molecule is documented in the lower panel of Fig. 4. The dissociation threshold is at an energy of 5.8 eV. For the weaker field, not enough energy is available so that the threshold is not approached. This is different if the field strength is increased by a factor of 4. Then, at a time of approximately 10 ps, enough energy is provided so that dissociation occurs (middle panel of the figure). In that case, the predissociation yield is lower asymptotically than in the zero-field limit (because all excited molecules will undergo predissociation). One might suspect that the described trends continue if the field strength is increased even more, which, however, is not the case. For example, it is found that the predissication yield is reduced for

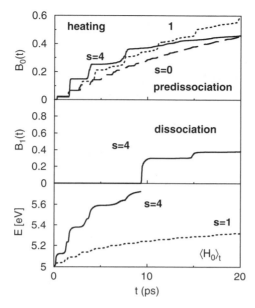

Figure 4. Dissociation and predissociation dynamics in a heating field. The lower panel shows the energy expectation value of the system for two values of the strength parameter ($\lambda = s \times 10^{-6}$ a.u.), as indicated. The middle and upper panels contain the excited-state dissociation ($B_1(t)$) and predissociation yields ($B_0(t)$), respectively. The dashed line curve in the upper panel represents the field free case.

a value of $\lambda = 10 \times 10^{-6}$ a.u.. This is due to the modification of the potential surfaces by the strong external fields—that is, the Stark shift [111]. In fact, such *field-dressed potentials* [118] have been discussed recently in the connection of laser control by Wollenhaupt et al. [119–122].

Next, we discuss the case of cooling, where the field strength parameter λ is chosen negative. Here, no excited-state dissociation takes place, so that only the energy expectation values (lower panel) and the predissociation yields (upper panel) are shown in Fig. 5. The same parameters as in the heating case are employed. It is seen that the LCT field takes energy out of the system and that the efficiency of this process is proportional to the field strength. Note that on the present energy scale, ground-state atoms with zero energy correspond to a value of 4.1 eV. The effect of the field on the predissociation is remarkable: In the shown energy interval, the predissociation yield is decreased dramatically, which is in accord with Landau–Zener theory. Thus, it is possible to stabilize the predissociating molecule with the help of control fields derived from local control theory. Again, for even higher field strengths, Stark effects are present, which strongly influence the yield [111].

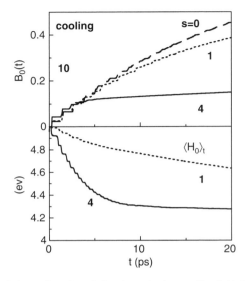

Figure 5. Dissociation and predissociation dynamics in a cooling field. The lower panel shows the energy expectation value of the system for two values of the strength parameter $\lambda = -s \times 10^{-6}$ a.u., as indicated. The predissociation yield $B_0(t)$ is shown in the upper panel, which also contains the field free case.

To conclude this subsection, we demonstrated that heating and cooling LCT fields can be employed to modify the branching ratio of ground-state predissociation and excited-state dissociation. Furthermore, a stabilization of predissociating molecules can temporarily be maintained.

C.   Photodissociation via Selective Mode Excitation

The last subsections treated an example of a diatomic molecule where a single degree of freedom is present. For larger molecules, in general, several fragmentation channels with different particle configurations exist. Here we apply LCT to a model where the field is constructed to selectively excite a determined vibrational mode, which then leads to dissociation into a target fragmentation channel. As an example, we use the water isotope HOD. This water isotope has been investigated as a model system to prove that the control schemes of *mode selective chemistry* [19] and also *vibrationally mediated photodissociation* [123] work successfully. Opposite to other studies [124–126], only the electronic ground state is taken into account. Fixing the bending angle to its value at equilibrium, the potential energy surface $V(R_H, R_D)$ depends on the distances of H (R_H) and D (R_D) to the O atom. A contour plot of a parameterized surface [127, 128] is shown in Fig. 6. One clearly distinguishes the two exit channels H + OD and D + OH.

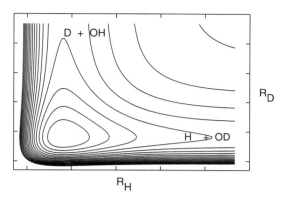

Figure 6. Potential surface $V(R_H, R_D)$ for the ground state of the HOD molecule. Contours are shown in increments of 0.05 a.u. starting at 0.05 a.u. Dissociation can occur into the exit channels corresponding to the fragment configurations H + OD and D + OH, as indicated.

We now employ the ideas described in Section III.A. For example, if one aims at a preferential dissociation into the products H + OD, a heating in the O–H bond should be triggered. Within LCT this can be achieved if one couples the field to the momentum of the H atom, so that (in analogy to the one-dimensional case, Eq. (20), and for a linear dipole moment) one has

$$E(t) = \lambda \frac{\langle P_H \rangle_t}{m_H} \tag{22}$$

(for a more detailed discussion see Refs. 129 and 130). Likewise, a selective population of the D + OH channel is triggered by a field of the same form but with $\langle P_H \rangle_t$ and m_H replaced by $\langle P_D \rangle_t$ and m_D, respectively. Figure 7 collects a typical set of curves for the local control of selective bond breaking. The molecule-field coupling was chosen such that the field couples equally strong to both bonds.

The left-hand panels illustrate the D + OH breakup. The LCT field (lower panel) oscillates with decreasing frequency (down chirp), just like in the one-dimensional example of the last subsection. The OH distance R_D exhibits the same oscillation behavior (middle panel), and thus the D atom is resonantly driven in such a way that the bond length increases as a function of time. Then, around a time of 60 fs, the distance becomes larger than 3 a.u., indicating that some molecules undergo dissociation. This is reflected by the increasing dissociation yield in the D + OH channel which is displayed in the upper left panel of Fig. 7. At later times the yield shows a further stepwise increase which shows that parts of the remaining bound-state wavepacket are forced into the

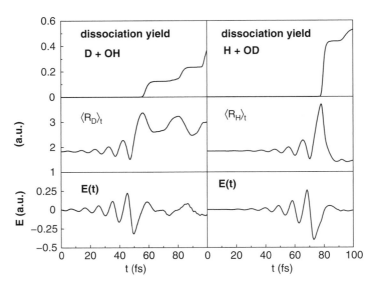

Figure 7. Selective bond breaking of HOD. Dissociation yields (upper panels), bond-length expectation values (middle panels), and LCT fields (lower panels) are shown for two objectives. The left-hand panels correspond to the case where the kinetic energy of the D atom is steadily increased, leading exclusively to D + OH dissociation. The results for the H + OD selective excitation and fragmentation are shown on the right-hand side of the figure.

exit channel. In the observed time interval, a yield of about 40% is obtained which, at later times, settles to a value of 60%. We note that no H+OD products are built, so that the dissociation is perfectly selective.

The forced dissociation process for a selective H + OD configuration is illustrated in the right-hand panels of Fig. 7. The same trends as in the D + OH case are found, and the final yield is about 50%. Again, only the target channel is populated, indicating a 100% selectivity. In the two cases, different strength parameters are used, which explains the fact that the times where dissociation starts differ. The curves in the figure document that the system dynamics (exemplified by $\langle R_{H/D} \rangle_t$) and the properties of the LCT field $E(t)$ are directly entangled.

IV.   SELECTIVE MODE EXCITATION

The recent availability of ultrafast and intense mid-IR laser pulses has opened the way of controlling nuclear motion in the electronic ground state by multiphoton vibrational excitation. In order to access higher vibrational levels, anharmonicities have to be taken into account. This is commonly done by using chirped laser pulses, which change the instantaneous frequency during their duration, thus being able to keep the resonance condition while exciting a specific mode to high-lying states. We already showed that, posing a heating condition, the fields emerging from LCT have this property, see Section III.A. Thus, the physics of "vibrational ladder climbing," which has successfully been demonstrated experimentally in systems like NO [131], $W(CO)_6$ [132, 133], CO adsorbed at a Ru(001) surface [134], $Cr(CO)_6$ [110], $Mo(CO)_6$, $Fe(CO)_5$ [133] and CH_2N_2 [135], emerges directly from the construction scheme. In some of the above-mentioned cases, even dissociation was observed [110, 135] which relates to the results presented in Section III.C. Recently, vibrational ladder climbing was observed in biological systems like carbon monoxyhemoglobin (HbCO) [136]. Within this context, the interesting question arises as to which extent vibrational ladder climbing can be achieved in complex molecules, like in biological systems, and which pulse shapes are best adapted to achieve this task. Calculation on infrared selective mode excitations have been extensively performed for model systems using optimal control theory (OCT) [137–139], adiabatic Floquet theory [140], or techniques of multidimensional wavepacket propagations [87, 141–143].

In this section, the question raised above will be addressed by applying the local control methodology to a two-dimensional quantum model of the active site of hemoglobin. This system has been studied extensively for many years, both theoretically and experimentally, using a wide range of theoretical methods or experimental techniques. For a recent review of the theoretical approaches, see Ref. 144 and references therein. Here, we choose the six-coordinated iron–porphyrin–imidazole–CO (FeP(Im)–CO) as active site model, which recently has also been used for the study of the excited states [145]. In this complex, the imidazole mimics the proximal histidine, which binds to the central Fe atom; and a second imidazole is placed in the proximity of the complex, to include the influence of the distal histidine. The respective configuration is sketched in Fig. 8.

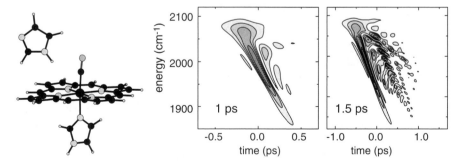

Figure 8. Model of the active site of carboxy hemoglobin (left) and Wigner plots of optimized laser pulses with different pulse durations, as indicated.

Inspired by recent experimental results on multiphoton IR excitation of the C–O stretch [136], we are specifically interested in the anharmonicity of this mode as well as its coupling to the other modes. The vibrational dynamics is investigated within a 2D model with the Hamiltonian

$$H_0(R_0, R_1) = -\frac{\omega_0}{2}\frac{\partial^2}{\partial R_0^2} - \frac{\omega_1}{2}\frac{\partial^2}{\partial R_1^2} + V(R_0, R_1) \tag{23}$$

comprising the CO vibration (R_0) with harmonic energy ω_0 and the most strongly coupled mode, the Fe–CO stretch vibration (R_1, energy ω_1). For a detailed description of the Hamiltonian, we refer to Ref. 146. Consistently, the dipole moment $\mu(R_0, R_1)$ is also expressed as a function of these two-mode displacements. To selectively excite the CO stretch, the operator A entering into the local control scheme is chosen as a one-dimensional Hamiltonian ($A = H^{(1D)}(R_0)$), depending exclusively on the CO stretch coordinate R_0; that is, we couple the field to the mode we want to be excited. We determine the heating field as before by evaluating the expression

$$E(t) = \lambda(t)\langle [H^{(1D)}(R_0), \mu(R_0, R_1)]\rangle_t \tag{24}$$

with a slight modification. Here, $\lambda(t)$ is not a mere parameter but a time-dependent function which ensures an overall pulse duration of 1 ps and 1.5 ps, respectively.

The pulses obtained led to a monotonic increase of the vibrational energy in the CO stretch, up to a total energy of $\sim 12,000\,\text{cm}^{-1}$ ($15,000\,\text{cm}^{-1}$) for the 1-ps (1.5-ps) pulse (vibrational levels of up to $\sim 13(\sim 15)$) [146]. The electric fields constructed in this way are further analyzed and shown as Wigner distributions [147] in Fig. 8. In the case of a 1-ps pulse, one clearly sees how the optimized pulse form requires a change in the excitation frequency from about $2080\,\text{cm}^{-1}$ at early times to $1900\,\text{cm}^{-1}$ at later times, thus resulting in a down-chirp as already discussed in Section III.A.

This result shows that for sufficiently short pulses, the local control algorithm automatically finds the well-known chirped pulse strategy of multiphoton vibrational excitation. This is indeed the strategy that has been used experimentally for a wide range of molecular systems (see, e.g., Refs. 131–133), including the HbCO considered here [136]. For longer pulses, (keeping the same overall laser intensity), we find a completely different behavior: The beginning of the pulse is very similar to the result obtained when imposing the short pulse considered before. However, in contrast to the previous situation, for this longer pulse, the algorithm does not continue this initial chirp to lower energies but creates new spectral components at energies of $\sim 2080\,\text{cm}^{-1}$, forming a chain parallel to the initial chirp. This structure can be viewed as a second chirped pulse, more structured than the initial one, but with the same chirp parameter, determined by the slope of the structures in the time–frequency plane. At this point, it is important to remember that the local control scheme aims to increase the total *energy* in the CO mode at any instant in time, and not the population in individual high-lying vibrational states. Since the total energy is the sum over the populations multiplied with their vibrational energies, the control algorithm has to find a compromise between a small population in high levels and a large population in fairly low levels. The multiple sweep strategy is also reflected in the final vibrational distributions, which, as compared to the simple chirp excitation, now shows a less uniform distribution, with strong population inversion. This suggests that in this case it is more efficient to excite the "leftovers" in fairly low-lying states in a second sweep than to continue exciting to even higher levels. With the latest pulse-shaping techniques, the experimental verification of this *multiple sweep* strategy for efficient energy pumping in specified modes should be possible.

It should be noted that for longer pulses, vibrational dephasing becomes important [136, 148, 149] and might modify the control scenario. The effect of the remaining modes coupled to the CO stretch will be investigated in future studies. Several lines are envisaged, ranging from mixed quantum-classical methods to high-dimensional quantum approaches like the TDSCF (Time-Dependent Self-Consistent Field) [150] or the MCTDH (Multiconfigurational Time-Dependent Hartree) [151–153] methods.

V. PHOTOASSOCIATION

In the examples described in Section III it was shown that a particular choice of the LCT field is able to steadily increase the internal energy of molecules, which eventually leads to fragmentation. In what follows we address the question if a field is able to induce the opposite process, namely, to reduce the energy of two colliding atoms to build a stable molecule. Many theoretical studies on such photoassociation processes have been presented [154–165], mostly involving more than a single electronic state. Also, with the modern technology of cooled atoms and molecules, photoassociation has become an important issue (see, e.g., Refs. 161–164).

As an example for the local control of photoassociation, we first regard the s-wave scattering of H and F atoms, where the total orbital momentum is equal to zero so that, effectively, a problem in a single dimension is to be treated. For an inclusion of the rotational degree of freedom, see Section X.C. The scattering takes place in the electronic ground state, which is the only one taken into account. It is described by an incoming Gaussian wavepacket of the form

$$\psi(R, t = 0) = \sqrt[4]{\frac{2\beta_i}{\pi}} e^{-\beta_i(R-R_i)^2 - i\bar{P}R} \tag{25}$$

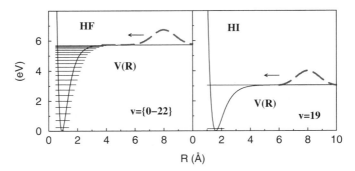

Figure 9. Potential energy curves $V(R)$ for the HF and HI molecule, as indicated. The collision process is described by an incoming Gaussian wavepacket (dashed lines). For the HF molecule the objective is to decrease the energy of the scattered particles in order that bound states are populated (vibrational energies E_v are indicated by the horizontal lines). For HI, the objective is to selectively populate the $v = 19$ state.

This packet, which is localized around $R_i = 15$ Å, has a full width at half-maximum (FWHM) of 5 Å. The mean momentum of the approaching particles is $\bar{P}$, which corresponds to an impact energy of $E_P = \bar{P}^2/2m$. Figure 9 exhibits the potential energy curve for the HF molecule [166] (left panel). Here, we present results for a linear dipole moment, for a more realistic description, see Ref. 107. As in the case of photodissociation, we use the energy rate $d\langle H_0\rangle_t/dt$, which leads to a field of the form Eq. (20) but now with a negative value of the parameter λ. This ensures that the internal energy is decreased at any instant of time.

In Fig. 10 we show results derived from a calculation with a parameter of $\lambda/m = -1.2 \times 10^{-3}$ a.u. and an impact energy of 0.1 eV. The LCT field (upper panel) is smoothly switched on with a Gaussian envelope function, which, upon reaching its maximum at 250 fs, remains at a constant value of 1. This avoids numerical instabilities caused by a sudden switch-on of the field. It can be seen that the field carries an up-chirp. Thus, here the behavior opposite to that present in the case of heating (see Section III) is found. In the latter case, the down-chirp of the field ensured that an effective *ladder climbing* from lower to higher vibrational states takes place. In the present case, on the other hand, one enforces an efficient *ladder descending* from higher to lower vibrational states. Because the level spacing increases, this requires a field frequency that increases with time—that is, an up-chirp. The *ladder descending* is clearly seen in the lower panel of Fig. 10, which contains the vibrational populations

$$B_v(t) = |\langle \varphi_v|\psi(t)\rangle|^2 \tag{26}$$

where $|\varphi_v\rangle$ is a vibrational eigenstate of energy E_v. The contour plot shows that the cooling takes place by a stepwise transfer from one to the next lower

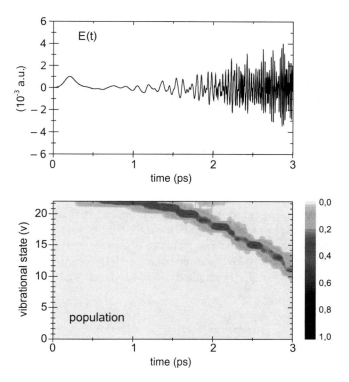

Figure 10. Control field (upper panel) and time-dependence of the population $B_v(t)$ in the vibrational states (lower panel) for the case of photoassociation of H and F atoms.

vibrational state. In the shown time interval the $v = 10$ vibrational state is reached. After 10 ps, the fraction of the photoassociated molecules (for the present parameter set around 60%) is transfered into the ground state (not shown).

The objective in the above example was to induce association by a field-induced decrease of the system energy. One might want to choose a particular target vibrational state $|\varphi_T\rangle$ and achieve photoassociation by populating this state selectively. In this case, the operator A in Eq. (4) may be defined as

$$A = T_\varphi = |\varphi_T\rangle\langle\varphi_T| \tag{27}$$

that is, it is the projector T_φ on the target state. The rate of change can be evaluated as

$$\frac{d}{dt}\langle T_\varphi\rangle_t = iE(t)\langle[-\mu, T_\varphi]\rangle_t = E(t)2\Im\{\langle\psi(t)|\mu|\varphi_T\rangle\langle\varphi_T|\psi(t)\rangle\}$$

where $\Im$ denotes the imaginary part. In order to keep this rate positive, so that the target state population increases steadily, a field of the form

$$E(t) = \lambda \Im \{ \langle \psi(t) | \mu | \varphi_T \rangle \langle \varphi_T | \psi(t) \rangle \} \qquad (28)$$

is chosen, where now λ is a positive number. From the latter equation it is clear that if the target state is not populated initially, the field, by construction, remains zero at all times. Therefore, a small fraction of population is deposited in the target state at the beginning of the scattering. Note, however, that this is done only to start the numerical algorithm and has no consequences if the derived control field would be applied in an experiment.

To illustrate the state-selective photoassociation, we employ a model of the HI molecule incorporating the molecular dipole moment [167] and the ground-state potential curve [168] which is shown in the right-hand panel of Fig. 9. The target state is chosen as the $v = 19$ vibrational state (indicated by the horizontal line in the figure). The results employing a value of $\lambda = -9.0$ a.u. are depicted in Fig. 11. Indeed, the population $B_{19}(t)$ increases (middle panel) and, for longer times,

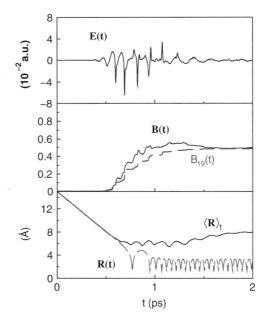

Figure 11. The LCT field inducing selective photoassociation of HI is shown in the upper panel. The bound-state population $(B(t))$ and the population in the target state $(B_v(t))$ with quantum number $v = 19$ are shown in the middle panel of the figure, as indicated. The lower panel compares the quantum mechanical coordinate expectation value $(\langle R \rangle_t)$ with a classical trajectory $(R(t))$.

becomes equal to the total fraction of associated molecules $B(t)$, which is also shown in the figure. A Fourier analysis of the LCT field (the time-dependence of the latter is shown in the upper panel of Fig. 11) reveals that the mean frequency equals the energy difference between the impact energy and the eigenenergy E_{19} of the target state. This again demonstrates the direct connection between the control field properties and the underlying dynamical process.

Let us briefly comment on the relation between the quantum mechanical field-assisted scattering process and its treatment within the classical limit. Therefore, a classical trajectory $(R(t))$ is determined in the presence of the LCT field-derived quantum mechanically, where the initial condition is defined by the average position and momentum of the initial wavepacket (Eq. (25)). The time evolution of this trajectory is compared to the coordinate expectation value in the lower panel of Fig. 11. It is seen that the trajectory is trapped by the field interaction, leading to a classical vibration at a smaller total energy (0.102 eV as compared to $E_{19} = 0.113$ eV). Deviations in the two curves are to be expected and arise from the spatial extent of the wavepacket. Here, we encounter a first example for the qualitative relation between quantum and classical dynamics in the case of local control.

We note that the efficiency of state-selective photoassociation decreases with decreasing quantum number and does not take place at all (quantum mechanically and classically) if one aims at a population of the vibrational ground state (see Ref. 107).

VI. MULTIPLE POTENTIAL WELLS

In Section III we showed how an LCT field is able to steer a bond-breaking by imposing the condition that the energy rate assumes only positive values (heating). Then, the opposite process of photoassociation (Section V) was discussed, where the field perturbation diminishes the energy of two colliding particles so that a bound molecule is formed (cooling). In what follows we combine heating and cooling fields in order to direct a particle between potential wells. This dynamics is typical for a charge transfer [169] and also for intra-molecular or intermolecular proton (or hydrogen) transfer processes [169–171]. A vast amount of theoretical studies is available (see Refs. 172–179).

A. Double Well Potentials

Let's discuss first an asymmetric potential curve in a single degree of freedom as is sketched in the lower part of Fig. 12. There, a wavepacket is shown, describing the localization of a particle (mass m) in the well at around 4 a.u.. The purpose now is to drive the wavepacket out of the well, over the barrier and trap it in the shallow well at negative values of the coordinate R [129]. Within the model

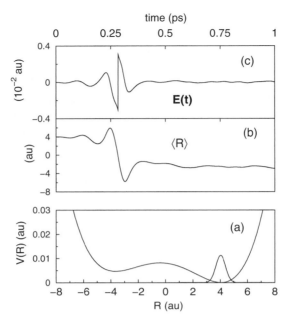

Figure 12. Particle transfer in a double-well potential. The initial wavepacket is localized in the potential well at positive distances (panel (a)). The LCT field induces a vibrational motion and drives the packet over the barrier, as can be taken from the coordinate expectation value displayed in panel (b). A phase jump is introduced in the field (panel (c)) at the time the barrier is passed, and afterwards a trapping in the second potential well takes place.

calculation, a linear dipole moment $\mu(R) = \mu_1 R$ is assumed. Then, employing Eq. (4) for the operator $A = H_0$, we define the control field as

$$E(t) = \lambda \frac{\langle R \rangle_t}{|\langle R \rangle_t|} \frac{\mu_1}{m} \langle P \rangle_t \qquad (29)$$

where the parameter λ is taken to be positive. This means that as long as the particle is localized, on the average, at positive distances, a heating field acts, whereas if the barrier is passed and the motion takes place in the region with negative positions, the field takes energy away from the system (cooling). As a consequence, at some time the field switches sign; that is, a phase jump takes place. The scenario described above, involving a control field depositing energy into, and then absorbing it from, a system is illustrated in the upper panels of Fig. 12. There, the time-dependence of the coordinate expectation value is shown together with the control field derived from Eq. (29). It is seen that the

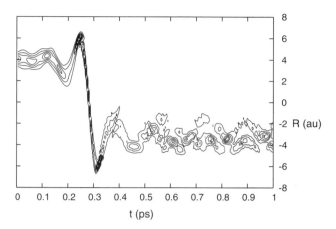

Figure 13. Transfer dynamics in a double-well potential: The time-dependent probability density illustrates the over-the-barrier and trapping dynamics.

perturbation induces a vibrational motion with increasing amplitude so that, around a time of 0.28 ps, the barrier is crossed. The control field that follows the wavepacket oscillations then exhibits a phase jump. Afterwards, the average coordinate takes a value close to -4 a.u. which corresponds to the minimum of the second potential well (at $R < 0$). A clearer picture can be obtained regarding the probability density $\rho(R, t) = |\psi(R, t)|^2$, which is displayed in Fig. 13. There, the initial vibrational motion, the barrier crossing, and the trapping can be seen.

The present example illustrates that fields derived from LCT are perfectly able to induce a particle transfer in a one-dimensional double-well potential. If the respective coordinate is regarded as a reaction coordinate, the question arises with regard to the extent to which other degrees of freedom have an influence on the transfer process. Of course, no general answer can be given, and we now present a discussion that rests on the inclusion of only a single additional coordinate.

A two-dimensional potential surface $V(R_1, R_2)$ is shown as a contour plot in Fig. 14. Here, a wavepacket describing a trapped particle is placed in the potential well at values of $R_1 > 0$. As in the one-dimensional case, the purpose is to induce a transfer into the second potential well by application of an LCT field [129]. In order to do so, we impose a heating condition for the positive values of the expectation value of the (reaction) coordinate $\langle R_1 \rangle_t$ and a cooling condition for $\langle R_1 \rangle_t < 0$. Although it is possible to exclusively incorporate the dynamics in the reaction coordinate into the construction of the field (as was done in the selective mode dissociation; see Section III.C), here we include both

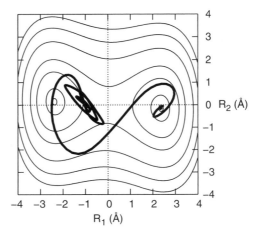

Figure 14. Transfer dynamics in a double-well potential in two degrees of freedom. Contour plots of a model potential surface are shown. The thick black line is the coordinate expectation value calculated for the first picoseconds of the interaction of the particle with the LCT field, starting in the potential well at positive values of R_1.

degrees of freedom. Employing a linear dipole moment $\mu = R_1 + R_2$, the energy rate is

$$\frac{d}{dt}\langle H_0 \rangle_t = \frac{1}{m} E(t)(\langle P_1 \rangle_t + \langle P_2 \rangle_t) \tag{30}$$

where the expectation values of the momentum operators appear. Then, the field is taken as

$$E(t) = \lambda \frac{\langle R_1 \rangle_t}{|\langle R_1 \rangle_t|} (\langle P_1 \rangle_t + \langle P_2 \rangle_t) \tag{31}$$

with $\lambda = 4 \times 10^{-4}$ a.u., and the proton mass is used in the numerical example. The transfer dynamics is illustrated in Fig. 15, which contains the probability densities

$$\rho(R_1, t) = \int dR_2 |\psi(R_1, R_2, t)|^2 \tag{32}$$

$$\rho(R_2, t) = \int dR_1 |\psi(R_1, R_2, t)|^2 \tag{33}$$

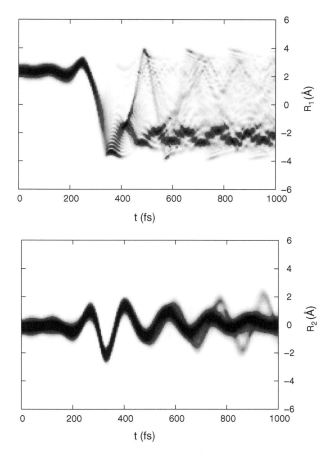

Figure 15. Transfer dynamics in a double-well potential in two degrees of freedom. Shown are the probability densities in the reaction coordinate ($\rho(R_1, t)$, upper panel) and the additional vibrational degree of freedom ($\rho(R_2, t)$, lower panel).

for times up to 1 ps. The reaction coordinate density $\rho(R_1, t)$ starts to oscillate as a consequence of the LCT-field interaction. Then, around 300 fs, the barrier is crossed and most of the density is trapped in the target potential well. However, it is seen that other parts of the density move over larger distances and become rather delocalized. By construction, the field also steers a vibrational motion in the other degree of freedom (R_2). There, most of the respective density $\rho(R_2, t)$ (lower panel of Fig. 15) remains at values close to $R_2 = 0$, but a smaller fraction performs a more extended vibrational motion.

A clearer picture of the complicated density dynamics can be obtained regarding the coordinate expectation value $\vec{R}_t = (\langle R_1 \rangle_t, \langle R_2 \rangle_t)$. This trajectory is

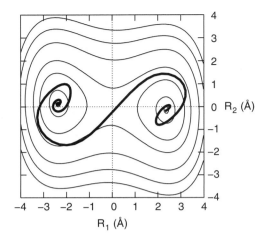

Figure 16. Classical transfer dynamics in a double-well potential. A trajectory is shown which is perfectly driven from one potential well to the other by a field that is constructed classically.

shown in Fig. 14 as a solid thick line). It starts at time $t = 0$ and ends at a value of $t = 1$ ps. Again, one recognizes the over-the-barrier motion accompanied by a vibrational excitation in the (R_2) degree of freedom. At later times, the quantum orbit does not end up in the target potential well, which is a reflection of the delocalization of the densities shown in Fig. 15.

For comparison, we now study the classical control problem and determine the field from Eq. (31), where the expectation values are replaced by the values of the classical coordinates and momenta, respectively. This yields the classical trajectory shown in Fig. 16, which is superimposed on the potential energy contours. Here, a perfect transfer is found where the particle stops in the minimum of the target potential well. A comparison with the trajectory derived from the quantum calculation (Fig. 14) shows that the classical orbit follows the quantum orbit closely until the reaction barrier is passed. At later times, due to the missing *dispersion* in the classical treatment, deviations are found.

The resemblance of the classical and quantum dynamics suggests that the LCT field derived from one or the other treatment might be similar. This is indeed the case, as is documented in Fig. 17. The classically and quantum mechanically derived fields shown in the figure are almost indistinguishable for times before the barrier crossing takes place (~ 300 fs), and deviate not essentially later on. Again, here we emphasize the close relationship between the LCT-field properties, the system dynamics, and also the classical limit, which is closest to our intuition.

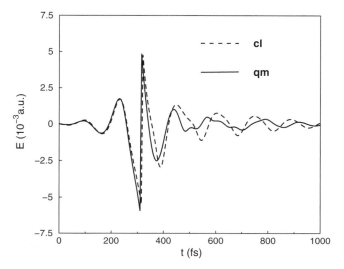

Figure 17.   Local control fields for the transfer dynamics in a double-well potential. Shown are fields constructed quantum mechanically and classically, as indicated.

B.   Multiple Transfer Processes: Triple-Well Potentials

In the previous subsection the particle transfer in double-well potentials was discussed. Many problems, however, involve a transfer over a more extended system. For a recent study, see the hydrogen transfer along an ammonia wire cluster [180]. These types of, more complicated dynamical processes can also be influenced by properly designed LCT fields. To hint at possible applications, we here treat a triple-well potential along a single coordinate. Figure 18 shows such a potential curve with wells numbered as (A), (B), and (C), respectively. Two scenarios are sketched starting from the same initial configuration, namely that a wavepacket (here, describing a proton) is localized in the (A)-potential well. The control pathway indicated in Fig. 18a leads to a first transfer and trapping in well (C), followed by a second transfer into well (B). On the other hand, the two-step transfer sketched in panel (b) involves an initial transfer between wells (A) and (B), along with a successive move into well (C). From what has been discussed above, it emerges that, in order to steer this process, a combination of heating and cooling fields has to be used. Details of the parameterization of these fields can be found in Ref. 181. Here, we only discuss the form of the control fields and their relation to the coordinate expectation values reflecting the average particle position. The latter quantities are shown in Fig. 19.

Let us first regard the field inducing the (B) ← (C) ← (A) process. The constructed field exhibits phase jumps at those times when the heating and

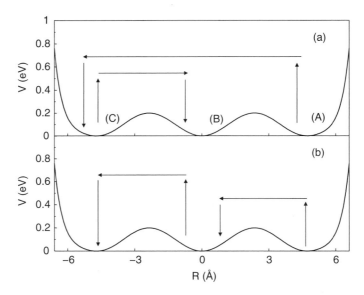

Figure 18. Transfer dynamics in a triple-well potential. Two transfer processes are sketched which involve stepwise transfers between the potential wells (A), (B), and (C).

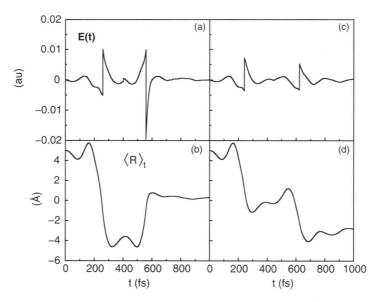

Figure 19. Transfer dynamics in a triple-well potential. Control fields (panels (a) and (c)) and mean positions (panels (b) and (d)) are shown for the two transfer processes sketched in Fig. 18.

cooling conditions are switched. Despite the somehow erratic form of the control field, the expectation value $\langle R \rangle_t$ indicates that the objective is perfectly obtained: Vibrations in the well (A) are excited until a motion over the barrier becomes possible. A temporal confinement in well (C) is achieved and, upon imposing another heating/cooling condition, the particle finally is kept in the middle potential well (B), at least for the time-interval displayed in the figure. The second multiple transfer process is also successfully performed by the LCT field (see the right-hand panel of Fig. 19). Vibrations take place first in well (A), then in well (B), and finally in well (C), which is the target region for the particle to be localized in.

In Fig. 20 we display the wavepacket dynamics for the two transfer pathways, taking place in the LCT field. It is remarkable that the depicted densities

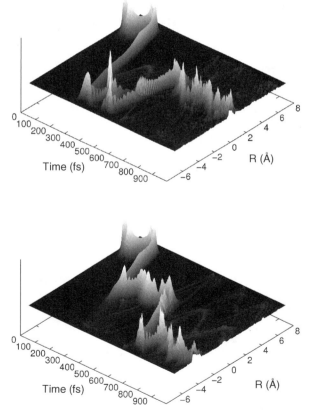

Figure 20. Transfer dynamics in a triple-well potential: The wavepacket dynamics $(|\psi(R,t)|^2)$ is shown for the processes indicated in Fig. 18.

remain strongly localized so that the two-step transfer from the initial to the respective target state is steered efficiently. What can be taken from the density dynamics is close to the intuitive picture of the sequential transfer of a classical particle in the multiple potential well curve.

VII. ELECTRONIC EXCITATION

Until now we have discussed the dynamics in single electronic states, that is, translational and vibrational nuclear motion. We now extend the treatment and include the possibility of an electronic excitation. This is done within the Born–Oppenheimer approximation where the nuclear dynamics in different electronic states is coupled exclusively by the external field. The system that serves as an example is the Na_2 molecule, which has been studied extensively employing femtosecond spectroscopy [182–188].

In Fig. 21, an excitation scheme including three electronic states is sketched. The respective potential curves $V_n(R)$ belong to to the $X^1\Sigma_g^+(|0\rangle)$, the $A^1\Sigma_u^+(|1\rangle)$,

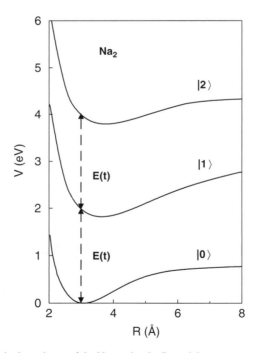

Figure 21. Excitation scheme of the Na_2 molecule. Potential energy curves of three electronic states are displayed. The arrows indicate a control field that induces a selective excitation of a target electronic state.

and the $2^1\Pi_g(|2\rangle)$ excited states of the sodium dimer [189–191]. Below, we discuss the selective population transfer between the three electronic states. They are decoupled from the ionization continuum, which can be excessed during the excitation process. The Hamiltonian for the unperturbed molecular system is

$$H_0 = \sum_{n=0}^{2} |n\rangle (T(P) + V_n(R))\langle n| \tag{34}$$

The interaction term reads

$$W(R, t) = -(|0\rangle \mu_{01} E(t)\langle 1| + |1\rangle \mu_{12} E(t)\langle 2|) + c.c. \tag{35}$$

where $c.c.$ denotes the conjugate complex and μ_{nm} is the transition dipole moment connecting the states $|m\rangle$ and $|n\rangle$. The state vector has three components:

$$|\psi(R, t)\rangle = \sum_{n=0}^{2} \psi_n(R, t)|n\rangle \tag{36}$$

where the nuclear motion in state $|n\rangle$ is described by the wavefunction $\psi_n(R, t)$. The rotational motion is not taken into account in what follows.

Here, we are interested in a selective population of a target electronic state $|k\rangle$. Within the LCT approach (Eq. (4)), we define the operator $A = |k\rangle\langle k|$ and note that the expectation value of this projector equals the population $B_k(t)$ in the electronic target state. The time derivative of this expectation value is evaluated as [43, 45, 47]

$$\frac{dB_k(t)}{dt} = -2E(t) \sum_{m} Im\langle \psi_k(t)|\mu_{km}|\psi_m(t)\rangle \tag{37}$$

From this equation it is obvious that a field chosen as

$$E_k(t) = -\lambda \sum_{m} Im\langle \psi_k(t)|\mu_{km}|\psi_m(t)\rangle \tag{38}$$

will keep the population change at a positive rate—that is, will steadily increase the target state population (for a positive strength parameter λ). The expression for the field contains the overlap of the target state nuclear wavefunction with those of all other states.

As a first example, we treat the selective population of the first excited state $|1\rangle$, with the ground state being populated at the beginning (for numerical details see Ref. 192). The results, for two different field strengths, are collected

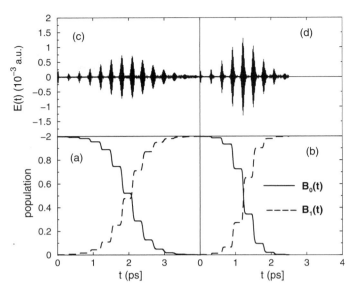

Figure 22. Selective population of the first electronic state of Na_2 achieved by LCT fields. The population in the ground and excited states are shown in the lower panels and the respective fields in the upper panels, as indicated. A weaker (left) and stronger (right) field strength was employed in the calculation.

in Fig. 22. There, the ground-state population $(B_0(t))$ and the excited-state population $(B_1(t))$ are displayed in panels (a) and (b), as indicated. The LCT fields are shown in the upper panels (c) and (d). Here, we employ a seed pulse to start the numerical algorithm (seen in the figure at very early times) which, as was stated above, is irrelevant for an experimental implementation. We note that the population in the second excited state $|2\rangle$ is negligible at all times (and thus is not visible in the figure). The target state population increases in steps which, because of norm conservation, are accompanied by a decrease of the ground-state population. The control process is finished after 4 ps for the weaker field; at that time, all of the ground-state population has been transferred to the excited state; that is, a 100% efficiency is found. Increasing the field strength has the effect that the process is faster, but the stepwise population change and also the structure of the control field are similar. The latter consists of a train of subpulses that are separated by about 300 fs, being the characteristic time for the vibrational wavepacket dynamics in the Na_2 excited state $|1\rangle$ [183]. Reasons why this particular field property arises are given below.

The seed pulse prepares a small-amplitude wavepacket $\psi_1(R, t)$ in the excited state which moves out of the Franck–Condon (FC) region so that the overlap with the ground-state wavefunction approaches zero. As the field is constructed from this overlap, the field vanishes as well. Then, after one vibrational period, the excited-state wavepacket returns to the Franck–Condon region so that the LCT field deviates from zero and is able to transfer population to the excited state. Afterwards, the scenario repeats itself: As a consequence of the outward motion of $\psi_1(R, t)$, the field approaches zero. Thus, the subpulse envelopes directly map the wavepacket motion in and out of the FC region. One notes that the peak height of the subpulses first increases, assuming a maximum at the time the ground- and excited-state populations are equal (see Fig. 22). This can be rationalized, taking into account that only two electronic states participate in the transfer process. It is reasonable (although not completely correct) to assume that the modulus of the overlap integral entering into the field (Eq. (38)) is proportional to $\sqrt{B_0(t)B_1(t)}$. Then, the field intensity is expressed as

$$|E(t)|^2 \sim B_0(t)B_1(t) = B_0(t)(1 - B_0(t)) \tag{39}$$

This function has a maximum at a value of $B_0(t) = B_1(t) = 0.5$, explaining the overall intensity pattern of the control field (Fig. 22).

Until now, we documented how the systems vibrational dynamics influences the LCT field—namely, that it is responsible for the encountered pulse sequence. The question remains as to why no transfer to the higher electronic state $|2\rangle$ occurs. To answer this question, a detailed analysis of the spectral properties of the subpulses is necessary. This was performed in Ref. 192 and is not repeated here. What is found can be explained on intuitive grounds. First, within the FC region (for the $|1\rangle \leftarrow |0\rangle$) it is necessary that the driving field be on resonance with the $|1\rangle \leftarrow |0\rangle$, but not with the $|2\rangle \leftarrow |1\rangle$ transition. This, indeed, is valid and the fast oscillations of the subpulses correspond to the $|1\rangle \leftarrow |0\rangle$ transition frequency. Nevertheless, when the wavepacket $\psi_1(R, t)$ moves outward, the resonance condition for the transition to the second excited state changes, which follows from a difference-potential analysis [184, 193] and can be employed for femtosecond spectroscopy with chirped pulses [194]. Then, eventually, the LCT field becomes resonant with the $|2\rangle \leftarrow |1\rangle$ transition and a transfer is possible so that the population in the target state is diminished. This, however, does not happen. It is found that the algorithm is smart enough to shift the carrier frequency in such a way that the second excited state is not populated.

To summarize, the field exhibits slow and fast oscillations that correspond to the timescales of vibrational and electronic motion, respectively. It is interesting to change the objective and aim at a selective population of the second excited state $|2\rangle$. This is a more demanding problem because the ground state is not directly coupled to the target state and thus transitions have to proceed via the intermediate state. The LCT field and the population dynamics for the selective

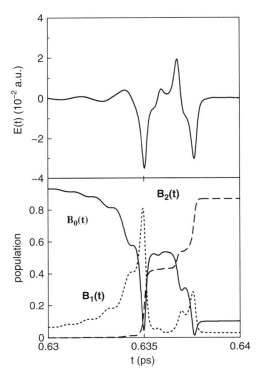

Figure 23. Selective population of the second electronic state of Na_2. The populations in the different states are shown in the lower panel, and the control field in the upper panel.

excitation of the target state $|3\rangle$ is displayed in Fig. 23. Note that only a small time interval is shown. For times outside of this interval, the populations change only insignificantly. For a long time, small amounts of population are transferred to the first excited state. Then, within a period of 10 fs, Rabi-like oscillations [26] are seen with the effect that the target-state population settles to a value of almost 90%. Thus, the transfer is effective but not to a 100%. The critical time interval of 10 fs is much smaller than any vibrational period in the system. This means that the timescale of the population transfer is determined exclusively by the electronic level spacing. We note that a neglect of the vibrational kinetic energy operators in a calculation results in a similar population dynamics as shown in Fig. 23, which emphasizes that the nuclear motion is of no importance here. This is supported by a Fourier analysis of the LCT field which identifies a main frequency component corresponding to an energy of ~ 2 eV, which is consistent with the separation of the potential curves (see Fig. 21).

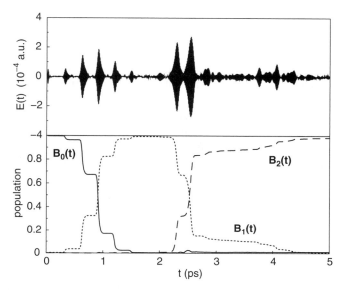

Figure 24. Selective population of the second electronic state $|2\rangle$ by a two-step process. The target state is state $|1\rangle$ until it is populated to 100%. After that (around 2 ps), the target state is redefined as state $|2\rangle$. In this way, the control field (upper panel) induces a perfect population transfer (lower panel).

There exists a different pathway for a selective population of state $|2\rangle$. In a first step, a selectively transfer from the ground to the first excited state is performed. We already showed that this is possible with a 100% yield. The second step then involves a change of the target state from $|1\rangle$ to $|2\rangle$. A numerical example for this successive excitation process is shown in Fig. 24. Until a time of 2 ps, the same features as already discussed above (i.e., Fig. 22) are found, namely, the stepwise increase of the population $B_1(t)$ and a pulse-train structure of the field. Afterwards, a more complex time-dependence of the field is encountered. This is because now the vibrational dynamics in the intermediate as well as the target electronic state enters into the construction scheme for the field. The control in this two-step process is more effective if compared to the direct transfer (Fig. 23). Here, we achieve an almost complete transfer of population into the target state $|2\rangle$.

VIII. OPTICAL PARALYSIS

A further application of local control theory is the so-called optical paralysis scheme [26, 43, 45, 47]. The basic idea is to manipulate ground-state properties while suppressing unwanted excitations to high-lying states, an important issue for strong-field coherent control. As will be shown below, only a condition on the phase of the electric field is required, leaving enough freedom to achieve further control goals, such as vibrational heating or cooling. In this section we show how ground-state vibrational excitation can be achieved while effectively locking the population on the ground and first excited electronic states only.

As before, we consider a molecular system interacting with an external field $E(t)$. At this stage, we assume that the field is complex, which is equivalent to adopting the rotating wave approximation [26]. Complex fields ($E(t)$) are necessary at the *design* stage, and subsequently the real fields are constructed by taking $E^{(r)}(t) = E(t) + E^*(t)$. To demonstrate the main principle, the vibronic manifold of the Na$_2$ dimer is chosen, including $n = 14$ electronic states together with all transition dipole moments between them. These curves have been calculated by a two-electron full configuration interaction method using effective core potentials and core polarization potentials [195, 196], similar to the method described in Ref. 197.

In order to have a fairly realistic picture, we do not invoke the Condon approximation, but use transition dipole moments which are a function of the internuclear distance.

Within the Born–Oppenheimer approximation, the time-dependent Schrödinger equation for the nuclear wavefunctions $\psi_j(r, t)$ in the different electronic states $|j\rangle$ using complex fields is

$$i\hbar \partial_t \psi_j(R, t) = H_j \psi_j(R, t) - \sum_{k<j} E^*(t) \mu_{kj} \psi_k(R) - \sum_{k>j} E(t) \mu_{kj} \psi_k(R) \quad (40)$$

where H_j is the field-free Hamilton operator for the jth electronic state. In the above equations, we have assumed an ordering of the electronic states with increasing energy at the equilibrium distance.

In order to keep the population $N_j(t) = \langle \psi_j | \psi_j \rangle_t$ in the ground-state and first-excited-state constants, we start from the condition

$$\frac{d}{dt}(N_1(t) + N_2(t)) = -2Im\left[E(t) \sum_{j=3}^{n} (\langle \psi_1 | \mu_{j1} | \psi_j \rangle_t + \langle \psi_2 | \mu_{j2} | \psi_j \rangle_t) \right] = 0 \quad (41)$$

This condition is satisfied if one chooses for the complex field [47]

$$E(t) = C(t) \sum_{j=3}^{n} (\langle \psi_j | \mu_{j1} | \psi_1 \rangle_t + \langle \psi_j | \mu_{j2} | \psi_2 \rangle_t) \quad (42)$$

where $C(t)$ can be any real-valued function.

Defining the ground-state energy to be

$$E_g(t) = \frac{\langle \psi_1 | H_1 | \psi_1 \rangle_t}{\langle \psi_1 | \psi_1 \rangle_t} \quad (43)$$

one finds for its temporal variation the following:

$$\frac{dE_g}{dt} = \frac{2}{\langle \psi_1 | \psi_1 \rangle_t} C(t) Im \left[\sum_{k=3}^{n} (\langle \psi_j | \mu_{j1} | \psi_1 \rangle_t + \langle \psi_j | \mu_{j2} | \psi_2 \rangle_t) \sum_{j=1}^{n} \langle \psi_1 | (H_1 - E_g) \mu_{1j} | \psi_j \rangle_t \right] \quad (44)$$

The sign of $C(t)$ can now be adjusted such that the vibrational energy in the ground state increases (or decreases) monotonically, while the sum of the populations on the ground and first excited states is kept constant. As in the general formulation of local control theory, in order to avoid unphysical divergences, one choice for $C(t)$ is

$$C(t) = Im \left[\sum_{k=3}^{n} [\langle \psi_j | \mu_{j1} | \psi_1 \rangle_t + \langle \psi_j | \mu_{j2} | \psi_2 \rangle_t] \sum_{j=1}^{n} \langle \psi_1 | (H_1 - E_g) \mu_{1j} | \psi_j \rangle_t \right] \quad (45)$$

which goes to zero smoothly when the terms in brackets in Eq. (5) approaches zero. It is intriguing that the design procedure only works with complex fields; as a consequence, the obtained fields need to be checked to assure that the neglect of the counter-rotating terms is justified.

In practice, the scheme proceeds as follows: Initially, the molecule is in its electronic and vibrational ground state. A "seed" pulse transfers a part of the ground-state population to higher-lying states. The overlap of the ground-state wavepacket with nuclear wavepackets in the excited states creates a instantaneous transition moment, which is used to locally construct an electric field satisfying the locking and heating (cooling) condition. The physical mechanism that underlies this scheme is a resonant stimulated Raman scattering process (RISRS) [198–200]. In Ref. 43, where a weak field version of this scheme was given, it was found that creating an instantaneous transition dipole moment without population transfer can be viewed in the frequency domain as excitations *between* the vibrational transitions—that is, in resonance with electronic transitions but off-resonant with vibrational transitions.

The 14 electronic states of the sodium dimer which are considered are shown in Fig. 25 (left panel). A selection of transition dipole moments are displayed in the right panel for illustration. Note, however, that all transition dipole moments

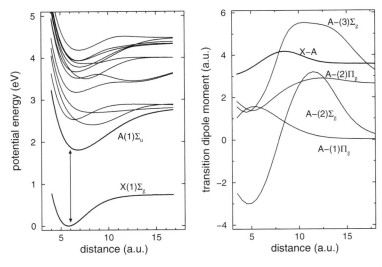

Figure 25. Potential energy surfaces of Na_2 (left) and a selection of transition dipole moments. Note that in the calculations all R-dependent transition dipole moments between all states depicted in the left panel are included.

were included in the calculations. In what follows, only linear polarizations are considered, and the rotational motion is frozen.

Figure 26 summarizes the results of the simulations. In panel (a) the result of the design step yielding a complex field is shown (real and imaginary part). The complex field is created according to Eqs. (42) and (45) by calculating the various overlaps at any instant in time, and the fields determined are immediately re-injected into the time-dependent Schrödinger equation within the spirit of local control theory. As stated above, the paralysis principle rests on the rotating wave approximation, and it is thus important to clarify the effects of counter-rotating contributions, especially in the high intensity regime. The locking and heating process thus needs to be confirmed by a second calculation based on a real field obtained by $E^r(t) = E(t) + E^*(t)$. Panel (b) shows the heating process on the electronic ground state. The results using the complex fields are shown (dotted line), together with the results using the real field $E^{(r)}$ (solid line). One clearly sees that the heating is well reproduced by the real field.

Panels (c) and (d) of Fig. 26 contain the populations of the ground (X) and first excited state (A), together with its sum (dashed line) to assess the locking mechanism. The result in panel (c) is obtained during the design step—that is, the complex fields shown in panel (a). One clearly sees how the algorithm works: After the seed pulse has transferred $\sim 40\%$ of the ground-state population onto the first excited state, and about 25% to the rest of the electronic states, the

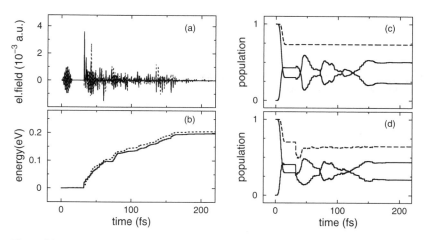

Figure 26. (a) Complex field obtained by the local control procedure. (b) Vibrational energy of the ground-state wavepacket, calculated with a complex (dotted line) and real field (solid line), respectively. (c,d) Time evolution of the X- and A-state population, along with their sum (dashed line) for the complex design field (c) and for the corresponding real-valued field $E^{(r)}$ (d).

control procedure sets it at 25 fs. As a consequence, the X and A states exchange population, but the sum remains constant over time (dashed line in (c)). When using $E^{(r)}$, the population in the X and A state together are not rigorously constant any more (d), indicating that the counter-rotating terms induce variations of a few percent. However, around 50 fs, the control procedure produces an extremely peaked pulse structure (see panel (a)) at high intensities. In this temporal region, the rotating wave approximation breaks down and the population locking is violated by $\sim 10\%$.

To summarize, the optical paralysis scheme can be viewed as local control with complex fields. If the processes are sufficiently well described within the rotating wave approximation—that is, counter-rotating terms do not play a crucial role—the real fields that can be produced in the laboratory can be successful in locking populations in selected electronic states while other strong field control objectives can be achieved, which might be much more complex than the heating considered in this section.

IX. PHOTODISSOCIATION AND ELECTRONIC TRANSITIONS, REVISITED

We now combine various aspects of the problems discussed in the previous sections, taking the NaI dynamics as an example. In Section III.A, the excited-state dissociation of this molecule was treated, excluding the possibility of predissociation. This approximation was removed in Section III.B, allowing for the population of exit channels corresponding to ground- and excited-state fragments. Then, the population transfer between different electronic states of Na_2 was discussed in Section VII. Regarding the excitation scheme of NaI (Fig. 1) we may now ask for the possibilities of a *complete local control of the molecular excited-state photofragmentation* [201], where the excitation step from the ground state is properly taken into account. Accordingly, the Hamiltonian is of the form

$$
H = \sum_{n=0}^{1} |n\rangle \{T(P) + V_n(R) + W_{nn}(R,t)\} \langle n|
$$
$$
+ |0\rangle \{V_c(R) + W_{01}(R,t)\} \langle 1| + |1\rangle \{V_c(R) + W_{10}(R,t)\} \langle 0|
$$
(46)

where $W_{nm}(R,t)$ are the dipole interaction terms. The objective, in what follows, is to control excited-state dissociation (i.e., to achieve a large value of the dissociation yield $B_1(t)$). Therefore, it is necessary to depopulate the ground state as much as possible and, furthermore, heat all excited molecules to break apart into the dissociation channel. In order to put this control scheme into action, two pathways can be followed. One possibility is to construct a field $E_1(t)$ to trigger the electronic transition (from Eq. (38)) and then afterwards set up a heating field (from Eq. (20)) to induce the excited-state dissociation. Alternatively, one could construct the two field components serving the described purposes simultaneously. Both scenarios are illustrated in what follows.

First, the successive excitation pathway is discussed. In Fig. 27 we show the two control fields that act after each other (lower panel). Scaling parameters of $\lambda_1 = 5 \times 10^{-3}$ a.u. and $\lambda_2 = 2 \times 10^{-3}$ a.u. are employed. As in the case of the Na_2 population transfer (Section VII), the field $E_1(t)$ consists of a pulse sequence (lower panel), which leads to a stepwise increase of the excited-state population. The field is switched off at 5 ps, and the heating field $E_2(t)$ starts to interact with the system. Oscillations characterized by the excited-state vibrational period are seen. A fast dissociation (seen in the increase of $B_1(t)$, upper panel of Fig. 27) is triggered, but also a population transfer between the two electronic states and predissociation takes place. For the present pulse parameters we find a total dissociation yield $B_1(t \rightarrow \infty)$ of about 30%.

The second interaction scheme incorporates a field being the sum $(E(t) = E_1(t) + E_2(t))$ of the two field components which are constructed as before. The structure of the obtained field then, of course, is more complex. In Fig. 28, we show the single field components in the lower panel (LCT field for the population transfer) and the middle panel (LCT field for the heating). It is worth mentioning that each field component fulfills the posed objective but the sum of the

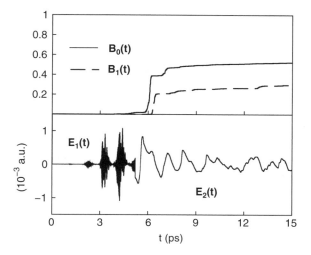

Figure 27. Successive interaction of two LCT fields where the first $(E_1(t))$ induces a population transfer and the second acting after 5 ps $(E_2(t))$ triggers excited-state dissociation. These fields are shown in the lower panel, whereas the predissociation $(B_0(t))$ and excited-state fragmentation $(B_1(t))$ yields are contained in the upper panel.

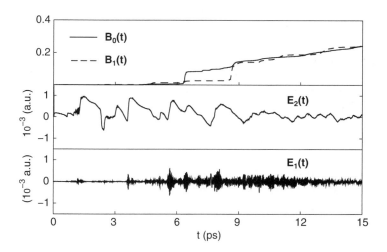

Figure 28. Simultaneous interaction of two LCT fields $(E_1(t) + E_2(t))$ constructed to induce population transfer and excited-state dissociation, respectively. The two field components are displayed separately in the lower panels. The predissociation $(B_0(t))$ and excited-state $(B_1(t))$ fragmentation yields are shown in the upper panel.

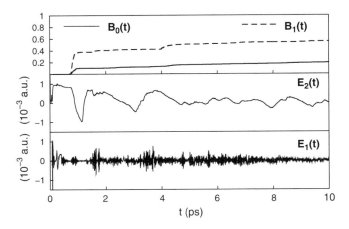

Figure 29. Same as Fig. 28 but for a different combination of field strength parameters. In this way, the branching ratio between the two exit channels can be varied.

two fields does not necessarily fulfill both objectives simultaneously. For example, it is found that the excited-state population does not increase monotonically. In the present case, the yield $B_1(t)$ increases at the same rate as the predissociation yield. At longer times (not shown) it levels to a value of ~ 0.3, which is the same as was found in the case of the subsequent interaction of the two fields.

Within the LCT approach, only the strength parameters λ_1 and λ_2 are at our disposal. Changing these numbers allows for a modification of the branching ratio of the two exit channels. An example is given in Fig. 29, where we modify the parameter for the field $E_1(t)$ to a value of $\lambda_1 = 1 \times 10^{-2}$ a.u., which enhances the total field strength. This amounts to a more effective excited-state fragmentation where the yield approaches a value of 0.50 asymptotically. The same is found for successive interactions of the field (not shown).

To conclude this section, it emerges that LCT offers an excellent approach to the control of excited-state dynamics. Here, the total electric field is composed out of two components where the first transfers population from the ground to the excited state, and the second steers the excited-state energy absorption. Concerning the efficiency and for the parameters regarded, it is not important if the fields interact successively or simultaneously.

X. ROTATIONAL MOTION

In the preceding sections, the rotational degrees of freedom were not included in the treatment of the coherent control processes. In what follows, a discussion of various aspects of rotational motion is presented. Several authors addressed the influence of rotations in the connection with laser control [192, 202, 203].

Nevertheless, the coupled vibrational–rotational-field dynamics has not been investigated in much detail. An exception is the work of Hornung and de Vivie-Riedle [204], who studied this interplay in connection with optimal control theory (OCT) [17, 42]. With respect to experiment, the new technique of polarization shaping [205–208] is directly connected to the angular motion.

A. Control of Directional Rotational Motion

We start with an example of a rotational motion in a single angular coordinate q. Although this is very similar to some of the above-presented one-dimensional vibrational problems, additional aspects arise. The example is a simplified treatment of what is called a "molecular motor." Such systems have been discussed vividly in connection with the building of nanoscale devices [209].

Here, we employ a model developed by Fujimura and co-workers which describes the rotation of a CHO functional group [210–212]. As in the latter work, we aim at the preparation of a unidirectional motion with the help of control fields. These will emerge from our classical intuition and the simple LCT construction scheme. A different local control approach to the motor motion was employed in Ref. 212 (see also Ref. 53).

The Hamiltonian is of the form

$$H_0 = \frac{-\hbar^2}{2I} \frac{d^2}{dq^2} + V(q) \tag{47}$$

where I is the moment of inertia and the potential curve is shown in Fig. 30. The latter has several extrema. Two excitation pathways are shown in panels (a) and (b). Adopting a classical point of view, a particle that is initially localized in the deepest well absorbs energy from a heating field (see Section III) until it acquires enough energy so that an unhindered periodic rotational motion becomes possible. The probabilities for a clockwise or counterclockwise rotation to take place are equal, and the actual outcome depends on the number of oscillations performed by the trapped particle. This is indicated in the figure by the arrows, where in panel (a) and panel (b) a rotation in one or the other direction is triggered.

In order to control the molecular motion, the heating field is taken as

$$E(t) = -\lambda \Im \langle \psi(t) | [\mu(q), T(P)] | \psi(t) \rangle \tag{48}$$

where the commutator has to be evaluated because the dipole moment is a complicated function of the angular variable q [210] (see Fig. 30). Note that the expectation value is purely imaginary (because it contains an anti-hermitian operator [213]) so that this particular choice of the control field again ensures a positive energy rate. The initial state is the ground state of the system and, to start

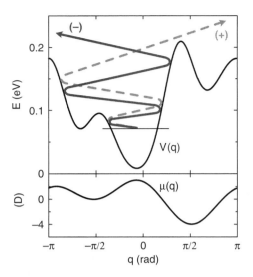

Figure 30.   Potential energy curve for the rotational motion of a molecular motor (upper panel). A hindered rotation takes place in a triple potential well. Starting from the bottom of the lowest potential well, the absorption of energy from a control field leads to free rotational motion. The case where a clockwise directional motion is obtained is indicated as (+). For a weaker field, the motion is forced to take place in the counterclockwise direction (−). The lower panel shows the dipole moment entering into the construction of the control field.

the algorithm, a seed pulse of small intensity is applied to the system [48]. To identify the direction of the rotational motion, we determined the time- and space-integrated flux

$$F(t) = \frac{1}{2Ii} \int dq \int_0^t dt' \left\{ \psi^*(q,t') \frac{d}{dq} \psi(q,t') - \left(\frac{d}{dq} \psi^*(q,t') \right) \psi(q,t') \right\} \quad (49)$$

The flux is displayed in the lower panel of Fig. 31. Also shown is the energy in the system. This energy is calculated quantum mechanically as the expectation value of the system Hamiltonian, and also classically as the average of an ensemble of trajectories sampled from the quantum wavefunction [181]. The dynamics of these trajectories is obtained employing the identical LCT field as in the quantum case. It is seen that at some time before 4 ps, the energy exceeds the barrier at 0.18 eV and the flux starts to decrease, indicating a net rotation in the counterclockwise direction. This time-dependence becomes linear after 4 ps because the control field is switched off (see the upper panel of Fig. 31). The classical and quantum energies show a similar time-dependence which suggests that the classical picture developed above (Fig. 30) is appropriate. The control

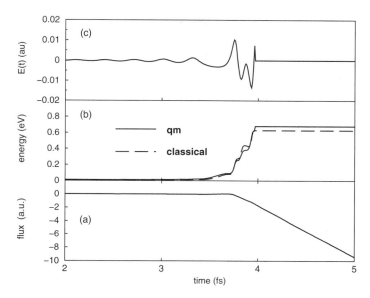

Figure 31. Directional motion of a molecular motor. The interaction of a field (panel (a)) leads to a heating so that the internal energy increases. This is seen in panel (b), which compares classically and quantum mechanically calculated energies, as indicated. Upon reaching the continuum, a net flux (panel (c)) in the counterclockwise direction is obtained. The control field is turned off at $t = 4$ ps.

field shows the systems period of the hindered rotation but here is not directly proportional to the angular momentum because the dipole moment is not a linear function of the variable q and also exhibits a sign change [210].

The quantum mechanical probability density is compared to an ensemble of classical trajectories in Fig. 32. Both quantities carry the same dynamical features, which, again, supports the intuitive picture evolving from LCT theory. It is seen that a bifurcation occurs, where two wavepackets move out of phase with each other (and likewise do sets of classical trajectories). This then has the consequence that, upon reaching the continuum, the rotational motion is not directional to 100%. Rather, for the present parameters, one finds a ratio of 2.2 in favor of the counterclockwise rotation.

In order to eliminate the bifurcation of the densities, it is sufficient to start from an initial state that is displaced from the equilibrium position. This can be achieved numerically without any problem. Nevertheless, we hint at a preparation of such a displaced state with the help of a static electric field $\vec{E}_s$. This amounts to the addition of the term

$$W_s(q) = -\vec{\mu}(q)\vec{E}_s = \mu(q)E_s \tag{50}$$

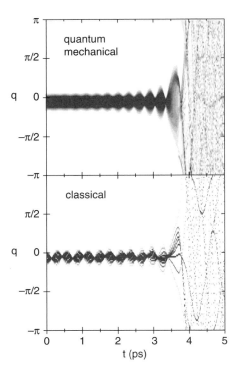

Figure 32. Field-driven rotational motion: Quantum mechanical and classical dynamics for an excitation from the ground state. The quantum mechanical probability density is compared to an ensemble of classical trajectories. Here, a preferential rotation in the counterclockwise direction is found.

to the Hamiltonian. Here, we assumed that the static field with field strength E_s acts antiparallel to the direction of the dipole moment. The additional interaction induces a Stark shift of the potential, and the field-dressed ground-state eigenfunction (for a strength of $E_s = 10^9$ V/m) is displaced by a value of $\Delta q = 0.07$ rad as compared to the field-free function. If then the static field is turned off, the LCT field starts interacting with the system and no additional seed pulse is needed. The density dynamics incorporating an initial field dressed state (and a value of $\lambda = 6 \times 10^{-5}$ a.u.) is shown in Fig. 33.

It is seen that the bifurcations are nearly removed and that now the motion is much more in favor of a counterclockwise rotation. In fact, one finds that 83% of the rotational wavepackets move in this direction. The classical trajectories exhibit a dynamics similar to that of the quantum densities. This is true for the ones moving freely but also for the trapped trajectories which do not have enough energy to escape the potential wells; for a discussion of similar features and their relation to femtosecond pump-probe signals, see Ref. 214.

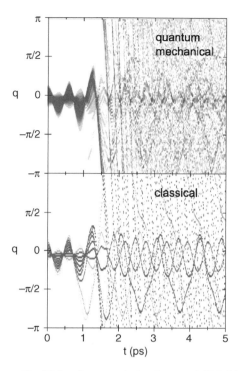

Figure 33. Same as Fig. 32 but for an excitation from an initial field dressed state being displaced in comparison to the field free state. An effective selective counterclockwise rotation is induced.

According to the classical picture suggested in Fig. 30, an increase of the field strength will accelerate the energy absorption and, eventually, prepare a clockwise rotation. That this is indeed the case is illustrated in Fig. 34. Due to the stronger molecule-field coupling ($\lambda = 10.5 \times 10^{-5}$ a.u.), the continuum is reached earlier, leading to the desired directional motion.

To summarize this subsection, we have shown that, in principle, a unidirectional motion of a functional molecular group can be induced by LCT fields. Quantum calculations revealed that it is possible to trigger a rotational dynamics in one or the other direction. The quantum dynamics is very much alike to that of a swarm of classical trajectories. This proves that the intuitive picture of a single-particle orbit driven out of the potential wells is valid. In particular, a change of the field strength triggers a hindered rotation that can be timed to move the then freely rotating particle selectively in the counterclockwise or clockwise direction. We have not addressed the role of the overall molecular orientation with respect to the direction of the external field

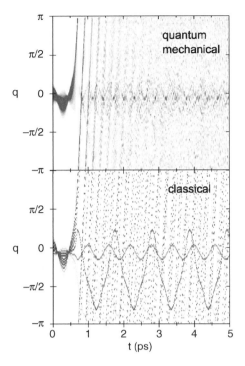

Figure 34. Same as Fig. 33 but for a stronger field strength. In this case rotation in the clockwise direction is triggered.

very carefully. However, the calculations illustrate that if the system is placed in a static field, the then-achieved forced orientation enhances the efficiency of the control process.

B.   Excited-State Fragmentation

We now come back to the initially studied problem of NaI fragmentation where the predissociation channel is not taken into account (Section III.A). As is sketched in Fig. 1, we aim at a field-induced excited-state dissociation after an initial femtosecond excitation from the ground state, but now take the rotational degree of freedom into account. Thus the molecular Hamiltonian in the electronic state $|n\rangle$ is written explicitly as

$$H_{0,n} = -\frac{1}{2m}\frac{\partial^2}{\partial R^2} + \frac{\vec{J}^2}{2mR^2} + V_n(R) = T(P) + T(J) + V_n(R) \qquad (51)$$

Here, $\vec{J}$ is the angular momentum operator and the Hamiltonian is understood to act on the wavefunction $\psi(R, \theta) = R\Phi(R, \theta)$, where $\Phi(R, \theta)$ is the complete wavefunction. We employ the expansion

$$\psi(R, \theta, t) = \sum_J \psi_J(R, t) Y_{J0}(\theta, 0) \tag{52}$$

where $Y_{J0}(\theta, 0)$ are spherical harmonics, i.e. eigenfunctions of the angular momentum operator. For linearly polarized light that we treat here, only the angle θ is of relevance. The field is polarized along the z-axis, so that the interaction is

$$W(t) = -\mu(R) \cos(\theta) E(t) \tag{53}$$

with the dipole moment $\mu(R)$ in the NaI excited state $|1\rangle$.

The fragmentation will proceed if enough energy is pumped into the radial degree of freedom. Therefore, we choose the operator $A = H_{vib} = H_{0,1} - T(J)$ in the local control scheme to arrive at a rate

$$\frac{d\langle H_{vib}\rangle_t}{dt} = i\langle \psi(t)|[T(J), T(P)]|\psi(t)\rangle - iE(t)\langle \psi(t)|[\mu(R)\cos(\theta), T(P)]|\psi(t)\rangle \tag{54}$$

where the brackets now denote integration over the vibrational and rotational variables. The first term in Eq. (54) contains the commutator between the kinetic energy operators corresponding to the radial and angular motion. It scales with $1/m^2$ and involves terms scaling with R^{-3} and R^{-4}, so that it safely can be ignored in the construction of the control field. Because the dipole moment is linear over most of the regarded bond-length region (see Section III.A), we then arrive at the rate expression

$$\frac{d\langle H_{vib}\rangle_t}{dt} = \frac{E(t)}{m} \langle \psi(t)|P\cos(\theta)|\psi(t)\rangle \tag{55}$$

which is similar to the one-dimensional case and one would anticipate that, in analogy, a control field could be constructed from the expectation value $\langle \psi(t)|P\cos(\theta)|\psi(t)\rangle$. This, however, is not possible for times when the control field has not already interacted sufficiently long with the system, a fact that can be related to the appearance of the $\cos(\theta)$ term in the expectation value. The initial excitation step (from $|0\rangle$ to $|1\rangle$), originating from an initial rotational state

Y_{J0}, prepares a linear combination of states corresponding to angular momenta with $J = J \pm 1$. Then, because of the property

$$\cos(\theta)Y_{J0}(\theta,0) = c_J^+ Y_{(J+1)0}(\theta,0) + c_J^- Y_{(J-1)0}(\theta,0) \tag{56}$$

where $c_J^\pm$ are known coefficients [215], the expectation value vanishes identically. We note that this is not only the case in the weak field limit but also for intense pulses [216], and also if a thermal distribution of initial states is considered. The reason is that, starting from an odd (even) initial state (in $|0\rangle$), the excited-state rotational manifold is always of even (odd) symmetry. According to the property Eq. (56), the operation of $\cos(\theta)$ on the wavefunction $\psi(R,\theta,t)$ changes its rotational symmetry, so that the expectation value is identically zero. If, on the other hand, a control field interacts with molecules being in the excited state, rotational states with even and odd symmetry are populated (see below).

In a first approach, we consider control fields constructed as in the rotationless case (or for fixed orientation) and take

$$E(t) = \lambda\langle P\rangle_t = \lambda\sum_J \langle\psi_J(R,t)|P|\psi_J(R,t)\rangle \tag{57}$$

Note that now, for a positive value of λ, the energy rate (Eq. (55)) will not necessarily be always positive. In Fig. 35, we collect results for a control field interaction with a field constructed from the linear momentum only. The

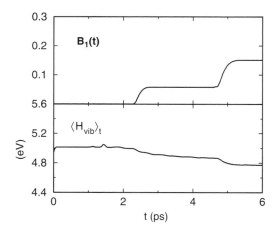

Figure 35.   Field-induced excited-state dissociation of NaI including the rotational degree of freedom. The upper and lower panels contain the dissociation yield and the expectation value of the vibrational Hamiltonian, respectively.

calculation employs the ro-vibrational ground state as initial state. The figure contains the fragmentation yield ($B_1(t)$, upper panel) and the expectation value of the vibrational Hamiltonian (lower panel). Although some dissociating molecules are produced (seen in the stepwise growth of the yield), it is curious that the internal energy decreases instead of showing the expected increase. Thus, the constructed LCT field, on the average, acts more like a cooling than a heating field.

The reasons for the above found trends are revealed by inspection of the density dynamics. Therefore, we consider the radial and angular densities separately. They are obtained by integration over one or the other degree of freedom:

$$\rho(R, t) = 2\pi \int d\theta \sin(\theta) |\psi(R, \theta, t)|^2 \tag{58}$$

$$\rho(\theta, t) = \int dR \sin(\theta) |\psi(R, \theta, t)|^2 \tag{59}$$

The latter are displayed in Fig. 36. At first, the radial density is very compact and performs one vibrational oscillation. Then, at a time of about 1.5 ps, a bifurcation takes place where one part of the former localized wavepacket moves into the fragmentation channel (which is reached, by definition, for distances larger than

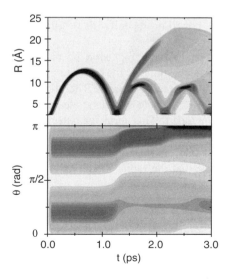

Figure 36. Density dynamics in the NaI molecule. The upper panel shows the radial density that bifurcates at $t \sim 1.5$ ps. The angular density, initially exhibiting a $|\cos(\theta)|^2$ distribution, accumulates around a value of $\theta = \pi$ at later times.

22.5 Å), whereas the other part remains bound and oscillates further. It is seen that the amplitude of the vibration decreases, which is consistent with the cooling taking place in the system.

An explanation of the bifurcation can be found from the angular density (lower panel of Fig. 36). Because we start from the rotational ground state, the first excitation step prepares a wavepacket with the rotational quantum number $J = 1$. Then, the density, initially, is proportional to $|Y_{10}(\theta, 0)| \sim |\cos(\theta)|^2$. It is seen that the density changes with time and that a depletion at angles smaller than $\pi/2$ occurs, which goes in hand with a concentration of density at a value of π. It is now straightforward to find a classical interpretation of the (radial) density bifurcation in regarding the classical force which stems from the external field interaction and acts in the radial direction:

$$F_R(t) = -\frac{\partial W(t)}{\partial R} = \cos(\theta(t))P(t) \qquad (60)$$

where the linear property of the dipole moment (i.e., $d\mu(R)/dR = 1$ (in atomic units)) is taken into consideration. A sampling of the initial probability distribution will result in trajectories starting at angles smaller and larger than $\pi/2$. Those which initiate from smaller angles, where the cos function assumes positive values, experience a positive force from the field which will, eventually, drive them into the exit channel. On the contrary, the orbits starting at larger angles are subject to a negative additional force that decelerates them. For longer times, when only these bound orbits exit, this results in a cooling as seen in the quantum result shown in Fig. 35.

So far, we have seen that the inclusion of the angular motion critically influences the energy transfer and that LCT is not applicable in a straightforward manner to induce a fragmentation process. Therefore, we now follow another strategy and place the NaI molecule in a static electric field, i.e. an extra term of the form Eq. (50) is included. This term leads to an initial orientation in the electronic ground state [217], that is, a *pendular state* [218]. In this way, states with various, even and odd, rotational quantum numbers are populated in the initial state (in the sense that the radial functions in the expansion Eq. (52) are nonzero). Consequently, we now may employ a field of the form

$$E(t) = \lambda \langle \psi(t)|P\cos(\theta)|\psi(t)\rangle \qquad (61)$$

which ensures that the energy rate can be influenced to be positive at all times. This is illustrated in Fig. 37, which contains the fragmentation yield and also the expectation value of the internal energy. Curves for three different field strengths of the static field are included, as indicated. For the weakest field, the fragmentation yield settles to 10%. This is enhanced to over 70% if the static

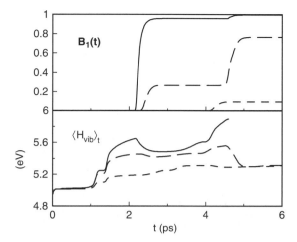

Figure 37. Same as Fig. 35 but for molecules in a static electric field. Curves are shown for different static field strengths of $E_s = 10^5$ V/m (short dashed line), 10^6 V/m (long dashed line), and 10^7 V/m (solid line).

field is more intense. Finally, for the strongest field, all molecules undergo dissociation. The internal energies now clearly show the heating effect. As a result, in a strong external static field the simple one-dimensional case, where the rotational motion is not included, is recovered. This comes as no surprise because the molecules are, to a large extent, forced to orient along the direction of the static field, as can be taken from the probability densities $|\psi(R, \theta, t)|^2$ displayed in Fig. 38 for a fixed time of 2 ps. Whereas the field free case (upper panel) exhibits the initial $|\cos(\theta)|^2$ distribution, an accumulation of density at small angles is found with increasing field intensity. Simultaneously, the mean position in the radial density shifts outward, which supports the classical picture developed above.

We have thus learned that the inclusion of the rotational degree of freedom in the simple one-state dissociation problem differs from the restricted one-dimensional vibrational problem. However, if an orientation of the molecule caused by an additional static field is present, the results of the simpler model are recovered.

C. Photoassociation: The Role of Orientation

In Section V, we have applied LCT to photoassociation processes including only s-wave scattering. Although this is the dominant event in ultra-cold collisions [159], the rotational degree of freedom cannot be ignored for higher impact energies. To get some insight into the influence of rotations on the yield of

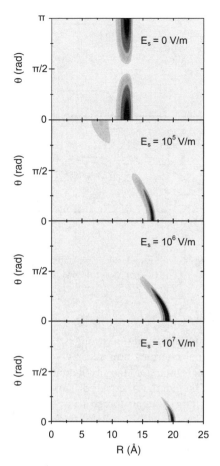

Figure 38. Probability density $|\psi(R, \theta, t)|^2$ at a time of 2 ps obtained for the interaction of the NaI molecule with a LCT field and an additional static field E_s. Plots are shown for the case where no static field is present and for different field strengths, as indicated.

photoassociation, we re-address the H + F scattering problem. Therefore, the initial state of angular momentum J_i is taken as

$$\psi_{J_i}(R, \theta) = \psi_i(R) Y_{J_i 0}(\theta, 0) \qquad (62)$$

where $\psi_i(R)$ is the Gaussian defined in Eq. (25).

We reexamine the example of Section V for incoming states which differ in their initial angular momentum, where we exemplarily choose $J_i = 1$ and $J_i = 6$. The LCT field is taken from the s-wave calculation (see Fig. 10). As a

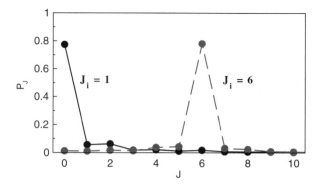

Figure 39. Photoassociation including rotations: shown is the rotational state distribution at a time of 3 ps for two different values of the initial angular momentum J_i, as indicated. It is seen that this value is approximately conserved.

first question we consider the rotational distribution prepared by the field in the limit of long times. This distribution is calculated from the expansion Eq. (52) as

$$P_J(t) = \langle \psi_J(R,t) | \psi_J(R,t) \rangle \tag{63}$$

Figure 39 shows the rotational state distribution for a time of $t = 3$ ps (where, afterwards, no further association takes place). The results document that the initial value of J is approximately conserved. The same is found for other values of the J_i. This means that the control field is such that no significant excitation or de-excitation in the rotational degree of freedom takes place. Thus we conclude that a single LCT field interacts similarly with scattered atoms belonging to different values of the orbital angular momentum.

Next, we regard the association yield, which is calculated as

$$B(t) = \sum_v \sum_J |\langle \varphi_{vJ}(R) | \psi_J(R,t) \rangle|^2 \tag{64}$$

where the states $|\varphi_{vJ}(R)\rangle$ are eigenfunctions characterized by the vibrational (v) and rotational quantum numbers (J). The yield is determined with the same field as above (determined from the pure s-wave calculation (Section V)) and is shown in Fig. 40 as a function of the value of the quantum number J_i. It is important to notice that the yield $B(t)$ is nearly independent of the initial rotational state. This means that it is not necessary to perform a thermal average in a calculation to predict association yields. These yields, however, are drastically reduced as compared to the s-wave scattering. This is to be expected because we employed a

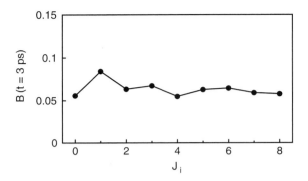

Figure 40. Photoassociation yield for different initial rotational states with quantum number J_i, determined at a time of 3 ps.

field that is not state selective, i.e. is taken from the restricted one-dimensional calculation.

What emerges from the obtained results is that the basic physical picture obtained from a simple calculation which only incorporates the radial degree of freedom (as is discussed in Section V) does not change significantly if the angular dynamics is taken into account.

XI.   COUPLED ELECTRONIC AND NUCLEAR MOTION

A.   A Useful Model

All of the problems we have addressed so far were based on a theoretical description resting on the Born–Oppenheimer approximation, which separates electronic and nuclear degrees of freedom. The usual approach to this problem is to calculate potential energy surfaces and, if necessary, nonadiabatic coupling elements employing methods of quantum chemistry [219]. These functions then enter into the Schrödinger equation for the nuclear motion. This procedure often involves the problem of "diabatization" [112, 113], where the kinetic coupling elements are replaced by potential couplings [114]. We will partly follow the outlined approach and compare the results to a calculation involving the coupled electronic–nuclear dynamics in control fields. This is a very demanding problem, and only a few model studies have been presented where both degrees of freedom are treated on the same footing (see Refs. 220–223). Below, we employ a simple model for a combined electron–nuclear dynamics which was introduced by Shin and Metiu [224, 225]. The model was later applied to illustrate basic features of non Born–Oppenheimer wavepacket dynamics [226–228] and also, in an extended version, was used to generalize the electron localization function (ELF)

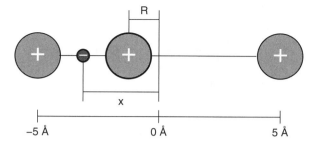

Figure 41. Model system for the investigation of coupled electron-nuclear dynamics in a control field. Three ions and an electron are arranged in one dimension, where only the middle ion (coordinate R) and the electron (coordinate x) are allowed to move.

[229] to include time and nuclear motion [230]. Following the former work, we consider a particle configuration as sketched in Fig. 41. There, two fixed ions of unit positive charge are placed at a distance of $L = 10$ Å. Another ion with proton mass and an electron move in one dimension. The charged particles interact via screened Coulomb forces which are parameterized such that it is easy to switch from the Born–Oppenheimer case to a situation where nonadiabatic coupling becomes important. A contour plot of the potential is shown in Fig. 42. It exhibits a double minimum structure where the minima correspond to two stable configurations of the ion–electron system. The objective, in what follows, is to induce a field-assisted transfer between these two configurations.

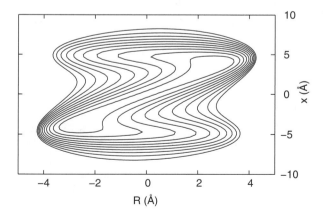

Figure 42. Contour plot displaying the model potential for a coupled electronic-nuclear motion. Contours are drawn in increments of 1 eV ranging from 1 eV to 10 eV as a function of the electronic (x) and nuclear coordinate (R).

The system Hamiltonian is

$$H^s(x, R) = -\frac{\hbar^2}{2}\frac{\partial^2}{\partial x^2} - \frac{\hbar^2}{2m}\frac{\partial^2}{\partial R^2} + V(x, R) \qquad (65)$$

and the dipole interaction term is

$$W(x, R, t) = -E(t)\mu(x, R) = -E(t)(R - x) \qquad (66)$$

Following the usual procedure, adiabatic potential curves $V_m(R)$ are determined from the electronic Schrödinger equation

$$\left\{ -\frac{\hbar^2}{2}\frac{d^2}{dx^2} + V(x, R) \right\}\varphi_m(x, R) = V_m(R)\varphi_m(x, R) \qquad (67)$$

where the $\varphi_m(x, R)$ are the electronic eigenfunctions depending parametrically on the nuclear coordinate R. Potential curves for the electronic states $|n\rangle (n = 0, 1, 2, +)$ are shown in Fig. 43 (upper panel), where $|+\rangle$ corresponds to the ionic

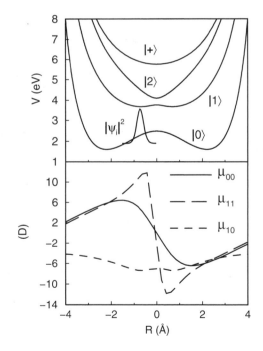

Figure 43. The upper panel shows adiabatic potential energy curves $V_n(R)$ for several electronic states. The $|+\rangle$ state is obtained if the single electron is detached from the system. Dipole matrix elements are shown in the lower panel of the figure.

state where the electron is detached from the system. The lower panel of the figure contains the dipole moments defined as

$$\mu_{nm}(R) = \int dx \; \varphi_n(x, R)\mu(x, R)\varphi_m(x, R) \tag{68}$$

We will now address the control problem adopting several levels of approximation.

B.   Transfer in the Adiabatic Ground State

We first employ the adiabatic approximation and restrict the theoretical treatment to the electronic ground state. This is equivalent to the (one-dimensional) double minimum problem discussed in Section VI.A. Thus, the control field is determined as

$$E(t) = -\lambda \Im \langle [T(P), \mu_{00}(R)] \rangle_t \tag{69}$$

Note that here the ground-state dipole moment $\mu_{00}(R)$ enters. As before, the parameter λ is adjusted such that, upon passing the barrier, the heating is changed to the cooling condition. From Fig. 43 it is clear that $\mu_{00}(R)$ is not a linear function of the nuclear coordinate R as was assumed in Section VI.A.

As an initial wavefunction we employ a Gaussian that is located in the potential well at negative distances, as indicated in Fig. 43. The interaction of the LCT field induces an average motion over the reaction barrier where afterwards a cooling takes place. This is demonstrated in Fig. 44 which contains the coordinate expectation value as a function of time (upper panel). Also shown is the target-state population, which we define as

$$B_T(t) = \int_0^\infty dR |\psi_0(R, t)|^2 \tag{70}$$

This quantity measures the norm of the wavefunction for positive values of the coordinate R. It is seen that the control field is not able to completely confine the wavepacket to the region right of the barrier. Regarding the quantum mechanical probability density, one finds that this loss of efficiency is due to the spreading of the wavepacket which becomes essential at later times and thus poses problems to the LCT procedure [87]. The control field is rather complicated (not shown; see Ref. 231) which stems from the coordinate dependence of the dipole moment. Nevertheless, here we find that the transfer is effective with about 80%.

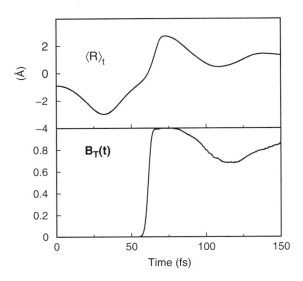

Figure 44. Control of particle transfer in an adiabatic treatment involving only the electronic ground state of the system. The coordinate expectation value (upper panel) and the population which is found at positive distances R are shown as a function of time.

C. Transfer in a Coupled Two-State Model

We next address the question if the inclusion of another electronic state influences the control yield. Therefore, the first excited state $|1\rangle$ is incorporated in the theoretical description. From Fig. 43 it can be taken that the respective potential curve is well separated from the ground state and thus nonadiabatic couplings are small (this, however, does not hold for the coupling between the states $|1\rangle$ and $|2\rangle$). We therefore employ the Hamiltonian

$$H^s(R) = \sum_{n=0,1} \{|n\rangle(T(P) + V_n(R) - \mu_{nn}E(t))\langle n|\} - E(t)(|1\rangle\mu_{10}\langle 0| + |0\rangle\mu_{01}\langle 1|)$$

$$(71)$$

where the potential energies $V_n(R)$ and dipole matrix elements μ_{nm} (see Fig. 43) appear. Thus, the field now not only perturbs the motion in the ground and excited state but also is able to induce a population transfer between the two electronic states.

Employing the same field as above leads to the results depicted in Fig. 45. It is seen that as long as the nucleus is confined to negative distances, a similar behavior as before is found (compare to Fig. 44). However, at the time the barrier crossing occurs, the ground-state population $P_0(t)$ is diminished to about

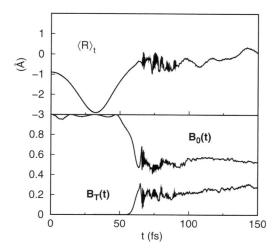

Figure 45. Same as Fig. 44, but for a model incorporating the ground and first excited state. The LCT field derived from the 1-state calculation is employed.

50% (lower panel). This, of course, stems from a population transfer to the excited state. Fast (Rabi-like [26]) oscillations in the population are observed which are due to the strong coupling [231]. Concerning the transfer efficiency, the target-state population approaches a value of ~ 0.3. This means that the yield is drastically reduced by the inclusion of the additional electronic state.

It is difficult to predict what will happen if other excited states and nonadiabatic couplings are taken into account. Nevertheless, the simple model employed here allows for a (numerically) exact treatment of the control problem, where the electron–nuclear Schrödinger equation is solved. The results are presented in the next subsection.

D.   Control of Combined Electron and Nuclear Dynamics

In the coupled problem, the initial wavefunction is the product

$$\psi(x, R, t = 0) = \psi_i(R)\varphi_0(x, R) \tag{72}$$

where $\psi_i(R)$ is the Gaussian used in the 1-state calculation (see Fig. 43). The time-dependent Schrödinger equation is now solved with the field derived from the adiabatic treatment of Section XI.B. The results of this calculation are collected in Fig. 46. There, the ground-state population defined as

$$B_0(t) = \left| \int dx \int dR \; \varphi_0(x, R)\psi(x, R, t) \right|^2 \tag{73}$$

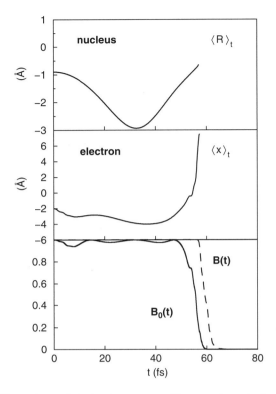

Figure 46. The nuclear and electronic average positions are shown in the upper and middle panel, respectively. The lower panel contains the ground state ($B_0(t)$) and total population ($B(t)$). It is seen that after 60 fs, the electron is detached from the system.

is shown. Also included is the population

$$B(t) = \int^{x_c} dx \int dR \, |\psi(x, R, t)|^2 \tag{74}$$

where x_c is set to a value of 19 Å. This quantity measures the total population in the system. The parts of the wavepacket that move toward larger values of the electronic coordinate x describe electron detachment and are removed by an optical potential [232–235] in the numerical calculation. From a comparison of the nuclear and electron coordinate expectation values with the populations, it emerges that as soon as the potential barrier is approached, the system is decomposed, leading to an electron promoted into the continuum. Thus, no control can be performed.

So far, we have demonstrated that a field derived from an adiabatic one-state model, if applied to the coupled dynamical problem, fails to induce a transfer between two stable particle configurations, so that the adiabatic approximation breaks down. We now consider an LCT field that is obtained directly from the exact treatment. In analogy to the particle control in the two-dimensional double well (see Section VI.A) the energy rate is

$$\frac{d}{dt}\langle H\rangle_t = E(t)\left\{-\langle P_x\rangle_t + \frac{\langle P_R\rangle_t}{m}\right\} \tag{75}$$

where the electronic and nuclear momentum operators P_x and P_R appear, and we used the fact that the dipole moment is linear in both x and R. Keeping in mind that the nuclear degree of freedom is to be controlled, the field is determined exclusively from the nuclear momentum as

$$E(t) = \lambda \frac{\langle P_R\rangle_t}{m} \tag{76}$$

In Fig. 47, we show the expectation values of the ion (upper panel) and the electron (lower panel) for two values of the field strength parameter where in

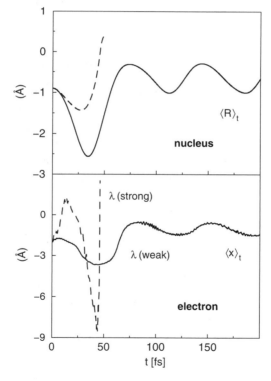

Figure 47. Coordinate expectation values in the nuclear (upper panel) and electronic (lower panel) degrees of freedom. Two cases, where a smaller (weak, solid lines) and larger (strong, dashed lines) field strength is employed, are illustrated.

one case (dashed lines) the field is to be considered as *strong*, while in the other case (solid lines) it is to be considered as *weak* (for details, see Ref. 231). In the latter case, the nucleus oscillates in the potential well localized at negative values of R. The electron adiabatically follows this vibrational motion. The fast, small-amplitude oscillations belong to a wiggling motion of the electron around the position of the nucleus. It is thus found that the control field does not induce a transfer between the wells.

It could be expected that an increase of the field strength will result in a successful transfer. This, however, is not the case as is documented in Fig. 47. In the *strong* field situation, the electron is accelerated fast and, at about 50 fs, leaves the system. Thus the attempt to control the particle transfer fails also in the present case. If the strength parameter is carefully tuned, it is found that the two situations shown in the figure are generic: Either the system cannot be effectively heated and remains in the initial potential well, or, for a more intense field, heating is effective and results in an electron detachment. This situation, although studied within a very simple model, suggests that the adiabatic approximation is not always appropriate in treating control processes in molecules.

XII.   SUMMARY AND OUTLOOK

In this chapter we have reviewed selected applications of *Local Control Theory* (LCT) to the control of processes taking place in molecules. The scheme rests on the construction of electric fields taking the instantaneous response of a perturbed system into account. In the simplest version, this is done by calculating the rate of an observable and the fields are adjusted to either increase or decrease this rate, depending on which objective is chosen.

From the given examples, it is revealed that the construction scheme is successful and easy to implement in a calculation. We have emphasized that the emerging fields can be interpreted in a straightforward way, often leading to a simple classical understanding of the underlying physics which determines the properties of the field and thus the outcome of the control process.

These obvious advantages suggest that other applications will be presented in the future. For example, preliminary work has shown that LCT can be applied to problems in quantum computing [51]. Here, the present approach could open new perspectives. Also, it is interesting to address the question of how the scheme can be realized experimentally. We are optimistic that there are many more interesting problems that can be treated using the local control scheme, and we hope that the present compilation of work will stimulate research in that direction.

Acknowledgments

We thank our co-workers and collaborators for their input on this work. In particular, V. E. thanks M. Erdmann, S. Graefe, N. E. Henriksen, P. Marquetand, and K. B. Møller.

References

1. A. H. Zewail, *Femtochemistry*, Vols. I and II, World Scientific, Singapore, 1994.

2. L. Wöste and J. Manz, eds., *Femtosecond Chemistry*, Vols. I and II, VCH, Weinheim, 1995.

3. M. Chergui, ed., *Femtochemistry—Ultrafast Chemical and Physical Processes in Molecular Systems*, World Scientific, Singapore, 1996.

4. V. Sundström, ed., *Femtochemistry and Femtobiology: Ultrafast Reaction Dynamics at Atomic-Scale Resolution*, Imperial College Press, London, 1996.

5. F. D. Schryver, S. DeFeyter, and G. Schweitzer, eds., *Femtochemistry*, Wiley-VCH, Weinheim, 2001.

6. A. Douhal and J. Santamaria, eds., *Femtochemistry and Femtobiology*, World Scientific, Singapore, 2002.

7. M. Martin and J. T. Hynes, eds., *Femtochemistry and Femtobiology: Ultrafast Events in Molecular Science*, Elsevier, Oxford, 2004.

8. M. Dantus and A. H. Zewail (eds.), *Chem. Rev.* **104**, 1717 (2004).

9. A. H. Zewail, *J. Phys. Chem.* **104**, 5660 (2000).

10. A. H. Zewail, *Angew. Chem. Int. Ed. Engl.* **39**, 2586 (2000).

11. A. M. Weiner, *Rev. Sci. Instrum.* **71**, 1929 (2000).

12. M. Wollenhaupt, A. Assion, and T. Baumert, in *Springer Handbook of Optics*, F. Träger, ed., Springer, New York, 2007, pp. 937–983.

13. D. J. Tannor and S. A. Rice, *J. Chem. Phys.* **83**, 5013 (1985).

14. D. J. Tannor, R. Kosloff, and S. A. Rice, *J. Chem. Phys.* **85**, 5805 (1986).

15. D. J. Tannor and S. A. Rice, *Adv. Chem. Phys.* **70**, 441 (1988).

16. P. Brumer and M. Shapiro, *Annu. Rev. Phys. Chem.* **43**, 257 (1992).

17. A. P. Peirce, M. A. Dahleh, and H. Rabitz, *Phys. Rev. A* **37**, 4950 (1988).

18. W. S. Warren, H. Rabitz, and M. Dahleh, *Science* **259**, 1581 (1993).

19. R. N. Zare, *Science* **279**, 1875 (1998).

20. S. A. Rice, *Nature* **403**, 496 (2000).

21. H. Rabitz, R. de Vivie-Riedle, M. Motzkus, and K.-L. Kompa, *Science* **288**, 824 (2000).

22. S. A. Rice, *Nature* **409**, 422 (2001).

23. N. E. Henriksen, *Chem. Soc. Rev.* **31**, 37 (2002).

24. M. Shapiro and P. Brumer, *Principles of Quantum Control of Molecular Processes*, Wiley, New York, 2003.

25. S. A. Rice and M. Zhao, *Optical Control of Molecular Dynamics*, Wiley, New York, 2000.

26. D. J. Tannor, *Introduction to Quantum Mechanics: A Time-dependent Perspective*, University Science Books, Sausalito, CA, 2007.

27. S. A. Rice, *Adv. Chem. Phys.* **101**, 213 (1997).

28. R. J. Gordon and S. A. Rice, *Annu. Rev. Phys. Chem.* **48**, 601 (1997).

29. K. Bergmann, H. Theuer, and B. W. Shore, *Rev. Mod. Phys.* **70**, 1003 (1998).

30. T. Brixner, N. H. Damrauer, and G. Gerber, *Adv. At. Mol. Opt. Phys.* **46**, 1 (2001).

31. S. A. Rice and S. P. Shah, *Phys. Chem. Chem. Phys.* **4**, 1683 (2002).

32. T. C. Weinacht and P. H. Bucksbaum, *J. Opt. B: Quantum Semiclass. Opt.* **4**, R35 (2003).

33. T. Brixner and G. Gerber, *Chem. Phys. Chem.* **4**, 418 (2003).

34. T. Brixner, N. H. Damrauer, G. Krampert, P. Niklaus, and G. Gerber, *J. Mod. Opt.* **50**, 539 (2003).

35. D. Goswami, *Phys. Rep.* **374**, 385 (2003).

36. M. Shapiro and P. Brumer, *Rep. Prog. Phys.* **66**, 859 (2003).

37. M. Dantus and V. V. Lozovoy, *Chem. Rev.* **104**, 1813 (2004).

38. T. Brixner, T. Pfeifer, G. Gerber, T. Baumert, and M. Wollenhaupt, in *Femtosecond Laser Spectroscopy*, P. Hannaford, ed., Springer, New York, 2005, pp. 225–266.

39. P. Nuernberger, G. Vogt, T. Brixner, and G. Gerber, *Phys. Chem. Chem. Phys.* **9**, 2470 (2007).

40. J. Werschnik and E. K. U. Gross, *J. Phys. B: At. Mol. Opt. Phys.* **40**, R175 (2007).

41. D. J. Tannor, R. Kosloff, and A. Bartana, *J. Chem. Soc., Faraday Trans.* **113**, 365 (1999).

42. R. Kosloff, S. A. Rice, P. Gaspard, S. Tersigini, and D. J. Tannor, *Chem. Phys.* **139**, 201 (1989).

43. R. Kosloff, A. D. Hammerich, and D. J. Tannor, *Phys. Rev. Lett.* **69**, 2172 (1992).

44. A. Bartana, R. Kosloff, and D. J. Tannor, *J. Chem. Phys.* **99**, 196 (1993).

45. D. Tannor, in *Molecules in Laser Fields*, A. D. Bandrauk, ed., Marcel Dekker, New York, 1994.

46. H. Tang, R. Kosloff, and S. A. Rice, *J. Chem. Phys.* **104**, 5457 (1996).

47. V. Malinovsky, C. Meier, and D. J. Tannor, *Chem. Phys.* **221**, 67 (1997).

48. V. Malinovsky and D. J. Tannor, *Phys. Rev. A* **56**, 4929 (1997).

49. J. Vala and R. Kosloff, *Opt. Express* **8**, 238 (2001).

50. A. Bartana, R. Kosloff, and D. J. Tannor, *Chem. Phys.* **276**, 195 (2001).

51. S. E. Sklarz and D. J. Tannor, *Chem. Phys.* **322**, 87 (2006).

52. N. Vitanov, T. Halfmann, B. W. Shore, and K. Bergmann, *Annu. Rev. Phys. Chem.* **52**, 763 (2001).

53. Y. Ohtsuki, H. Kono, and Y. Fujimura, *J. Chem. Phys.* **109**, 9318 (1998).

54. M. Sugawara, S. Yoshizawa, and S. Yabashita, *Chem. Phys. Lett.* **350**, 253 (2001).

55. M. Sugawara, *Chem. Phys. Lett.* **358**, 290 (2002).

56. M. Sugawara, *J. Chem. Phys.* **118**, 6784 (2003).

57. M. Sugawara, *Chem. Phys. Lett.* **378**, 603 (2003).

58. M. Sugawara and Y. Fujimura, *J. Chem. Phys.* **100**, 5646 (1994).

59. M. Sugawara and Y. Fujimura, *J. Chem. Phys.* **101**, 6586 (1994).

60. Y. Ohtsuki, Y. Yahata, H. Kono, and Y. Fujimura, *Chem. Phys. Lett.* **287**, 627 (1998).

61. H. Umeda and Y. Fujimura, *Chem. Phys.* **274**, 231 (2001).

62. P. Gross, H. Singh, H. Rabitz, K. Mease, and G. M. Huang, *Phys. Rev. A* **47**, 4593 (1993).

63. Z.-M. Lu and H. Rabitz, *Phys. Rev. A* **52**, 1961 (1995).

64. Y. Chen, P. Gross, V. Ramakrishna, and H. Rabitz, *J. Chem. Phys.* **102**, 8001 (1995).

65. W. Zhu, M. Smit, and H. Rabitz, *J. Chem. Phys.* **110**, 1905 (1999).

66. W. Zhu and H. Rabitz, *J. Chem. Phys.* **119**, 3619 (2003).

67. M. Mirrahimi, G. Turinici, and P. Rouchon, *J. Phys. Chem. A* **109**, 2631 (2005).

68. A. Rothman, T.-S. Ho, and H. Rabitz, *J. Chem. Phys.* **123**, 134104 (2005).

69. T. Tung Nguyen-Dang, C. Chatelas and D. Tanguay, *J. Chem. Phys.* **102**, 1528 (1995).

70. G. Katz, M. A. Ratner, and R. Kosloff, *Phys. Rev. Lett.* **98**, 203006 (2007).

71. M. Sugawara and Y. Fujimura, *Chem. Phys.* **196**, 113 (1995).

72. Y. Watanabe, H. Umeda, Y. Physuki, H. Kono, and Y. Fujimura, *Chem. Phys.* **217**, 317 (1997).

73. K. Hoki, Y. Ohtsuki, H. Kono, and Y. Fujimura, *J. Phys. Chem.* **103**, 6301 (1999).

74. K. Hoki, Y. Ohtsuki, and Y. Fujimura, *J. Chem. Phys.* **114**, 1575 (1999).

75. Y. Fujimura, L. Gonzalez, K. Hoki, J. Manz, and Y. Ohtsuki, *Chem. Phys. Lett.* **306**, 1 (1999).

76. J. Salomon and G. Turinici, *J. Chem. Phys.* **124**, 074102 (2006).

77. P. A. M. Dirac, *The Principles of Quantum Mechanics*, 4th ed., Oxford, Oxford Science Publications, 1958.

78. C. Cohen-Tannoudji, B. Diu, and R. Laloë, *Quantum Mechanics, Vol. I* (Wiley, New York, 1977).

79. C. D. Schwieters and H. Rabitz, *Phys. Rev. A* **44**, 5224 (1991).

80. M. H. Lissak, J. D. Sensabaugh, C. D. Schwieters, J. G. B. Beumee, and H. Rabitz, *Chem. Phys.* **174**, 1 (1993).

81. C. D. Schwieters and H. Rabitz, *Phys. Rev. A* **48**, 2549 (1993).

82. M. Demiralp and H. Rabitz, *J. Math. Chem.* **16**, 185 (1994).

83. J. Botina, H. Rabitz, and N. Rahman, *Phys. Rev. A* **51**, 923 (1995).

84. J. Botina, H. Rabitz, and N. Rahman, *J. Chem. Phys.* **102**, 226 (1995).

85. J. Botina, H. Rabitz, and N. Rahman, *J. Chem. Phys.* **103**, 6637 (1995).

86. Y. Chen, P. Gross, V. Ramakrishna, H. Rabitz, K. Mease, and H. Singh, *Automatica* **33**, 1617 (1997).

87. Y. Zhao and O. Kühn, *J. Phys. Chem. A* **104**, 4882 (2000).

88. H. Risken, *The Fokker–Planck Equation*, 2nd ed., Springer, Berlin, 1989.

89. R. Loudon, *The Quantum Theory of Light*, 2nd ed., Clarendon Press, Oxford, 1983.

90. R. Schinke, *Photodissociation Dynamics*, Cambridge University Press, Cambridge, 1993.

91. S. Gräfe, P. Marquetand, N. E. Henriksen, K. B. Møller, and V. Engel, *Chem. Phys. Lett.* **398**, 180 (2004).

92. V. Engel and H. Metiu, *J. Chem. Phys.* **90**, 6116 (1990).

93. P. Davidovits and D. L. McFadden, eds., *Alkali Halide Vapors* (Academic, New York, 1979).

94. T. S. Rose, M. J. Rosker, and A. H. Zewail, *J. Chem. Phys.* **88**, 6672 (1988).

95. T. S. Rose, M. J. Rosker, and A. H. Zewail, *J. Chem. Phys.* **91**, 7415 (1989).

96. J. L. Herek, A. Materny, and A. H. Zewail, *Chem. Phys. Lett.* **228**, 15 (1994).

97. H. Kono and Y. Fujimura, *Chem. Phys. Lett.* **184**, 497 (1991).

98. T. Taneichi, T. Kobayashi, Y. Ohtsuki, and Y. Fujimura, *Chem. Phys. Lett.* **231**, 50 (1994).

99. C. J. Bardeen, J. Che, K. R. Wilson, V. V. Yakovlevi, P. Cong, B. Kohler, J. L. Krause, and M. Messina, *J. Phys. Chem. A* **101**, 3815 (1997).

100. E. Charron and A. Giusti-Suzor, *J. Chem. Phys.* **108**, 3922 (1998).

101. M. Grønager and N. E. Henriksen, *J. Chem. Phys.* **109**, 4335 (1998).

102. B. H. Hosseini, H. R. Sadeghpour, and N. Balakrishnan, *Phys. Rev. A* **71**, 023402 (2005).

103. P. Cong, A. Mokhtari, and A. H. Zewail, *Chem. Phys. Lett.* **172**, 109 (1990).

104. S. Chapman and M. S. Child, *J. Phys. Chem.* **95**, 578 (1991).

105. C. Meier, V. Engel, and J. S. Briggs, *J. Chem. Phys.* **95**, 7337 (1991).

106. G. H. Peslherbe, R. Bianco, J. T. Hynes, and B. M. Ladanyi, *J. Chem. Soc., Faraday Trans.* **93**, 977 (1997).

107. P. Marquetand and V. Engel, *J. Chem. Phys.* **127**, 084115 (2007).

108. J. V. José and E. J. Saletan, *Classical Dynamics*, Cambridge University Press, Cambridge, 1998.

109. S. Chelkowski, A. D. Bandrauk, and P. B. Corkum, *Phys. Rev. Lett.* **65**, 2355 (1990).

110. T. Witte, T. Hornung, L. Windhorn, D. Proch, R. de Vivie-Riedle, M. Motzkus, and K. L. Kompa, *J. Chem. Phys.* **118**, 2021 (2003).

111. P. Marquetand and V. Engel, *Chem. Phys. Lett.* **407**, 471 (2005).

112. V. Sidis, *Adv. Chem. Phys.* **82**, 73 (1992).

113. T. Pacher, L. S. Cederbaum, and H. Köppel, *Adv. Chem. Phys.* **84**, 293 (1993).

114. G. Stock and W. Domcke, *Adv. Chem. Phys.* **100**, 1 (1997).

115. V. Engel, H. Metiu, R. Almeida, R. A. Marcus, and A. H. Zewail, *Chem. Phys. Lett.* **152**, 1 (1988).

116. L. D. Landau, *Phys. Z.* **2**, 46 (1932).

117. L. D. Landau, *Proc. R. Soc. London A* **137**, 696 (1932).

118. A. D. Bandrauk, ed., *Molecules in Laser Fields*, Marcel Dekker, New York, 1994.

119. M. Wollenhaupt, A. Präkelt, C. Sarpe-Tudoran, D. Liese, and T. Baumert, *J. Opt. B Quantum Semiclass. Opt.* **7**, S270 (2005).

120. M. Wollenhaupt, A. Präkelt, C. Sarpe-Tudoran, D. Liese, and T. Baumert, *Appl. Phys. B* **82**, 183 (2006).

121. M. Wollenhaupt, D. Liese, A. Präkelt, C. Sarpe-Tudoran, and T. Baumert, *Chem. Phys. Lett.* **419**, 184 (2006).

122. M. Wollenhaupt and T. Baumert, *J. Photochem. Photobiol. A* **180**, 248 (2006).

123. F. F. Crim, *Annu. Rev. Phys. Chem.* **44**, 397 (1993).

124. B. Armstrup and N. E. Henriksen, *J. Chem. Phys.* **97**, 8285 (1992).

125. S. Meyer and V. Engel, *J. Phys. Chem.* **101**, 7749 (1997).

126. N. Elgobashi, P. Krause, J. Manz, and M. Oppel, *Phys. Chem. Chem. Phys.* **5**, 4806 (2003).

127. J. R. Reimers and R. O. Watts, *Mol. Phys.* **52**, 357 (1984).

128. J. Zhang, D. G. Imre, and J. H. Frederick, *J. Phys. Chem.* **93**, 1840 (1989).

129. S. Gräfe, C. Meier, and V. Engel, *J. Chem. Phys.* **122**, 184103 (2005).

130. S. Gräfe, P. Marquetand, and V. Engel, *J. Photochem. Photobiol. A* **180**, 271 (2006).

131. D. J. Maas, D. I. Duncan, R. B. Vrijen, W. J. van der Zande, and L. D. Noordham, *Chem. Phys. Lett.* **290**, 75 (1998).

132. V. D. Kleiman, S. M. Arrivo, J. M. Melinger, and E. J. Heilweil, *Chem. Phys.* **233**, 207 (1998).

133. L. Windhorn, T. Witte, J. S. Yeston, D. Proch, M. Motzkus, K.-L. Kompa, and W. Fuss, *Chem. Phys. Lett.* **357**, 85 (2002).

134. M. Bonn, C. Hess, and M. Wolf, *Phys. Rev. Lett.* **85**, 4341 (2000).

135. L. Windhorn, J. S. Yeston, T. Witte, W. Fuss, M. Motzkus, D. Proch, K.-L. Kompa, and C. B. Moore, *J. Chem. Phys.* **119**, 641 (2003).

136. C. Ventalon, J. M. Fraser, M. H. Vos, A. Alexandrou, J.-L. Martin, and M. Joffre, *Proc. Natl. Acad. Sci. USA* **101**, 13261 (2004).

137. W. Jakubetz, B. Just, J. Manz, and H.-J. Schreier, *J. Chem. Phys.* **94**, 2294 (1990).

138. W. Jakubetz, J. Manz, and H.-J. Schreier, *Chem. Phys. Lett.* **165**, 100 (1990).

139. M. V. Korolkov, J. Manz, and G. K. Paramonov, *Adv. Chem. Phys.* **101**, 327 (1997).

140. H. P. Breuer, K. Dietz, and M. Holthaus, *Phys. Rev. A* **45**, 550 (1992).

141. A. Orel, Y. Zhao, and O. Kühn, *J. Chem. Phys.* **97**, 94 (2000).

142. H. Naundorf, G. A. Worth, H.-D. Meyer, and O. Kühn, *J. Phys. Chem. A* **102**, 719 (2000).

143. M. Petkovic and O. Kühn, *J. Phys. Chem. A* **107**, 8458 (2003).

144. C. Rovira, *J. Phys. Condens. Matter* **15**, S1809 (2003).

145. A. Dreuw, B. D. Duniez, and M. Head-Gordon, *J. Am. Chem. Soc.* **124**, 12070 (2002).

146. C. Meier and M.-C. Heitz, *J. Chem. Phys.* **123**, 044504 (2005).

147. C. Bardeen, J. Cau, F. L. H. Brown, and K. R. Wilson, *Chem. Phys. Lett.* **302**, 405 (1999).

148. R. B. Williams, R. F. Loring, and M. D. Fayer, *J. Phys. Chem. B* **105**, 4086 (2001).

149. J. C. Owrutsky, M. Li, B. Locke, and R. M. Hochstrasser, *J. Phys. Chem.* **99**, 4842 (1995).

150. A. E. Roitberg, R. B. Gerber, and M. A. Ratner, *J. Phys. Chem. B* **101**, 1700 (1997).

151. M. H. Beck, A. Jäckle, G. A. Worth, and H.-D. Meyer, *Phys. Rep.* **324**, 1 (2000).

152. U. Manthe, H.-D. Meyer, and L. S. Cederbaum, *J. Chem. Phys.* **97**, 3199 (1992).

153. H.-D. Meyer, U. Manthe, and L. S. Cederbaum, *J. Chem. Phys.* **97**, 3199 (1992).

154. H. R. Thorsheim, J. Weiner, and P. S. Julienne, *Phys. Rev. Lett.* **58**, 2420 (1987).

155. M. Machholm, A. Giusti-Suzor, and F. H. Mies, *Phys. Rev. A* **50**, 5025 (1994).

156. M. V. Korolkov, J. Manz, G. K. Paramonov, and B. Schmidt, *Chem. Phys. Lett.* **260**, 604 (1996).

157. P. Backhaus, B. Schmidt, and M. Dantus, *Chem. Phys. Lett.* **306**, 18 (1999).

158. E. Luc-Koenig, R. Kosloff, F. Masnou-Seeuws, and M. Vatasescu, *Phys. Rev. A* **70**, 033414 (2004).

159. C. P. Koch, R. Kosloff, E. Luc-Koenig, F. Masnou-Seeuws, and A. Crubellier, *J. Phys. B: At. Mol. Opt. Phys.* **39**, S1017 (2006).

160. A. Fioretti, D. Comparat, A. Crubellier, O. Dulieu, F. Masnou-Seeuws, and P. Pillet, *Phys. Rev. Lett.* **80**, 4402 (1998).

161. U. Marvet and M. Dantus, *Chem. Phys. Lett.* **245**, 393 (1995).

162. F. Fatemi, K. M. Jones, H. Wang, I. Walmsley, and P. D. Lett, *Phys. Rev. A* **64**, 033421/1 (2001).

163. B. L. Brown, A. J. Dicks, and I. A. Walmsley, *Phys. Rev. Lett.* **96**, 173002/1 (2006).

164. W. Salzmann, U. Poschinger, R. Wester, M. Weidemüller, A. Merli, S. M. Weber, F. Sauer, M. Plewicki, F. Weise, A. M. Esparza, and L. Wöste, A. Lindinger, *Phys. Rev. A* 73, 023414/1 (2006).

165. M. Shapiro and P. Brumer, *Phys. Rep.* **425**, 195 (2006).

166. A. Guldberg and G. D. Billing, *Chem. Phys. Lett.* **186**, 229 (1991).

167. M. A. Buldakov and V. N. Cherepanov, *J. Phys. B: At. Mol. Opt. Phys.* **37**, 3973 (2004).

168. Y. Niu, S. Wang, and S. Cong, *Chem. Phys. Lett.* **428**, 7 (2006).

169. V. May and O. Kühn, *Charge and Energy Transfer Dynamics in Molecular Systems*, Wiley-VCH, Berlin, 2000.

170. S. Lochbrunner and E. Riedle, *Recent Res. Devel. Chem. Physics* **4**, 31 (2003).

171. E. T. J. Nibbering and T. Elsaesser, *Chem. Rev.* **104**, 1887 (2004).

172. F. Grossmann, T. Dittrich, P. Jung, and P. Hänggi, *Phys. Rev. Lett.* **67**, 516 (1991).

173. M. Holthaus, *Phys. Rev. Lett.* **69**, 1596 (1992).

174. R. Bavli and H. Metiu, *Phys. Rev. Lett.* **69**, 1986 (1992).

175. N. Došlić, O. Kühn, J. Manz, and K. Sundermann, *J. Phys. Chem. A* **102**, 9645 (1998).

176. H. Naundor, K. Sundermann, and O. Kühn, *Chem. Phys.* **240**, 163 (1999).

177. J. Karczmarek, M. Stott, and M. Ivanov, *Phys. Rev. A* **60**, R4225 (1999).

178. Y. Ohta, T. Bando, T. Yoshimoto, K. Nishi, H. Nagao, and K. Nishikawa, *J. Phys. Chem. A* **105**, 8031 (2001).

179. A. Matos-Abiague and J. Berakdar, *Phys. Rev. B* **69**, 155304 (2004).

180. C. Tanner, C. Manca, and S. Leutwyler, *J. Chem. Phys.* **122**, 204326 (2005).

181. P. Marquetand, S. Gräfe, D. Scheidel, and V. Engel, *J. Chem. Phys.* **124**, 054325 (2006).

182. T. Baumert, M. Grosser, R. Thalweiser, and G. Gerber, *Phys. Rev. Lett.* **67**, 3753 (1991).

183. V. Engel, T. Baumert, C. Meier, and G. Gerber, *Z. Phys. D: At. Mol. Clusters* **28**, 37 (1993).

184. T. Baumert and G. Gerber, *Adv. At. Molec. Opt. Phys.* **35**, 163 (1995).

185. T. Frohnmeyer, M. Hofmann, M. Strehle, and T. Baumert, *Chem. Phys. Lett.* **312**, 447 (1999).

186. A. Assion, T. Baumert, U. Weichmann, and G. Gerber, *Phys. Rev. Lett.* **86**, 5695 (2001).

187. M. Wollenhaupt, A. Assion, O. Bazhan, C. Horn, D. Liese, C. Sarpe-Tudoran, M. Winter, and T. Baumert, *Chem. Phys. Lett.* **376**, 457 (2003).

188. M. Wollenhaupt, V. Engel, and T. Baumert, *Annu. Rev. Phys. Chem.* **56**, 25 (2005).

189. P. Kusch and M. M. Hesse, *J. Chem. Phys.* **68**, 2591 (1978).

190. G. Gerber and R. Möller, *Chem. Phys. Lett.* **113**, 546 (1985).

191. A. J. Taylor, K. M. Jones, and A. L. Schawlow, *J. Opt. Soc. Am. B* **73**, 994 (1983).

192. S. Gräfe, M. Erdmann, and V. Engel, *Phys. Rev. A* **72**, 013404 (2005).

193. R. S. Mullikan, *J. Chem. Phys.* **55**, 309 (1971).

194. A. Assion, T. Baumert, J. Helbing, V. Seyfried, and G. Gerber, *Chem. Phys. Lett.* **259**, 488 (1996).

195. F. Spiegelman and M. Gross, *J. Chem. Phys.* **108**, 4148 (1998).

196. F. Spiegelman, private communication (2007).

197. S. Magnier, P. Millie, O. Dulieu, and F. Masnou-Seeuws, *J. Chem. Phys.* **98**, 7113 (1993).

198. S. Ruhman, A. G. Jdy, and K. A. Nelson, *J. Chem. Phys.* **86**, 6563 (1987).

199. J. Chesnoy and A. Mokhtari, *Phys. Rev. A* **38**, 3566 (1988).

200. A. M. Weiner, D. E. Leaird, G. P. Wiederrecht, and K. A. Nelson, *J. Opt. Soc. Am. B* **8**, 1264 (1991).

201. P. Marquetand and V. Engel, *Chem. Phys. Lett.* **426**, 263 (2006).

202. D. Kröner and L. Gonzalez, *Chem. Phys.* **298**, 55 (2004).

203. G. Turinici and H. Rabitz, *Phys. Rev. A* **70**, 063412 (2004).

204. T. Hornung and R. de Vivie-Riedle, *Europhys. Lett.* **64**, 703 (2003).

205. T. Brixner and G. Gerber, *Opt. Lett.* **26**, 557 (2001).

206. T. Brixner, G. Krampert, P. Niklaus, and G. Gerber, *APB* **74**, 133 (2002).

207. T. Brixner, G. Krampert, T. Pfeifer, R. Selle, G. Gerber, M. Wollenhaupt, O. Graefe, C. Horn, D. Liese, and T. Baumert, *Phys. Rev. Lett.* **92**, 208301 (2004).

208. M. Aeschlimann, M. Bauer, D. Bayer, T. Brixner, F. J. G. de Abajo, W. Pfeiffer, M. Rohmer, C. Spindler, and F. Steeb, *Nature* **446**, 301 (2007).

209. V. Balzani, M. Venturi, and A. Credi, *Molecular Devices and Machines*, Wiley-VCH, Weinheim, 2003.

210. K. Hoki, M. Yamaki, S. Koseki, and Y. Fujimura, *J. Chem. Phys.* **118**, 497 (2003).

211. K. Hoki, M. Yamaki, S. Koseki, and Y. Fujimura, *J. Chem. Phys.* **119**, 12393 (2003).

212. M. Yamaki, K. Hoki, Y. Ohtsuki, H. Kono, and Y. Fujimura, *J. Am. Chem. Soc.* **127**, 7300 (2005).

213. J. J. Sakurai, *Modern Quantum Mechanics*, Benjamin Cummings Publishing, Menlo Park, CA, 1985.

214. V. A. Ermoshin, V. Engel, and C. Meier, *J. Chem. Phys.* **113**, 5770 (2000).

215. E. Merzbacher, *Quantum Mechanics*, Wiley, New York, 1998.

216. P. Marquetand, C. Meier, and V. Engel, *J. Chem. Phys.* **123**, 204320 (2005).

217. P. Marquetand and V. Engel, *Phys. Chem. Chem. Phys.* **7**, 469 (2005).

218. J. M. Rost, J. C. Griffin, B. Friedrich, and D. R. Herschbach, *Phys. Rev. Lett.* **68**, 1299 (1992).

219. P. von Ragué Schleyer, ed., *Encyclopedia of Computational Chemistry*, Vols. 1–5, Wiley, New York, 1998.

220. S. Chelkowski, T. Zuo, O. Atabek, and A. D. Bandrauk, *Phys. Rev. A* **52**, 2977 (1995).

221. S. Chelkowski, C. Foisy, and A. D. Bandrauk, *Phys. Rev. A* **57**, 1176 (1996).

222. S. Chelkowski, M. Zamojski, and A. D. Bandrauk, *Phys. Rev. A* **63**, 023409 (2001).

223. M. Lein, E. Gross, T. Kreibich, and V. Engel, *Phys. Rev. A* **65**, 033403 (2002).

224. S. Shin and H. Metiu, *J. Chem. Phys.* **102**, 9285 (1995).

225. S. Shin and H. Metiu, *J. Phys. Chem.* **100**, 7867 (1996).

226. M. Erdmann, P. Marquetand, and V. Engel, *J. Chem. Phys.* **119**, 672 (2003).

227. M. Erdmann and V. Engel, *J. Chem. Phys.* **120**, 158 (2004).

228. M. Erdmann, S. Baumann, S. Gräfe, and V. Engel, *Eur. Phys. J. D* **30**, 327 (2004).

229. A. D. Becke and K. E. Edgecombe, *J. Chem. Phys.* **92**, 5397 (1990).

230. M. Erdmann, E. K. U. Gross, and V. Engel, *J. Chem. Phys.* **121**, 9666 (2004).

231. S. Gräfe and V. Engel, *Chem. Phys.* **329**, 118 (2006).

232. D. Neuhauser and M. Baer, *J. Chem. Phys.* **90**, 4351 (1989).

233. U. V. Riss and H.-D. Meyer, *J. Chem. Phys.* **105**, 1409 (1996).

234. B. Poirier and T. Carrington, *J. Chem. Phys.* **118**, 17 (2003).

235. B. Poirier and T. Carrington, *J. Chem. Phys.* **119**, 77 (2003).

DYNAMICS OF DOUBLE PHOTOIONIZATION IN MOLECULES AND ATOMS

JOHN H. D. ELAND

Physical and Theoretical Chemistry Laboratory, University of Oxford, Oxford, OX1 3QZ, United Kingdom

CONTENTS

Advances in Chemical Physics, Volume 141, edited by Stuart A. Rice
Copyright © 2009 John Wiley & Sons, Inc.

103

I. INTRODUCTION

The impact of a single sufficiently energetic photon on a molecule or atom frequently results in the ejection of two or even more electrons, even though the basic interaction discovered by Hertz [1] and interpreted by Einstein [2] is of one photon with one electron. Immense theoretical and experimental efforts have been devoted to this topic in recent decades, partly because the effect is intimately connected with the question of electron correlation. In the orbital model of atomic and molecular electronic structure, double photoionization (DPI) is effectively forbidden as a direct process, so its observed extent, where simple two-step processes are excluded, is taken to be a gauge of the real matter's divergence from the simple model. Every one of the several recent reviews of this topic [3–9] restates this connection with electron correlation, which is expressed as a motivation for the work, but it cannot (yet) be said that experimental results on DPI have improved our understanding of correlation in neutral species. This review is mainly concerned with two new techniques that can, in principle, provide results capable of giving direct and relevant information on electron correlation in molecules. There are considerable caveats, however, because the new data also cast some doubt on the concept of initial state electron correlation as the major factor in the extent of double photoionization for most species and energy ranges.

All the previous reviews [3–9] have been concerned mainly with what is called *direct* double photoionization in atoms. This is a process in which a single photon ejects two electrons simultaneously, without the intervention of any real intermediate state of either the neutral molecule or the singly ionized cation. It can be written

$$M + h\nu \rightarrow M^{2+} + e_1^- + e_2^- \tag{1}$$

Its chief characteristic is that the excess energy of the photon above the double ionization threshold is shared as a continuous monotonic distribution between the two outgoing electrons. This direct double photoionization is contrasted with several *indirect* processes, of which by far the most important can be written

$$M + h\nu \rightarrow M^{+*} + e_1^- : \qquad M^{+*} \rightarrow M^{2+} + e_2^- \tag{2}$$

Here the intermediate monocation M^{+*} may have an inner shell (core) hole, in which case we have the classical Auger effect, or it may be in a valence two-hole one-particle state (or an even more distant state) reached by a multielectron transition from the neutral ground state. The characteristic of either process is that both ejected electrons have distinct fixed energies, though their sharpness will depend on the lifetime of the intermediate state. The new data discussed in this review confirm and quantify the known [10] but seldom acknowledged truth that in the EUV (extreme ultraviolet) and soft X-ray region, most double photoionization of both atoms and molecules is in fact *indirect*, even below inner-shell thresholds. In addition to the major indirect process (2), a number of other related mechanisms have been discovered recently, all stemming from the capability of intermediate superexcited cations M^{+*} to decay in a variety of ways. Dissociation or fluorescence photon emission from the intermediate may intervene distinctly before, or on the same timescale as, the second electron ejection. Because the direct and indirect pathways, or even two different indirect pathways to double ionization may have the same initial and final states, the possibility of quantum mechanical interference also arises, but this aspect has not yet been explored in detail.

The structure of this chapter is as follows. First, a brief review of the characteristics of direct double ionization of atoms is presented as a background to the recent results on indirect processes and on molecules. Second, the new experimental methods based on threshold electron detection and on the magnetic bottle time of flight (TOF) technique are described. The main body of the chapter then presents representative results on selected atoms and molecules. Finally the relevance of the experimental findings to the characterization of electron correlation is discussed qualitatively.

II. DIRECT DOUBLE PHOTOIONIZATION OF HELIUM

Because He^+ or any helium-like ion has only one electron, there are no states at all above the double ionization potential, so no indirect double ionization is possible at any energy. This has made helium a happy hunting ground for both experimentalists and theoreticians concerned with direct double photoionization. There are also some multielectron closed-shell atoms, particularly the rare gases, where all excited electron configurations of the dication are at much higher energies than the lowest energy configuration. For formation of the highest energy level belonging to the lowest configuration, there are then no accessible higher energy states over a wide range of energy. Double ionization to these specific states is therefore mainly direct over a certain photon energy range. In other atomic and molecular cases, there may exist a narrow range immediately above threshold where only a direct process is possible, but this is of negligible practical importance. The following remarks therefore refer to the whole process of double photoionization of helium and formation of a few specially selected dication states of other atoms. The majority of other double photoionization is indirect. Some probable exceptions in molecular double photoionization have been discovered recently and are discussed later. Even where the overall DPI process is mainly *indirect*, however, the ever-present direct contribution may be selectable by experimental choice of a particular energy sharing between the two electrons, where indirect processes do not contribute.

The starting point in all discussion of direct DPI is the purely classical theory of Wannier [11], which is about the need for correlation in the final state of two free electrons if double escape is to succeed. If one electron is much slower than the other and lags behind, it will screen the positive nuclear charge from its faster partner, which gains energy and escapes. The slow electron loses energy and remains trapped. To avoid this failure of escape at threshold, the two electrons need to leave the doubly positive core maintaining sensibly equal radial distances, and preferably in opposite directions. The need for such strict correlation diminishes as the excess energy above threshold increases, so the probability of double ionization has an energy-dependent effect on the cross section. Wannier [11] showed that the cross section for direct DPI has the form

$$\sigma = \sigma_0 E^m \tag{3}$$

where $m = 1.056$ for this process and E is the excess energy above threshold. The final-state correlation causes the exponent m to depart from unity, the value given by the statistical phase space available as a function of energy [12].

The threshold law (3) has been confirmed by quantum theory calculations [13, 14] and also by experiment [15]. In the energy range where the Wannier theory applies, the energy distribution of the emitted electrons is classically predicted to be flat; this has also been confirmed [16, 17], but has been found to hold true over a much wider energy range than the threshold law. Adherence to the threshold law in helium double photoionization has been observed in both real experiments and accurate numerical simulations [14] over an energy range up to about 2 eV above threshold, but the electron energy distribution remains flat within 20% over almost 20 eV. At higher energies the energy distribution is U-shaped, with equal energy sharing becoming less probable than emission of one high-energy electron and one slow one. This form of energy distribution, calculated by Chang and Poe [18] without explicit inclusion of the final-state correlation, was found by Wehlitz et al. [17] to fit the observed distributions in double photoionization of helium right down to the lowest excess energy measured (5 eV). The *flatness* or concavity of the electron pair energy distribution is therefore very insensitive as a criterion of final-state electron correlation. Flatness of the distribution may be an indicator of direct double ionization near threshold, but is not a sufficient indicator. The vital characteristic of the direct process, critically important for the interpretation of the new data discussed in this review, is that the pair distribution is a smooth *monotonic* function of energy, with no structure besides its U-shaped concavity.

Final-state correlation in the threshold region has a very strong effect on the angular distributions of the ejected electrons, both relative to the polarization of the light and relative to each other. There has been a great deal of experimental and theoretical work in this area, but because it is largely irrelevant to the main topic of this review, only a brief overview is given here. The reviews by Bolognesi et al. [7] and by Avaldi and Huetz [8] should be consulted for a full account up to 2005. If only one of the two electrons is measured relative to the polarization vector, the distribution has the familiar form

$$I(\theta) = \sigma/4\pi\{1 + \beta/4(1 + 3P\cos 2\theta)\} \qquad (4)$$

where P is the degree of linear polarization of the light and θ is the angle from the polarization direction. Exactly at threshold the parameter β should be -1 in DPI, corresponding to perpendicular alignment [19, 20], but it rises extremely rapidly away from threshold, in a manner that may depend on the overall symmetry of the electron pair [21–23]. This symmetry also has a notable effect on the cross section for double photoionization, because the electron pair symmetries can be classified as "favorable" or "unfavorable." The overall process is treated as a dipole transition from the neutral ground state, $^1S^e$, for a closed-shell atom. The dication and two electrons must then together have $^1P^0$ symmetry, which is also the symmetry of the electron pair if the dication is in a $^1S^e$ state, as with He. If the dication has another symmetry, such as $^3P^e$ (ground state of rare gas dications), the electron pair symmetry must be complementary (here $^3P^0$ or $^3D^0$). Of the possible electron pair symmetries, those in which spin, parity, and orbital angular momentum are either all even or all odd are the favorable ones, for which the electrons can depart from the core at 180° to each other as in the best Wannier geometry. This geometry is not available for the unfavorable pair symmetries, essentially because of the requirement that the electron pair wavefunction be antisymmetric to exchange. In helium the only possible outgoing electron pair wave has an unfavorable symmetry, a circumstance to which the low cross section is partly attributed.

It is evidently a stringent test of theory and a great challenge to experiment to measure both of the two electrons in coincidence with energy and angle analysis. Each such measurement, of which a great many have now been carried out, is a determination of a triply differential cross section (TDCS). The cross sections and angular distributions depend on the total electron energy $E = E_1 + E_2$ and on the ratio E_1/E_2 ; the limiting cases $E_1 = E_2$ and $E_1 = 0$ are particularly attractive for comparison with theory. The form of the TDCS depends on both geometrical and dynamical factors, and three complete

schemes for its parameterization have been derived [20, 24–26]. The scheme of
Huetz et al. [20, 26] is most easily related to experimental measurements and
has been widely used. The TDCS is expressed as

$$\text{TDCS} = |a_g(E_1, E_2, \theta_{12})(\cos\theta_1 + \cos\theta_2) + a_u(E_1, E_2, \theta_{12})(\cos\theta_1 + \cos\theta_2)|^2$$

(5)

where the two functions a_g and a_u are *gerade* and *ungerade* complex
amplitudes which express the ionization dynamics. The angles θ_1 and θ_2 are
measured from the light polarization direction. In the case of equal energy
sharing $(E_1 = E_2)$ the ungerade term vanishes, while for $\theta_1 = 0$ and $\theta_2 = 180°$
only the scalar square of the ungerade term remains. The contributions of the
two terms to experimental data can thus be separated and can be compared with
theory. Three forms of theory have been particularly successful, namely
convergent close coupling (CCC, [27, 28]), time-dependent close coupling
(TDCC, [29, 30]), and hyperspherical R-matrix with semiclassical outgoing
waves (HRM-SOW [31, 32]). In the energy range up to 40 eV above threshold,
all three theoretical models give good agreement with experimental results.
The angular distributions for He DPI generally have two lobes, whose direction
and widths are well predicted, and a node for back-to-back ejection. The widths
of the angular distributions show the strength of angular correlation and are
an aspect of the DPI process where the extended Wannier law [33] fails. The
Wannier prediction for the characteristic width of $91E^{0.25}$ (angle in degrees,
E in electronvolts) turns out to be too large except in the very close vicinity
(0.1 eV) of threshold [8].

The double photoionization of helium has recently been studied experimentally and theoretically at 529 eV photon energy [34], and in an accurate
numerical experiment using the TDCC method up to the same excess energy,
450 eV above threshold [35]. These very useful studies demonstrate that two
processes play major roles in the double photoionization of helium, *direct
knockout* and *shake-off*. In direct knockout (also called TS1, "two-step-one"),
a primary photoelectron of sufficiently high energy collides with another
electron and ejects it from the ion, losing energy itself in the process. The two
electrons can end up with similar energies. In shake-off, a primary electron
ejection changes the local electric field very suddenly; the resulting
perturbation causes a second electron to be ejected. Alternatively, we can say
that the first electron ejection leaves the residual ion in a state that is not an
eigenstate, but contains some continuum character. At all events, the result is
that one electron (the primary) has high energy and the other (the shaken-off
electron) has near-zero energy. As the excess energy of the photon above the
ionization threshold is increased, shake-off becomes more and more dominant,

contributing to deeper and deeper concavity of the electron distribution. The experimental results of Knapp et al. [34] show that at 450-eV excess energy the shake-off mechanism produces electrons with the most unequal energy sharing (2 eV and 448 eV), but for slightly less extreme sharing (30 eV and 420 eV) direct knockout is more important. The overall process is a coherent sum of the two mechanisms.

The direct knockout mechanism had earlier been evoked by Samson [36, 37], who pointed out that the variation of some double photoionization cross sections as a function of energy closely resembles the form of electron impact ionization cross sections of ions. This is, of course, the expected behavior if direct knockout dominates, which is expected in the energy range relatively near threshold. Further consideration of these model mechanisms is left for a later section where the role of initial state electron correlation is discussed.

III.   ANGULAR CORRELATIONS IN DIRECT DPI OF COMPLEX ATOMS

Because of the higher cross sections, experiments on complex atoms are somewhat less demanding than those on helium, and several measurements have been made using selected energy sharings to isolate the direct component. On the other hand, the larger number of accessible states can make the interpretation more complicated. Where the symmetries are exactly the same as in the helium case, as in removal of two s electrons from an inner shell, similar two-lobe angular structures are found with interesting differences of detail [7, 38]. Where other unfavored electron pair symmetries are involved, multilobe structures avoiding the favored back-to-back emission are found [39–41], and this is true even where both favored and unfavored symmetries are in play for the same final dication state. The simplest angular distributions, totally dominated by a single lobe of back-to-back electron emission, are found where only favored symmetries of the electron pair are available, as in formation of a $^1P^o$ state of an Rg^{2+} ion by $s^{-1}p^{-1}$ ionization [42].

IV.   INDIRECT PROCESSES IN ATOMS

The prototypical indirect double photoionization process is the classical Auger effect, where the single-hole state created by ejection of an inner-shell electron exists for a significant time before ejection of a second electron. Alternatively, an intermediate singly charged state may exist as a multiply excited valence state or a Rydberg state, but whatever its nature, a lifetime of at least a few femtoseconds will normally allow the whole process to be considered as a sequence of two distinct steps. A sharp peak in the photoelectron spectrum and also, therefore, in the Auger spectrum is the obvious experimental signature of a separable process. Measurement of the photoelectron and the Auger electron in coincidence, with angular resolution of both over a full range, presents the possibility of what is called a *complete experiment*. By this is meant that the amplitudes and relative phases of all partial waves describing each electron emission can, in principle, be determined. In practice there is often more than one solution of the equations linking the observed angular distributions to the theoretical parameters, so an ambiguity may remain. Nevertheless, a complete solution has been found in at least a few cases [43–46].

Even where a sharp peak appears in the photoelectron spectrum, divergences from the two-step picture can arise for several reasons. First, if the direct double photoionization is not of negligible relative amplitude at the same photon energy, interference is possible. Peaks with Fano profiles should be seen and the angular distributions will be complicated. Second, if a photon energy is chosen such that the photoelectron and an Auger electron have the same kinetic energy, the two become indistinguishable and strong interference effects are expected [47]. The effects present an interesting challenge to theory, and they have been examined in detail for several atoms [48–50]. Third, if the photoelectron is of lower energy than the Auger electron, the electron emitted later may overtake the one emitted first and can exchange energy with it by the Coulomb interaction. The faster electron gains energy by being shielded from the core by the slower one, which loses energy. This process is called post-collision interaction (PCI), and it has been known and studied for many years in connection with Auger spectra. Its effect on the electron energies can be understood semiclassically [51] as well as by full quantum mechanical treatments [52], but the angular effects are expected to be complex and are not yet well known. The newer experiments discussed below are particularly well suited to the study of the energetic effects of PCI, and relevant results are mentioned in a later section.

V.   EXPERIMENTAL TECHNIQUES

A.   The New Experiments with Angular Integration

The main focus of this review is on double photoionization experiments carried out using two rather new techniques that accept electrons over a wide range of angles, approximating 4π solid angle, simultaneously. Because of their high collection efficiency, they are very sensitive and gather spectra fast. The integration over all angles means, of course, that they are not suited to study of the finest details of the phenomena, but their sensitivity makes them ideal for exploration of a wide range of effects in atoms and molecules. These experimental methods can be broadly divided into two classes: (a) those that specifically detect one or more electrons of near zero energy and (b) those in which electrons of all energies are detected concurrently. Readers should be aware that there is another group of techniques based on COLTRIMS (cold target recoil momentum spectroscopy) and the "reaction microscope" which share many of the characteristics of the techniques covered here. These COLTRIMS techniques have been reviewed exhaustively elsewhere [53, 54] and are not included in this review, though some of their results have already been mentioned [34]. Because they require cold molecular beams, they have not yet been applied to a very wide range of molecular and atomic targets.

B.   Threshold Electron Selection: TPEsCO and Its Derivatives

To study double ionization, one must detect and analyze two electrons from each ionization event. The very first experiments in which two zero-energy photoelectrons were detected in coincidence were essentially demonstrations of principle, because they accepted electrons only from small angular ranges [55, 56]. A significant advance was made with the application of the penetrating field electron extraction technique [57] to the study of double photoionization, with the technique that Hall and his collaborators named threshold photoelectrons coincidence (TPEsCO) [58]. Ionization takes place in a volume that would be field-free, except for the penetrating field from a positively biased electrode placed behind a narrow aperture. Electrons of near zero energy are drawn through the aperture with almost 100% efficiency and are accelerated and imaged by an electrostatic lens onto a second small aperture. The chromatic aberration of the lens diminishes the fraction of fast electrons that pass into the apertures following emission into the small solid angle subtended. The fraction of fast electrons eventually detected is reduced further by passage at low energy through a cylindrical or hemispherical condenser and in some cases by selection on the basis of total flight time relative to the ionizing light pulse [58]. The very high detection efficiency and specificity of this penetrating field technique for threshold electrons makes it ideal for coincidence experiments.

In order to detect the electron pairs emitted in double photoionization, the first TPEsCO apparatus used two penetrating field analyzers mounted back-to-back [59]. The exact field configuration was not discussed, but the apparatus worked well after careful balancing of count rates in the two channels. Since the two electrons detected in this way are both of nominally zero energy, the spectrum of double ionization is scanned by varying the photon energy. Because the photon energy required is outside the range of any suitable laboratory light source, the experiments could only be done at a synchrotron radiation source. First results were obtained at the SRS, Daresbury, UK, where the work has been continued; further developments were made by Hall's group in Paris, using the synchrotron radiation source SACO at Orsay. One major

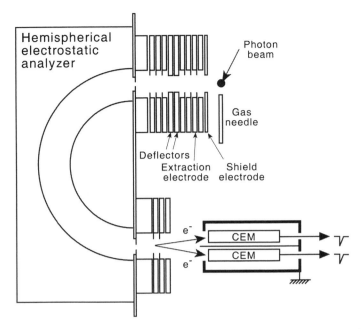

Figure 1. TPEsCO apparatus using two-channel electron multiplier (CEM) electron detectors. Potentials on the several lens elements and deflectors are tuned manually for optimum electron pair detection and resolution.

change was to use a single penetrating field threshold electron selector, with a detector for electron pairs consisting first of a microchannel plate and later (Fig. 1) of two-channel electron multipliers mounted side-by-side [60, 61]. Careful shielding of the two detectors from each other to eliminate cross-talk was essential; and after tuning of steering electrodes, the fraction of electron pairs detected could be made comparable to, though always less than, the fraction expected on the basis of the detection efficiency for single electrons. In a later refinement a detector with four individual channeltrons was used, giving a further improvement in sensitivity for electron pairs. The resolution of the apparatus has also been improved; in the latest version a resolution of 10 meV is achieved [62].

The great advantage of using a single penetrating field analyzer for the electron pairs is that the ionization volume is accessible for the simultaneous observation of ions, fluorescence photons, or metastable atoms, any of which can be detected in coincidence. The combination of a time-of-flight mass spectrometer (TOF-MS) with the penetrating field threshold electron detector produced a very powerful apparatus [61] (Fig. 2) capable of measuring branching ratios, kinetic energy releases, and dissociation rate constants in decay of specific dication states.

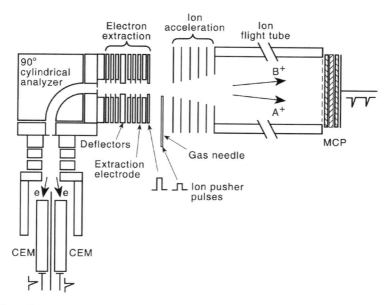

Figure 2. Combination of a TPEsCO spectrometer with a TOF mass spectrometer, to allow multiple coincidence experiments detecting up to two electrons and two ions from individual events. The mass spectrometer uses second-order time focus.

C.   Energetic Electron Pair Analysis: Magnetic Bottle TOF Methods

The earliest experiments on double photoionization with energy analysis of both photoelectrons were done using twin hemispherical electrostatic analyzers and a continuous high-energy photon source [63]. These were very lengthy experiments even at low resolution, because only a single energy of each electron could be sampled at a time. An early conceptual advance was the use of a dual magnetic bottle time-of-flight (TOF) system with pulsed synchrotron radiation from a small storage ring [64], but the full power of this technique could not be realized at the time because of the limited capability of the timing electronics and the short interpulse period. The recent efflorescence of double and multiple ionization studies has stemmed from improved timing electronics, the development of a pulsed light source and long magnetic bottle TOF spectrometer [65], and the use of synchrotron radiation from large storage rings in single-bunch mode.

The success of the magnetic bottle TOF electron–electron coincidence technique, which is the main topic of this review, arises from two basic ideas. First, a solenoid can be made into an electron trap or "magnetic bottle" by arranging a strongly divergent axial magnetic field at each end [66]. In a spectrometer, one end has no divergent field but has an electron detector instead, while ionization occurs within the divergent field at the closed end. By suitable choice of field strengths (typically about 0.5 T at the source, 10^{-3} T in the solenoid), all electrons emitted into more than 90% of the 4π solid angle from a point source on axis can be made to travel to the detector on nearly parallel paths. The theory of this device was given by Kruit and Read [67] in the original article describing its use as a TOF analyzer. Second, in simple TOF energy analysis of electrons without retardation, electrons of all energies are detectable all the time, so long as multihit detection and recording electronics are available. The combination of very high collection efficiency with complete energy multiplexing is absolutely ideal for coincidence experiments, where the signal depends on the product of overall detection–collection efficiencies for all the particles involved.

The initial development of the TOF-PEPECO (TOF photoelectron–photoelectron coincidence) technique as a laboratory experiment also depended on the development of a suitable pulsed light source. The lamp developed for this purpose is a gas discharge, usually in He or Ne, where each flash is powered by a small (50–200 pF) capacitor, charged to about 10 kV. The design of the lamp itself was inspired by the short-pulse lamp of Choi and Favre [68] and combines the character of a capillary channel with the hollow cathode effect. In the most highly developed form, the lamp (Fig. 3) firing is controlled by a fast hydrogen thyratron and the condenser is charged through a high-voltage switch, to ensure that the thyratron conductivity is extinguished between pulses. Pulse lengths of about 3 ns at repetition rates around 10 kHz are achieved at photon energies from 15 eV to 50 eV on selected atomic lines [69]. The lines most frequently used are at photon energies 26.91 (46.1 nm, NeII), 32.69 eV (37.9 nm, NeIII), 40.81 eV (30.4 nm, HeII), 48.37 eV (25.6 nm, HeII), and 51.02 eV (24.3 nm, HeII); these are referred

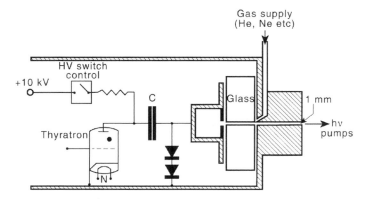

Figure 3. Scheme of a fast-pulsed discharge lamp for TOF-PEPECO spectroscopy. After the condenser is charged to 10 kV the HV control switch opens and the thyratron fires, driving a rapid rise of negative potential at the anode which caused spontaneous breakdown. Pulse lengths less than 5 ns at repetition rates of 10 kHz are possible.

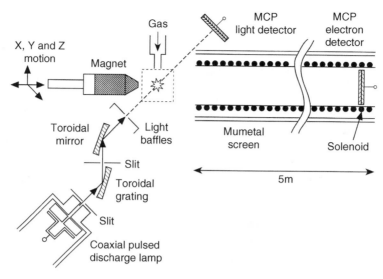

Figure 4. Elements of a TOF-PEPECO apparatus using the pulsed lamp. The magnetic field provided by the shaped permanent magnet at the ionization point is about 0.5 T, while it is about 10^{-3} T in the long solenoid. Improvements over the original form of the apparatus [65] include a truncated cone magnet and a cage around the ionization zone to provide a uniform electrical potential.

to below as the laboratory lines. The overall scheme of the apparatus with this lamp is indicated in Fig. 4; essential points are the provision of a high permeability (mumetal) screen to remove the geomagnetic field and an X, Y, Z manipulation stage to position the truncated cone-ended permanent magnet which produces the strong divergent field. The original apparatus has a total electron flight path of 5.6 m, as long as physically possible in the space available, to allow good energy resolution even with long (50–100 ns) light pulses. Later work, particularly using synchrotron radiation pulses of 1 ns or shorter duration, has been successful with much shorter (0.25–2 m) electron flight paths. As with all analyzers dealing with low-energy electrons, careful attention must be paid to the quality of surfaces "seen" by the electrons, because serious surface potential changes can occur. Most workers cover all metal surfaces near the ionization region with graphite to minimize surface charging, and they also provide electrodes or a surrounding "cage" to control the electrical potential of the ionization zone. It is helpful to accelerate all electrons to an energy of a few tenths of an electronvolt to make the longest flight times, for electron of zero initial energy, fit a time range of 20 μs or less, according to the repetition rate of the light source in use.

The very high detection efficiency of this apparatus, which normally exceeds 50% as directly determined from the coincidence counting statistics, invites attempts to detect other particles in coincidence in order to learn more about the mechanism and consequences of ionization. Ions [70], vis–UV photons [71], and EUV photons [72] have all been measured in coincidence with the electrons by appropriate modifications and additions to the apparatus. Its most important characteristics from the point of view of the present review are that electron detection is essentially independent of emission angle and of electron energy up to at least 100 eV. The energy resolution is a strong function of the energy and also depends on the conditions, particularly the magnetic field strengths, but a numerical resolution $E/\Delta E$ of 50–100 gives the correct order of magnitude. At the lowest energies a fixed energy width of 15–50 meV is usually found, presumably because of stray fields in the source region. For high-energy electrons the peaks may be split into doublets by a difference in total flight time between electrons that start out in the direction toward the detector and those whose initial velocity must be reversed by the divergent magnetic field. This is most serious in short instruments, but the doubling can usually be eliminated, at least over a certain energy range, by careful tuning of electrical potentials in the source.

D.   Characteristics of the TPEsCO and TOF-PEPECO Techniques

These two modern techniques for the study of double and multiple photoionization are complementary in a number of ways. In TPEsCO, where the photon energy is scanned and the electron analysis method is always the same, good resolution is available at all photon energies. On the other hand, only a small fraction of the total photoionization process is sampled, and the relative intensities of different peaks and bands are extremely susceptible to effects of autoionization from levels of singly ionized or even neutral species which happen to be at the same energy as the dication states being sought. TOF-PEPECO allows the whole double photoionization process to be studied and can distinguish between direct and indirect ionization, so its relative intensities are meaningful. On the other hand, the energy resolution is good only for low electron energies and degrades sharply at high energies. Attempts to alleviate this problem by retardation are under way, but always come at the cost of limiting the range of observable energies.

VI.   FORMS OF DATA PRESENTATION

Double photoionization spectra from TPEsCO are simple one-axis graphs, but complete data sets from double photoionization are inherently three-dimensional: intensity as a function of two electron energies. Presentation is in the form of two-axis "maps," with grayscale or color points to indicate intensity, or by projection into normal spectra of intensity against a single variable. Many choices of axes are possible, and different authors prefer

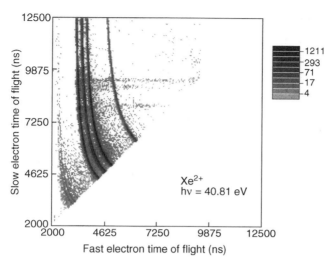

Figure 5. Raw TOF-PEPECO data from ionization of Xe at 30.4 nm (40.81 eV photon energy) in the instrument of Fig. 4. The strong curved lines show coincident electron pair formation, while all the other weak background is due to scattering.

different conventions. The TOF technique using a single detector imposes an unavoidable classification of the first and second electrons to arrive as the "fast" and "slow" electrons, even though this classification may have no physical significance. The most basic form of data is a raw map of intensity against the times of arrival of the two electrons; Fig. 5 is the only example of such a plot in this review, because energy is usually the physically relevant parameter. However, the TOF peak width is more nearly constant in time than in energy, so details of processes at low electron energy are often more clearly seen in the time maps.

Figure 5 shows the raw time map for double photoionization of xenon at 30.4 nm, with symmetrical axis scales. The intensity coding (grayscale) is highly nonlinear (here logarithmic) in order to show very weak features in the presence of the strong ones. Each major curve represents pairs of electrons with a fixed energy sum, corresponding to a final state of Xe^{2+}. The much weaker curved lines at the lower left of the figure represent electron pairs of fixed total energy, but these arise from a secondary process in which a primary photoelectron from formation of one Xe^+ collides with another atom, producing a second Xe^+ ion. The secondary process is distinguishable by its quadratic pressure dependence, and usually the pressure used in the experiments is so low that such processes are of negligible intensity. Weak horizontal lines in the figure arise from inelastic collisions of one electron of a true pair, with consequent energy loss. These are also negligible in most spectra.

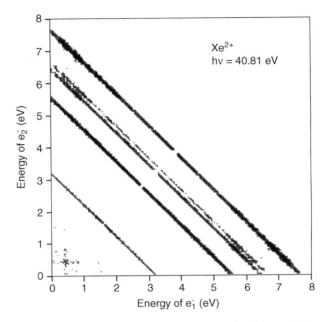

Figure 6. Double photoionization map of Xe on energy scales with some background and the weak secondary processes suppressed. The gap on the diagonal is caused by a dead time, during which almost simultaneous electron arrivals are not recorded as pairs.

Figure 6 shows data from a similar Xe experiment, with times converted to energies and the figure symmetrized by duplication about the line of equal electron energies. There is a narrow dead zone on the diagonal of the figure, because electron pairs with exactly equal energies, which would arrive simultaneously at the detector, are not recorded. Some authors prefer a diagram of this triangular form but with the lines of constant energy sum running horizontally. Finally, Fig. 7 shows the same data plotted with the summed energy as the vertical scale and one electron energy as the horizontal scale, which is the form of presentation used in this review. The vertical scale is equally a scale of final-state energies, obtained by subtracting the summed electron energy from the photon energy. Horizontal lines in such figures denote final dication energies, whereas vertical lines or alignments of peaks denote fixed electron energies, which represent intermediates in two-step double ionizations.

Single-axis spectra can be made as projections of maps such as Fig. 7 in the vertical or horizontal directions. A projection integrating intensity onto the vertical axis gives the spectrum of final dication states populated by the totality of double photoionization processes; the Xe example is shown in Fig. 8. Projection for a selected range of one electron energy emphasizes the states

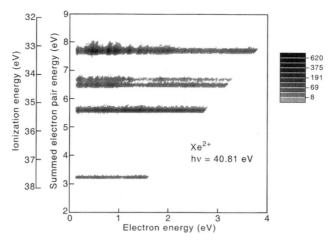

Figure 7. Double photoionization map for Xe at 40.81 eV in the form of representation used throughout the remainder of this review. The vertical scale of summed electron energies is equivalent to a scale of (double) ionization energy, as shown.

produced through a particular intermediate. Projection of the whole map onto the horizontal axis gives an overall view of the energy distribution and of any intermediate state energies; it is usually more useful to select a final-state energy and plot the spectrum of intermediates which contribute to it, as illustrated in Fig. 9. In this case the horizontal scale can be one of ionization energy (binding energy), which is appropriate where the main process is intermediate single photoionization following Eq. (2). Although these different spectra can be visualized as projections of the maps, they are derived in practice directly from

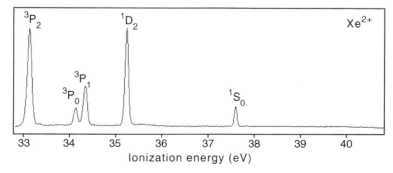

Figure 8. Spectrum of Xe^{2+} states populated in double photoionization at 40.81 eV photon energy, equivalent to projection of Fig. 7 onto the vertical axis. Note the evident conformity of the peak heights to the statistical weights of the populated levels.

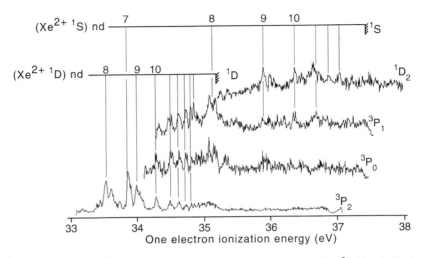

Figure 9. Electron distributions for formation of the individual levels of Xe^{2+}. The distributions are effectively intensities along the horizontal lines in Fig. 7.

the raw data, which gives the best possible resolution and avoids problems of pixellation.

VII. DOUBLE PHOTOIONIZATION OF COMPLEX ATOMS

For all atoms except helium and isoelectronic ions (H^-, Li^+, Be^{2+}...) there exist indirect paths for double photoionization, in addition to the direct path. Excited states of the singly charged cation can be found above the double ionization threshold, particularly as inner-shell hole states or Rydberg states of series converging on higher double ionization limits. If such a state is formed, it must decay by either photon emission or electron emission. Electron emission from a "superexcited" cation state is variously called Auger decay or autoionization, and it dominates over photon emission at all the relatively low energies (soft X-ray, <1000 eV) discussed here. The lifetimes of the emitting states can vary over a very wide range, from tens of picoseconds to femtoseconds, but are generally much shorter than typical radiative lifetimes. They have hitherto been determined only from line widths, which depend on the lifetimes of both initial and final states, which cannot be distinguished by this method. We shall see that the coincidence methods sometimes allow intermediate level lifetimes to be determined independently. A new technique using high-harmonic generation of ultrashort light pulses has recently demonstrated direct determination of femtosecond lifetimes in shake-up states of Ne and of Xe ions [73].

Work on energy distributions in double photoionization of complex atoms has so far touched only closed-shell species, the rare gases, mercury, and cadmium. Results are discussed in the following sections according to the energy regime covered.

A.   Valence DPI below Deep Inner-Shell Thresholds

The example of Xe, already illustrated in Figs. 5–9, is representative of observations on all the complex rare gases studied so far in the valence energy range. From the electron energy distributions contributing to the formation of the different final Xe^{2+} states (Fig. 9), it appears that double photoionization to the ground state at 30.4 nm is dominated by an indirect sequential process, passing through Rydberg states of the monocation as intermediates. The many peaks in the distribution are visually dominant, but when the summed area of the peaks (indirect) is compared with the area of the underlying continuum (direct process), the indirect process is found to contribute less than half the total intensity [68]. At shorter wavelengths (higher photon energy) the continuum becomes relatively stronger and the indirect ionization becomes weaker. In formation of the excited states of Xe^{2+}, the continuum is clearly the dominant contributor, with the indirect component diminishing as higher states are chosen and almost completely vanishing for the highest state from the p^4 configuration, $^1S_0$. This is simply because there are fewer and fewer excited states of the monocation available, because no higher excited states of the dication itself are within range and capable of supporting intense Rydberg series to act as intermediates in the double ionization process. The states of Xe^{2+} arising from the configuration $5s5p^5$ start above 45 eV and have been observed both in TPEsCO experiments [74] and by TOF-PEPECO [75] at shorter wavelengths (Fig. 10). In addition to the main lines $^3P_{2,1,0}$ and $^1P_1$ from the $5s5p^5$ configuration, many "satellite lines" from three-hole one-particle configurations $5s^25p^3nl$ appear with comparable strength, in spectra taken by either method. The electron energy distributions in forming the states shown in Fig. 10 are rather smooth, giving evidence of only very minor population by indirect double ionization. The distributions, like the underlying continua in Fig. 9, are also rather flat; this confirms the notion that flatness of the distributions at 10–20 eV above threshold is an indicator of direct double photoionization.

The relative intensity of the $5s^{-1}5p^{-1}$ ionization compared with $5p^{-2}$ ionization is about 1/20, at both wavelengths. Within the groups of lines of the p^4 configuration the relative intensities in complete photoionization spectra seem to agree roughly with the statistical weights of 5:3:1:5:1 for $^3P_{2,1,0}$, $^1D_2$, and $^1S_0$, with minor deviations. The relative contributions of direct and indirect photoionization can sometimes be separated [65], and where this is done the continuum (direct) part approximately follows the statistical weights at the

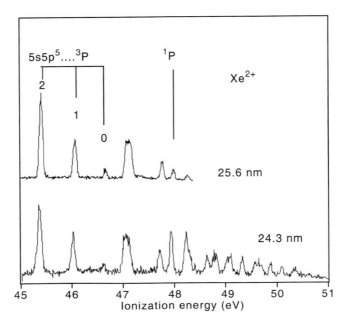

Figure 10. Spectra of Xe^{2+} states from removal of electrons from the $5p$ and $5s$ orbitals, at two shorter wavelengths equivalent to 48.37 eV and 51.02 eV photon energy. The lines without spectral identifications are mostly blends from the configurations $5s^2 5p^3 5d$ and $5s^2 5p^3 6s$.

lower photon energies, while indirect ionization is stronger in populating the lowest energy state than any higher one. This result may seem surprising, because Wannier theory for direct ionization near threshold indicates that only the $^3P^e$ term is favored by the symmetry rule. However, the energy range over which this "near threshold" conclusion is valid may be a narrow one, similar to the range where the threshold law is obeyed. At the energies where the $5s5p^5$ configuration is accessible, there is again no extra intensity for the symmetry-favored $^1P^o$ term (Fig. 10).

Many of the above remarks apply to Ar and Kr double photoionization as much as to Xe; the electron energy distributions for formation of the $^3P_2$ ground states of Ar^{2+} and Kr^{2+}, for instance, are shown in Fig. 11. Indirect processes are visually dominant at photon energies up to 50 eV, but the underlying continuum processes are at least as strong and become relatively stronger at the higher photon energies. In both Kr and Ar double photoionization at these energies the overall (direct + indirect) relative intensities of the terms from the p^4 configuration are again in rough agreement with the statistical weights. However, if the continuum parts are taken, states of the ^3P term are less intense than expected relative to $^1D_2$, which seems to be particularly favored. At

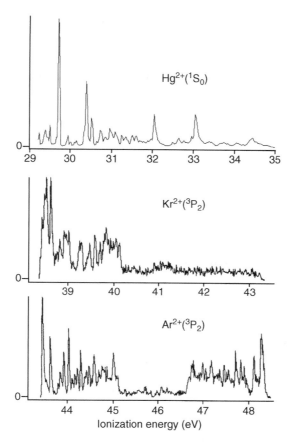

Figure 11. Electron distributions from double photoionization producing the ground states of Ar^{2+} and Kr^{2+} at 48.37 eV and of Hg^{2+} at 40.81 eV. The whole distribution is shown for Ar, and it illustrates the lower resolution of TOF-PEPECO for higher-energy electrons.

considerably higher energies (80–150 eV, synchrotron light), this trend continues for Ar [76]; the intensity ratio $\{^3P_{2,1,0}\}/\{^1D_2\}$, which should be 1.8 statistically, is 1.2 at 50 eV photon energy, 0.8 at 80 eV, and 0.75 at 90 eV.

Neon has not been examined this way in the laboratory, because the existing fast pulsed lamps are not capable of doubly ionizing it. Using the TPEsCO method with synchrotron radiation, Avaldi et al. [77] obtained a double photoionization spectrum of Ne over the whole energy range up to the Ne^{3+} threshold. Two characteristics of this spectrum that match the observations on Xe and other inert gases are that the intensities of the Ne^{2+} states from the $2s2p^5$ configuration are much less (where not evidently distorted by autoionization) than those from $2p^4$ and that satellites of $2p^3nl$ configuration are rather strongly populated. Most recently, Kaneyasu et al. [78] have examined the double photoionization of Ne over a wide energy range using TOF-PEPECO with synchrotron radiation. The full electron energy analysis allowed direct and indirect processes to be separated, and absolute cross sections were obtained for comparison with theory. The relative intensities of the $2p^4$ states are far from statistical except for total (direct + indirect) intensities in a narrow energy range near threshold, where any agreement may be accidental. At all higher energies the triplet terms are much weaker than statistical relative to the singlets, both for the $2p^4$ states and the $2s2p^5$ states. In the purely direct cross sections the singlets are stronger than the triplets at all energies despite the contrary statistical weights; it seems that the apparent statistical intensities at low energy are entirely due to extra population of the lowest terms (triplets) by indirect double ionization. It is not clear whether shake-off (or even direct knockout) has a predilection for spin conservation because this question does not arise for He, but the question may be interesting for future enquiry.

Complete double photoionization spectra of atomic mercury [79] and cadmium [75] have been examined by TOF-PEPECO and show characteristics similar to those of the rare gases. The ground states of the doubly charged ions are of $^1S_0$ symmetry, reached by s^{-2} ionization, and the first excited states are $^3D_{3,2,1}$ and $^1D_2$ reached by $d^{-1}s^{-1}$ processes. For both elements the formation of ground-state dications at the laboratory wavelengths is dominated by indirect ionization; the direct part is essentially undetectable in spectra at 40.81 eV photon energy (Fig. 11) and remains undetectable for Cd even at 48.37 eV. The photoionization map for Cd at 40.81 eV, showing this unusual signature, is given in Fig. 12, and the final-state spectra of Cd^{2+} at 40.81 and 48.37 eV are in Fig. 13. In addition to the main lines, the spectrum at 48.37 eV contains a strong contribution from levels of the configuration $4d^9 5p$, which can be reached by shake-up accompanying the two-electron ejection. In the equivalent spectra of Hg^{2+} at 48.4 eV [79], the similar terms and levels arising from $5d^9 6p$ are visible, but are interspersed with levels from $5d^8 6s^2$.

In both Cd and Hg double photoionization, the electron distributions show some involvement of indirect processes in formation of essentially all the doubly charged states populated. The relative populations within levels of each configuration are nevertheless not wildly different from the relative statistical weights. A common feature is that singlet terms ($^1D_2$) are more favored at higher photon energies, a tendency also noted in the rare gases and one that goes in the direction of classical Auger spectra where singlets are more strongly populated than triplets in ionization from a closed shell neutral.

B.   Inner-Shell DPI of Atoms

The combination of magnetic bottle spectrometers with synchrotron radiation from storage rings has opened a huge energy range but also posed a problem.

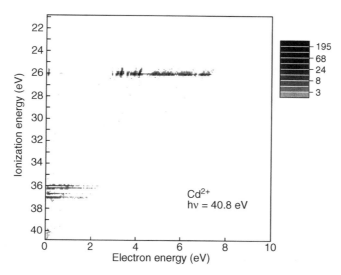

Figure 12. Double photoionization map for Cd at 40.81 eV. The complete absence of direct double photoionization to the ground state is notable.

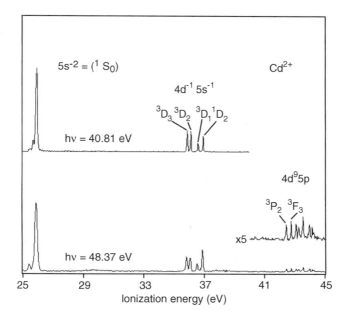

Figure 13. Spectra of Cd^{2+} from photoionization at two photon energies. The increased relative intensity of Cd^{2+} $(\cdots 4d^9 5s(^1D_2))$ at the higher energy is notable. The small peaks just above 25 eV in both spectra are artifacts of the lamp pulse structure.

Even at the largest storage rings operating in single-bunch mode, the interval between light pulses is less than $1\mu s$. The time of flight of an electron of initially zero energy in a 2-m magnetic bottle is typically arranged to be 5–$10\ \mu s$, but at a storage ring the flight time can be measured only modulo the pulse period. An ingenious solution to this problem, first introduced by Penent et al. [80], is to choose a photon energy at which the cross section is dominated by an inner-shell ionization and where the photoelectron is faster than any subsequent Auger electrons. The first electron of each coincident group is thus of known time origin, and all later electrons can be referred to the first. This technique, together with data filtering based on energy conservation, has enabled detailed studies on many inner-shell effects.

The first studies by the new technique [80] were on the decay of $4d$ core-hole states in Xe. An important advantage of the multielectron coincidence method is that it allows complete separation of Auger spectra into the parts due to each initial level, where the core-hole state is split by spin–orbit coupling. For instance, separate spectra are obtained for the $^2D_{3/2}$ and $^2D_{5/2}$ states of the Xe^+ core hole [81, 82]. The relative intensities of the Xe^{2+} final states in this Auger decay are not at all statistical, but show the singlet states $^1D_2$ and $^1S_0$ of the ground configuration much more prominently than the triplet $^3P_{2,1,0}$. This is, of course, normal behavior in the Auger effect. The electron spectra taken in coincidence with electrons from each core hole also have a continuous part, which is due to a double Auger effect in which two more electrons are ejected, leaving Xe^{3+}. In this double Auger effect, as in double photoionization, the electron distribution can be continuous, denoting a direct process, or structured when an intermediate Xe^{2+} state is involved. The distributions are found to be highly structured, a major part of the double Auger intensity going through two particularly favored intermediate levels [80]. Very similar behavior has been found in double Auger decay from Ar $2p$ hole states, where the intermediate states could be identified as $Ar^{2+}\ 3s^{-2}$ correlation satellites [83]. A triple Auger decay leading to Ar^{4+} could also be observed and measured. The lifetime of the intermediate states in these cases can be determined from the linewidths in the final electron emission; for instance, an intermediate Ar^{2+} state lifetime was found to be more than 22 fs, in contrast to the 5.5 fs lifetime of the $Ar^+\ 2p$ hole. The first electron in the cascade process gives a broader peak, because both the hole-state lifetime and the intermediate lifetime affect it. The fact that the two peaks in cascade double Auger decay can have different widths contrasts with the situation in double photoionization following Eq. (1), where the first and second electron peaks in a cascade pathway both have exactly same intrinsic width, because only one lifetime is involved. Similarly, in a process that goes through an initial neutral excited state, such as resonant double Auger (below), the two electron peaks of a sequential pathway could again have different widths.

The availability of a complete range of photon energies from storage ring synchrotron emission has allowed the formation and decay of some exotic inner-shell hole states to be studied. The formation of double core-hole states, where two electrons are removed from the same inner shell by direct photoionization is expected to be weaker by several orders of magnitude than the formation of doubly charged ions with two vacancies in the valence shell [84]. Nevertheless, the $4d^{-2}$ double inner shell hole states of Xe^{2+} could be studied in detail, because of a strong configuration mixing with the single electron ejection leading to a $4p_{3/2}$ hole with an εf electron in the continuum [85]. The relative populations of the Xe^{2+} $4d^{-2}$ levels produced in this process are governed by the symmetry of the mixing state. Once formed, the double hole states decay by a double Auger process, again involving a cascade through Xe^{3+} intermediate states, to final states of Xe^{4+}.

Another unusual type of highly excited doubly charged ion has one hole in a deep inner shell and one in the valence shell. Such states are of interest theoretically because of (a) the presence of distinct electron vacancies and (b) several possible pathways for their production [86]. The first observation of direct formation of such states in Ne and N_2 was recently achieved by the magnetic bottle TOF method [87]. The cross section for their formation is very low, and the electron distribution in the Ne case shows both a direct contribution and an indirect pathway. For the final states from $1s^{-1}2p^{-1}$ ionization the intermediate states are satellites of the $1s2s(^{1,\,3}S)ns$ series whose lines show asymmetric Fano profiles. The presence of these profiles demonstrates that the matrix elements representing direct double ionization, formation of the intermediate states, and interaction of those states with the continuum are of similar magnitudes.

Line shapes in electron spectra from double photoionization are often asymmetric for another reason, namely *post-collision interaction* (PCI) between electrons in the final state. This is particularly important if one or more of the electrons are slow ($<1eV$, say), and therefore concerns TPEsCO results strongly, but may also affect the overall outcome of double photoionization. In the classical PCI effect, a fast electron is emitted after a slow one and overtakes it; the change in residual ion charge state then retards the slow electron and accelerates the fast one. PCI's effect on the production of different final dication states in coincidence with threshold electrons has been investigated by the new coincidence techniques in the double ionization of Xe at energies near the opening of the $4d$ inner-shell ionization—that is, by the Auger and "resonant" Auger effects [88]. A conclusion of these investigations is that at a photon energy just above a threshold, the slow electron may sometimes be so retarded as to be recaptured into a quasi-bound state of the singly charged ion, which then autoionizes to a final dication. The same quasi-bound monocation states may also be formed directly, so interference is possible. The full extent of PCI in the Auger effect near inner-shell thresholds thus involves interactions and possible energy exchange between two or three (in double Auger) electrons and can affect the partial cross sections into all accessible channels. It is a strong manifestation of final-state electron correlation. In double photoionization without inner-shell involvement the same process of slow electron capture into short-lived superexcited states which reemit an electron must be taken into account in full theoretical treatments.

Besides forming inner-shell hole states directly, photons of energy just below and just above the ejection thresholds can populate states of the neutral atom with an inner-shell hole and one electron in a normally unoccupied orbital. Such states form the well-known "near-edge structure" in absorption spectra. They can decay by one- or two-electron emission in what are called *resonant* Auger effects. With two-electron emission, this is another form of indirect double photoionization. The electron pair emission may again be simultaneous and produce a monotonic electron distribution, or may occur by a cascade through singly charged states. Both types of route are followed. The spectrum of final singly charged ions populated by the resonant Auger process generally extends to higher energy than the normal photoelectron spectrum at a similar photon energy, because the initially excited electron can act as a "spectator" and can also be "shaken" to a new orbital in the emission process [89]. The same thing happens in resonant double Auger formation of doubly charged ion states, but has not been studied in detail by the new coincidence methods, largely because of the low resolution in TOF-PEPECO for the high-energy electrons usually involved.

VIII.   DOUBLE PHOTOIONIZATION OF MOLECULES

A.   TPEsCO and TOF-PEPECO Spectra

Until very recently, the spectra of doubly ionized molecules were almost completely unknown. The existence of many "stable" doubly charged molecules was familiar from mass spectrometry, and some appearance energies had been measured by electron impact and other collisional methods, but with low accuracy. Previous coincidence methods added some new spectral information from measurements of kinetic energy releases in dissociative double ionization, eventually with excellent resolution on a few molecules [90]. The situation was completely changed by the advent of the TPEsCO and TOF-PEPECO techniques, and now (mid-2007) reliable spectra of more than 50 molecular dications have been published. Table I gives references to all the molecular spectra, but very few are reproduced here because this review is focused instead on the mechanisms of final dication formation. Where the same molecules have been investigated by both techniques, the positions of sharp peaks in the spectra agree in almost all cases within 20 meV or better; this figure gives a safe guide to the absolute energy accuracy. The relative intensities of different final states agree in a few cases, but often differ because of the different roles of autoionization in the two techniques.

TABLE I   References to Double Photoionization Spectra of Molecules

Molecule(s)	TPEsCO	TOF-PEPECO
HCl	113, 114	
HBr	115, 116	117
HI		118
N_2	59, 62, 119	62, 117
CO	59, 61, 119	91, 117
NO	59	94, 117
O_2	59, 120, 121	106, 117
Cl_2	113	
Br_2	122	107
I_2		123
ICl		136
HCN		124
H_2O		98
H_2S	125	117
N_2O		71, 117
CO_2	126	93, 117
OCS	126	117
CS_2	127	117, 135
SO_2		117, 128
BrCN, ICN		70
C_2H_2	126, 129	129
H_2CO		100
NH_3		98
CH_4		98
GeH_4, $GeBr_4$, $GeCl_4$		99
CF_4	126	130
SF_6	131	132
CH_3I, CF_3I		118
RI: R = C_nH_{2n+1}, $n = 1, 2, 3, 6$		133
$I(CH_2)_nI$: $n = 1, 2, 3$		133
$Br(CH_2)_nBr$: $n = 1, 2, 3$		133
DABCO, dithiane, diamines		133
Ethylene, 1,3-butadiene		134
Cyclooctatetraene		134
Benzene	126	134
Naphthalene		134
Toluene		95

The TPEsCO and TOF-PEPECO spectra of CO^{2+} [91, 92], N_2^{2+} [62] and CO_2^{2+} [93] have been compared in detail. The intensity differences could be attributed to the special sensitivity of TPEsCO to indirect double ionization through states of the singly charged ions which accidentally coincide in energy with dication final states. In molecules, indirect pathways through monocation states according to Eq. (2) give access to dication states outside the Franck–Condon zone for vertical double ionization from the neutral molecules, and such effects must be expected in TOF-PEPECO as well as in TPEsCO. In fact, TOF-PEPECO, where all electron energies are accepted, could involve more indirect pathways than TPEsCO, but in practice the influence of any such individual pathway on the spectra is diminished by dilution. The evidence for this assertion is empirical; vibrational intensity distributions in TOF-PEPECO spectra closely match the calculated Franck–Condon factors [94], whereas in TPEsCO this is not usually so. In many cases the molecular ion geometries are not well enough known to allow reliable calculation of Franck–Condon factors, but the TOF-PEPECO resolved vibrations have monotonic intensity distributions that would be compatible with vertical transitions from the neutral ground state, whereas TPEsCO spectra show distributions with irregular intensities or excitation of normally forbidden modes. An interesting example is acetylene (Fig. 14), where the TPEsCO spectrum shows a whole vibrational progression that is absent from the TOF-PEPECO spectrum. The new vibration is a bending motion, whose excitation in a vertical transition between the linear neutral and linear dication is forbidden by symmetry; it may become allowed by intervention of a nonlinear intermediate monocation state. Clearer evidence for and against indirect pathways in double photoionization comes from two sources, detailed in the following sections.

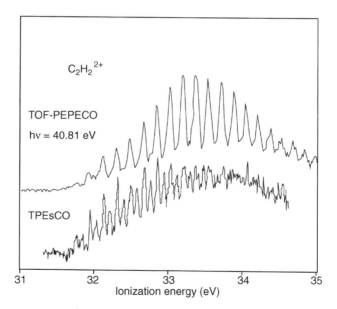

Figure 14. Comparison between the TOF-PEPECO and TPEsCO spectra of acetylene (ethyne). The intermediate peaks in the TPEsCO spectrum and its very low intensity of resolved bands above 33 eV are both attributed to autoionization from energy-coincident monocation states, populating the ground state $(^3\Sigma_g^-)$ of $C_2H_2^{2+}$.

B. Double Ionization Quantum Yields

Whereas in an atom the only possible fates of superexcited states are to emit light or electrons (or both), for a molecule the possibility of dissociation always exists in addition. In single ionization this leads to the definition of the quantum yield of ionization as the number of ions produced per photon absorbed. An identical definition is possible for double photoionization, but as yet there is no experimental means of measuring the quantity so defined. Instead we can define a new quantity, easily derived from TOF-PEPECO data, for which no suitable distinctive name has yet been proposed [95, 96]. It is the probability that after one electron has been ejected by a single photon, a second one will follow. It can be determined because in TOF-PEPECO using the magnetic bottle technique, the electron detection efficiency f is known and is independent of electron energy or emission angle. The true coincident electron pair detection rate C^{2+} (after any necessary background subtraction) depends only on the true electron pair emission rate $N^{2+}(E_1, E_2)$ for each particular pair of energies:

$$C^{2+}(E_1, E_2) = N^{2+}(E_1, E_2)f^2 \tag{6}$$

The coincidence rates can be integrated in the data over E_2, giving a sum that represents the total double photoionization with one electron at E_1:

$$C^{2+}(E_1) = N^{2+}(E_1)f^2 \tag{7}$$

The single electron detection rate C^+ at the same electron energy depends on both the pair emission rate and on the single ionization rate $N^+(E_1)$:

$$C^+(E_1) = N^{2+}(E_1)f(1 - f) + N^+(E_1)f \tag{8}$$

The ratio $N^{2+}(E_1, E_2)/N^+(E_1)$ is the desired quantum yield at "first" electron energy E_1, and it can be derived from the data as

$$QY = C^{2+}(E_1)/\{C^+(E_1)f - C^{2+}(E_1)(1 - f)\} \tag{9}$$

In future paragraphs this quantity QY is referred to as "quantum yield" with inverted commas to avoid any possible confusion. The choice of E_1 is not arbitrary; it must be the higher of the two ejected electron energies. This allows the quantum yield to be determined from the double ionization threshold up to an excess energy of half the difference between the photon energy and the double ionization energy. For any lower electron energy E_1 (i.e., a higher ionization energy), the complementary electron can have an energy overlapping the range already covered, confusing the data. Some examples of these "quantum yields" are given in Fig. 15, plotted on a scale of single ionization energy (binding energy, hv-E_1).

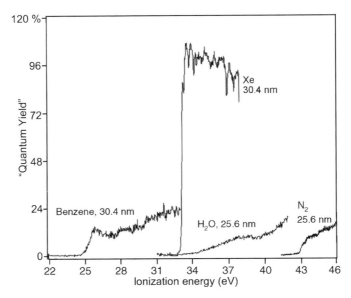

Figure 15. "Quantum yields" of double photoionization, as defined in the text, for a selection of molecules and for atomic Xe, with the wavelengths used.

From Eq. (9) it is clear that correct determination of QY depends on accurate knowledge of the overall detection efficiency f, which is generally about 0.5. It can be determined in two ways. One is to compare the measured ratio C^{2+}/C^+ with the calculated ratio for an atomic rare gas such as Xe at a wavelength where the true relative production ratio N^{2+}/N^+ is known from photoionization mass spectrometric measurement of the ion yields. This method has been used in deriving the Xe data of Fig. 15. As expected, the "quantum yield" in the atomic case is unity (100%) within experimental error; this must be generally true because deactivation of superexcited states by light emission is very rare. Thus the second and quicker method is simply to measure the apparent QY for an atomic gas and determine f accordingly.

The striking fact about the measured "quantum yields" for molecular double ionization is that they are never unity near threshold; they rise slowly at higher energies, and hardly reach unity at even 20 eV above threshold. The interpretation is very simple: At the binding energies $h\nu$-E_1 intermediate monocation states are populated, but instead of ejecting a second electron, they dissociate to singly ionized products. This should certainly be expected, because exactly the same thing happens in single ionization with neutral superexcited states, as has been well known for many decades and is clearly summarized in Berkowitz' more recent book [97]. The curves shown in Fig. 15, which are absolutely typical, demonstrate two basic facts:

i. Superexcited states of the monocations can always be populated at energies above the double ionization energy.

ii. The great majority of such states can dissociate to singly charged products in competition with second electron ejection.

Any deviation from this behavior would be very surprising, and none has yet been found. Unity "quantum yield" in molecular double photoionization is therefore not expected except at such high energy that direct double ionization completely dominates, overwhelming the cross section for intermediate monocation formation. The highest "quantum yields" may be expected for small molecules with very strong bonding; this is borne out by the examples in the figure. One extreme form of behavior, which may be approached in some cases discussed below, would be that all populated monocation states might be fully dissociated and not autoionizing. In such a case the double photoionization that remains would be pure direct double ionization.

"Quantum yields" in double photoionization involving deep inner shells of molecules have not yet been determined, but are expected to show the same general pattern of behavior.

C.   Double Photoionization Maps and Electron Distributions

The signature of direct double photoionization to a single state of an atomic dication is a purely monotonic electron energy distribution, while a structured distribution unambiguously denotes indirect pathways. At photon energies not too far ($< 20\,\mathrm{eV}$?) above threshold, distributions in direct double ionization are also rather flat. These statements are also true for molecules, but need careful reservations. Every molecular electronic transition may be accompanied by vibrational energy changes; if all the molecules do not start in the vibrational ground state, the Franck–Condon envelope may not be monotonic. In any indirect process, two sets of vibrational overlap integrals are implicated, and there is also the possibility of interference with a direct pathway. The resulting intensity distribution, even for a single state, can be highly complex. Spectral congestion, where several indirect pathways lead to a single final electronic state, could give the appearance of a continuous monotonic distribution. Thus the observation of structured electron distributions in double photoionization proves the existence of indirect pathways, but a lack of it does not exclude them.

In the following sections, we first consider molecules for which direct double photoionization seems to dominate, then we present clear examples of molecular Auger effects, and finally we discuss the dissociative double ionization mechanisms that have no counterpart in atoms.

D.   Mainly Direct Double Photoionization

For some molecules the "quantum yield" of double photoionization is low and the electron energy distributions are both flat and relatively smooth. This

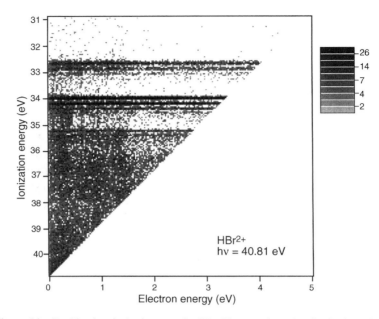

Figure 16. Double photoionization map for HBr. The even intensity distributions along the horizontal lines for each final vibration level suggest mainly direct double photoionization.

evidence suggests that their singly charged superexcited states, although formed, mostly dissociate to singly charged products. Examples are the small molecular hydrides H_2O, NH_3, CH_4 [98], and HBr and molecular nitrogen, N_2 [92], The electron distribution maps for photoionization of each of these also show the existence of more complex processes of the types discussed below, but they are not the major processes at the laboratory wavelengths ($hv < 50$ eV). The map for HBr is shown in Fig. 16 as an example, with projections giving the electron energy distributions in forming some of the final HBr^{2+} states in Fig. 17. Figure 18 shows the spectrum of HBr^{2+} and the "quantum yield" of double ionization, which has clear steps at the positions of each final electronic state. The visibility of such steps is another proof that the monocation states, which could fill in the steps, are in fact almost fully dissociated rather than ionized. It is a small irony that apart from He, it is easier to find almost pure direct double photoionization in small molecules, which physicists often eschew, rather than in atoms.

E. Auger Effects

The advantage of the magnetic bottle TOF-PEPECO technique relative to classical Auger spectroscopy, that it allows separate examination of the Auger

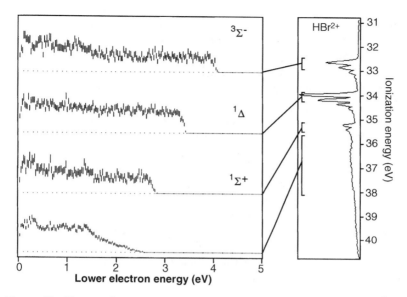

Figure 17. Electron distributions for production of different final states of HBr^{2+} from projections of Fig. 16 over the ranges shown.

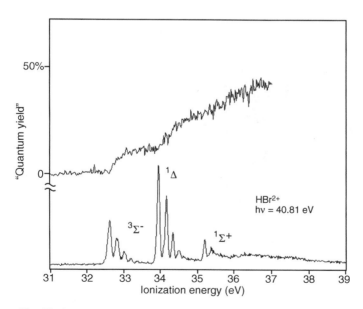

Figure 18. Final states and "quantum yield" in double photoionization of HBr. The step structure of the yield is a further indication that most superexcited states of HBr^{+*} dissociate rather than autoionize to the doubly charged cation.

spectra from each initial hole state, is no less valuable for molecules than for atoms. On the debit side, the resolution is generally poor at the high Auger energies released in the decay of deep inner shell holes by single electron emission. Shallow core holes, which are nevertheless distinct from the valence shell and inactive in molecular bonding, can, however, be studied with advantage. Particularly interesting are the $2p$ shells of Si, P, and S near 100, 135, and 163 eV, $3d$ shells from Ge to Br, and $4d$ shells of Te and I. Metallic elements offer many more possibilities; because their compounds are generally involatile, they will not be studied in this way until suitable oven sources are developed.

Germanium compounds have an inner $3d$ shell which lies conveniently within the range of laboratory light sources, but above the double ionization energy. By choice of compound the energy of the $3d$ hole can be varied; electron-donating substituents on the Ge atom lower it (GeH_4, $Ge(CH_3)_4$), while electron-withdrawing substituents raise it ($GeCl_4$, $GeBr_4$). When the resulting $3d$ photoelectron energy is low, the double photoionization map demonstrates the effect of PCI and allows a test of the theory of this final-state electron interaction [99]. The lines of nominally fixed (photo)-electron energy, which are the strongest features in the electron pair distribution map, become curved for low Auger electron energies, as illustrated in Fig. 19.

For the majority of molecules without accessible shallow inner-shell hole states, two distinguishable and much more widespread forms of the Auger effect

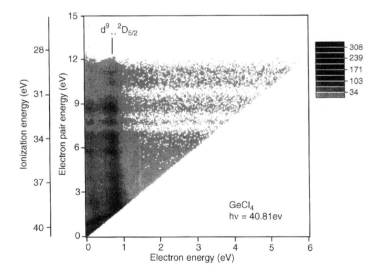

Figure 19. Photoionization map for $GeCl_4$. The inner shell hole states corresponding to atomic $^2D_{5/2}$ produces the curved but roughly vertical concentrations of intensity for electron energies below 1 eV. A similar line for the $^2D_{3/2}$ hole is present but hardly visible in the figure. Curvature of the lines is caused by post-collision interaction (PCI).

exist at low and intermediate energy. First, just as in atoms, there are Rydberg states of the singly charged ions in series converging on every higher double ionization threshold. The continuity of oscillator strengths as a function of energy ensures that if higher doubly ionized states can be populated directly, the related Rydberg series states are also populated and can autoionize to any lower dication state. This could be called an *inner Rydberg* Auger effect. Surprisingly, in view of its prevalence in atoms, evidence for this expected inner Rydberg Auger effect in molecules is scarce. We should expect to detect it most easily in cases where there are resolved and strongly bound excited states of the dication, a requirement that points to small and multiply bonded molecules. The process should be revealed by structured energy distributions, particularly in formation of dication ground states, in cases where strongly populated bound dication states exist at high energy with sharp spectra dominated by (0,0) transitions. The excited dication states should best be far above the ground state, because the Z^2 term in the Rydberg energy makes the low $n*$ Rydberg states, which are the most intense, lie 5 or 6 eV below their convergence limit. No unambiguous example of this process is yet known. Rapid dissociation of Rydberg states based on highly excited dication cores may make the signals weak, and congestion may conceal them.

Second, in molecular photoelectron spectra at energies above about 20 eV, there are states and groups of states classified as "inner valence" (IV) states. They are normally numerous states of mixed (often distant) electronic configurations, gaining intensity in the photoionization process from one-electron ionization out of molecular orbitals based on deeper valence shell atomic orbitals, such as the $2s$ orbitals of carbon. The relative intensity of the IV bands in photoelectron spectra generally increases at higher photon energies and becomes comparable to that of the main valence bands at X-ray wavelengths. If the double ionization threshold is lower in energy than an IV band, an *inner valence* Auger effect can and does occur. For carbon compounds the energy range of $2s$-based IV states is about 20–25 eV; as the molecules get bigger, especially if they are unsaturated, the double ionization threshold easily descends into this zone. Because the IV bands are generally broad, and not far above the double ionization limit, this Auger process produces electron distributions with pronounced intensity peaks near zero energy. The double photoionization maps look typically "hollow", with very little intensity for equal energy sharing in formation of the lowest dication states. To illustrate this, the map for toluene [95] is shown as Fig. 20 in the alternative representation where equal energy sharing corresponds to the central vertical axis. For the other important first-row atoms N and O the inner valence ionizations lie deeper, sometimes well above the double ionization limit. The double photoionization electron pair distribution and "quantum yield" may then show a peak at the IV band position, as illustrated by the

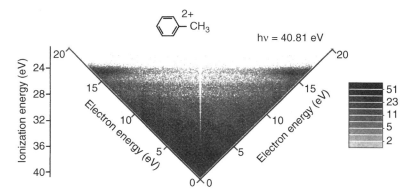

Figure 20. Photoionization map for toluene, in the triangular representation. The low ionization energy states of the dication are weakly populated by direct photoionization but strongly populated by the inner valence Auger effect, giving deeply hollow electron distributions. Valence photoionization maps for large molecules are generally like this, with the hollowness becoming more extreme for the largest molecules.

double photoionization of formaldehyde [100] in Fig. 21. Another consequence of the inner valence Auger effect from a deep inner valence band is seen in the double photoionization map for DABCO (1,4-diazobicyclo[2,2,2] octane) in Fig. 22. Here and in the maps for several related diamines we see diagonal bands in the region of the lowest double ionization energy, as plotted in the usual representation where the lower electron energy defines the horizontal scale. If the map were plotted against the higher electron energy of the two, the same events would form a vertical bar, as in the classical Auger effect.

The intensity distribution in the DABCO map illustrates two other basic phenomena. First, the DABCO^{2+} dication ground state is populated much more strongly by the Auger effect than by the direct processes visible as horizontal lines. The ground state has its charges on the two nitrogen atoms, but removal of two electrons from separate distant atoms simultaneously by direct ionization is inherently improbable, in line with simple prediction of the direct knockout model. The two-step process can facilitate it, since the intermediate excitation can be effectively delocalized. This phenomenon is quite general in the double photoionization of large molecules, where the charges are far distant from one another in the ground-state dication. Second, the spectrum of the dication from the Auger process shows a broader band for the ground state than does the direct process. This is the result of a change in internuclear distances in the intermediate cation, made possible by an appreciable lifetime, expressed as a wide spread of significant Franck-Condon factors. Such change by broadening (or conceivably, narrowing) of vibrational envelopes in molecular indirect

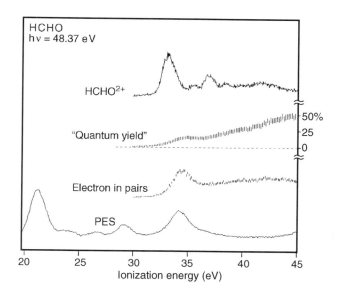

Figure 21. The inner valence Auger effect in formaldehyde (methanal). The broad peak in the regular photoelectron spectrum near 34 eV is an inner valence (IV) band based on $C2s^{-1}$ ionization. It recurs as a peak in the electron distribution for double photoionization and in the "quantum yield." The uppermost curve shows the spectrum of $HCHO^{2+}$, where the ground state is strongly populated by this effect.

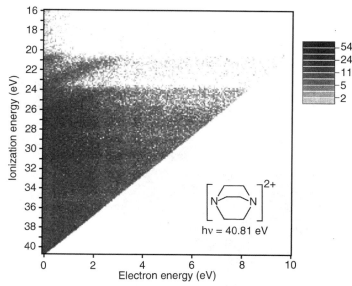

Figure 22. Photoionization map for DABCO (1,4-diazabicyclo[2,2,2]octane). The sloping feature at the upper left is a product of this form of representation; it really represents electrons of fixed energy of 17.3 eV, from photoionization to an inner valence band of the singly charged ion, which autoionizes to the dication ground state. The extreme hollowness of the distribution for forming the dication in its lowest state (charges on the two N atoms) is also apparent.

139

double ionization is always to be expected. From the magnitudes of the quantum yields in inner valence Auger effects, it is apparent that in all cases so far studied, the superexcited IV states dissociate to singly charged products in competition with the autoionization.

F. Dissociative Double Photoionization

The first sight of the double photoionization map for molecular oxygen (Fig. 23) was a considerable (though perhaps unnecessary) surprise to the researchers. The map is dominated not by horizontal lines for dication states (though these also exist) but by vertical lines of fixed single electron energy, lines which cannot be attributed to any molecular Auger effect. They are caused by autoionization of oxygen *atoms* produced by dissociation of superexcited singly charged molecular states. The process

$$M + h\nu \rightarrow M^{+*} + e_1^- : \qquad M^{+*} \rightarrow m_1^+ + m_2^* : \qquad m_2^* \rightarrow m_2^+ + e_2^- \quad (10)$$

had actually been discovered by conventional electron spectroscopy some time previously [101], but its extent was not appreciated. In fact this process involving oxygen atoms dominates double ionization at laboratory wavelengths

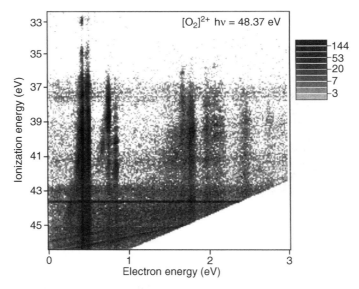

Figure 23. Partial photoionization map for O_2 at 48.37 eV. The many vertical lines represent states of atomic oxygen O^* which autoionize to ground state O^+ (^4S). Many of them extend to energies below the lowest level of molecular O_2^{2+} near 36.5 eV. Sloping lines in the lower part of the figure, which join on to the vertical lines, are again an effect of this representation.

in O_2, CO, NO, and NO_2 and is prominent for CO_2, SO_2, and N_2O (references in Table I). The same atomic autoionizations are also visible after inner-shell hole creation in several of these molecules. The same process but involving nitrogen atoms is visible (but not prominent) for valence double photoionization of N_2 and N_2O and involving sulfur atoms in H_2S and OCS. Faint indications of the same processes are seen for some halogen compounds at laboratory wavelengths, and they can be stronger when inner shells are excited [102, 103]. In many cases the atomic autoionization extends to total double ionization energies well below the molecular double ionization threshold, almost down to the thermodynamic limit for cation pair production. This means that the auto-ionization step of reaction (7) must sometimes take place when the fragments are far apart, at a minimum separation easily calculable as e^2/KER, where KER is the maximum possible kinetic energy release allowed by conservation of energy, $KER = hv - (E_1 + E_2)$. Such large distances (often 1 nm or more) also imply a long lifetime for the separating excited states, ranging up to picoseconds.

The nature of the atomic autoionizing states of O, N, and S can be discovered with fair certainty from the electron energies and the known optical spectra of the atoms. They are Rydberg states of low effective principal quantum number ($n^* = 3$ or 4) from series converging on the excited states of the atomic ions in their ground configurations. The nature of intermediate molecular monocation states that dissociate to produce them is much harder to discern, even though the "action spectrum" for production of each atomic autoionization is easy to extract from the data. Many electronic states seem to be involved, each with its attendant vibrational structure adding to a highly congested picture. The fullest, but not necessarily correct, identifications have been made in the case of CO [104] and O_2 [105, 106]. In general, they are probably Rydberg states belonging to series converging on higher molecular dication states that correlate to excited atomic dissociation products. The density of electronic states and the multiplicity of curve crossings in both the singly and doubly ionized manifolds make identifications beyond this general statement very difficult at present. The reasons why the phenomenon of atomic autoionization in DPI is so marked for oxygen above all other atoms and why it is so restricted to particular compounds, all small molecules, are also entirely obscure.

In the double photoionization maps of a select subset of small molecules, another phenomenon appears. It is a stepwise increase in intensity in the electron energy distribution as lower energies are approached. The "curtain edge" structures are usually seen above the minimum molecular double ionization energies, but in two cases they extend below them (N_2O, NO_2). Examples of the maps are shown in Figs. 24 and 25, and the same phenomenon can be recognized in HBr (Fig. 16). It has recently been recognized [107] that the positions of the edges always match excitation energies of possible atomic ion fragments, usually

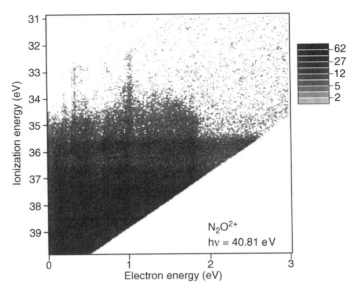

Figure 24. Photoionization map for nitrous oxide, N_2O. Vertical lines show states of N^* and O^* which autoionize, as in O_2. The threshold for ionization to molecular N_2O^{2+} is at 35.5 eV; all the intensity at lower ionization energies represents formation of dissociation products carrying two separate positive charges; the broad "curtain" structure is attributed to autoionization from high Rydberg states. A similar map is found for NO_2.

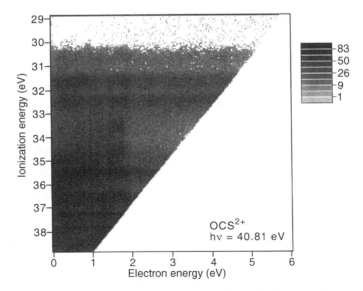

Figure 25. Photoionization map for OCS. Atomic autoionization is seen as sharp vertical lines and a "curtain structure," but in this case they do not extend below the molecular double ionization threshold of 30 eV. Similar maps are found for CS_2, Br_2, CH_3I, ICl, and others.

in their first or second excited states. In Br_2 and HBr double photoionization, for instance, two edges appear at 0.4 and 1.4 eV corresponding to Br^+, $^3P_{1,0}$, and $^1D_2$, respectively. In N_2O the edge is at 1.9 eV (N^+ $^1D_2$), in NO_2 it is at 3.3 eV (O^+ ^2D), in CS_2 and OCS it is at 1.8 eV (S^+ ^2D), and in I_2 and CH_3I it is at 0.8 eV (I^+ ^3P). For IC1 only the Cl^+ ^1D edge appears, while in CH_3Cl we see both Cl^+ ^1D and Cl^+ ^3P. Below the edges the extra intensity is a smooth and nearly constant or slowly falling function of energy, right down to zero. Because these edge energies are characteristic of the free atoms, we can safely conclude that the process involved is a dissociation. Because the intensity extends closely up to the edges, a reasonable interpretation is that intermediate singly charged states dissociate, producing an ion and an atom in a high Rydberg state. Such states have long lifetimes and might ionize (possibly under the influence of the other charge) at a range of internuclear distances, producing a continuous range of energies. This proposed mechanism is the same as reaction (7), the only difference being the identity of the autoionizing atom state as a high Rydberg rather than a low one. Unanswered, however, are the questions of why the electron distributions spread over a wide energy range and why the phenomenon is observed only in selected molecules. In the maps for N_2O, OCS, and NO_2 we see autoionization from both low Rydbergs and (putatively) high ones; the combination is particularly clear for OCS (Figs. 25 and 26), where the autoionizing states of S^* can be identified from the spectra of Gibson et al. [108].

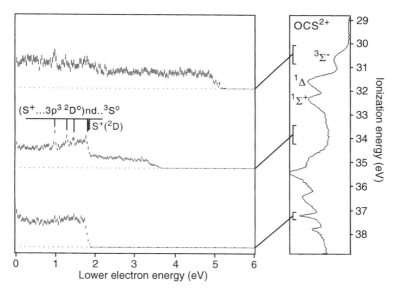

Figure 26.   Spectrum and electron distributions extracted from Fig. 25. The "curtain edge" is particularly clearly seen at ionization energies between real dication states; in this as in all other cases the edge position exactly matches an atomic ion excited level.

IX. RESONANT AUGER PHENOMENA IN MOLECULES

Near-edge structures appear as prominently in the X-ray absorption spectra of molecules as in atoms, or even more prominently. The resonant Auger decay of molecular single-hole one-particle states by single electron emission has been studied intensively by noncoincidence methods, particularly in small molecules [109]. Resonant *double* Auger decay, producing doubly charged final states, is seen prominently in synchrotron-radiation–excited TOF-PEPECO spectra of the common atmospheric molecules at their resonant photon energies, but is difficult to study in detail because the electron energies are high even for C K-shell holes, so resolution is poor. There are indications that dissociative double ionization and stepwise cascades occur as contributors to the overall decay mechanisms, as expected [110]. There will be studies of resonant Auger processes in Si, P, and S compounds, where convenient low-energy inner shells exist, in the very near future.

X. RELEVANCE TO CORRELATION

Because direct double photoionization is a prototypical two-electron process, it is natural to think that correlation between pairs of electrons may be of great importance for its occurrence. The relevance of double photoionization studies to the understanding of final-state electron correlation is indisputable. The form of the cross-section law, angular distributions, and PCI effects all relate directly to specific interactions between the almost free electrons. Whether double photoionization studies have contributed or will contribute to the understanding of *initial state* correlation in neutral atoms and molecules is less clear, however. Much double photoionization is indirect, taking place by two one-electron steps; initial correlation has no more relevance to such processes than to individual one-electron steps, which are more easily studied. Direct double photoionization, where correlation might be directly relevant, is rather well described as involving two model mechanisms, shake-off and direct knockout [34, 111], neither of which seems to involve initial correlation directly in the ionization process itself. These are, however, only models of the quantum mechanical description, and it is clear that the cross sections are underestimated in calculations using the sudden approximation if correlated initial wavefunctions are not used [112]. To go further, we need a more detailed understanding of the term electron correlation, which has been used loosely hitherto in this review.

The most precisely defined quantity is the *correlation energy*, the difference between the energy of a molecule calculated at the Hartree–Fock limit and the energy that would result from an exact solution of the nonrelativistic Schrödinger equation. This energy difference corresponds to *dynamic correlation*, which may be included in an accurate wavefunction by use of terms in specific interelectron distances, r_{ij}, or by equivalent techniques. It relates to instantaneous, rather than time-averaged, pairwise interaction. Some correlation (sometimes *Fermi correlation*), corresponding to the effect on each electron of the average coordinates of the others, is already included in the Hartree–Fock treatment. The magnitude of the correlation energy is typically 1 eV per valence electron pair, which is to be contrasted with the typical double ionization energy for such a pair of about 25 eV. From this comparison it seems unlikely that dynamic correlation could be the dominant factor determining double ionization probabilities.

Full accounting for all forms of correlation is necessary for accurate calculation of double ionization energies, but it is transition amplitudes, not state energies, that are the special characteristics of photoionization. In single photoionization of closed-shell atoms, the population of some satellite states of cations can be related to the magnitude of excited configuration contributions to CI (configuration interaction) descriptions of the neutral species. In mercury, for instance, the appearance of three-hole two-particle states of Hg^+ of configuration $5d^9 6p^2$ could be related to the coefficient of the corresponding term $(5d^{10} 6p^2)$ in the CI expansion for the ground-state neutral [95]. Similar deductions from spectral intensities of satellite bands should be possible in *direct* double photoionization, but in practice only qualitative observations have been made. The observation of $d^9 p^1$ shake-up satellites in double photoionization spectra of Hg and Cd (Fig. 12), for instance, confirm the importance of $d^{10} p^2$ configuration mixing in the CI description of the neutral atoms.

If pairwise electron correlation is really an important progenitor of direct two-electron photoionization, as the usual introductory statements imply [3–9], it should be interesting to discover experimentally which particular pairs of electrons are removed most effectively in direct double photoionization transitions. Is removal of electrons from the same shell or the same orbital favored relative to removal of electrons from different shell or orbitals? How about electrons of the same spin relative to pairs of different spin? The TOF-PEPECO method, with its lack of bias on the basis of energy or emission angle, is ideally suited to make such comparisons and thus in principle to contribute direct measurements of electron correlation. Other known factors thought to affect transition strengths such as photon energy or electron pair symmetry should be allowed for, and direct photoionization must be isolated. So far, there are very few examples where this can be done, so the jury is still out. One possibly indicative observation is that in double photoionization of small molecules with a full π-orbital, the $^1\Delta$ term is always the most intense (e.g., HBr, Fig. 18, OCS, Fig. 26). To form $^1\Delta$, two electrons must be removed from exactly the same spatial orbital. The same intensity pattern is found in every full π^{-2} ionization, and it may be remembered that for small molecules, the process is mainly direct.

Finally, it is not entirely obvious from an abstract point of view whether strong pairwise dynamic correlation would increase or decrease the probability of ejection of particular electron pairs. Where electrons are strongly correlated, as for instance in the He atom as represented by the Hyleraas model [4], correlation always has the effect of keeping the affected electrons apart, rather than together. On the model of direct knockout, this correlation would apparently diminish rather than increase the probability of double ejection near threshold. Probably this argument takes the models beyond their range of validity, but a clear theoretical explanation on this point would be most welcome. Spatial proximity may be a more relevant parameter, as suggested by the very low cross section for direct ionization to states with charges on distant atoms. The best hope for a direct experimental approach to electron correlation in molecules may be new pump-probe experiments involving direct double ionization, but with attosecond time resolution. Such capabilities are promised by forthcoming coherent soft X-ray sources based at "fourth-generation" synchrotron radiation facilities.

XI. SUMMARY AND CONCLUSIONS

Several mechanisms contribute to the overall effect of double photoionization in molecules, and this review attempts to extend the conventional classification. It may be helpful to summarize the different pathways and their characteristics very briefly.

1. *Direct Double Photoionization.* This is the sole mechanism for He, and it is present (though sometimes only weakly) for other species. At high energies it occurs mainly by shake-off, which gives deeply concave electron distributions. Nearer threshold, direct knockout is the main mechanism and gives flatter distributions. The very low intensity of direct ionization to molecular dication states with separated charges on distant atoms accords with the knockout model.

2. *Inner-Shell Auger Effect.* This is dominant for photon energies not far above the inner-shell thresholds. Singlet dication states are favored in ionization from closed shells. PCI effects are important at low photoelectron energies. Closely related, but relatively unstudied as yet by modern methods, is the resonant double Auger effect, where a near-edge neutral state is initially populated.

.3. *Inner Rydberg Auger Effect.* Autoionization from Rydberg states converging on higher double ionization limits is very important or dominant in low-energy (valence) double photoionization of atoms and must also happen in molecules. It has not yet been definitely observed there, because dissociation competes strongly. Similar cascade pathways are very important in atomic double Auger processes.

4. *Inner Valence Auger Effect.* Autoionization from inner valence states of the singly charged ions occurs very widely in organic molecules. For carbon compounds this produces deeply hollow electron distributions emphasizing unequal energy sharing. Dissociation to singly charged products competes strongly again.

5. *Double Photoionization by Dissociation with Autoionization.* In small molecules, particularly those with terminal O atoms, atomic autoionization may follow dissociative ionization. This can be from low Rydberg states or very high Rydbergs converging on excited levels of the atomic ions. Strongly structured electron distributions are produced, with peaks or edges covering broad ranges of final dication energy, often below the molecular double ionization potential.

Acknowledgments

I am particularly grateful to Pascal Lablanquie for reading an early version of the manuscript and for many helpful comments and to the Leverhulme Fund for support.

References

1. H. Hertz, *Ann. Phys. (Wiedermannsche Annalen)* **31**, 983 (1887).

2. A. Einstein, *Ann. Phys.* **17**, 132 (1905).

3. A. Huetz, L. Andric, A. Jean, P. Lablanquie, P. Selles, and J. Mazeau, *AIP Conf. Proc.* **360**, 139 (1995).

4. V. Schmidt, *Electron Spectrometry of Atoms Using Synchrotron Radiation*, Cambridge University Press, New York, 1997.

5. J. S. Briggs and V. Schmidt, *J. Phys. B: At. Mol. Opt. Phys.* **33**, R1 (2000).

6. G. C. King and L. Avaldi, *J. Phys. B: At. Mol. Opt. Phys.* **33**, R215 (2000).

7. P. Bolognesi, G. C. King and L. Avaldi, *Radiat. Phys. Chem.* **70**, 207 (2004).

8. L. Avaldi and A. Huetz, *J. Phys. B: At. Mol. Opt. Phys.* **38**, S861 (2005).

9. U. Kleiman, J. Colgan, M. S. Pindzola, and F. Robicheaux, *AIP Conf. Ser.* **183**, 131 (2005).

10. M. Ya. Amusia, in *VUV and Soft X-Ray Photoionization*, U. Becker and D. A. Shirley, eds., Plenum, New York, 1996, p. 1.

11. G. H. Wannier, *Phys. Rev.* **90**, 817 (1953); ibid. **100**, 1180 (1955).

12. S. Geltman, *Phys. Rev.* **102**, 171 (1956).

13. R. Peterkop, *J. Phys. B: At. Mol. Opt. Phys.* **4**, 513 (1971).

14. C. Bouri, L. Malegat, P. Selles, and M. G. Kwato Njock, *J. Phys. B: At. Mol. Opt. Phys.* **40**, F51 (2007).

15. D. B. Thompson, P. Bolognesi, M. Coreno, R. Camilloni, L. Avaldi, K. C. Prince, M. deSimone, J. Karvonen, and G. C. King, *J. Phys. B: At. Mol. Opt. Phys.* **31**, 2225 (1998).

16. P. Lablanquie, K. Ito, P. Morin, I. Nenner, and J. H. D. Eland, *Z. Phys.* **D16**, 77 (1990).

17. R. Wehlitz, F. Heiser, O. Hemmers, B. Langer, A. Menzel, and U. Becker, *Phys. Rev. Lett.* **67**, 3764 (1991).

18. T. N. Chang and R. T. Poe, *Phys. Rev.* **A12**, 1432 (1975).

19. C. Greene, *J. Phys. B: At. Mol. Opt. Phys.* **20**, L357 (1987).

20. A. Huetz, P. Selles, D. Waymel, and J. Mazeau, *J. Phys. B: At. Mol. Opt. Phys.* **24**, 1917 (1991).

21. F. Maulbetsch and J. S. Briggs, *Phys. Rev. Lett.* **68**, 2004 (1992).

22. R. I. Hall, L. Avaldi, G. Dawber, M. Zubek, K. Ellis, and G. C. King, *J. Phys. B: At. Mol. Opt. Phys.* **24**, 115 (1991).

23. A. Huetz and J. Mazeau, *Phys. Rev. Lett.* **85**, 530 (2000).

24. H. Klar and M. Fehr, *Z. Phys.* **D23**, 295 (1992).

25. L. Malegat, P. Selles, and A. Huetz, *J. Phys. B: At. Mol. Opt. Phys.* **30**, 251 (1997).

26. A. Huetz, P. Lablanquie, L. Andric, P. Selles, and J. Mazeau, *J. Phys. B: At. Mol. Opt. Phys.* **27**, L13 (1994).

27. A. S. Kheifets and I. Bray, *J. Phys. B: At. Mol. Opt. Phys.* **31**, L447 (1998).

28. A. S. Kheifets and I. Bray, *Phys. Rev.* **A62**, 065402 (2000).

29. J. Colgan, M. S. Pindzola, and F. Robicheaux, *J. Phys. B: At. Mol. Opt. Phys.* **34**, L457 (2001).

30. J. Colgan and M. S. Pindzola, *Phys. Rev.* **A65**, 032709 (2002).

31. L. Malegat, P. Selles, and A. K. Kazansky, *Phys. Rev. Lett.* **85**, 4450 (2000).

32. L. Malegat, P. Selles, and A. K. Kazansky, *Phys. Rev.* **A65**, 032711 (2002).

33. J. M. Feagin, *J. Phys. B: At. Mol. Opt. Phys.* **17**, 2433 (1984).

34. A. Knapp, A. Kheifets, I. Bray, Th. Weber, A. L. Landers, S. Schössler, T. Jahnke, J. Nickles, S. Kammer, O. Jagutzki, L. Ph. H. Schmidt, T. Osipov, J. Rösch, M. H. Prior, H. Schmidt-Böcking, C. L. Cocke, and R. Dörner, *Phys. Rev. Lett.* **89**, 033004 (2002).

35. J. Colgan and M. S. Pindzola, *J. Phys. B: At. Mol. Opt. Phys.* **37**, 1153 (2004).

36. J. A. R. Samson, *Phys. Rev. Lett.* **65**, 2861 (1990).

37. J. A. R. Samson, R. J. Bartley, and Z. X. He, *Phys. Rev.* **A46**, 7277 (1992).

38. P. Bolognesi, R. Flammini, A. Kheifets, and L. Avaldi, Abstract contribution at ICPEAC XXIII, 2003.

39. M. Y. Krässig, S. J. Schaphorst, O. Schwarzkopf, N. Scherer, and V. Schmidt, *J. Phys. B: At. Mol. Opt. Phys.* **29**, 4255 (1996).

40. A. Huetz, L. Andric, A. Jean, P. Lablanquie, P. Selles, and J. Mazeau, *AIP Conference Proc.* **260**, 139 (1995).

41. L. Malegat, P. Selles, P. Lablanquie, J. Mazeau, and A. Huetz, *J. Phys. B: At. Mol. Opt. Phys.* **30**, 263 (1997).

42. J. Mazeau, P. Lablanquie, P. Selles, L. Malegat, and A. Huetz, *J. Phys. B: At. Mol. Opt. Phys.* **30**, L293 (1997).

43. S. J. Schaphorst, Q. Qian, B. Krässig, P. van Kampen, N. Scherer, and V. Schmidt, *J. Phys. B: At. Mol. Opt. Phys.* **30**, 4003 (1997).

44. K. Ueda, Y. Shimmizu, H. Chiba, Y. Sato, M. Kitajima, H. Tanaka, and N. Kabachnik, *Phys. Rev. Lett.* **83**, 5463 (1999).

45. G. Turri, L. Avaldi, P. Bolognesi, R. Camilloni, M. Coreno, A. Rocco, R. Colle, S. Simonucci, and G. Stefani, *J. Electron Spectrosc. Related Phenom.* **114**, 199 (2001).

46. A. DeFanis, H.-J. Beyer, and J. B. West, *J. Phys. B: At. Mol. Opt. Phys.* **34**, L99 (2001).

47. L. Vegh and J. H. Macek, *Phys. Rev.* **A50**, 4036 (1994).

48. S. A. Scheinerman and V. Schmidt, *J. Phys. B: At. Mol. Opt. Phys.* **30**, 1677 (1997).

49. N. Scherer, H. Lörch, T. Kerkau, and V. Schmidt, *J. Phys. B: At. Mol. Opt. Phys.* **34**, L339 (2001).

50. S. Rioual, B. Rouvellou, A. Huetz, and L. Avaldi, *Phys. Rev. Lett.* **91**, 173001 (2003).

51. A. Russek and W. Mehlhorn, *J. Phys. B: At. Mol. Opt. Phys.* **16**, L745 (1983).

52. M. Y. Kuchiev and S. A. Scheinerman, *Sov. Phys. JETP* **63**, 986 (1986).

53. R. Dörner, V. Mergel, O. Jagutzki, L. Spielberger, J. Ullrich, R. Moshammer, and H. Schmidt-Böcking, *Physics Reports* **330**, 95 (2000).

54. J. Ullrich, R. Moshammer, A. Dorn, R. Dörner, L. Ph. H. Schmidt, and H. Schmidt-Böcking, *Rep. Prog. Phys.* **66**, 1463 (2003).

55. J. H. D. Eland and D. Mathur, *Rapid Commun. Mass Spectrom.* **5**, 475 (1991).

56. B. Krässig and V. Schmidt, *J. Phys. B: At. Mol. Opt. Phys.* **22**, L153 (1992).

57. S. Cvejanovic and F. H. Read, *J. Phys. B: At. Mol. Opt. Phys.* **7**, 1180 (1974).

58. R. I. Hall, A. McConkey, K. Ellis, G. Dawber, L. Avaldi, M. A. MacDonald and G. C. King, *Meas. Sci. Technol.* **3**, 316 (1992).

59. G. Dawber, A. G. McConkey, L. Avaldi, M. A. MacDonald, G. C. King, and R. I. Hall, *J. Phys. B: At. Mol. Opt. Phys.* **27**, 2191 (1994).

60. M. Hochlaf, H. Kjeldsen, F. Penent, R. I. Hall, P. Lablanquie, M. Lavollée, and J. H. D. Eland, *Can. J. Phys.* **74**, 856 (1996).

61. F. Penent, R. I. Hall, R. Panajotović, J. H. D. Eland, G. Chaplier, and P. Lablanquie, *Phys. Rev. Lett.* **81**, 3619 (1998).

62. M. Ahmad, P. Lablanquie, F. Penent, J. G. Lambourne, R. I. Hall, and J. H. D. Eland, *J. Phys. B: At. Mol. Opt. Phys.* **39**, 3599 (2006).

63. S. D. Price and J. H. D. Eland, *J. Phys. B: At. Mol. Opt. Phys.* **23**, 2269 (1990).

64. K. Okuyama, J. H. D. Eland, and K. Kimura, *Phys. Rev.* **A41**, 4930 (1990).

65. J. H. D. Eland, O. Vieuxmaire, T. Kinugawa, P. Lablanquie, R. I. Hall, and F. Penent, *Phys. Rev. Lett.* **90**, 053003 (2003).

66. T. Hsu and J. L. Hirschfield, *Rev. Sci. Iustrum.* **47**, 236 (1976).

67. P. Kruit and F. H. Read, *J. Phys. E: Sci. Instrum.* **16**, 313 (1983).

68. P. Choi and M. Favre, *Rev. Sci. Instrum.* **69**, 3118 (1998).

69. J. H. D. Eland, *J. Electron Spectrosc. Related Phenom.* **144–147**, 1145 (2005).

70. J. H. D. Eland and R. Feifel, *Chem. Phys.* **327**, 85 (2006).

71. S. Taylor, J. H. D. Eland, and M. Hochlaf, *J. Chem. Phys.* **124**, 204319 (2006).

72. P. Lablanquie and F. Penent, personal communication.

73. M. Uiberacker, Th. Uphues, M. Schultze, et al., *Nature* **446**, 627 (2007).

74. P. Bolognesi, S. J. Cavanagh, L. Avaldi, R. Camilloni, M. Zitnik, M. Stuhec, and G. C. King, *J. Phys. B: At. Mol. Opt. Phys.* **33**, 4723 (2000).

75. Unpublished work, this laboratory.

76. R. Feifel, E. Andersson, J.-E. Rubensson, and J. H. D. Eland, in preparation.

77. L. Avaldi, G. Dawber, N. Gulley, H. Rojas, G. C. King, R. I. Hall, M. Stuhec, and M. Zitnik, *J. Phys. B: At. Mol. Opt. Phys.* **30**, 5197 (1997).

78. T. Kaneyasu, Y. Hikosaka, E. Shigemasa, F. Penent, P. Lablanquie, T. Aoto, and K. Ito, *Phys. Rev.*, **A76**, 012717 (2007).

79. J. H. D. Eland, R. Feifel, and D. Edvardsson, *J. Phys. Chem. A* **108**, 9721 (2004).

80. F. Penent, J. Paladoux, L. Andric, P. Lablanquie, R. Feifel, and J. H. D. Eland, *Phys. Rev. Lett.* **95**, 083002 (2005).

81. F. Penent, P. Lablanquie, J. Paladoux, L. Andric, T. Aoto, K. Ito, Y. Hikosaka, R. Feifel, and J. H. D. Eland, *Ionization, Correlation and Polarization in Atomic Collisions*, A. Lahmann-Bennani and B. Lohmann, eds., AIP, New York, 2006.

82. F. Penent, P. Lablanquie, R. I. Hall, J. Paladoux, K. Ito, Y. Hikosaka, T. Aoto and J. H. D. Eland, *J. Electron Spectrosc. Related Phenom.* **144–147**, 7 (2005).

83. P. Lablanquie, L. Andric, J. Paladoux, U. Becker, M. Braune, J. Viefhaus, J. H. D. Eland, and F. Penent, *J. Electron Spectrosc. Related Phenom.* **156–158**, 51 (2007).

84. T. Mukoyama and K. Taniguchi, *Phys. Rev.* **A36**, 693 (1987).

85. Y. Hikosaka, P. Lablanquie, F. Penent, T. Kaneyasu, E. Shigemasa, J. H. D. Eland, T. Aoto, and K. Ito, *Phys. Rev. Lett.* **98**, 183002 (2007).

86. H. D. Schulte, L. S. Cederbaum and F. Tarantelli, *J. Chem. Phys.* **105**, 11108 (1996).

87. Y. Hikosaka, T. Aoto, P. Lablanquie, F. Penent, E. Shigemasa, and K. Ito, *Phys. Rev. Lett.* **97**, 053003 (2006).

88. S. Scheinerman, P. Lablanquie, F. Penent, J. Paladoux, J. H. D. Eland, T. Aoto, Y. Hikosaka, and K. Ito, *J. Phys. B: At. Mol. Opt. Phys.* **39**, 1017 (2006).

89. H. Aksela, S. Aksela, and N. Kabachnik, in VUV and Soft X-Ray Photoionization, U. Becker and D. A. Shirley, eds., Plenum, New York, 1996, Chapter 11.

90. M. Lundqvist, P. Baltzer, D. Edvardsson, L. Karlsson, and B. Wannberg, *Phys. Rev. Lett.* **75**, 1058 (1995).

91. J. H. D. Eland, M. Hochlaf, G. C. King, P. S. Kreynin, R. J. LeRoy, I. R. McNab, and J.-M. Robbe, *J. Phys. B: At. Mol. Opt. Phys.* **37**, 3197 (2004).

92. Y. Hikosaka and J. H. D. Eland, *Chem. Phys.* **299**, 147 (2004).

93. A. E. Slattery, T. A. Field, M. Ahmad, R. I. Hall, J. Lambourne, F. Penent, P. Lablanquie and J. H. D. Eland, *J. Chem. Phys.* **122**, 084317 (2005).

94. J. H. D. Eland, S. S. W. Ho, and H. L. Worthington, *Chem. Phys.* **290**, 27 (2003).

95. R. D. Molloy and J. H. D. Eland, *Chem. Phys. Lett.* **421**, 31 (2006).

96. J. H. D. Eland, *J. Electron Spectrosc. Related. Phenom.* **144–147**, 1145 (2005).

97. J. Berkowitz, *Atomic and Molecular Photoabsorption*, Academic Press, San Diego, 2002, Chapter 3.

98. J. H. D. Eland, *Chem. Phys.* **323**, 391 (2006).

99. J. H. D. Eland, *Chem. Phys.* **409**, 245 (2005).

100. M. Hochlaf and J. H. D. Eland, *J. Chem. Phys.* **123**, 164314 (2005).

101. U. Becker, O. Hemmers, B. Langer, A. Menzel, and R. Wehlitz, *Phys. Rev. A* **45**, R1295 (1992).

102. P. Morin and I. Nenner, *Phys. Rev. Lett.* **56**, 1913 (1986).

103. O. Björneholm, S. Sundin, R. R. T. Marinho, A. Naves de Brito, F. Gel'mukhanov, and H. Ågren, *Phys. Rev. Lett.* **79**, 3150 (1997).

104. Y. Hikosaka and J. H. D. Eland, *Chem. Phys.* **299**, 147 (2004).

105. P. Bolognesi, D. B. Thompson, L. Avaldi, M. A. MacDonald, C. A. Lopes, D. R. Cooper, and G. C. King, *Phys. Rev. Lett.* **82**, 2075 (1999).

106. R. Feifel, J. H. D. Eland, and D. Edvardsson, *J. Chem. Phys.* **122**, 144308 (2005).

107. T. Fleig, D. Edvardsson, S. B. Banks, and J. H. D. Eland, *Chem. Phys.*, **343**, 270 (2008).

108. S. T. Gibson, J. P. Greene, B. Ruščić, and J. Berkowitz, *J. Phys. B: At. Mol. Opt. Phys.* **19**, 2825 (1986).

109. S. Svensson, *J. Phys. B: At. Mol. Opt. Phys.* **38**, S821 (2005).

110. P. Lablanquie, personal communication.

111. T. Patard, T. Schneider, and J. M. Rost, *J. Phys. B: At. Mol. Opt. Phys.* **36**, L189 (2003).

112. T. Mukoyama and K. Taniguchi, *Phys. Rev. A* **36**, 693 (1987).

113. A. G. McConkey, G. Dawber, L. Avaldi, M. A. MacDonald, G. C. King, and R. I. Hall, *J. Phys. B: At. Mol. Opt. Phys.* **27**, 271 (1994).

114. A. D. J. Critchley, G. C. King, P. Kreynin, M. C. A. Lopez, I. R. McNab, and A. J. Yencha, *Chem. Phys. Lett.* **349**, 79 (2001).

115. A. J. Yencha, S. P. Lee, A. M. Juarez, and G. C. King, *Chem. Phys. Lett.* **381**, 609 (2003).

116. M. Alagia, B. G. Brunetti, P. Candori, S. Falcinelli, M. M. Teixidor, F. Pirani, R. Richter, S. Stranges, and F. Vecchiocattivi, *J. Chem. Phys.* **120**, 6980 (2004).

117. J. H. D. Eland, *Chem. Phys.* **294**, 171 (2003).

118. A. Pilcher-Clayton and J. H. D. Eland, *J. Electron Spectrosc. Related Phenom.* **142**, 313 (2005).

119. M. Hochlaf, R. I. Hall, F. Penent, H. Kjeldsen, P. Lablanquie, M. Lavollée, and J. H. D. Eland, *Chem. Phys.* **207**, 159 (1996).

120. R. I. Hall, G. Dawber, A. McConkey, M. A. MacDonald, and G. C. King, *Phys. Rev. Lett.* **68**, 2751 (1992).

121. P. Bolognesi, D. B. Thompson, L. Avaldi, M. A. MacDonald, M. C. A. Lopez, D. R. Cooper, and G. C. King, *Phys. Rev. Lett.* **82**, 2075 (1999).

122. A. J. Yencha, S. P. Lee, and G. C. King, XXIV ICPEAC Conference, 2005.

123. D. Edvardsson, A. Danielsson, L. Karlsson, and J. H. D. Eland, *Chem. Phys.* **324**, 674 (2006).

124. M. Hochlaf, A. Pilcher-Clayton, and J. H. D. Eland, *Chem. Phys.* **309**, 291 (2005).

125. J. H. D. Eland, P. Lablanquie, M. Lavollée, M. Simon, R. I. Hall, M. Hochlaf, and F. Penent, *J. Phys. B: At. Mol. Opt. Phys.* **30**, 2177 (1997).

126. R. I. Hall, L. Avaldi, G. Dawber, A. G. McConkey, M. A. MacDonald, and G. C. King, *Chem. Phys.* **187**, 125 (1994).

127. M. Hochlaf, R. I. Hall, F. Penent, J. H. D. Eland, and P. Lablanquie, *Chem. Phys.* **234**, 249 (1998).

128. M. Hochlaf and J. H. D. Eland, *J. Chem. Phys.* **120**, 6449 (2004).

129. T. Kinugawa, P. Lablanquie, F. Penent, J. Paladoux, and J. H. D. Eland, *J. Electron Spectrosc. Related Phenom.* **141**, 143 (2004).

130. R. Feifel, J. H. D. Eland, L. Storchi, and F. Tarantelli, *J. Chem. Phys.* **125**, 194318 (2006).

131. A. J. Yencha, M. C. A. Lopes, D. B. Thompson, and G. C. King, *J. Phys. B: At. Mol. Opt. Phys.* **33**, 945 (2000).

132. R. Feifel, J. H. D. Eland, L. Storchi, and F. Tarantelli, *J. Chem. Phys.* **122**, 144309 (2005).

133. R. D. Molloy, A. Danielsson, L. Karlsson, and J. H. D. Eland, *Chem. Phys.* **335**, 49 (2007).

134. J. H. D. Eland, *Inst. Phys. Conf. Series* **183**, 115 (2005).

135. S. Taylor, J. H. D. Eland, and M. Hochlaf, *Chem. Phys.* **330**, 16 (2006).

136. D. Edvardsson, A. Danielsson, L. Karlsson, and J. H. D. Eland, *Chem. Phys.* **332**, 249 (2007).

THE ELECTRIFIED LIQUID–LIQUID INTERFACE

R. A. W. DRYFE

School of Chemistry, University of Manchester, Manchester M13 9PL, United Kingdom

CONTENTS

I. INTRODUCTION

Electrified interfaces continue to play a central role in condensed-phase physical chemistry. The last couple of decades have seen enormous strides in our understanding of these interfaces at scales approaching the molecular and, for certain systems, the atomic. Over a similar period, there have also been significant developments in our understanding of the liquid–liquid interface. However, without the benefit of such high-resolution experimental probes of interfacial structure as scanning probe microscopy, other theoretical and experimental methods must be developed to build a framework to understand the structure and dynamics of this interface. This chapter assesses the recent advances in the understanding of the liquid–liquid interface, concentrating particularly on interfaces between immiscible electrolyte solutions.

An interfacial potential difference generally develops on contact between two immiscible liquid phases. The physical origin of this potential was the

Advances in Chemical Physics, Volume 141, edited by Stuart A. Rice

subject of debate for a good deal of the 20th century. In addition to the surface potential arising from the reorientation of solvent dipoles, for oil–water interfaces where the relative permittivity of the oil phase is significant (ε_{org} in the range $\sim 10 - 30$), the introduction of an electrolyte will lead to a distribution potential due to the partition equilibrium of the electrolyte between the two phases. Generally, therefore, a potential difference will exist between electrolyte phases formed from immiscible liquids of moderate polarity, analogous to that arising for electrolyte phases in contact with metallic phases. Classical electrochemical techniques provide a ready way of varying the potential difference at the electrode–electrolyte interface. Similarly, in the presence of appropriate electrolytes, the behavior of the "interface between two immiscible electrolyte solutions" (ITIES) as a function of potential can be investigated. This is a fruitful approach that, in spite of the complexity of the interface, has yielded quite considerable information on interfacial structure. There is also a broader link to colloidal systems (specifically, emulsions) and the physical chemistry of liquid membranes, since double-layer and charge transfer properties are often central to the stability of these systems. In earlier times, there was considerable overlap between the electrochemical, colloid, and membrane areas [1], however in recent years the study of liquid–liquid (L–L) electrochemistry has tended to be pursued in isolation from the aforementioned areas. There are some signs of a renewed overlap between the electrochemical study of the L–L interface and colloid science, as will be discussed at the end of the chapter.

This chapter will focus on the physicochemical properties of the ITIES, but will not be restricted to the study of these systems using electrochemical methods. The electrochemical approach was used widely from the end of the 1970s onwards, but the last decade or so has witnessed important contributions to the understanding of ITIES structure and dynamics from the application of molecular dynamic and Monte Carlo simulations to this interface, and through the application of scattering and spectroscopic experiments to the interface, most recently with simultaneous electrochemical (potential) control. The application of the above new methods, along with parallel improvements in electrochemical techniques, has transformed our understanding of the ITIES over the last 10–15 years. It therefore seems timely to appraise the most recent experimental and theoretical work in this area, as well as to suggest promising directions for the study of this important class of phase boundary.

The historical development of this area is worthy of brief consideration, to place the present review in context. For more detail, there are a number of earlier reviews on the electrochemistry of the L–L interface which cover the general developments in the field up to the time of their publication [2–5]. Brief summaries of the pertinent electrochemical literature have also been reported [6–9]. Other reviews deal with the miniaturization of the L–L interface and electroanalytical applications of this interface [10–12]. Early work devoted to the double-layer structure at the interface has been summarized [13], while a more recent review discusses the insights that molecular scale simulations have yielded on interfacial structure and charge transfer [14].

Although the area has a long pedigree, its modern development was catalyzed by (a) the electrocapillary measurements performed by Gavach and co-workers [15] and (b) the introduction of the four-electrode potentiostat methodology by the Prague group [16], during the 1970s. The 1980s and early 1990s saw the electrochemical study of interfacial ion transfer, and to a lesser extent electron transfer, under potential control. Thermodynamic parameters and insights into double-layer structure were obtained, although attempts to extract dynamic information (specifically, on rates of charge transfer; see Sections III and IV) from classical electrochemical techniques proved to be contentious. The thermodynamic basis of the electrochemical experiments is summarized briefly here to aid the reader who is unfamiliar with this field. More detailed treatments can be found in the reviews cited above [2–5].

The thermodynamic treatment of the ITIES starts with the equality of electrochemical potential for ion i in the aqueous (w) and organic (org) phases. This equality leads directly to a version of the Nernst equation for the ITIES:

$$\varphi_w - \varphi_{org} = \frac{\mu_{i,org} - \mu_{i,w}}{z_i F} \tag{1}$$

where the φ terms represent the Galvani (inner) potential of each solution phase, the μ terms describe the chemical potential of ion i (of charge number z_i) in that phase, and F is the Faraday constant. Consequently, at an activity ratio of unity, one can define a standard Galvani potential difference for i, related to the Gibbs energy of transfer of the ion:

$$\Delta_{org}^{w} \varphi_i^0 = \frac{\Delta G_i^{0,w \to org}}{z_i F} \tag{2}$$

where the Galvani potential difference in Eq. (2) signifies the Galvani potential of the water relative to that of the organic phase. The directly accessible quantity is the standard Gibbs energy of transfer of a salt, given by the sum of the transfer energies of the constituent ions: "Extra-thermodynamic" approaches have been taken to resolve the contributions of the individual ions [4], which in turn allows formulation of scales of Gibbs energy of ion transfer. For the general case of multi-ionic partition equilibria, the electrochemical equilibrium condition combined with the assumption of electroneutrality in each solution phase leads to [17]

$$\sum_i \frac{z_i c_i^0}{\left\{1 + \frac{\gamma_i^w}{\gamma_i^{org}} \exp\left(\frac{z_i F(\Delta_{org}^w \varphi - \Delta_{org}^w \varphi_i^0)}{RT}\right)\right\}} = 0 \qquad (3)$$

for equal phase volumes. (Here c_i represents total ion concentration across both phases, the γ_i terms are activity coefficients in each phase, and R and T have their usual significance.) The general form of this equation requires numerical solution to determine the equilibrium Galvani potential difference resulting from the partition of a multi-ionic system. For the simple case of a monovalent salt (AB) referred to above, with both ions able to partition freely between the *org* and *w* phases, Eq. (3) reduces to

$$\Delta_{org}^w \varphi = \frac{1}{2}(\Delta_{org}^w \varphi_{A^+}^0 + \Delta_{org}^w \varphi_{B^-}^0) \qquad (4)$$

This "distribution potential," expounded by Randles, defines a nonpolarizable ITIES [18].

Two further limiting cases are encountered experimentally, particularly in electrochemical approaches to measurement of charge transfer kinetics (Sections III and IV). The first is the case of an interface with a single potential determining ion. This corresponds to the addition of salts AX and CX to the system (i.e., a common ion (X) with hydrophilic and hydrophobic counterions), such that the distribution of X is controlled and can in turn be used to control the Galvani potential difference via the Nernst–Donnan equation:

$$\Delta_{org}^w \varphi = \Delta_{org}^w \varphi_{X^-}^0 + \frac{RT}{z_i F} \ln\left(\frac{a_{X^-}^{org}}{a_{X^-}^w}\right) \qquad (5)$$

The ion activities employed in Eq. (5) should be those established after equilibration of the two phases: It is generally assumed that the equilibrium distribution values can be approximated by the initial ionic concentrations. The validity of this approximation is dependent on the standard transfer potentials of the counterions, and in practical cases it will only hold over a restricted range ($\pm 0.1 - 0.2$ V [19]). The variance of such a system means that its Galvani potential difference is defined, over the above working range, through the ratio of activities of X.

By contrast, the Galvani potential difference will be undefined, and the interface will therefore be "ideally polarizable," if no ion readily transfers across the interface (i.e., individual salts AX and CY are present in w and org, respectively) and the following inequalities are met:

$$\Delta_{org}^{w}\varphi_{A^+}^{0} \gg 0 \quad \text{and} \quad \Delta_{org}^{w}\varphi_{X^-}^{0} \ll 0$$
$$\Delta_{org}^{w}\varphi_{C^+}^{0} \ll 0 \quad \text{and} \quad \Delta_{org}^{w}\varphi_{Y^-}^{0} \gg 0 \tag{6}$$

A potential window, where negligible ion re-partitioning occurs, then exists, with limits defined by the lowest absolute values of the standard Galvani potentials. Experimentally, this window is on the order of 0.5–1.0 V, for an ITIES composed of aqueous electrolyte phases and solvents such as 1,2-dichloroethane (DCE) or nitrobenzene (NB). In this type of system, the interfacial potential difference can be imposed externally using potentiostatic control.

II. THE STRUCTURE OF THE ELECTRIFIED LIQUID–LIQUID INTERFACE

The state of understanding of liquid-phase structure in general, and interfaces in particular, has been transformed by the application of computational (molecular dynamics, Monte Carlo) and experimental (scattering of X rays and neutrons, nonlinear optical spectroscopy) techniques to these systems. Liquid interfaces are complex environments, but for the experimental methods at least, the presence of the interface can confer local selectivity on spectroscopic or scattering-based techniques. Treatments of "neat" (i.e., electrolyte free) L–L interfaces have also been hampered by the dynamic nature of the interface, and details concerning interfacial structure on a molecular scale have only been clarified during the last decade or so. The second layer of complexity is the effect of the electrolytes' presence (aqueous and organic) on the interfacial structure: The properties of the resultant double layer have to be considered. Finally, there is the issue of the charge transfer process itself, in addition to its effect on the interface. The following section will attempt to summarize the current state of knowledge pertaining to interfacial structure. The focus of this chapter is on the interface between immiscible electrolyte solutions, but much of the recent progress has been based on developments in interfaces between *neat* liquids so some general remarks will be made concerning the significant progress in understanding the interfacial structure of the latter. The main results, rather than the details of the theoretical and experimental methods, will be presented: The aim is to define the current view of interfacial structure, before considering the behavior of the case of immiscible electrolytes.

The most basic question concerning the L–L interface is its *width*, leading to the closely related question of the *density profile* as one traverses the interface, from the interior of one liquid phase to the interior of the second phase. The dynamic nature of the interface, specifically the existence of capillary waves, was generally overlooked in earlier electrochemical investigations of electrolyte solutions (i.e., the ITIES; see Sections III and IV). The theoretical treatment of these thermal fluctuations is well established: The upper and lower wavelength limits are taken to be the macroscopic length of the interface (L) and the correlation length of the liquid, l_b, respectively [20]. Note that the correlation lengths of water and the organic phases typically used in electro-chemical experiments (see below) are relatively similar. The mean-square interfacial displacement due to the capillary waves, ξ^2, can be determined from [21, 22]:

$$\xi^2 \approx \frac{k_B T}{2\pi \gamma_{org/w}} \ln \frac{L}{l_b} \tag{7}$$

where k_B is the Boltzmann constant and $\gamma_{org/w}$ denotes the tension of the L–L interface. Typical tensions for the interfaces between water and the pure solvents used at the ITIES give ξ^2 values on the order of 1 nm; this value has been borne out by molecular dynamic (MD) simulations of the water–DCE interface [23]. Note that Eq. (7) would predict a decrease in the amplitude of fluctuations as the polarity of the solvent (for example, by lengthening the carbon chain in a homologous series) is decreased. MD calculations applied to the neat L–L interface have provided tremendous insight, which has guided subsequent experimental work. Use of a simple point charge model and pairwise Coulombic and Lennard-Jones intermolecular interactions, allowing for molecular vibration, was sufficient to reproduce the bulk behaviour of water and DCE. Moreover, the interfacial properties were also satisfactory since a water–DCE interfacial tension in reasonable agreement with the experimental value was recovered (via Eq. (7)). The water–DCE interface emergent from MD was sharp on a molecular scale and, on the aqueous side of the interface, possessed a lower number of H-bonds per molecule compared to the bulk. Subsequent correlation analysis determined the persistence of the H-bond network in the vicinity of the interface, which was found to be enhanced over the bulk value [24]. Aqueous protrusions into the organic solvent underlie the persistence of the H-bond network: The resultant isolation of the water molecules from the bulk enhances the lifetime of existing H-bonds. Widths of the w–CCl₄, w–DCE, and w–NB interfaces were on the sub-nanometer scale (quoted as 0.35 nm, 0.54 nm, and 0.53 nm, respectively, on the basis of the variation of the aqueous density from 90% to 10% of its bulk value). This interfacial width is attributed to the "averaging" effects of the capillary fluctuations.

There has been some debate over the existence of an "intrinsic" interfacial thickness or "mixed region" (see below), ξ_0, onto which the capillary waves are assumed to be superimposed; the square of the total width (denoted ξ_{tot}) is then given by [25]

$$\xi_{tot}^2 = \xi_0^2 + \xi^2 \tag{8}$$

A thermodynamic analysis of two model immiscible liquid phases, able to interact via dipolar and Yukawa interactions, has been reported [26]. The Yukawa component of the potential, u_{ij}, is characterized by an interaction parameter, ψ_{ij}, which was varied (for separations, R_{ij}, exceeding the molecular radius, r):

$$u_{ij}(R_{ij}) = -\psi_{ij}\frac{2r}{R_{ij}}\exp\left[-\zeta\left(\frac{R_{ij}}{2r}-1\right)\right] - \frac{m_i m_j}{R_{ij}^3}J(i,j) \tag{9}$$

where the ζ, m and $J(i,j)$ terms represent the range of the Yukawa interaction, the dipole moment and the dipolar interaction, respectively. Calculation of the Helmholtz energy and thus the Gibbs energy of the system, via a perturbation approach, spontaneously gave a two-phase system when the cross-interaction parameter, ψ_{12}, was less than the self-interaction parameters ψ_{11} and ψ_{22}. For such cases, the Gibbs energy was minimized along the coordinate normal to the interface, to determine the equilibrium mole fraction in each phase along the normal. These compositional profiles indicate that the interfacial width thickens as the cross-interaction term ψ_{12} is increased—that is, as the phases become more miscible (see Fig. 1). A finite interfacial width, approaching 1 nm, was obtained via this method for interfacial tensions close to those of typical ITIES systems. The two-phase system was also analyzed through a lattice gas treatment, where simple pairwise interactions ($\pm w$) were introduced between nearest neighbors for molecules of type "1" and "2" [25]. Monte Carlo simulation of the system as a function of the interaction energy revealed results consistent with the analytical study (namely, an overall increase in interfacial width with decreasing $|w|$); however, the width was in the range 0.2–0.3 nm for a typical ITIES-like interface (see Fig. 2). The power spectra of the capillary waves were found to reproduce the classical result [20] down to wavelengths of 2 nm. Moreover, the simulated interfacial widths were found to agree closely with the predictions of Eq. (7), indicating that the intrinsic component to the interfacial width (Eq. (8) is indeed negligible (0.2 nm or less)).

How do the predictions of capillary wave persistence to the molecular scale correlate with experiment? Scattering techniques are detailed probes of interfacial structure: The quasi-elastic scattering of light, the reflection of X-rays, and the reflection of neutrons have all been used as experimental probes of liquid interfacial structure. Schlossman and co-workers [27] have employed

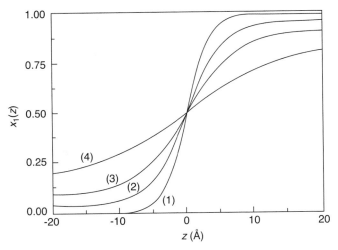

Figure 1. Mole fraction as a function of position along the interfacial normal for $\varepsilon_{12}/k_B T = (1)$ 0.5, (2) 0.75, (3) 0.8 and (4) 0.83. (Reprinted with permission from Ref. 26; copyright 1996, Royal Society of Chemistry.)

the specular reflection of synchrotron X rays to determine total (intrinsic, plus roughening by the capillary contribution) interfacial thickness for water–alkane interfaces. The technique rests on the exponential dependence of the reflectivity on the interfacial electron density profile. Interfacial widths in the

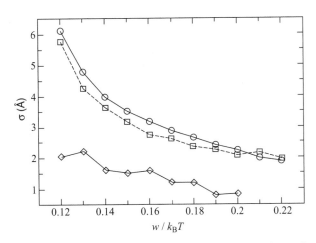

Figure 2. Simulated interfacial width as a function of the lattice gas interaction parameter, w (circles). Comparison with capillary wave expression (squares, Eq. (7)). The residual intrinsic width is shown (diamonds). (Reprinted with permission from Ref. 25; copyright 2004, Elsevier SA.)

range 0.35–0.60 nm were observed, larger than the predictions of the capillary wave theory, and accordingly a finite intrinsic width was suggested. Moreover, an increase in interfacial thickness with increasing carbon chain length was noted, contrary to expectations based on (a) Eq. (7) and (b) the earlier quasi-thermodynamic study where decreasing mutual interactions were reflected in a narrower interfacial region [26]. A larger interfacial thickness (1–1.5 nm) was initially determined, albeit with lower precision, for the water–hexadecane interface using neutron reflection [28]. This value was subsequently reevaluated to be 0.6 nm, in agreement with the X-ray work [29], while a thickness of less than 1 nm was quoted for the water–DCE interface in a separate study [30]. The quasi-elastic scattering of light from water–alcohol and water–carboxylic acid interfaces has also been interpreted in terms of a lower interfacial density, relative to the bulk solvents; preliminary data in this area have described the interface as "porous" [31]. We also note that the interface between electrolyte solutions of water and NB was shown to be sharp on the submicrometer scale through the application of the scanning electrochemical microscope (see Sections III and IV) to this interface [32].

Further structural information has been derived from the recent use of optical spectroscopy as a probe of structure at the L–L interface; the two general approaches pursued are (i) total internal reflectance (TIR) fluorescence spectroscopy of adsorbed solute species and (ii) nonlinear spectroscopic methods to generate interface-sensitive spectra of the solvent molecules in the interfacial region. The former approach uses the polarization of the emitted signal, from an s-polarized excitation, to determine the effective dimensionality of the interface. The anisotropy of the fluorescent response from an interface that is flat on the length-scale of the adsorbed probe ($\sim 1 \mathrm{nm}$) is [33]

$$\sigma(t) = \frac{F_{\parallel}(t) - F_{\perp}(t)}{F_{\parallel}(t) + F_{\perp}(t)} = \sigma(0) \exp\left(\frac{-t}{\tau_r}\right) \qquad (10)$$

where the fluorescent intensity terms (F) are parallel and perpendicular to the plane of polarization of the incident beam, $\sigma(0)$ is the initial anisotropy, and τ_r can be equated with a rotational correlation time. This expression leads to a "magic angle" (isotropic response) when the emission polarizer is measured at $45°$ with respect to excitation. By contrast, an interface that is corrugated with respect to the probe dimensions gives an anisotropic response that reverts to the bulk expression:

$$\sigma(t) = \frac{F_{\parallel}(t) - F_{\perp}(t)}{F_{\parallel}(t) + 2F_{\perp}(t)} = \sigma(0) \exp\left(\frac{-t}{\tau_r}\right) \qquad (11)$$

yielding a magic angle of 54.7°. Analysis of the experimental data for the adsorption of a rhodamine derivative suggests that the interfaces between water and various phthalate esters and water–CCl_4 are flat (i.e., Eq. (10) holds). By contrast, the water–DCE interface was found to be rough, since a good fit of the fluorescent decay to Eq. (11) was seen at an emission polarization of 54.7° [33, 34]. Further information on the water–CCl_4 and water–DCE interfaces was obtained through analysis of the fluorescent lifetimes in the presence of a second adsorbate, able to interact with the rhodamine solute via resonant energy transfer [34, 35]. The characteristic one-sixth power law dependence of the donor–acceptor lifetime term is multiplied by an exponent, d, reflecting the effective dimensionality of the interface, thus

$$F_d(t) = A \exp(-(t/\tau_D) - A'(t/\tau_D)^{d/6})$$ (12)

where A is a preexponential factor and A' describes the probability of donor–acceptor colocation within the critical distance for energy transfer. For aqueous interfaces with nonpolar phases (cyclohexane, CCl_4), d was equal to 2, but interfacial corrugation increased with an increase in organic polarity ($d \approx 2.2$ for water–toluene and water–chlorobenzene; $d \approx 2.3$ for water–1,2-dichlorobenzene; $d \approx 2.5$ for water–DCE). These findings are consistent with the classical picture of capillary waves, being driven thermally and suppressed by increasing interfacial tension (and gravity).

In the second spectroscopic approach, vibrational sum-frequency spectroscopy has been used to shed light on the molecular structure of water–organic interfaces [36, 37]. Interrogation of the water–CCl_4 interface by this approach revealed ordering, with water molecules "straddling" the interface. There is also evidence of similar ordering in the organic phase, the latter presumably induced by the dipole of the oriented interfacial water molecules [38]. The profile of the sum frequency amplitude, arising from the species located along the interface normal, has been simulated; a full spectral width at half-maximum (effectively, a measure of interfacial width) of 0.45 nm was calculated for the water–hexane interface [37] (cf. 0.35 nm reported from X-ray reflectivity [27]).

Of more relevance to ITIES work is the interrogation of the interface between pure water and DCE via the same nonlinear spectroscopic techniques: The less distinct sum-frequency spectral features were taken as evidence of a rougher, less structured interface compared to interfaces between water and nonpolar organic solvents, consistent with the fluorescent anisotropy work [39]. The transition from a sharp to a "blurred" interface could be induced by a progressive increase in the mole fraction of DCE in a CCl_4–DCE mixture. Subsequent MD calculations have been used to gain structural information on the water–DCE interface, since direct spectral interpretation is inherently harder to obtain, due to the less distinct spectral features [40]. The MD work was used to reproduce experimental spectra by calculating vibrational frequencies and nonlinear susceptibility terms. The comparison of experimental and calculated intensities was used, as above, to estimate the net interfacial width contribution. The calculations successfully reproduced the weaker spectral features observed at the water–DCE interface, suggesting that the structural information yielded by the MD approach (a wider interface, 0.8 nm compared to 0.6 nm for water–CCl$_4$, and greater mutual interpenetration of liquids in the water–DCE case) was valid. This compares against the slightly narrower interfacial region (0.5 nm), reported in a separate MD simulation of the water–DCE interface [24]. X-ray reflectivity has also been applied to the interfaces between water and more polar organic solvents: The width of the water–NB interface was found to be lower than the predictions of the capillary wave model (Eq. (7)) [41, 42]. This finding suggests that this interface possesses zero "intrinsic" width, somewhat at odds with the findings for DCE [40]. Importantly, the electron density profile determined by the X-ray work has been compared directly with MD simulation, by convoluting the calculated interfacial density with a Gaussian function representing the capillary wave contribution. Interfacial bending and/or layering of the organic molecules [42] was invoked to explain the damping of the capillary waves, the former leading to a replacement of the correlation length term in Eq. (7) with a factor dependent on the interfacial bending rigidity. The water–2-heptanone interface yielded a width comparable to the predictions of the capillary wave model [41], and MD calculations at this interface support this conclusion [43]. Therefore the picture emerging is one of a vanishingly small intrinsic interfacial width (substantially less than 1 nm) for the interface of water with polar organic liquids, whereas the X-ray reflectivity studies suggest that a finite intrinsic width exists when the polarity of the organic phase decreases. The roughening results from the action of capillary waves, which is more pronounced for more polar solvents [24].

The second general parameter of relevance to L–L studies, and particularly those concerned with charge transfer, is the interfacial *polarity*. The structural work referred to above gives length-scales over which the two liquids mix; hence the length-scales over which a measure of polarity such as the relative permittivity, ε_w or ε_{org}, would be anticipated to vary can be estimated. Polarity scales based on solvatochromic shifts of absorption maxima have been constructed and shown, at least in some cases, to be independent of probe molecule identity [44]. This approach has been extended to interfacial polarity scales, detected using nonlinear spectroscopy (in this case, the optical second harmonic signal, SHG, from adsorbed species) to relate the solvatochromic shift to the local polarity experienced by the adsorbate. The interfacial polarity was found to be well-described by the arithmetic mean of the bulk polarities for the water–chlorobenzene and water–DCE interfaces, that is,

$$P_{int} = \frac{(P_{org} + P_w)}{2} \tag{13}$$

where P represents polarity as determined by the solvatochromic scale employed and the "int" subscript refers to the interfacial environment. This observation is interpreted in terms of long-range dipolar forces, from both phases, contributing equally to the environment experienced by the interfacial species and hence to the effective polarity of Eq. (13). Kitamura et al. used the dynamic fluorescence approach (q.v.) to measure the lifetimes of the adsorbates' electronically excited states. An empirical exponential dependence of the nonradiative decay constant of the excited state on the solution polarity was exploited; this analysis was extended to the interfacial relaxation process to probe local polarity [35]. Additive polarity (Eq. (13)) was found for the interface with water and nonpolar organic phases; but the polarity of interfaces between water and more polar organic solvents, such as 1,2-dichlorobenzene and DCE, was found to be lower than predicted from Eq. (13). This nonadditivity of polarity has been investigated in greater detail by Walker and co-workers using SHG. Adsorbates with varying numbers of methylene units between the lipophilic chromophore and the hydrophilic "head" group are treated as direct probes of the interfacial "dipolar width," by determining the relationship between the observed solvatochromic shift of the absorbance maximum and the carbon chain length [45]. The conclusion was that interfacial polarity at "weakly interacting" organic phases (water–alkane) obeyed Eq. (13) [45, 46]. Contrasting behavior was observed for "strongly interacting organic phases" (water/alcohol) where, as Kitamura observed for water–DCE [35], an interfacial polarity substantially *lower* than that of the bulk organic phase was noted [45, 47]. The interfacial width, as probed through polarity changes, of the water–alkane interface was found to be less than 1 nm [46]. The nonadditivity of the water–alcohol polarity was attributed to an ordering of the alkyl fragment of the alcohols, driven by the incorporation of the alcoholic hydroxyl into the hydrogen bond network of the water [47]. MD simulations of the two-phase system have supported this interpretation [48]: the effective polarity of the interface was calculated, along the interfacial normal, by simulating the electronic absorbance of a model chromophore at a variety of positions. The MD method was similar to that described earlier [23], recovering an interfacial tension in reasonable agreement with experiment [48]. Consequently, the picture emerging for interfacial polarity at present is that an arithmetic mean value characterizes interfaces between water and nonpolar liquids, while a nonadditive value applies to interfaces with liquids capable of H-bonding. The physical origin of this difference is not immediately apparent, particularly in view of the rather loose definition of interfacial polarity. However, the insight gleaned through MD would suggest that orientational ordering, induced in the organic phase, is responsible for the deviation from Eq. (13).

The foregoing discussion deals with interfaces between neat liquids, whereas the structure of interfaces between electrolyte solutions has been a topic of much debate over the past three decades [3, 5, 13]. The presence of electrolytes allows a variable potential to be imposed on the interface, via either the common-ion or external potentiostatic approach (see Section I). An important parameter arises for the electrolyte case, namely, what is the *potential distribution* at the interface between the two electrolyte phases? The excess charge present on either side of the interface can be probed directly via macroscopic measurements of interfacial tension or capacitance and can thereby be used to infer structural information, albeit lacking in molecular detail. The developments along these lines up to the late 1980s/early 1990s have been reviewed [3, 5, 13, 49]; hence only a brief outline of the "bulk" approaches to this problem will be presented here.

Verwey and Niessen treated the interfacial ion distribution as two "back-to-back" double layers, each described using the Poisson–Boltzmann approach developed by Gouy and Chapman for electrode–electrolyte interfaces [13]. For a 1:1 electrolyte, the excess ionic charge density on the aqueous side of the interface, q^w, is given by

$$q^w = -\sqrt{8RT\varepsilon_w\varepsilon_0 c^w}\,\sinh\left[\frac{F(\varphi(0) - \varphi^w)}{2RT}\right] \tag{14}$$

where $\varphi(0)$ is the Galvani potential at the interface and ε_0 is the vacuum permittivity. An analogous expression applies to the organic phase. The interfacial capacitance is related to the interfacial charge density and tension via the Lippmann equation [50]:

$$C_{GC} = \frac{\partial q}{\partial \varphi} = -\frac{\partial^2 \gamma_{o/w}}{\partial \varphi^2} \tag{15}$$

Hence Eq. (14) can be differentiated with respect to the potential drop within the aqueous phase, $\varphi(0)\text{-}\varphi^w$, to give

$$C_{GC}^w = -\sqrt{\frac{2F^2\varepsilon_w\varepsilon_0 c^w}{RT}}\,\cosh\left[\frac{F(\varphi(0) - \varphi^w)}{2RT}\right] \tag{16}$$

Equations (14) and (16) assume that the potential distribution is entirely due to the back-to-back ionic space charge regions in each phase. In the early experimental work on ITIES capacitance, the existence of a "compact layer" of oriented solvent molecules (analogous to the inner layer found at the electrode–electrolyte interface) was introduced [15]. This approach is known as the modified Verwey–Niessen model. The interfacial charge density was measured as a function of potential using the common-ion and the potentiostatic methods to control potential [15, 51–53]. To a first approximation, reasonable agreement with the modified Verwey–Niessen model was seen for the water–NB interface for a number of electrolyte systems. A major achievement was the validation of Eq. (15) at the ITIES, with the charge density calculated from capacitance data, obtained from a.c. impedance, shown to give good agreement with that calculated from interfacial tension (electrocapillary) data [51, 52, 54]. However, the structure of this "compact layer," which is clearly key to understanding of the ITIES, has been a matter of considerable debate. Measurements of the surface excess of water at various aqueous electrolyte–organic interfaces suggested that less than one monolayer of interfacial water was present in the case of more polar organic phases [55]. Similarly, capacitance measurements can probe the contribution due to the compact layer, since it exists in series with the ionic space-charge regions; hence

$$\frac{1}{C_d} = \frac{1}{C_{GC}} + \frac{1}{C_{CL}} \tag{17}$$

where C_{GC} is the Gouy–Chapman capacitance (given by Eq. (16) with an equation of the same form describing the organic phase contribution) and C_{CL} is the contribution from the ion-free compact layer where charging results from the orientation of solvent dipoles. The simplest approach to the compact layer, treating it as two parallel monolayers of solvent, yielded capacitances that were substantially lower than those observed experimentally [13]. Moreover, the potential drop across the inner layer was found to be small at the potential of zero charge (the electrocapillary maximum) [3]. These findings were taken as evidence of the existence of a "mixed solvent" layer, which can be partially penetrated by ions from either phase, at the plane separating the two diffuse double layers [7].

In addition to the need to reconcile the inner-layer structure inferred from the macroscopic electrochemical data with the microscopic structural information gleaned to date for neat interfaces (see above), the following issues have recurred on comparison of experimental data from different laboratories:

1. *The Role of Ion Identity.* Some families of ions (e.g., alkali metals) show considerable differences in their capacitance response for a given concentration, particularly for potentials away from the potential of zero charge, with shifts of the potential of zero charge being observed in some

cases [56]. The conclusion drawn from the macroscopic measurements is that substantial interfacial ion-pairing occurs between the aqueous cations and organic anions [54, 57], although this has been contested [58].

2. *Enhancement of Capacitance.* The agreement between (modified) Verwey–Niessen models and experiment is less satisfactory for lower-polarity organic media (e.g., DCE, as opposed to NB) and for lower electrolyte concentrations [13]. What is the physical origin of the higher experimental capacitances seen for these conditions? As noted by Schmickler and co-workers [59], this enhancement of capacitance at the ITIES relative to the classical model stands in contrast to the response of electrode–electrolyte interface, where the capacitance is often found to be lower than the Gouy–Chapman function.

3. *The Limits of Validity of the Experimental Data.* Interfacial capacitance is generally measured through perturbation methods (pulse or a.c. methods). An earlier review has been strongly critical of the use of these experimental techniques because of the tendency of artifacts (such as stray capacitances and the high resistance of the organic phase) to affect the data and because of the mechanical/instrumental instability such experiments often induce [49, 58].

4. Finally, a general limitation is the applicability of mean-field theories, based on the Poisson–Boltzmann equation, to the ITIES. It has been noted that such theories do not generally give a good description of the double layer at electrode–electrolyte interfaces, where the electrolyte is present in a medium of low relative permittivity [60]; this problem will arise at the ITIES, given that the organic phase generally falls into this category.

We now consider the recent (mid-1990s onwards) developments in understanding of the double layer structure at the ITIES and consider how the theoretical developments can rationalize some of the trends observed experimentally. Three distinct approaches have been taken to improve the description of the ITIES double layer:

1. Those using analytical methods (often via perturbation approaches, which necessitate some level of approximation such as linearization of the Poisson–Boltzmann equation) to incorporate increasing complexity into the model of interfacial structure.

2. A lattice-gas approach to ion–ion and ion–solvent interactions (treated via a quasi-chemical approach or Monte Carlo simulations).

3. MD simulations: Statistically meaningful numbers of individual ions are not included within the calculations, but an electric field can be imposed on the system, and the resultant effect of the field on interfacial structure has been investigated.

There have been considerable efforts to move beyond the simplified Gouy–Chapman description of double layers at the electrode–electrolyte interface, which are based on the solution of the Poisson–Boltzmann equation for point charges. So-called modified Poisson–Boltzmann (MPB) models have been developed to incorporate finite ion size effects into double layer theory [61]. An early attempt to apply such "restricted primitive" models of the double layer to the ITIES was made by Cui et al. [62], who treated the problem via the MPB4 approach and compared their results with experimental data for the more "problematic" water–DCE interface. This work allowed for the presence of the compact layer, although the potential drop across this layer was imposed, rather than emerging as a self-consistent result of the theory. The expression used to describe the potential distribution across this layer was

$$\varphi(z) = \varphi(0) + z\frac{d\varphi(0)}{dz} + \frac{z\delta c_i^0 q^w}{\bar{\varepsilon}} \tag{18}$$

where z represents the interfacial normal, the relative permittivity of the layer is described by the mean of the bulk solvent values and δ is a term introduced to represent the penetration of the ions into the compact layer. The authors concluded that a substantial part (up to half) of the Galvani potential difference at the ITIES was due to the potential drop across the inner layer, although the δ factor appears to have a solely empirical basis and the assumption about the dielectric behavior of the interface is suspect, in view of subsequent work (see above).

An entirely different analytical approach has been pursued by Urbakh and co-workers, who have sought to incorporate the effects of interfacial roughness and compact layer structure on the capacitance of the ITIES. Essentially, mean-field models of increasing complexity have been developed. The initial approach considered only the effect of capillary waves on the capacitance of an interface, which was assumed to be sharp on the molecular scale (i.e., assuming that the fluctuations themselves were independent of the electric field) [63]. The linearized Gouy–Chapman equation, valid for $\Delta\varphi < RT/F$, was solved via a perturbation technique to determine the potential distribution normal to the interface as a function of the capillary wavevector. A pair of characteristic inverse length parameters, for each phase, determine the capacitative enhancement in this analysis, namely the inverse Debye length (κ) and the capillary wavenumber (k); for increasing Debye lengths, the potential distribution will be less sensitive to the interfacial fluctuation and the capacitance will reduce to the Gouy–Chapman value, C_{GC}. The form of the amplification factor, the roughness function $R(\kappa_{org}, \kappa_w)$, where

$$C = R(\kappa_{org}, \kappa_w)C_{GC} \tag{19}$$

was determined in this analysis to be

$$R(\kappa_{org}, \kappa_w) = 1 + 2C_{GC} \int_0^\infty dk g(k) \frac{k}{\varepsilon_{org} q_{org} + \varepsilon_w q_w}$$

$$\left\{ \begin{array}{l} (Q_{org} - \kappa_{org}) \left[(\kappa_{org} + \kappa_w) + \dfrac{Q_w}{\varepsilon_{org}} (\varepsilon_w - \varepsilon_{org}) \right] \\[3mm] + (Q_w - \kappa_w) \left[(\kappa_{org} + \kappa_w) + \dfrac{Q_{org}}{\varepsilon_w} (\varepsilon_{org} - \varepsilon_w) \right] \end{array} \right\} \quad (20)$$

where $g(k)$ is the Fourier transform of the wave's height correlation function and the Q terms for each phase combine the characteristic inverse length-scales:

$$Q = \sqrt{\kappa + k} \quad (21)$$

Equation (20) collapses to give $R(\kappa_{org}, \kappa_w)$ equal to one for an infinite interfacial tension. A more complete analysis allows for the aforementioned effect of the electric field on the capillary waves [22]. The free energy functional describing the capillary waves was solved, including an electrostatic potential term and an entropic term, for the nonlinear Poisson–Boltzmann equation (i.e., dispensing with the low potential approximation given above). Thus a more cumbersome roughness function is obtained, which is now a function of the inverse lengths and the normalized Galvani potential difference, V, defined as $F\Delta\varphi/RT$ (i.e., $R(V, \kappa_{org}, \kappa_w)$). $R(\kappa_{org}, \kappa_w)$ is plotted as a function of the inverse aqueous Debye length in Fig. 3 and 4, which show the effect of organic electrolyte concentration

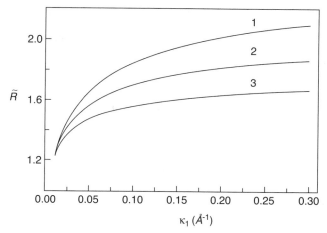

Figure 3. Interfacial roughness as a function of Debye length with $\gamma_{org-w} = 15\,\text{mN m}^{-1}$ and $\varepsilon_{org} = 10$ with organic electrolyte concentration of (1) 0.1 M, (2) 0.05 M, and (3) 0.025 M. (Reprinted with permission from Ref. 22; copyright 2000, Elsevier SA.)

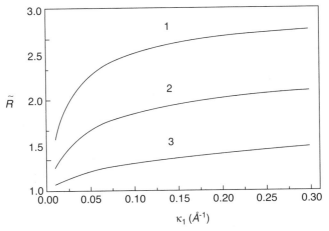

Figure 4. Roughness as a function of Debye length with $\gamma_{org-w} = 15\,\mathrm{mN\,m^{-1}}$ and organic electrolyte concentration of 0.1 M with ε_{org} of (1) 4, (2) 10, and (3) 30. (Reprinted with permission from Ref. 22; copyright 2000, Elsevier SA.)

and organic phase relative permittivity on this function. Increasing the former and decreasing the latter have the effect of enhancing the roughness function since more charges are present and their screening with distance is decreased. The most interesting consequence of the solution of the nonlinear equation is that substantial increases in the roughness function are found as a function of V: Essentially a critical potential difference exists, beyond which the interface becomes unstable. The onset and amplitudes of oscillation are increased by an increase of organic electrolyte concentration as Fig. 5 shows; physically this change arises because the double layer is compressed, and thus the field is acting on a narrower interfacial region. The critical potential for potential-induced oscillation is on the order of 0.2 V for system parameters close to those found experimentally, which lies well within the range of potential windows accessible potentiostatically (see Section I). Two shortcomings with the above model were noted by its authors. First, the model does not allow for the potential induced transfer of the ions across the ITIES (i.e., the interface is treated as infinitely polarisable). Second, it is physically unrealistic to assume that the constituent ions of the electrolyte will respond to the highest capillary wave frequencies, particularly for the bulky molecular ions normally employed as organic phase electrolytes; thus an upper frequency cutoff should be imposed on the correlation function of the capillary waves.

The first of these two shortcomings has been addressed using the free-energy functional approach by assuming a physical picture where a static interface of smoothly varying dielectric properties is considered (effectively, a "mixed

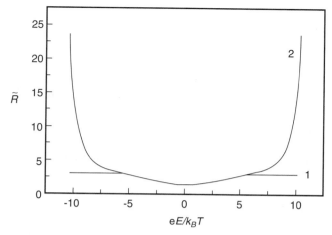

Figure 5. Roughness as a function of potential difference with $\gamma_{org-w} = 30\,\text{mN}\,\text{m}^{-1}$ and organic electrolyte concentration of 0.1 M with ε_{org} fixed at 10. (Reprinted with permission from Ref. 22; copyright 2000, Elsevier SA.)

solvent" layer) and finite ionic penetration (effectively, finite polarization) of the interface is allowed [64]. A perturbation approach was used once more to solve the Poisson–Boltzmann equation, subject to the assumption that the thickness of the interfacial region is small with respect to the thickness of the double layer. This approach yields three corrective terms to the classical Gouy–Chapman capacitance, each of which depends on a characteristic length, related to the variation of the relative permittivity and the integral of the Gibbs energy profile of the ions along the interfacial normal:

$$L_1 = \int_{-\infty}^{\infty} dz \left[\exp\left(\frac{-G_{org}^{C+}(z)}{RT} \right) - \exp\left(\frac{-G_{org}^{Y-}(z)}{RT} \right) \right] \tag{22}$$

$$L_2 = \int_{-\infty}^{\infty} dz \left[\exp\left(\frac{-G_{w}^{A+}(z)}{RT} \right) - \exp\left(\frac{-G_{w}^{X-}(z)}{RT} \right) \right] \tag{23}$$

$$L_3 = \int_{-\infty}^{\infty} dz \left[\frac{1}{2} \left(\frac{1}{\varepsilon_{org}} \exp\left(\frac{-G_{org}^{C+}(z)}{RT} \right) + \frac{1}{\varepsilon_{org}} \exp\left(\frac{-G_{org}^{Y-}(z)}{RT} \right) \right. \right.$$
$$\left. \left. + \frac{1}{\varepsilon_w} \exp\left(\frac{-G_{w}^{A+}(z)}{RT} \right) + \frac{1}{\varepsilon_w} \exp\left(\frac{-G_{w}^{Y-}(z)}{RT} \right) \right) - \frac{1}{\varepsilon(z)} \right] \tag{24}$$

The integral parameters therefore allow for the effects of ion penetration (L_1 and L_2) and finite interfacial thickness (L_3) on the capacitance. Note that each integral can assume positive or negative values, depending on the Gibbs energy of transfer of the cation with respect to the anion; hence the capacitance can show either positive or negative deviations from the Gouy–Chapman response in this analysis. The results of this analysis on the calculated capacitance are shown in Figs. 6 and 7, which again confirm the general finding that corrections to the classical picture are more susceptible to changes in the organic phase due to the reduced screening of the ions intrinsic to this phase. Essentially, this method of analysis allows for the typical experimental findings (asymmetric capacitance, shifts of the potential of zero charge) to be reproduced by adjustment of the length parameters specified above. An explicit comparison with experiment has been made [65], where the capacitance of various combinations of electrolytes $AX_{(aq)} \| CY_{(org)}$ (see Section I) was found for various aqueous–organic combinations [56]. In an approach similar to that pursued by Schmickler (see below) [66, 67], the ion penetration parameter was generalized into an ion–ion interaction parameter, a finite interfacial width, and an ion–interface interaction parameter, to account for deviations from the classical "noninteracting" double layer picture. Adjustment of these parameters gave good agreement with experimental data, with different parameters found to dominate for different ITIES systems (see Figs. 8–10). A drawback is that the physical reason for the dominance of a particular factor within a given system is not clear; therefore, it is hard to impart predictive properties to this approach. Finally, a numerical

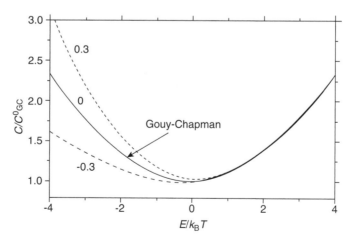

Figure 6. The effect of hydrophobic cation penetration on interfacial capacitance, normalized to the Gouy–Chapman behavior. The Gouy–Chapman response is calculated for $L_1 = L_2 = L_3 = 0$, with L_2 set to ± 0.3 nm in the broken lines. ε_{org} is fixed at 10 and the inverse Debye lengths are set to $0.3\,nm^{-1}$ for both phases. (Reprinted with permission from Ref. 64; copyright 2001, Elsevier SA.)

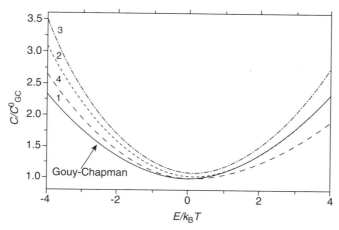

Figure 7. The effect of ion penetration (Fig. 6) coupled to smearing of the dielectric profile. Lines 1 and 2 correspond to the Gouy–Chapman behavior and $L_2 = 0.3$ nm from Fig. 6, respectively. Lines 3 and 4 are identical to line 2, except L_3 is set to 0.02 nm and -0.02 nm, respectively. (Reprinted with permission from Ref. 64; copyright 2001, Elsevier SA.)

solution to the potential distribution was obtained for the case where the interpenetration factor of the four electrolyte ions ($A^+_{(aq)}$, $X^-_{(aq)}$), $C^+_{(org)}$, and $Y^-_{(org)}$) was described as a potential-dependent adsorption equilibrium; thus,

$$B_{A+} = K_{A+}B_0 \exp\left(-\frac{Z_A + F}{RT}\Delta\varphi_w\right) \tag{25}$$

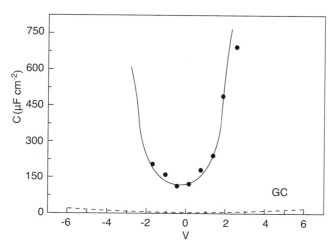

Figure 8. Comparison of experimental (dots) and calculated dimensionless capacitance curves for the water–2-heptanone interface: GC represents the predicted Gouy–Chapman response. (Reprinted with permission from Ref; 65; copyright 2003, Elsevier SA.)

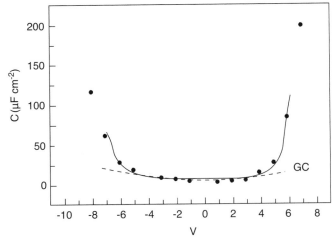

Figure 9. Comparison of experimental (dots) and calculated dimensionless capacitance curves for the water–2-octanone interface. (Reprinted with permission from Ref. 65; copyright 2003, Elsevier SA.)

(and analogous expressions hold for the other ions) where the Galvani potential term in the above case is the potential drop within the aqueous phase, B_0 term represents the fractional surface coverage of available sites at the ITIES, and the B_{A+} term is the fractional coverage of ion A^+ [68]. The four ion adsorption terms

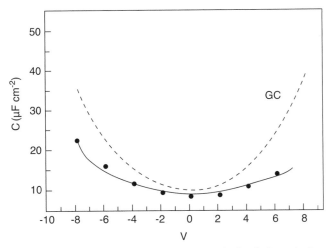

Figure 10. Comparison of experimental (dots) and calculated dimensionless capacitance curves for the water–DCE interface. (Reprinted with permission from Ref. 65; copyright 2003, Elsevier SA.)

are coupled via a Langmuir model of maximum (monolayer) adsorption. The advantage of this approach is that the potential dependence of the adsorption process can be related to an independently measurable parameter, namely the standard Gibbs energy of transfer of each ion (see Section I). Treating the interface as a mixed layer, the Gibbs energy profile is assumed to vary linearly as the ion traverses this mixed zone (increasing solvation by the second phase); the only adjustable parameter is the fraction, β, relating the Gibbs energy of transfer to the Gibbs energy of transfer to the interfacial site, that is,

$$\Delta G_{A+}^{\text{int}} = \beta \Delta G_i^{w \to org} \tag{26}$$

β was assumed to be constant for all four ions to a first approximation. Again this type of analysis can be used to demonstrate that interfacial capacitance increases (for all potential differences, including the potential of zero charge with increasing ionic interpenetration; see Fig. 11) and that substantial differences in the degree of penetration (due to dissimilar Gibbs energies of transfer) induce asymmetry in the capacitance curves, with a more marked dependence for the organic electrolyte. Essentially, the correct quantitative dependence can be reproduced phenomenogically, but none of the above approaches have combined mixed solvation with interfacial distortion due to capillary waves. Moreover, the assumptions of the models are somewhat at odds with recent findings on interfacial polarity, specifically the observation for the pure (electrolyte-free) case that dielectric properties may not be additive (see above).

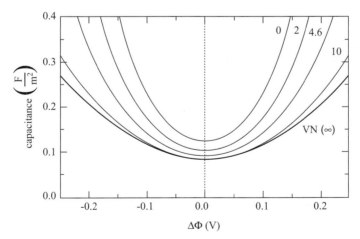

Figure 11. Interfacial capacitance as a function of potential difference for varying $\Delta G_{A+}^{\text{int}}$ values (kJ mol^{-1}). VN represents the Verwey–Niessen (compact layer) response. (Reprinted with permission from Ref. 68; copyright 2005, Elsevier SA.)

Monte Carlo simulation methods have also been employed to describe potential distribution at the ITIES. An early simulation of this type used a restricted primitive approach where ion-size effects are considered but each solvent is treated as a dielectric continuum [60]. The ITIES was treated as being infinitely sharp, and the ions were assumed to interact via Coulombic interactions, modified to include the image charges arising from the consequent dielectric discontinuity at the interface. Ion penetration of the interface was included by adjusting the distance of closest approach of the ions to the interface. The influence of ion–ion correlations was clear, particularly in the organic phase, serving to contract the organic diffuse layer. Image forces also have a more pronounced effect on the organic diffuse layer (being inversely proportional to the relative permittivity), and they act to contract this layer. This work undermines the entire basis of the Verwey–Niessen appoach, since it shows that the ion interactions in each phase are correlated and the diffuse layers cannot, therefore, be treated independently of one another. Nearest-neighbor ion–ion and ion–solvent interactions have been treated via a lattice gas approach developed by Schmickler and co-workers. In the initial work, which was complemented by experimental data, a quasi-chemical (equilibrium) model was developed to describe the ion–solvent interaction terms [59]. The Poisson–Boltzmann equation was solved numerically to obtain the potential distribution. The main improvement over classical theories was that the interfacial width, characterized by a parameter d_{eff}, such that the fraction of sites along the interfacial normal occupied by solvent 1 is given by

$$f_1(z) = \frac{1}{2}\exp\left(\frac{-z}{d_{eff}}\right) \tag{27}$$

was treated as a variable. The effect of varying d_{eff} relative to the Debye length is shown in Fig. 12. A clear increase in capacitance is observed as the interfacial width is increased for fixed values of the ion–solvent interaction parameters. Physically, this arises because more ions (and hence more charge) accumulate in the space-charge region, because the double-layer interpenetration partially counteracts their presence. Similarly, as Fig. 13 shows, increase of the ion–solvent interaction parameters for a given interfacial width increases the capacitance; hence the enhanced capacitance can be related to the Gibbs energy of transfer of the ion. Asymmetry in the capacitance–voltage response can be induced through asymmetry in the ion–solvent interaction energies as Fig. 13 also demonstrates; thus many of the deviations from classical theory observed in experiment could be successfully reproduced. This is, effectively, an alternative route to incorporate the factors introduced by Urbakh et al. [66]. For the specific case of the alkali metal ions for example, the increase in capacitance is found to

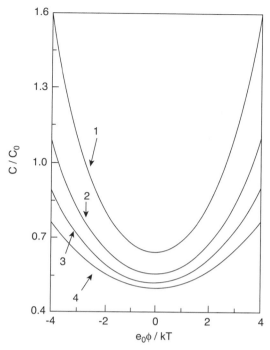

Figure 12. Normalized capacitance-potential response for varying interfacial widths with symmetrical interaction parameter: d_{eff} relative to Debye length as (1) 0.2, (2) 0.1, (3) 0.05, and (4) Gouy–Chapman response. (Reprinted with permission from Ref. 59; copyright 1997, Elsevier SA.)

be less pronounced with decreasing Gibbs transfer energy. Note that the existence of the interpenetrating solvent layers, effectively a mixed layer to use Girault's terminology [7], removes the dielectric discontinuity and hence the need to consider image forces. This approach was subsequently re-cast with Monte Carlo simulations, obviating the need for the rather arbitrary quasi-chemical formalism [66], with interfacial ion-pairing introduced as a further parameter contributing to the capacitive enhancement [67]. The strength of these works is their ability to produce a capacitive response that can reproduce experimental trends, with appropriate adjustment of parameters, but the predictive power is not clear. The reliance on a finite (static) interfacial width, rather than incorporating capillary fluctuations, is a further weakness.

To close the discussion of analytical and/or computational approaches to potential distribution at the ITIES, the results of MD simulations of the effect of electric fields on interfacial structure are worthy of consideration. MD was performed using the potentials briefly discussed earlier [23], with external electric fields on the order of 1 V nm^{-1} (comparable with the experimental

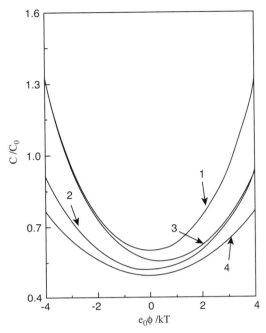

Figure 13. The effect of asymmetry of interaction parameters on the normalised capacitance–potential response with the dimensionless interfacial width fixed at 0.1. (Reprinted with permission from Ref. 59; copyright 1997, Elsevier SA.)

polarization window) imposed on the system. A significant change in the interfacial density profile as these fields were imposed on the interface was not found [21]. On a macroscopic level, imposition of an electric field is expected (see the Lippmann equation, Eq. (15)) to reduce the interfacial tension and hence increase the interfacial width; MD offers an insight into the microscopic structural changes associated with the latter effect. The molecular level analysis of the interfacial height and width distributions suggested that the interfacial corrugation was enhanced as the electric field was increased. Molecular trajectories can also be used to determine the mean square amplitude of the interfacial fluctuations and hence to recover the interfacial tension, assuming Eq. (7) is valid. This exercise gave a zero-field tension for the water–DCE interface of 32 mN m^{-1}, compared to an experimental value at ambient temperature of 28.43 mN m^{-1} [69]. Classically, an inverse quadratic dependence of $\gamma_{org/w}$ on the electric field is predicted [70]; this approximate trend was recovered from the MD approach. The picture of fewer, yet stronger, hydrogen bonds for interfacial water (as compared to the bulk) is largely unaffected by the imposition of an electric field [21], although the lack of ions in the system makes it difficult to extend the conclusions to the ITIES. For example, in the absence of an external field, the average orientation of the aqueous dipoles is parallel to the interface with DCE; but as the field increases, the dipoles tend to orient parallel to the interfacial normal [21]; a key question would be how this picture is modified by the presence of electrolyte in the system.

More recent experimental studies have addressed the question of double-layer structure at immiscible electrolyte interfaces using the techniques discussed earlier in this section for electrolyte-free phases. X-ray reflectivity, along with SHG spectroscopy and quasi-elastic light scattering (QELS), have been applied in the case where the Galvani potential difference is controlled either via an external potentiostat [41] or via a single potential determining ion (tetrabutyammonium at the water–NB interface) [71, 72]. The X-ray studies have determined the interfacial width, a function of the common-ion concentration (Eqs. (5) and (9)), which varies between 0.6 and 0.8 nm. These values are ~ 0.1 nm greater than the capillary wave prediction, in contrast to the behavior found with electrolyte-free NB [42]. Poor agreement was found between the classical Verwey–Niessen-based double-layer models and the ion distributions determined experimentally through reflectivity. Similarly, the modified Verwey–Niessen approach, incorporating the ion-free inner layer, did not give good agreement with experimental reflectivity data, even when the thickness of the ion-free layers was treated as an adjustable parameter. Discrepancies with experiment were also seen when a MPB-based approach (MPB5) was pursued [61]. Parallel MD simulations of the interfacial potential distribution were performed. The Boltzmann energy term (E_i) giving the potential-dependent ion distribution as a function of the interface normal (z) was determined from

$$E_i(z) = z_i e \varphi(z) + G_i(z) \tag{28}$$

(cf. Eq. (1)), where G_i is the local Gibbs energy determined from the potential of mean force on a single ion in the MD simulations and e is the elementary charge. Introducing this functional dependence into the recalculated double-layer profiles essentially presents a route to combine the local details of the electric field variation with the dynamics of the solvents, once the profile was "roughened" by a Gaussian function to account for the capillary wave contribution. Good agreement with the experimental reflectivity data was seen in this case. The obvious objection to this procedure is the inability of the calculation to include interionic interactions; however, an iterative treatment to include an ion correlation term in Eq. (28) suggested that this factor was not significant under the conditions explored.

At the present time there is, however, a need for more detailed nonele-ctrochemical experimental probes of double-layer structure. Spectroscopic experiments performed to date have not yielded the resolution required to probe molecular-level interfacial details. Ellipsometric experiments under external potential control have been reported at the ITIES; the analysis of the refractive index variation was sufficient to detect the interpenetration of the interfacial region by the supporting electrolyte ions [73]. The *in situ* SHG studies reported previously for the water–DCE interface under potentiostatic control are in reasonable agreement with early (modified Verwey–Niessen) electrochemical reports of double layer structure [52], but to date have not been sufficiently sensitive to yield the level of discrimination which more recent theories have revealed about interfacial potential distribution [74]. Other resonant SHG studies have been performed on surfactants adsorbed on the water–DCE interface, for example, and used to determine the potential dependent adsorption of the probe molecules [75]. Similarly, ionic surface excesses have been probed through QELS, which have yielded results consistent with electrochemical studies of capacitance, but more work is required to provide detailed new insights into interfacial structure [76]. Further developments in understanding ITIES structure at the molecular level will rest on (a) including finite numbers of ions within computational methods and (b) extending the "dipolar width" work to the case of electrolyte solutions: the aim would be to provide local information on interfacial potential distribution via the measurement of the solvatochromic shift as a function of applied potential. Similarly the extension of the SFG (of solvent vibration) work to cases of variable interfacial potential would be very welcome as an experimental test of double-layer structure on the molecular scale.

III.   ION TRANSFER KINETICS

Ion transfer (IT) across the ITIES is readily studied via electrochemical methods. In classical voltammetric experiments at the electrode/electrolyte interface, charge transfer may be limited by (i) the rate at which the active species is transported to the interface or (ii) an intrinsic charge transfer rate constant, which presupposes that there is a considerable $(>k_B T)$ energetic barrier between the reduced and oxidised form of the active molecule. Theories relating charge transfer activation barriers to the driving force for charge transfer are well established for the electrode–electrolyte interface [77]. With the development of the four-electrode potentiostat to allow reliable study of IT as a function of potential, those working in the area started to ask whether there was a measurable barrier to IT at the L–L interface, meaning a barrier between the two solvated states of the ion.

Clearly the experimental measurement of apparent IT kinetics requires that the rate constant describing the process (i), k_{MT}, must be significantly higher than the rate constant of process (ii), k_{CT}. The pertinent question here is the time constant of the electrochemical techniques used to probe the IT process. Here we will summarize the experimental data, concentrating primarily on developments in IT over the past 10 years or so, since earlier work has been reviewed [3, 78], before describing some of the theoretical developments and suggesting how these data in this area can be interpreted in the light of the structural studies summarized in Section II.

IT processes are generally classed as "simple" (unassisted) or "facilitated" (assisted), borrowing terminology from the literature on charge transfer through membranes [79]. Unassisted IT is generally denoted

$$A^+_{(aq)} \rightleftharpoons A^+_{(org)}$$

At the ITIES, numerous assisted transfers of hydrophilic ions have been reported, where the transfer is facilitated by a neutral ligand (or "carrier," to borrow membrane terminology) present within the organic phase. Facilitated transfer of cations has been reported most frequently [3], although a few reports of facilitated anion transfer also exist [80, 81]. In fact, this distinction can be viewed as rather artificial, since all transferring ions must interact with the phase to which they are transferred; the distinction is simply whether the agent is the major component (solvent) or the minor component (solute, the "carrier") of the receiving phase. Clearly the proportion of the minor component present in the receiving phase can be controlled, which adds another variable to the assisted transfer process; however, if an IT to a nonaqueous phase composed of two solvents (one strongly interacting, one weakly) were conceived, the relative proportions of the two solvents could be used as a variable in this nominally "unassisted" transfer process. The unassisted IT process is better represented as

$$A^+.nH_2O_{(aq)} + m.org \rightleftharpoons A^+.m \ org_{(org)} + nH_2O$$

although the discussion below suggests that n is not completely removed for hydrophilic ions.

By way of introduction, we note that the IT process has been treated as a quasi-reversible first-order process [82]:

$$I = z_i F(k_f c_i^w - k_b c_i^{org}) \tag{29}$$

where I is the current density, the k terms are the first-order heterogeneous rate constants (dimensions of velocity, analogous to k_{CT}) describing the interfacial charge transfer process in the forward and reverse directions and the c terms represent the concentrations of the ion on each side (w, org) of the interface. Note that I is conventionally defined as positive for a cation transfer from aqueous to organic (or an anion moving in the opposite direction) [82]. Following the approach for the electrode–electrolyte interface, the rate constants are assumed to be interrelated via a phenomenological Butler–Volmer potential dependence [3]; thus,

$$k_f = k_0 \exp\left(\frac{\alpha z_i F(\Delta_0^w \varphi - \Delta_0^w \varphi_i^0)}{RT}\right) \tag{30}$$

$$k_b = k_0 \exp\left(\frac{-(1 - \alpha) z_i F(\Delta_0^w \varphi - \Delta_0^w \varphi_i^0)}{RT}\right) \tag{31}$$

where k_0 is the standard rate constant, when the applied potential equals the standard transfer potential of the ion, and the charge transfer coefficient (α) describes the potential dependence of k_0 (effectively the position of the transition state with respect to the interface). Classical electrochemical methods, based on cyclic voltammetry or a.c. impedance at the polarized ITIES, were used during the 1980s and early 1990s to extract values of k_0 and α. A crucial experimental factor is the elimination of ohmic drop, the product of current flow (i) and uncompensated solution resistance (R); the effect of this feature is that the local potential at the interface differs from the applied potential by iR, and thus a low apparent rate constant will be extracted through Eqs. (29–31). The values of k_0 initially reported for the interfacial transfer of small ions were on the order of 10^{-4} to 10^{-3} cm s^{-1} [16, 83–85] due to the lack of ohmic compensation. Refinement of the instrumentation to give closer control over the ohmic drop yielded k_0 values in the range 10^{-2} to 0.1 cm s^{-1} [86–88]. Results obtained at the water–NB interface using a.c. impedance have suggested that the k_0 value is approximately invariant with ion identity (at ~ 0.1 cm s^{-1}) and hence independent of the ion's standard Gibbs energy of transfer [89]. A weak correlation of k_0 with supporting electrolyte concentration has also been noted from comparison of earlier data [90]. However, general problems persist with the use of classical electrochemical techniques: One is the fact that the rate of the kinetic processes was often close to the limit of the timescales of the technique (i.e., the rate of process (ii), as defined above, is approaching that of process (i); see below for a discussion of intrinsic experimental timescales). A further intrinsic time constant to consider with voltammetric approaches to measure charge transfer rates is that required to charge the electrical double layer at the ITIES; if the interfacial potential is varied rapidly, the mass-transport rate (process (i) above) may become comparable with RC_d, the time constant for double-layer charging. Consequently an alternative approach to resolve the charge-transfer flux from the background charging response was developed by Kakiuchi and co-workers, who investigated the interfacial transfer of fluorescent ions as a function of potential [91, 92]. The logic behind this approach is that the fluorescent response, unlike the current response, is not convoluted with charging effects. The experimental approach involved illumination of the interface, via the organic phase, at an angle greater than the critical angle; the potential-dependent transfer of a fluorescent ion from the aqueous phase was treated in terms of an approximate model (treating illumination as being along the interfacial normal), for small excursions in applied potential, where the fluorescent intensity is given by

$$F^{org}(t) = \Omega^{org} \Phi^{org} I_0 \int_0^\infty c_i^{org}(z, t)\, dz \tag{32}$$

where Ω^{org}, Φ^{org}, I_0, and c_i^{org} correspond to the organic phase molar absorption coefficient, fluorescent quantum yield, incident light intensity, and local fluorescent ion concentration. This model does not take the possibility of fluorescent quenching into account. Since, by Faraday's law, the total amount of ion transferred to the organic phase is related to the current via

$$\int_0^\infty c_i^{org}(z, t)\, dz = \frac{1}{z_i F} \int_0^t I(t)\, dt \tag{33}$$

Thus Eqs. (32) and (33) demonstrate that the fluorescent intensity is proportional to the integral of the current; hence "volt-fluorograms" (fluorescent analogues of voltammograms) can be constructed from the time derivative of the fluorescent response to potential [93]. This approach was used to consider the fluorescent response to applied potential for the transfer of the anionic dye Eosin Y; a standard rate constant of $\sim 10^{-2}$ cm s^{-1} was found, and notably the potential dependence of k_f did not accord with the model of Eq. (30) [94]. A further investigation of interfacial dye transfer kinetics, using an a.c. modulation of the interfacial potential (and hence of the fluorescent response), yielded a similar apparent k_0 value, although in this case the retardation was attributed to adsorption of the dye on the organic side of the interface [95]. An analogous approach, using the analysis of the visible absorbance (as opposed to fluorescent) response, has yielded k_0 values on the order of 10^{-2} cm s^{-1} for the transfer of the dyes methyl orange and ethyl orange across the water–DCE interface [96, 97].

An alternative approach to decrease the time constant of electrochemical systems rests on the use of the miniaturized ITIES. Micron-scale ITIES use a micropipette to support the aqueous phase [98]. Subsequently an alternative approach, using "micro-holes" formed by laser ablation of thin polymer films, was reported [99]. The advantage of the micron-scale approach is that the radial diffusive flux to such an inlaid interface reaches a steady state, so the problems of transient current methods due to the double-layer charging constant are avoided. The low currents measured at such interfaces, due to their small size, mean that distortions arising from ohmic drop are also minimized. Moreover, for an inlaid micron-scale ITIES, the steady-state current is given by the well-known equation describing the flux to an inlaid disc [100]:

$$i = 4z_i FDcr_d \qquad (34)$$

where r_d is the disc radius, and D is the diffusion coefficient of the transferring ion. The mass transport coefficient (k_{MT}), process (i) from the opening paragraph of this section, is given by

$$k_{MT} = \frac{4D}{\pi r_d} \qquad (35)$$

The earlier IT reports at microscopic ITIES used interfaces with a diameter on the order of 10 μm [98, 99]; Eq. (35) gives k_{MT} on the order of 0.01 cm s^{-1}, using typical solution phase diffusion coefficients of small molecular ions, meaning that k_{CT} values similar to, or in excess of, this value would not be measurable. The use of smaller ($\sim$ 1-μm diameter) pipettes, combined with analysis of the a.c. response of the IT current, led to a further increase in k_{MT}, which in turn led the authors to conclude that there was no effective interfacial barrier to IT [101–103]. Subsequently a further decrease of pipette radius to the order of 1–10 nm has been achieved [104], meaning that interfacial rate constants exceeding 1 cm s^{-1} should fall within the measurable domain for small molecular ions. When k_{MT} begins to exceed k_{CT} the voltammogram for ion transfer across the ITIES will start to deviate from the general form of the ideal (Nernstian) response for a reversible transfer, given by

$$i = i_{\lim} \left(\frac{\exp(\vartheta)}{1 + \exp(\vartheta)} \right) \qquad (36)$$

with $i_{\lim}$ as the mass-transport limited current and

$$\vartheta = \frac{z_i F}{RT} (\Delta_{org}^w \varphi - \Delta_{org}^w \varphi_i^0) \qquad (37)$$

The extent of distortion of the voltammogram from Eq. (36) can be measured and related to k_0 since the relationship between the two parameters has been determined by numerical solution of the mass-transport equation for microelectrode voltammetry under mixed kinetic–mass-transport control [105]. This approach has been used to determine k_0 values (~ 1.0 cm s^{-1}) for the transfer of the laurate anion across the water–n-octanol interface [106] and for the transfer of the tetramethylammonium ion (~ 1.5 cm s^{-1}) across the water–DCE interface

[107]. The rate of tetraethylammonium transfer has been measured in both directions at the water–DCE interface; consistent values of k_0 (~ 2 cm s^{-1}) were obtained in both cases, although the sum of the α values (Eqs. (30) and (31)) exceeded unity [107]. The facilitated transfer of alkali metal ions by the ligand dibenzo-18-crown-6 has been investigated using the micropipette approach at the water–DCE interface: k_0 values around 1.0 cm s^{-1} have been measured for the potassium ion [19, 104, 107], whereas a separate study reported k_0 values increasing in the series Li$^+$ (0.3 cm s^{-1}) < Na$^+$ (0.9 cm s^{-1}) < K$^+$(1.7 cm s^{-1}) [108].

An alternative electrochemical approach to the measurement of fast interfacial kinetics exploits the use of the scanning electrochemical microscope (SECM). A schematic of this device is shown in Fig. 14; the principle of the method rests on the perturbation of the intrinsic diffusive flux to the microelectrode, described by Eq. (34) above. A number of reviews of the technique exist [109, 110]. In the case of the L–L interface, the microelectrode probe is moved toward the interface; once the probe-interface separation falls within the diffusion layer, a perturbation of the current–distance response is seen, which can be used to determine the rate of interfacial processes, generally by numerical solution of the mass-transport equations with appropriate interfacial boundary conditions. The method has been

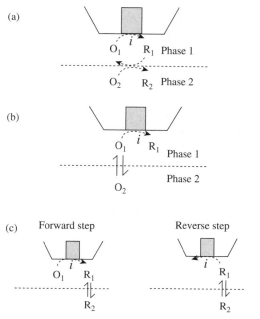

Figure 14. Schematic of operation of the SECM as applied to L–L interfaces. Interfacial ET is shown in (a). Interfacial IT, with the partition of the oxidized reduced forms, is shown in (b) and (c). (Reprinted with permission from Ref. 109; Copyright 1999, Elsevier SA.)

extensively used since the mid-1990s in the study of electron transfer kinetics at the ITIES (see Section IV). Its initial application to IT [111, 112] has been extended to kinetic studies by combining the SECM with the conventionally used four electrode potentiostat to polarize the ITIES externally [19]. Alternatively, control over the potential at the ITIES can be achieved through the Nernst–Donnan approach; the transfer of electrogenerated ferricenium across the L–L interface has been investigated as a function of the interfacial perchlorate ion ratio (Eq. (5)), but the transport was solely diffusion controlled (i.e., $k_{CT} > k_{MT}$, lower bound on k_{CT} estimated as $0.5 \, \text{cm s}^{-1}$) [113].

An analogous technique to the SECM involves the expansion of drops of one immiscible liquid into the other. In the method of microelectrochemical measurements at expanding droplets (MEMED), a microelectrode probe is placed within vicinity of the drop, such that the changes in the flux to the electrode due to drop expansion can be used to report on the process occurring at the L–L interface [114]. This expanding droplet approach has been applied to IT kinetics, specifically of the hexachloro-iridate(IV) ion, where a second partitioning ion (perchlorate) was used to control the interfacial potential via the Nernst–Donnan approach (Eq. (5)). Finite IT kinetics were shown to be measurable for certain common-ion concentrations, although no k_0 value was reported [115]. Hydrodynamic L–L electrochemical systems have also been developed which, essentially, increase k_{MT} by imposing a defined flow regime on the immiscible liquid system. Rotating diffusion systems have been used to measure IT (tetraethylammonium) and facilitated IT (Na$^+$ facilitated by dibenzo-18-crown-6) across the membrane-stabilized water–DCE interface: respective k_0 values of $1 \, \text{cm s}^{-1}$ and $0.4 \, \text{cm s}^{-1}$ were measured, in broad agreement with the micropipette data [116, 117].

The interfacial transfer of neutral molecules is also amenable to measurement by some of the above techniques, notably the SECM and MEMED, and by laser trapping of individual droplets. Clearly, no potential dependence of these interfacial phenomena can be probed, but for comparison with the IT work, the general conclusions are presented. The transfer of small, neutral solutes (molecular bromine and oxygen) across the water–DCE interface investigated via SECM revealed no measurable kinetic barrier, on the timescales accessible to the SECM [118, 119]. An alternative experimental configuration is the single microdroplet method, where continuous irradiation via an infrared laser source is used to "trap" a droplet of organic electrolyte solution against a microelectrode surface [120, 121]. A schematic of the experimental configuration is shown in Fig. 15; electrolysis of the solutes within the droplet perturbs the aqueous–organic partition equilibrium; hence for sufficiently small droplets, the charge–time response can be related to the transport of the solute across the L–L interface. The approach has been applied to various derivatives of ferrocene, which showed that in most cases transport was simply limited by

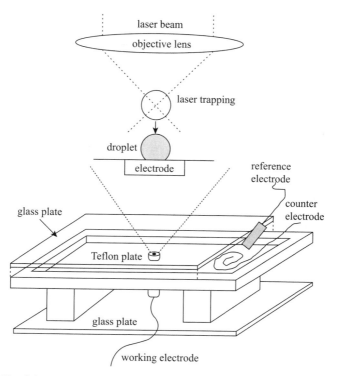

Figure 15. Schematic of the single-laser microdroplet experiment. (Reprinted with permission from Ref. 121; copyright 1998, American Chemical Society.)

diffusion [122]. However, for ferrocene derivatives containing groups capable of strong hydrogen bonding, mixed adsorption–transport control over the solute flux was invoked. Rate constants for the adsorption of these solutes at the water–NB interface were found (order of 10^{-5} cm s^{-1}). A droplet size dependence of both the adsorption rate constant and the normalized droplet area (ratio of effective area to actual area) was found for the transfer of 2-ferrocenyl-2-propanol, with the former parameter decreasing and the latter parameter approaching unity as the droplet size increased (see Fig. 16) [121]. The authors separate this effect from the increasing diffusive flux to a smaller surface (compare Eq. (34)); the parameters shown in Fig. 16 are normalized to the relevant interfacial area. Instead the size dependence of the adsorption rate was attributed to the effect of capillary waves on the droplet surface. A quasi-thermodynamic explanation was offered for this phenomenon: capillary fluctuations of smaller droplets were assumed, from the higher surface:volume ratio, to generate a greater perturbation, leading to stronger adsorption on such droplets. The rate of transfer of ionic solutes, namely ferricenium cation formed

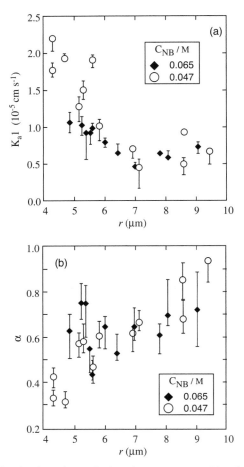

Figure 16. Droplet size dependence of adsorption rate constant (a) and effective area factors (b) for ferrocene derivatives on the water–NB interface. (Reprinted with permission from Ref. 121 copyright 1998, American Chemical Society.)

on the oxidation of ferrocene, was also extracted through the microdroplet approach. A rather low rate constant, of $\sim 10^{-5}$ cm s^{-1}, was measured at the water–NB interface [123].

The final experimental approach to IT kinetics reported recently involves the use of a thin film of organic solvent covering a graphitic electrode; the whole assembly is then immersed in aqueous solution, as depicted in Fig. 17 [124]. Potentiostatic control is maintained over the aqueous/electrode potential difference, which is the sum of the aqueous/organic and organic/electrode potential differences. The aqueous/organic term is assumed to be controlled by

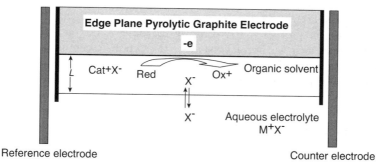

Figure 17. Schematic of the thin-layer configuration used to study IT and ET processes at the L–L interface. (Reprinted with permission from Ref. 124; copyright 2005, Elsevier SA.)

an excess of a common-ion (Eq. (5)) relative to a lipophilic electron donor (or acceptor). Oxidation (reduction) of the latter at the graphite surface is compensated by IT from the aqueous phase; a kinetic analysis is performed on the basis that IT is rate-limiting, an unusual assumption given the large excess of ion present. An initial report gave k_0 values on the order of 1×10^{-3} cm s^{-1} [124] for a variety of ions including tetramethylammonium, whereas values an order of magnitude higher were reported in a subsequent report, albeit using slightly different experimental conditions [125]. Anion transfer rates on the order of 1×10^{-2} cm s^{-1} have also been reported [126]. There is a clear discrepancy with the significantly larger rate constants measured at the externally polarized interface [19, 104, 107, 116].

To summarize, although rate constants (or, perhaps, apparent rate constants) for IT across the ITIES have been reported for more than 20 years, there is still controversy about the interpretation of this phenomenon, not least because the reported rate constants have "increased over the years as experimental measurements became more and more sophisticated" [127]. As noted above, a general problem has been that the characteristic timescales of the (apparent) kinetic process are often not markedly lower than the timescales of the experimental technique, a fact that has been remarked upon in the literature [128]. For example, in some of the recent data [94, 96] the time constants of the technique are frequently of the same order of magnitude as the timescales of the process they are purporting to measure. The question therefore becomes whether the rate constants reported for IT using nanopipettes [104, 107, 108] will increase in the future or whether these represent true values. Clearly at this point it is reasonable to ask whether theory predicts that a barrier to interfacial IT should exist and, if so, what the physical origin of such a barrier might be.

The observation that capillary waves exist, even on scales approaching the molecular as borne out by MD work, has prompted a reevaluation of the view of the IT process. Previously, theories had been developed which treated IT as an activated transfer process [87, 90]; however, these works do not include the molecular detail that has become available through the recent improvements in understanding of interfacial structure. Consequently, the spatial variation of the electrochemical potential could only be assumed, and kinetic models could be developed on the basis of these assumptions. For example, Kakiuchi assumed that the electrochemical potential varied linearly across a small layer on either side of the interface [129], that is,

$$\bar{\mu}(z) - \bar{\mu}(z_i) = [\bar{\mu}(z_f) - \bar{\mu}(z_i)]\frac{(z - z_i)}{(z_f - z_i)} = \Delta\bar{\mu}\frac{(z - z_i)}{(z_f - z_i)} \tag{38}$$

Thus from the integral form of the Nernst–Planck equation, the steady-state interfacial flux gives

$$k_f = \frac{1}{\int_{z_i}^{z_f}\left[\exp\left(\frac{\bar{\mu}(z)-\bar{\mu}(z_i)}{RT}\right)/D\right]dz} = \frac{\left(\frac{\Delta\bar{\mu}}{2RT}\right)\exp\left(\frac{\Delta\bar{\mu}}{2RT}\right)}{\sinh\left(\frac{-\Delta\bar{\mu}}{2RT}\right)}\frac{D}{(z_f - z_i)} \tag{39}$$

where k_f here relates the flux in the one direction (see Eq. (29)) to the interfacial concentration of the ion. In the limit of zero driving force, Eq. (39) reduces to

$$k_f = k_0 = \frac{D}{(z_f - z_i)} \tag{40}$$

an expression identical to one put forward by Girault [87]. The problem with Eq. (40) is that typical solution-phase diffusion coefficients (10^{-5} cm^2 s^{-1}) and realistic values of the interfacial width (1 nm at most; see Section II) give k_0 values two orders of magnitude higher than the highest experimental values measured (or indeed, accessible) to date.

A possible resolution to this problem is to allow for the dynamic properties of the interface. The dynamic nature of the interface has been incorporated into a protrusion–extrusion model of IT by Marcus [128]. In this model, IT (e.g., from water to organic) is viewed as being initiated by a "finger" of organic solvent attaching to the ion; the finger then moves back toward the bulk of the organic solvent with the ion, and a finger of aqueous solution then develops, which pushes the ion into the organic phase (see Fig. 18). Marcus expressed the overall process as a series of pseudo-elementary steps, involving attachment of the ion, its diffusion along the protrusion in solvational and real space [130], and

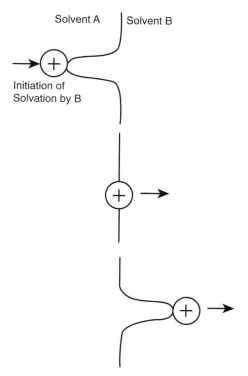

Solvent A | Solvent B

Initiation of
Solvation by B

Figure 18. Protrusion mechanism of IT. (Reprinted with permission from Ref. 128; copyright 2000, American Institute of Physics.)

the detachment process. Statistical thermodynamic arguments were used to give order of magnitude estimates of each of the three contributions to the IT process. Marcus concluded that the ion attachment/detachment terms were likely to be substantially higher than the observed rate constants; hence the only conceivable factor behind retardation of the IT process is the motion of the ion along the spatial-solvation coordinate. The concept of interfacial dynamics influencing the IT process has been considered in terms of a hydrodynamic analogy [127]. In this case, the motion of a micron-scale sphere toward a L–L interface, where the receptor phase had a higher viscosity, was determined from numerical solution of the Navier–Stokes equations. The frictional force on the sphere was found to increase over a larger ($\sim 10\,\mu$m) length-scale, as the sphere approached the interface. The coupling of interfacial and ion motion was taken further by Urbakh and co-workers, who viewed ion motion toward the interface as the cause of the protrusion, as the ion attempts to remain in its preferred state of solvation [131]. The Langevin equations, describing the coupling of the ionic and interfacial diffusive movement, are written

$$\eta_z \frac{dz}{dt} + \frac{\partial V(z,h)}{\partial t} = f_z(t) \tag{41}$$

$$\eta_h \frac{dh}{dt} + \frac{\partial V(z,h)}{\partial t} = f_z(t) \tag{42}$$

where V is the local potential, h is the position of the interface, the η terms are dissipation constants, and the f terms describe thermal motion. Equations (41) and (42) were solved analytically [131] (for the case of strong interface–ion coupling, i.e., high γ) and numerically [132] (the weak coupling, or low γ, case) and expressions for the IT rate constant were so obtained. Two limiting cases were considered in the strong coupling case: (a) one where the ion motion was faster than that of the interface and (b) another where the opposite condition held. The latter limit yields an expression identical to Eq. (40) above. The former condition makes more physical sense, since all but the heaviest ions should diffuse more rapidly than the collective motions of the interfacial molecules, yielding

$$k_{IT} = \frac{D_h}{(z_f - z_i)} \left(\frac{\Lambda}{h_{max}} \right)^2 \tag{43}$$

where Λ is the position where the ion begins to influence the motion of the interface. Interpolation between the limits represented by Eqs. (40) and (43) leads to an effective diffusion coefficient,

$$\frac{1}{D_{eff}} = \frac{1}{D_z} + \frac{1}{D_h \left(\frac{\Lambda}{h_{max}} \right)^2} \tag{44}$$

thereby providing a physical explanation of the slow apparent interfacial diffusion step observed experimentally. Rather severe approximations were made in the derivation of the rate expression, although the interpolation to give Eq. (44) was supported by the later numerical work [132]. The work predicts an increase in IT rate with increasing interfacial tension (since h_{max} in Eq. (43) should fall, according to Eq. (7)) although it does not allow for any detachment of the protrusion from its "mother" solvent along with the transferring ion (see below), nor are the parameters introduced in Eqs. (43) and (44) readily predictable.

As discussed in Section II, MD calculations of the structure of the ITIES have made an important contribution to the understanding of the field. The MD approach has been extended to the case of IT, with the particular aim of gaining evidence to support—or discredit—the experimental indications that a finite barrier to IT exists. As with any MD approach, the validity depends on the applicability of the potentials used; the general drawback for ionic systems is the difficult in including meaningful numbers of ions, which mean that local effects of the double layer structure on the IT dynamics cannot readily be incorporated. The Gibbs energy profile for the transfer of relatively hydrophilic ions (standard Gibbs energy of transfer a least 40 kJ mol^{-1}) across the water–DCE interface has been calculated, treating the liquids via the point charge model and including pairwise Coulombic and Lennard-Jones interactions [133]. The standard Gibbs energies of single ions of different type were found from the difference between their initial and final "bulk" solvation states: reasonable agreement (within 20%) with experiment was obtained. A nonequilibrium Gibbs energy profile along the interfacial normal was calculated from the potential of mean force of each hydrophilic ion as it traversed the interface; notably an interfacial barrier to IT was also suggested, and it was correlated by the form of the ionic trajectories in the vicinity of the interface [130]. Successful transfer events were found to occur when a "finger" of solvent attached to the ion. The calculations also suggested that, for hydrophilic ions transferring to DCE, part of the hydration shell remained intact within the organic phase although the restricted system size warrants caution in this interpretation. An exception was the transfer of the less hydrophilic tetramethylammonium ion from water to NB (standard Gibbs energy of transfer at this interface is reported as 2.9 kJ mol^{-1} [134]), which was found to lose most of its solvation shell on transfer to the organic phase [135]. Analysis of the trajectories showed that considerable interfacial roughening is associated with the transfer process, indicating that the "fingering" mechanism of transfer holds even for ions where the hydrated layer is lost, although no barrier to tetramethylammonium transfer was observed. MD calculations of iodide transfer across the water–organic interface, by evaluating the potential of mean force, have indicated that the extent of residual hydration is dependent on the identity of the organic solvent (as well as the ion) [43]. For iodide, an almost intact first hydration shell is found on transfer to 2-heptanone, whereas the retracting finger removes the entire shell on transfer to iso-octane. These conclusions on the extent of hydration within the organic phase have been supported by experimental studies of ion extraction at the water/NB interface, where the change in water content of the organic phase associated with ion extraction has been found [136, 137]. The results imply that first solvation shells are co-extracted almost intact with hydrophilic ions such as lithium and sodium (six and four water molecules, respectively), and substantial parts of the second shell are extracted for ions with higher charge (e.g., 12 water molecules per ion of calcium extracted), whereas no detectable hydration sphere was co-extracted with tetramethylammonium. The calculated hydration numbers on transfer from water to 2-heptanone [138] also correlated well with the measured residual hydration in NB [136, 137], in spite of the significant differences in relative permittivity [135, 139]. However, the longer timescale (several nanoseconds) MD calculations did not display any interfacial barrier to IT for iodide or more hydrophilic ions at the water–2-heptanone interface [138], a conclusion at variance with Benjamin's work [133]. Despite the general agreement with experiment (level of residual hydration, reasonable prediction of standard Gibbs energies of transfer), some caution is still needed with regard to the mechanistic conclusions drawn from these works. The length-scale of the fingers (in excess of 1 nm with divalent ions) will be reduced on introduction of a background electrolyte to form a double layer. Under experimental conditions, the presence of the supporting electrolyte within the organic phase means that the ions of the electrolyte could also play a role in assisting the IT process. Such a mechanism has been suggested recently [79, 140] where the use of nanopipettes has allowed IT to be studied in the absence of an organic electrolyte, due to the extremely low currents passed and, in turn, the ability to study IT in high-resistance media. Alkali metal cation transfer was found to be highly sensitive to the presence of organic-phase electrolytes, whereas this sensitivity did not extend to anions (chloride) with similar standard Gibbs energies of transfer, nor did it apply to less hydrophilic cations. A "shuttling" mechanism, due to interfacial ion pair formation between the cation and the anion of the supporting electrolyte, was put forward [79].

A very recent article has also suggested that the water content of the organic phase can play a similar role in facilitating the transfer of hydrophilic cations [141]; the suggestion is that hydrophilic cations are not transferred to the bulk organic solvent per se, but rather remain within an aqueous microenvironment, for water-saturated organic solvents. How this conclusion relates to the mechanism of IT, and why it should only hold for cations and not for anions of comparable hydrophilicity, where the number of co-extracted water molecules is similar [136], is not yet clear. Understanding of the mechanism of IT is intimately bound to knowledge of interfacial, and associated double-layer, structure; and until molecular level treatment of the ITIES can allow for these effects, the IT mechanism will be uncertain. The question of the size dependence of kinetic parameters is worthy of further consideration [121]. This phenomenon could be an artifact of the similarity of k_{CT} to k_{MT}, for the experimental techniques employed, as has been suggested [128]. The lack of agreement between kinetic parameters (see preceding text) reported with different techniques tends to support the artifact argument, although to date, experimental conditions often vary making direct comparisons difficult. Given that protrusion formation has been invoked as central to the IT mechanism, a systematic study of the dependence of kinetic parameters on interfacial tension (via surfactant addition) may provide a route to validating the mechanism of the IT process. The idea that interfacial size may influence fluctuation amplitude [121] is plausible (cf. Eq. (7)), although longer wavelength fluctuation should have a minimal influence on the formation of the fluctuations driving IT.

IV. ELECTRON TRANSFER KINETICS

Electron transfer (ET) at the ITIES can occur when an organic solution containing a hydrophobic redox couple is placed in contact with an aqueous solution containing a hydrophilic couple. Charge transfer can occur spontaneously, or can be controlled through the application of an external potential, depending on the relative reduction potentials of the two couples. ET under external potential control at the ITIES using four-electrode potentiostat methodology was first reported in 1979 [142], although an earlier, less refined experimental configuration was used to deposit copper and silver at the water–DCE interface [143]. The general case of spontaneous ET at the L–L interface dates back to the 19th century for the case of metal deposition [144]. The thermodynamics of the ET process have been described in earlier reviews; as with the IT case, the purpose of this section is to give only a brief recap of the older work and to discuss the most recent (past 10–15 years) progress in this area.

In a perceptive remark made in a review of the ITIES literature in the early 1990s, Girault noted that "experimental studies of ET reactions at L–L interfaces, during the coming decade, will be dedicated to test the theories available" [3]. The past decade or so has indeed seen a tremendous improvement in the understanding of ET kinetics at the ITIES, where the kinetic picture is now clearer than in the IT case. This improvement has been largely due to the application of two important experimental approaches to ET at the ITIES, namely SECM (discussed previously in the context of IT, in Section III) and intensity-modulated photocurrent modulation (IMPS) and related photoelectrochemical techniques for the study of light-induced ET. The general interest in understanding ET at the ITIES stems from its combination of features from homogeneous (solution phase) ET and heterogeneous (electrode/electrolyte) ET; that is, the process is bimolecular, as in the former case, yet the potential can be used as a rate-controlling variable, as in the latter case. The question suggested by Girault can be expressed as, What is the actual relationship between ET rate and driving force?

The general ET process can be written as

$$O_{1(aq)} + R_{2(org)} \rightleftarrows R_{1(aq)} + O_{2(org)}$$

For transfer of a single electron, the standard Gibbs energy change is

$$\Delta G^0 = -F(E_1^0 - E_2^0) \tag{45}$$

where the E^0 values are the standard reduction potentials of the redox couples in the appropriate (water, organic) phases. The Nernst equation for the ET process is therefore

$$\Delta_{org}^w \varphi = E_2^0 - E_1^0 + \frac{RT}{F} \ln \left[\frac{a_{R_1}^w a_{O_2}^{org}}{a_{O_1}^w a_{R_2}^{org}} \right] = \Delta_{org}^w \varphi_{ET}^{0'} + \frac{RT}{F} \ln \left[\frac{c_{R_1}^w c_{O_2}^{org}}{c_{O_1}^w c_{R_2}^{org}} \right] \tag{46}$$

where the a and c terms denote the thermodynamic activities and concentrations, respectively, of each component. (The prime symbol on the Galvani potential term indicates that the effects of nonideality have been absorbed into this term.) As with the IT process, the potential difference can be imposed externally via a potentiostat [142, 145] or via a common-ion (Eq. (5)) [146, 147], if it is assumed that none of the redox active species transfer and that they do not interfere with the partition equilibrium of the common-ion. If, following the convention used for IT, positive current density is defined as the flux of negative charge from the organic phase, then the current density for a single ET process is

$$I = F(k_f c_{O_1}^w c_{R_2}^{org} - k_b c_{R_1}^w c_{O_2}^{org}) \tag{47}$$

where, following the phenomenological Butler–Volmer formalism, the bimolecular rate constants for ET are written as

$$k_f = k_0 \exp \left(\frac{\alpha F(\Delta_{org}^w \varphi - \Delta_{org}^w \varphi_{ET}^{0'})}{RT} \right) \tag{48}$$

and

$$k_b = k_0 \exp \left(\frac{-(1-\alpha)F(\Delta_{org}^w \varphi - \Delta_{org}^w \varphi_{ET}^{0'})}{RT} \right) \tag{49}$$

where α again represents the phenomenological charge-transfer coefficient. Experimental systems often employ an excess of, for example, both forms of the aqueous phase redox couple, meaning pseudo-first-order rate constants describe the current density:

$$I = F(k_f' c_{R_2}^{org} - k_b' c_{O_2}^{org}) \tag{50}$$

where

$$k_f' = k_f c_{O_1}^w \quad \text{and} \quad k_b' = k_b c_{R_1}^w \tag{51}$$

In the cases where the potential is controlled via the common-ion, it has been widely assumed that a simple coupling of the potential established via the ion partition and ET equilibria occurs [146, 148]; hence combining Eqs. (5) and (48) yields a Tafel expression for ET coupled to the distribution of a common-ion (of charge number z_i) at the ITIES:

$$\ln\left(\frac{k_f}{k_0}\right) = \frac{\alpha F(\Delta_{org}^w \varphi_i^o - \Delta_{org}^w \varphi_{ET}^{0'})}{RT} + \frac{\alpha}{z_i} \ln\left(\frac{a_i^{org}}{a_i^w}\right) \tag{52}$$

This approach has been employed in experimental studies of ET at the ITIES using the SECM [148]; note that the definitions of the potential and/or directionality of rate constants above do not necessarily accord with those given in the literature, and thus the apparent dependence of kinetic parameters on ion activity may differ from that presented here. It is noted that a general experimental difficulty in identifying practical ET systems is that, ideally, none of the four components of the redox equilibrium should readily cross the interface and all four components should be stable at ambient conditions. These conditions have proven to be relatively challenging to meet in practice [149, 150].

Marcus has extended the semiclassical ET theory to the case of ET at the ITIES. The barrier to the ET reaction is given by [151, 152]

$$\Delta G^{\neq} = \frac{\lambda}{4}\left(1 + \frac{\Delta G^0}{\lambda}\right)^2 \tag{53}$$

an expression presented previously for ET across the metal–electrolyte interface [153], which reduces to a Butler–Volmer-type dependence for $\Delta G^0 < \lambda$. The reorganization energy, λ, is given by the sum of the outer-sphere (solvent continuum) and inner-sphere (vibrational) contributions [154]. The latter term can be evaluated from a normal mode analysis; the outer sphere expression at the ITIES is given by

$$\lambda_0 = \frac{N_A(\Delta e)^2}{4\pi\varepsilon_0} \left\{ \begin{array}{l} \left[\frac{1}{2r_w}\left(\frac{1}{\varepsilon_w^{op}} - \frac{1}{\varepsilon_w}\right)\right] + \left[\frac{1}{2r_o}\left(\frac{1}{\varepsilon_{org}^{op}} - \frac{1}{\varepsilon_{org}}\right)\right] - \left[\frac{1}{4Z_w}\left(\frac{\varepsilon_{org}^{op} - \varepsilon_w^{op}}{\varepsilon_w^{op}(\varepsilon_{org}^{op} + \varepsilon_w^{op})} - \frac{\varepsilon_{org} - \varepsilon_w}{\varepsilon_w(\varepsilon_{org} + \varepsilon_w)}\right)\right] \\ - \left[\frac{1}{4Z_o}\left(\frac{\varepsilon_w^{op} - \varepsilon_{org}^{op}}{\varepsilon_{org}^{op}(\varepsilon_{org}^{op} + \varepsilon_w^{op})} - \frac{\varepsilon_w - \varepsilon_{org}}{\varepsilon_{org}(\varepsilon_{org} + \varepsilon_w)}\right)\right] - \left[\frac{2}{\Theta}\left(\frac{1}{\varepsilon_{org}^{op} + \varepsilon_w^{op}} - \frac{1}{\varepsilon_{org} + \varepsilon_w}\right)\right] \end{array} \right\} \tag{54}$$

where the ε^{op} terms denote the optical dielectric constants of each phase, the r terms denote the radii of the reactants, the Z terms denote the separation of the reactants from the interface, and Θ is approximately twice the mean interfacial separation. Equations (53) and (54) neglect the work required to bring the reactants and products to/from the interface. Expressions have been given for the work terms using a dielectric continuum model [154]. For the reactant case, using the ET process above, the work for one mole of reactants is given by

$$w^r = \frac{-N_A}{4\pi\varepsilon_0}\left\{\left(\frac{(z_{O_1})^2}{4Z_w\varepsilon_w} - \frac{(z_{R_2})^2}{4Z_o\varepsilon_{org}}\right)\left(\frac{\varepsilon_{org} - \varepsilon_w}{\varepsilon_w + \varepsilon_{org}}\right) + \frac{2z_{O_1}z_{R_2}}{\Theta(\varepsilon_w + \varepsilon_{org})}\right\} \qquad (55)$$

where the z terms represent the charge number of the reactants. Although these terms are quite significant, their magnitude in typical experimental conditions with excess inert electrolyte is unknown; hence their effect on the overall driving force has generally been neglected [155, 156]. Marcus has provided expressions for the preexponential factors, to allow evaluation of the ET rate constant k_0 at a sharp L–L interface [152, 154]:

$$k_0 = N_A j \nu v \exp\left(\frac{-\Delta G^{\neq}}{RT}\right) \qquad (56)$$

where j denotes the Landau–Zener factor for ET, ν is a characteristic frequency of motion for the process and v is the product of the square area term, describing the effective reaction volume for the bimolecular process. For a sharp interface, the square area term has been approximated as

$$v \approx \frac{2\pi(r_w + r_{org})^3}{\Gamma} \qquad (57)$$

where Γ is the decay factor describing the distance sensitivity of solution phase ET. The level of approximation made above means the preexponential term in Eq. (56) can only be estimated to within an order of magnitude. Marcus extended this analysis to the case of an interface which is several solvent molecules thick [157], but this case will not be considered further here, given the more recent experimental evidence on interfacial width (Section II).

The experimental studies reported in the past 10–15 years have attempted to measure the rate of ET at the ITIES, along with its potential dependence. The Marcus approach predicts a potential dependent rate that results from the variation of the rate constant. By contrast, other theories have developed a Frumkin-type approach, where the potential drop between the planes of closest approach of the reactants is small; hence the potential dependence of the rate constant is also low [158]. An extension of the lattice gas model (see Section II) was used to determine concentration of the redox partners local to the interface; the weakness in this approach is its reliance on the Gouy–Chapman method to determine the local electrostatic potential. The potential dependence of the overall rate was derived from the interfacial concentrations of the reactants; given a rather diffuse double layer (large Debye lengths compared to the thickness of the reaction zone), the approach of charged reactants/products to the interface becomes a function of the local Galvani potential, and thus this local concentration change drives the change in reaction rate.

Classical electrochemical techniques (cyclic voltammetry, a.c. impedance) were employed by Cheng and Schiffrin to determine ET kinetics between an aqueous mixture of ferri/ferrocyanide and a variety of organic phase electron acceptors. Analysis of the problem was simplified by treating the ET reaction as pseudo-first order, hence the use of an excess of aqueous phase redox species. Pseudo-first-order rate constant (k_0') values of 10^{-3}–10^{-2} cm s^{-1} were obtained, corresponding to second-order k_0 values of ~ 0.1 cm s^{-1} M^{-1}, with α in the range 0.5–0.7 [159, 160]. The rapid development of the field in the intervening period has been due, in part, to the application of the SECM method to ET kinetics at the ITIES, first reported in 1995. An experimental configuration similar to that shown in Fig. 14 was used, where an excess of organic phase species is used to regenerate the aqueous phase redox couple (electrolyzed at the microelectrode surface) via interfacial ET [32]. Again, a pseudo-first-order kinetic model was used to estimate the interfacial ET rate constant; its potential dependence was probed via the common-ion approach in a subsequent publication, with α values around 0.5 obtained once more [148]. One intrinsic problem of the use of pseudo-first-order kinetic conditions is that the reaction rate will be maximized (see Eqs. (47) and (48)) and the process is therefore likely to become transport limited, following the arguments given at the beginning of Section III. One approach taken to slow down the ET rate has been the adsorption of phospholipids at the ITIES; monolayers have been assumed to form, with a decrease in ET rate constant assumed to follow from the resultant increase in separation of the redox partners (cf. Eqs. (56) and (57)) [160, 161]. Use of this approach in conjunction with the SECM has allowed a range of aqueous phase redox couples to be probed, without the hindrance of transport limitations [161]. This allowed the potential dependence of the ET rate to be varied through the change in the ΔE^0 term accorded; importantly, a decrease in rate constant was seen as the driving force increased, which was taken as evidence in support of the Marcus model with its prediction of an inverted rate-driving force dependence. The use of adsorbates to slow ET rates at solid electrode surfaces is well established [162]; its application to the ITIES is simpler to achieve experimentally, although it does raise other questions about the integrity of any monolayer structure and the positions of the reactants with respect to the adsorbate.

A further increase in the range of systems amenable to study was achieved by dispensing with the pseudo-first-order model referred to above. The deceleration of interfacial charge transfer rate relative to the intrinsic mass transport rate, brought about with conditions where neither redox couple is in excess, can extend the range of measurable kinetic parameters—with the proviso that a model exists to extract the parameters from the experimental data. Unwin and co-workers [163] obtained such a model via numerical solution of the coupled two-phase transport equations for the SECM as a function of interfacial rate constant, thus enabling comparison with experiment for second-order conditions. The more extensive data set available, with a larger range of ΔE^0 probed below the transport limit, provided further evidence for an inverted dependence of the rate constant on driving force as the latter increased; notably these measurements were performed in the absence of an interfacial phospholipid film (see Fig. 19) [155, 163]. ET kinetics were investigated in both directions, at low driving force, for the the tetracyanoquinodimethane (TCNQ)/ferrocyanide system and found to follow a Butler–Volmer dependence (Eqs. (48) and (49)) with the transfer coefficients for the forward and reverse ET summing to a value close to unity (see Fig. 20) [155, 164]. The value of k_0 obtained experimentally

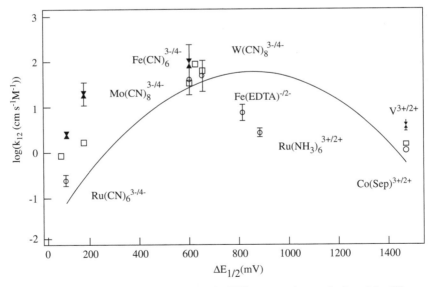

Figure 19. Dependence of the rate constant for ET between a zinc porphyrin and the difference in reduction potential for the aqueous redox couples shown. (Reprinted with permission from Ref. 155; copyright 2001, American Chemical Society.)

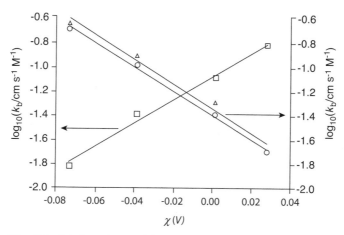

Figure 20. Potential dependence of the forward and reverse rate constants for electron transfer between TCNQ and ferrocyanide. (Reprinted with permission from Ref. 164; copyright 2001, Elsevier SA.)

for the TCNQ/ferrocyanide system, 0.1 cm s^{-1} M^{-1}, was compared against a value obtained from the Marcus approach (Eqs. (53 to 57)) of 0.08 cm s^{-1} M^{-1}, using previously quoted values for the inner-sphere reorganization energies of the reactants [165]. Similarly, ET between aqueous-phase ferrocyanide and organic-phase ferricinium ions was studied by a hybrid SECM-polarizable interface configuration, which allowed a larger potential window (see Section I) to be attained. This approach gave rate constants consistent with Eq. (52) (Tafel) and α values close to 0.4 [166].

The other notable experimental improvement in recent times has been the application of spectroscopic and/or spectro-electrochemical methods to ET at the ITIES. Photocurrents at the ITIES were reported some time ago, but the relatively crude experiments hindered the conclusive identification of the processes responsible [167, 168]. Pulsed laser spectroscopy was used in TIR mode to probe the photoinduced ET between organic-phase anthracene and aqueous phase europium ions, via the change in the lifetime of the europium excited state [169]. A TIR geometry was also used to study potential-dependent optical absorbance in the TCNQ/ferrocyanide ET system [97], analogous in conception to the transfer of fluorescent ions investigated by Kakiuchi (see previous section) [91–95]. The time dependence of the absorbance, along with the frequency response to potential modulation, was used to extract a rate constant for the interfacial ET process ($k_0 \approx 0.1$ cm s^{-1} M^{-1}), which revealed an unusual (non-exponential) dependence on applied potential. Subsequent investigations concentrated on photoinduced ET processes, using a porphyrin heterodimer adsorbed on the aqueous side of the ITIES. The photoinduced processes are readily interrogated using light intensity modulation (IMPS), permitting the resolution of kinetic processes with different time-scales. Photo-induced ET from the porphyrin dimer to/from TCNQ and decamethylferrocene, respectively, gave an exponential dependence of ET rate constant on applied potential, with α close to 0.5 [170], in broad agreement with the dark ET systems probed with the SECM. Extending the work to a series of ferrocene derivatives, analogous to Bard's SECM work, revealed an apparent levelling of the ET rate constant with driving force, consistent with the Marcus model (see Figs. 21 and 22) [156]. Similar behavior was observed for photoinduced reduction of a group of quinones with the porphyrin dimer [171]; the slow rate of the back (dark) ET, despite its considerable driving force ($\sim 1\,V$), was taken as further evidence in support of the existence of the Marcus "inverted" region at the ITIES. Stabilized gold nanoparticles have been employed as organic-phase electron donors and acceptors, due to the range of easily accessible oxidation states [172, 173]; potential-dependent ET rate constants for dark- and light-driven processes with k_0 approaching $100\,cm\,s^{-1}\,M^{-1}$ have been reported.

The body of experimental data summarized in the preceding paragraphs shows reasonable agreement with the Marcus model and suggests that the potential dependence of the ET rate at the ITIES results from a potential-dependent ET rate constant, rather than the response of the reactant concentrations to applied potential. However, the picture is not complete, as

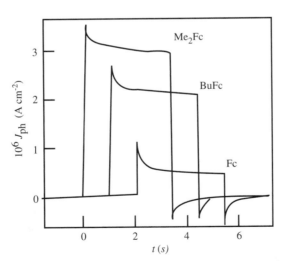

Figure 21. Temporal response of photocurrent density to illuminaion obtained with the zinc porphyrin heterodimer and dimethylferrocene (Me$_2$Fc), butylferrocene (BuFc), and ferrocene (Fc) dissolved in the organic phase. (Reprinted with permission from Ref. 156; copyright 2002, American Chemical Society.)

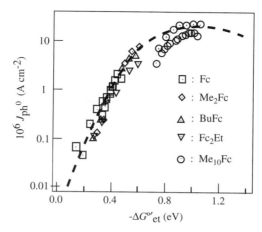

Figure 22. Dependence of the initial photocurrent density on the net driving force for electron transfer for the porphyrin heterodimer–ferrocene system. (Reprinted with permission from Ref. 156; copyright 2002, American Chemical Society.)

some aspects of the recent literature suggest. There is some disagreement concerning the dependence of the ET rate constant on driving force; notably the linear (and, ultimately, inverted quadratic) dependence is observed when the driving force has been altered by changing the redox couples [155, 156, 163, 171]. By contrast, a less clear-cut dependence of the ET rate constant on driving force emerges when the distribution of a common-ion has been used to control the Galvani potential difference at the ITIES. For example, when a thin-layer configuration (see Fig. 17) was used to investigate ET across the ITIES, the ET rate constant for the decamethylferrocene–ferricyanide system was reported to be independent of the common-ion ratio [174–176]. By contrast, the forward and reverse ET of the TCNQ–ferrocyanide was investigated using the same approach; both the forward and reverse rate constants varied in a manner consistent with Eq. (52). Similar behavior—that is, rate constant invariance with the common-ion ratio, but an exponential dependence on the difference in donor–acceptor reduction potential (shown in Fig. 23)—has been observed with the SECM [177]. The identity of the common-ion has also affected the kinetic data: An SECM study of the decamethylferrocene–ferricyanide ET system showed a reversal in apparent dependence of ET rate constant on driving force when the common-ion was changed from perchlorate to tetrabutylammonium [178]. Moreover, a marked sensitivity of kinetic parameters to the overall composition of the electrolyte has been noted, with the addition of supposedly inert aqueous-phase electrolyte retarding ET kinetics [155, 178]. Finally, a recent report has applied nanopipette voltammetry (see Section III) to ET systems [179]: Finite pseudo-first-order (k_0') rate constants were measured,

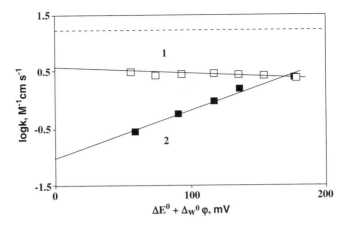

Figure 23. ET rate constant (k_f) as a function of potential for zinc porphyrin complex and hexacyanoruthenium (III), curve 1. The rate constant for the reverse process is shown in curve 2. (Reprinted with permission from Ref. 177; copyright 1999, American Chemical Society.)

although an unexplained dependence on the pipette radius was observed (shown in Fig. 24).

To summarize: The common-ion approach to potential control at the ITIES has given inconsistent results, with the driving force dependence of the ET process

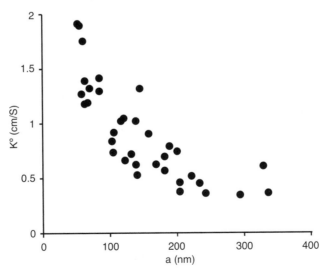

Figure 24. Apparent kinetic parameters obtained as a function of pipette radius for ET between TCNQ and the EDTA complex of Fe(II). (Reprinted with permission from Ref. 179; copyright 2006, American Chemical Society.)

varying from system to system. The validity of Eq. (52) should be addressed for each experimental ET system. Specifically, the assumption of driving force additivity implicit in Eq. (52) is only valid if the standard potentials of transfer of ionic species involved in the redox equilibria are far from the standard potential of the potential-determining ion (see Eq. (3)); coupling between IT and ET equilibria can give rise to intermediate potential values [180]. The general role of the supporting electrolyte deserves further investigation, given that it plays a "non-innocent" role in ET kinetics. The Marcus model appears to describe ET kinetics at the ITIES well, in certain cases. The model assumes a sharp interface exists between the two dielectrics, with an associated steep change in local potential. The ionic strength dependence of the apparent ET rate is worthy of exploration: As the potential distribution widens, the relaxation of the kinetic model to one closer to that described by Schmickler [158] would be anticipated. The initial report on the use of a potential-determining ion to control ET equilibria at the ITIES correlated the position of the redox equilibrium (measured via optical absorbance) with the potential difference [146]. It is also clear that nonelectrochemical methods are required as probes of ET at the L–L interface. Electrogenerated chemiluminescence, resulting from back ET between species located on either side of the ITIES, has been taken as further evidence for the existence of an inverted Marcus region at the ITIES [181]. Second-harmonic generation has recently been applied to ET at the water–dimethylaniline interface: A biexponential decay of the acceptor excited state was observed (picosecond and subpicosecond lifetimes), which was attributed to the ET process and excited-state solvation, respectively. The assignment was confirmed because a similar (picosecond) lifetime was observed in the electron donor signal [182]. Further spectroscopic studies of this type, under conditions of variable potential difference, are required to probe of the dynamics of ET at the ITIES.

V. INTERFACIAL DEPOSITION

Electrochemical methods offer controlled ways to study the nucleation, and subsequent growth, of metallic phases from solution phase precursors [183]. ET at the ITIES can be used as a route to the deposition of metals, if one of the reactants (e.g., O_1 in the reaction scheme of Section IV) is a metal precursor. The first report of electrolysis at the L–L interface, as noted above, involved the deposition of copper on passage of current across the water–DCE interface [143]. Spontaneous ET can also occur: The formation of gold films at the water–carbon disulfide interface was described by Faraday [144]. The interest in studying metal deposition at the ITIES is twofold. First, the interface in its initial state can be considered as a "perfect" surface, free of fixed defect sites, which are generally the sites of nucleation on solid surfaces. Thus the transposition of models describing nucleation and growth to the ITIES should represent a good testing ground for such theories. Second, the control (via the potential) available at the ITIES over the current (particle growth rate) means that quantitative studies of deposition at the ITIES could yield information on the mechanism of formation of nanoparticulate phases, given that these species are generally formed through solution-phase reduction processes, and little is known about the mechanism of their growth [184, 185]. A number of recent articles have continued Faraday's approach to particle formation, describing the formation of nanoparticulate films by spontaneous ET at the L–L interface [186, 187]. Aside from the preliminary study referred to above, the first deposition at the ITIES using the four-electrode potentiostat, of gold, was reported in 1996 [188]. ET-induced polymerization has also been reported at the ITIES using both potentiostatic and common-ion approaches to control the interfacial potential [189, 190].

Palladium deposition at the water–DCE interface has been described: Attempts have been made to correlate the experimental current response for deposition with nucleation models allowing for two-phase transport and interfacial nucleation [191, 192]. Relatively poor agreement was observed, a fact attributed to the growing particles' lateral motion and resultant tendency to aggregate. This problem led to the use of micropipette methods (see Sections III and IV), applied to silver deposition at the ITIES in an attempt to produce a finite number of particles [193]. Another approach to the restriction of lateral aggregation involved the use of porous membranes: palladium and platinum particles formed at the ITIES were ~ 10 nm in diameter [145, 194, 195]. The form of typical experimental current–time responses for the deposition process is consistent with the "classical" nucleation response seen for deposition on solid electrode surfaces (see Fig. 25) [50]. The quantitative analysis of this

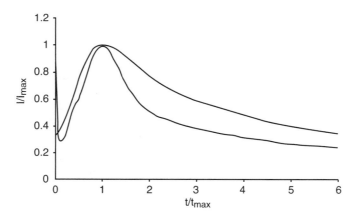

Figure 25. Normalized current–time response for palladium deposition (at constant applied potential) at the water–DCE interface.

process is in its infancy, and there are conflicting conclusions about the applicability of classical nucleation models to the ITIES [196]. One report has suggested that the nucleation process depends on a random number of atoms forming a critical nucleus, implying that no consistent nucleation parameters can be extracted from the experimental data [197].

Macroscopic thermodynamic models have been presented to give the Gibbs energy profile of the particle, as a function of its position along the interfacial normal [198, 199]. If the aqueous phase $(z < 0)$ is treated as the reference state and line tension at the three-phase boundary is neglected, then

$$\Delta G(z) = 0 \quad \text{for } -z > r \tag{58}$$

$$\Delta G(z) = 2\pi r(r+z)(\gamma_{org-m} - \gamma_{org-m}) - \pi(r^2 - z^2)\gamma_{org-w} \quad \text{for } -r \leq z \leq r \tag{59}$$

$$\Delta G(z) = 4\pi r^2(\gamma_{org-m} - \gamma_{w-m}) \quad \text{for } z > r \tag{60}$$

where the γ terms introduced above represent the tensions of the organic–metal $(org-m)$ and aqueous–metal $(w-m)$ interfaces, respectively. Combination with the Young–Dupré equation [1] provides the interrelation between the interfacial tension terms; thus Eqs. (59) and (60) become

$$\Delta G(z) = -2\pi r(r+z)\gamma_{org-w} \cos\phi_{org-m} - \pi(r^2 - z^2)\gamma_{org-w} \quad \text{for } -r \leq z \leq r \tag{61}$$

$$\Delta G(z) = -4\pi r^2 \gamma_{org-w} \cos\phi_{org-m} \quad \text{for } z > r \tag{62}$$

where ϕ is the contact angle of the organic phase on the metal surface. This analysis shows that the particle Gibbs energy profile can pass through a minimum when z is zero; that is, particle adsorption occurs. This phenomenon is manifest in the adsorption of the particles electrodeposited in situ at the ITIES [191–197], but can also be used to assemble preformed particles at this interface. Nanoparticle assembly at organic–water interfaces can be controlled via the particle surface charge density [200] or through specific chemical interactions mediated by the addition of surfactants [201]. More external control over assembly can be induced through the voltage-dependent, reversible assembly of stabilized gold particles at the ITIES, which has been reported recently [202]: Essentially the particle partition equilibrium is controlled through the Galvani potential (Eq. (1); see Fig. 26). A considerable body of (nonelectrochemical) literature has appeared in recent years describing size-dependent segregation phenomena of nanoparticles [203] at organic–aqueous interface and the spontaneous ordering of micron-scale particles at these interfaces [204, 205]. The physical origin of these effects is not well understood at present; however, the introduction of electrochemical methodology, with the degree of freedom induced by an externally variable potential, should shed some light on the interactions underlying these intriguing phenomena.

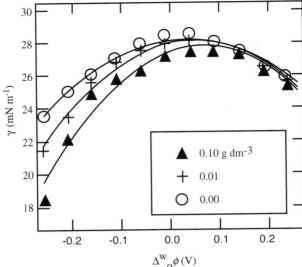

Figure 26. Electrocapillary curves obtained at the water–DCE interface in the presence of varying amounts surface active Au nanoparticles. (Reprinted with permission from Ref. 202; copyright 2004, American Chemical Society.)

VI.   CONCLUSIONS AND PERSPECTIVES

The transfer of charge at the ITIES is an intrinsically interesting phenomenon and, as many articles on the topic have noted, it has wider ramifications for charge transport within membranes and for two-phase chemical syntheses. The response of electrochemical cells based on the ITIES can give some information on the structure of the ITIES itself, but many of the conclusions drawn about the molecular level structure of the interface were misleading. The application of accurate MD methods to this interface has addressed some of the structural questions, which now provides a better foundation from which the dynamic phenomena can be interpreted. The difficulties associated with the inclusion of multiple charges and polarization terms into the MD calculations reported so far means that results from MD work must be treated with some care. However, the existence, and importance of, capillary waves on the microscopic scale has been an important finding from the MD work applied to this interface to date. Some progress has been made with regard to the quantitative description of the electrolyte double layer, moving beyond the simple Verwey–Niessen model. Measurements of the rate of charge transfer have given more consistent results for ET, compared to IT; in the latter case, neither the experimental nor computational work has yielded a consensus on the rate of IT (or the magnitude of the rate constant). Theory and experiment have reached agreement on the role of hydration: Some water molecules are co-extracted with hydrophilic ions, but many basic features of interfacial IT are not understood at the microscopic level. Perhaps one way forward may be to concentrate on the dynamics of the facilitated (assisted) ion transfer process, where the role of solvent (and/or supporting electrolyte) should be minimized. Spectroscopic methods, notably SFG, under potential control are required to improve understanding of ITIES structure. Similarly, the "dipolar width" approach [45–47] could be applied to the ET process by adsorption of species with donor and acceptor groups separated by rigid spacers, to understand the distance dependence (and hence potential distribution) of the interfacial ET process. EPR spectroscopy has been applied to ET at the ITIES [206], but no reports of the application of pulsed EPR to probe the ET process exist. The adsorption of lipid species at the ITIES has not been considered explicitly in this chapter [207]: Particle adsorption at the ITIES is a related phenomenon, which has emerged recently, although there is a link to particle-stabilized emulsion formation [205]. The ITIES is an attractive interface for the study of nucleation phenomena, given the lack of fixed defect sites and molecularly sharp nature. External control over the potential difference allows control over the supersaturation of the nucleating phase; hence quantitative investigations of phase formation and growth (metals or polymers) at this interface would aid the microscopic understanding of nucleation. Some initial studies toward this end have been reported [191, 196, 197], but much more work toward this goal is required. Some very interesting applications, exploiting double-layer properties of the ITIES to control electro-wetting and liquid injection, have recently been reported [208, 209]. These works highlight the overall need for an improved, molecular-level understanding of the structure and dynamics of this important class of interface.

Acknowledgments

The author gratefully acknowledges financial support from the U.K. Engineering and Physical Sciences Research Council (EP/C509773/1 and EP/E000665/1) and the donors of the American Chemical Society Petroleum Research Fund (PRF# 41667-AC-5).

References

1. J. T. Davies and E. K. Rideal, *Interfacial Phenomena*, Academic Press, London, 1961.

2. V. Mareček, Z. Samec, and J. Koryta, *Adv. Coll. Int. Sci.* **29**, 1 (1988).

3. H. H. Girault, in *Modern Aspects of Electrochemistry*, Vol. 25, J. O. Bockris, ed., Plenum, New York, 1993, pp. 1–62.

4. H. H. J. Girault and D.J. Schiffrin, in *Electroanalytical Chemistry*, Vol. 15, A. J. Bard, ed., Marcel Dekker, New York, 1989, pp. 1–141.

5. A. G. Volkov and D. W. Deamer, eds., *Liquid–Liquid Interfaces: Theory and Methods*, CRC Press, Boca Raton, FL, 1996.

6. J. Koryta, *Electrochim. Acta* **33**, 189 (1988).

7. H. H. Girault, *Electrochim. Acta* **32**, 383 (1987).

8. M. Senda, T. Kakiuchi, and T. Osakai, *Electrochim. Acta* **36**, 253 (1991).

9. F. Reymond, D. J. Fermín, H. J. Lee, and H. H. Girault, *Electrochim. Acta* **45**, 2647 (2000).

10. B. Liu and M. V. Mirkin, *Electroanalysis* **12**, 1433 (2000).

11. S. X. Guo, P. R. Unwin, A. L. Whitworth, and J. Zhang, *Prog. React. Kin. Mech.* **29**, 43 (2004).

12. P. Vanýsek, *Anal. Chem.* **62**, 827A (1990).

13. Z. Samec, *Chem. Rev.* **88**, 617 (1988).

14. I. Benjamin, *Annu. Rev. Phys. Chem.* **48**, 407 (1997).

15. M. Gros, S. Gromb, and C. Gavach, *J. Electroanal. Chem.* **89**, 29 (1978).

16. Z. Samec, V. Mareček, J. Koryta, and M. W. Khalil, *J. Electroanal. Chem.* **83**, 393 (1977).

17. L. Q Hung, *J. Electroanal. Chem.* **149**, 1 (1983).

18. F. M. Karpfen and J. E. B. Randles, *Trans. Farad. Soc.* **49**, 823 (1953).

19. P. Sun, Z. Q. Zhang, Z. Gao, and Y. H. Shao, *Angew. Chem. Int. Ed.* **41**, 3445 (2002).

20. J. S. Rowlinson and B. Widom, *Molecular Theory of Capillarity*, Clarendon Press, Oxford, 1989.

21. K. J. Schweighofer and I. Benjamin, *J. Electroanal. Chem.* **391**, 1 (1995).

22. L. I. Daikhin, A. A. Kornyshev, and M. Urbakh, *J. Electroanal. Chem.* **483**, 68 (2000).

23. I. Benjamin, *J. Chem. Phys.* **97**, 1432 (1992).

24. I. Benjamin, *J. Phys. Chem. B* **109**, 13711 (2005).

25. S. Frank and W. Schmickler, *J. Electroanal. Chem.* **564**, 239 (2004).

26. D. J. Henderson and W. Schmickler, *J. Chem. Soc. Farad. Trans.* **92**, 3839 (1996).

27. D. M. Mitrinovic, A. M. Tikhonov, M. Li, Z.Q. Huang, and M. L. Schlossman, *Phys. Rev. Lett.* **85**, 582 (2000).

28. J. Bowers, A. Zarbakhsh, J. R. P. Webster, L. R. Hutchings, and R. W. Richards, *Langmuir* **17**, 140 (2001).

29. A. Zarbakhsh, J. Bowers, and J. R. P. Webster, *Langmuir* **21**, 11596 (2005).

30. J. Strutwolf, A. L. Barker, M. Gonsalves, D. J. Caruana, P. R. Unwin, D. E. Williams, and J. R. P. Webster, *J. Electroanal. Chem.* **483**, 163 (2000).

31. I. Tsuyumoto, N. Noguchi, T. Kitamori, and T. Sawada, *J. Phys. Chem. B* **102**, 2684 (1998).

32. C. Wei, A. J Bard, and M. V. Mirkin, *J. Phys. Chem. B* **99**, 16033 (1995).

33. S. Ishizaka, K. Nakatani, S. Habuchi, and N. Kitamura, *Anal. Chem.* **71**, 419 (1999).

34. S. Ishizaka, S. Habuchi, H. B. Kim, and N. Kitamura, *Anal. Chem.* **71**, 3382 (1999).

35. S. Ishizaka, H. B. Kim, and N. Kitamura, *Anal. Chem.* **73**, 2421 (2001).

36. L. F. Scatena, M. G. Brown and G. L. Richmond, *Science* **292**, 908 (2001).

37. M. G. Brown, D. S. Walker, E. A. Raymond, and G. L. Richmond, *J. Phys. Chem. B* **107**, 237 (2003).

38. D. K. Hore, D. S. Walker, and G. L. Richmond, *J. Am. Chem. Soc.* **129**, 752 (2007).

39. D. S. Walker, M. G. Brown, C. L. McFearin, and G. L. Richmond, *J. Phys. Chem. B* **108**, 2111 (2004).

40. D. S. Walker, F. G. Moore, and G. L. Richmond, *J. Phys. Chem. C* **111**, 6103 (2007).

41. G. Luo, S. Malkova, S. Venkatesh Pingali, D. G. Schultz, B. Lin, M. Meron, T. J. Braber, J. Gebhardt, P. Vanýsek, and M. L. Schlossman, *Farad. Disc.* **129**, 23 (2005).

42. G. M. Luo, S. Malkova, S. V. Pingali, D. G. Schultz, B. H. Lin, M. Meron, I. Benjamin, P. Vanýsek, and M. L. Schlossman, *J. Phys. Chem. B* **110**, 4527 (2006).

43. P. A. Fernandes, M. N. D. S. Cordeiro, and J. A. N. F. Gomes, *J. Phys. Chem. B* **103**, 8930 (1999).

44. H. Wang, E. Borguet, and K. B. Eisenthal, *J. Phys. Chem. B* **102**, 4927 (1998).

45. W. H. Steel and R. A. Walker, *Nature* **424**, 296 (2003).

46. W. H. Steel, Y. Y. Lau, C. L. Beildeck, and R. A. Walker, *J. Phys. Chem. B* **108**, 13370 (2004).

47. W. H. Steel, C. L. Beildeck, and R. A. Walker, *J. Phys. Chem. B* **108**, 16107 (2004).

48. I. Benjamin, *Chem. Phys. Lett.* **393**, 453 (2004).

49. A. Watts and T. J. VanderNoot, Chapter 5 of Ref 5.

50. W. Schmickler, *Interfacial Electrochemistry*, Oxford University Press, Oxford, 1996.

51. T. Kakiuchi and M. Senda, *Bull. Chem. Soc. Jpn.* **56**, 1322 (1983).

52. T. Kakiuchi and M. Senda, *Bull. Chem. Soc. Jpn.* **56**, 1753 (1983).

53. T. Kakiuchi and M. Senda, *Bull. Chem. Soc. Jpn.* **56**, 2912 (1983).

54. H. H. Girault and D. J. Schiffrin, *J. Electroanal. Chem.* **170**, 127 (1984).

55. H. H. Girault and D. J. Schiffrin, *J. Electroanal. Chem.* **150**, 43 (1983).

56. C. M. Pereira, A. Martins, M. Rocha, C. J. Silva, and F. Silva, *J. Chem. Soc. Farad. Trans.* **90**, 143 (1994).

57. Y. Cheng, V. J. Cunnane, D. J. Schiffrin, L. Murtomäki, and K. Kontturi, *J. Chem. Soc. Farad. Trans.* **87**, 107 (1991).

58. Z. Samec, A. Trojanek, and J. Langmaier, *J. Electroanal. Chem.* **444**, 1 (1998).

59. C. M. Pereira, W. Schmickler, A. F. Silva, and M. J. Sousa, *Chem. Phys. Lett.* **268**, 13 (1997).

60. G. M. Torrie and J. P. Valleau, *J. Electroanal. Chem.* **206**, 69 (1986).

61. S. L. Carnie and G. M. Torrie, *Adv. Chem. Phys.* **56**, 141 (1984).

62. Q. Cui, G. Zhu, and E. Wang, *J. Electroanal. Chem.* **383**, 7 (1995).

63. L. I. Daikhin, A. A. Kornyshev, and M. Urbakh, *Electrochim. Acta* **45**, 685 (1999).

64. L. I. Daikhin, A. A. Kornyshev, and M. Urbakh, *J. Electroanal. Chem.* **500**, 461 (2001).

65. L. I. Daikhin, and M. Urbakh, *J. Electroanal. Chem.* **560**, 59 (2003).

66. T. Huber, O. Pecina, and W. Schmickler, *J. Electroanal. Chem.* **467**, 203 (1999).

67. S. Frank and W. Schmickler, *J. Electroanal. Chem.* **483**, 18 (2000).

68. C. W. Monroe, A. A. Kornyshev, and M. Urbakh, *J. Electroanal. Chem.* **582**, 28 (2005).

69. H. H. Girault, D. J. Schiffrin, and B. D. V. Smith *J. Colloid Interface Sci.* **101**, 257 (1984).

70. Y. A. Shchipunov and A. F. Kolpakov *Adv. Coll. Int. Sci.* **35**, 31 (1991).

71. G. M. Luo, S. Malkova, J. Yoon, D. G. Schultz, B. H. Lin, M. Meron, I. Benjamin, P. Vanýsek, and M. L. Schlossman, *Science* **311**, 216 (2006).

72. G. M. Luo, S. Malkova, J. Yoon, D. G. Schultz, B. H. Lin, M. Meron, I. Benjamin, P. Vanýsek, and M. L. Schlossman, *J. Electroanal. Chem.* **593**, 142 (2006).

73. R. D. Webster and D. Beaglehole, *Phys. Chem. Chem. Phys.* **2**, 5660 (2000).

74. J. C. Conboy and G. L. Richmond, *J. Phys. Chem. B* **101**, 983 (1997).

75. R. R. Naujok, D. A. Higgins, D. G. Hanken, and R. M. Corn, *J. Chem. Soc. Farad. Trans.* **91**, 1411 (1995).

76. A. Trojanek, P. Krtil, and Z. Samec, *Electrochem. Commun.* **3**, 613 (2001).

77. R. A. Marcus and N. Sutin, *Biochim. Biophys. Acta* **811**, 265 (1985).

78. Z. Samec, Chapter 8 of Ref 5.

79. F. O. Laforge, P. Sun, and M. V. Mirkin, *J. Am. Chem. Soc.* **128**, 15019 (2006).

80. T. Shioya, S. Nishizawa, and N. Teramae, *J. Am. Chem. Soc.* **120**, 11534 (1998).

81. R. A. W. Dryfe, S. S. Hill, A. P. Davis, J. B. Joos, and E. P. L. Roberts, *Org. Biomol. Chem.* **2**, 2716 (2004).

82. Z. Samec, *Pure Appl. Chem.* **76**, 2147 (2004).

83. C. Gavach, B. d'Eponoux, and F. Henry, *J. Electroanal. Chem.* **64**, 107 (1975).

84. R. P. Buck and W. E. Bronner, *J. Electroanal. Chem.* **197**, 179 (1986).

85. W. E. Bronner, O. R. Melroy, and R. P. Buck, *J. Electroanal. Chem.* **162**, 263 (1984).

86. Z. Samec, V. Mareček, and J. Weber, *J. Electroanal. Chem.* **100**, 841 (1979).

87. Y. Shao and H. H. Girault, *J. Electroanal. Chem.* **282**, 59 (1990).

88. T. Osakai, T. Kakutani, and M. Senda, *Bull. Chem. Soc. Jpn.* **58**, 2626 (1985).

89. T. Wandlowski, V. Mareček, K. Holub, and Z. Samec, *J. Phys. Chem.* **93**, 8204 (1989).

90. Z. Samec, *Electrochim. Acta* **44**, 85 (1998).

91. T. Kakiuchi, Y. Takasu, and M. Senda, *Anal. Chem.* **64**, 3096 (1992).

92. T. Kakiuchi and Y. Takasu, *Anal. Chem.* **66**, 1853 (1994).

93. T. Kakiuchi and Y. Takasu, *J. Electroanal. Chem.* **381**, 5 (1995).

94. T. Kakiuchi and Y. Takasu, *J. Phys. Chem. B* **101**, 5963 (1997).

95. N. Nishi, K. Izawa, M. Yamamoto, and T. Kakiuchi, *J. Phys. Chem. B* **105**, 8162 (2001).

96. Z. F. Ding, F. Reymond, P. Baumgartner, D. J. Fermín, P. F. Brevet, P. A. Carrupt, and H. H. Girault, *Electrochim. Acta* **44**, 3 (1998).

97. D. J. Fermín, Z. F. Ding, P. F. Brevet, and H. H. Girault, *J. Electroanal. Chem.* **447**, 125 (1998).

98. G. Taylor and H. H. J. Girault, *J. Electroanal. Chem.* **208**, 179 (1986).

99. J. A. Campbell and H. H. Girault, *J. Electroanal. Chem.* **266**, 465 (1989).

100. A. J. Bard and L. R. Faulkner, *Electrochemical Methods: Fundamentals and Applications*, 2nd ed., Wiley, New York, 2001.

101. P. D. Beattie, A. Delay, and H. H. Girault, *Electrochim. Acta* **40**, 2961 (1995).

102. P. D. Beattie, A. Delay, and H. H. Girault, *J. Electroanal. Chem.* **380**, 167 (1995).

103. B. Quinn, R. Lahtinen, and K. Kontturi, *J. Electroanal. Chem.* **436**, 285 (1997).

104. Y. H. Shao and M. V. Mirkin, *J. Am. Chem. Soc.* **119**, 8103 (1997).

105. M. V. Mirkin and A. J. Bard, *Anal. Chem.* **64**, 2293 (1992).

106. P. Jing, M. Q. Zhang, H. Hu, X. D. Xu, Z. W. Liang, B. Li, L. Shen, S. B. Xie, C. M. Pereira, and Y. H. Shao, *Angew. Chem. Int. Ed.* **45**, 6861 (2006).

107. C. X. Cai, Y. H. Tong, and M. V. Mirkin, *J. Phys. Chem. B*, **108**, 17872 (2004).

108. Y. Yuan and Y. H. Shao, *J. Phys. Chem. B* **106**, 7809 (2002).

109. A. L. Barker, M. Gonsalves, J. V. Macpherson, C. J. Slevin, and P. R. Unwin, *Anal. Chim. Acta* **385**, 223 (1999).

110. P. Sun, F. O. Laforge, and M. V. Mirkin, *Phys. Chem. Chem. Phys.* **9**, 802 (2007).

111. Y. H. Shao and M. V. Mirkin, *J. Electroanal. Chem.* **439**, 137 (1997).

112. Y. H. Shao and M. V. Mirkin, *J. Phys. Chem. B* **102**, 9915 (1998).

113. A. L. Barker and P. R. Unwin, *J. Phys. Chem. B* **105**, 12019 (2001).

114. C. J. Slevin and P. R. Unwin, *Langmuir* **15**, 7361 (1999).

115. J. Zhang and P. R. Unwin, *Langmuir* **18**, 2313 (2002).

116. B. Kralj and R. A. W. Dryfe, *J. Phys. Chem. B* **106**, 6732 (2002).

117. B. Kralj and R. A. W. Dryfe, *J. Electroanal. Chem.* **560**, 127 (2003).

118. C. J. Slevin, J. V. Macpherson, and P. R. Unwin, *J. Phys. Chem. B* **101**, 10851 (1997).

119. A. L. Barker, J. V. Macpherson, C. J. Slevin, and P. R. Unwin, *J. Phys. Chem. B* **102**, 1586 (1998).

120. K. Nakatani, T. Uchida, N. Kitamura, and H. Masuhara, *J. Electroanal. Chem.* **373**, 383 (1994).

121. K. Nakatani, M. Sudo, and N. Kitamura, *J. Phys. Chem. B* **102**, 2908 (1998).

122. K. Nakatani, M. Wakabayashi, K. Chikama, and N. Kitamura, *J. Phys. Chem.* **100**, 6749 (1996).

123. N. Terui, K. Nakatani, and N. Kitamura, *J. Electroanal. Chem.* **494**, 41 (2000).

124. V. Mirčeski, F. Quentel, M. L'Her, and A. Pondaven, *Electrochem. Comm.* **7**, 1122 (2005).

125. R. Gulaboski, Mirčeski, C. M. Pereira, M. N. D. S. Cordeiro, A. F. Silva, F. Quentel, M. L'Her, and M. Lovric, *Langmuir* **22**, 3404 (2006).

126. F. Quentel, V. Mirčeski, and M. L'Her, *Anal. Chem.* **77**, 1940 (2005).

127. R. Ferrigno and H. H. Girault, *J. Electroanal. Chem.* **496**, 131 (2001).

128. R. A. Marcus, *J. Chem. Phys.* **113**, 1618 (2000).

129. T. Kakiuchi, *J. Electroanal. Chem.* **322**, 55 (1992).

130. K. J. Schweighofer and I. Benjamin, *J. Phys. Chem.* **99**, 9974 (1995).

131. A. A. Kornyshev, A. M. Kuznetsov, M. Urbakh, *J. Chem. Phys.* **117**, 6766 (2002).

132. C. G. Verdes, M. Urbakh, and A. A. Kornyshev, *Electrochem. Commun.* **6**, 693 (2004).

133. I. Benjamin, *Science* **261**, 1558 (1993).

134. T. Wandlowski, V. Mareček, and Z. Samec, *Electrochim. Acta* **35**, 1173 (1990).

135. K. Schweighofer and I. Benjamin, *J. Phys. Chem. B* **103**, 10274 (1999).

136. T. Osakai, A. Ogata, and K. Ebina, *J. Phys. Chem. B* **101**, 8341 (1997).

137. T. Osakai and K. Ebina, *J. Phys. Chem. B* **102**, 5691 (1998).

138. P. A. Fernandes, M. N. D. S. Cordeiro, and J. A. N. F. Gomes, *J. Phys. Chem. B* **104**, 2278 (2000).

139. Y. F. Cheng and D. J. Schiffrin, *J. Electroanal. Chem.* **409**, 9 (1996).

140. P. Sun, F. O. Laforge, and M. V. Mirkin, *J. Am. Chem. Soc.* **127**, 8596 (2005).

141. P. Sun, F. O. Laforge, and M. V. Mirkin, *J. Am. Chem. Soc.* **129**, 12410 (2007).

142. Z. Samec, V. Mareček, and J. Weber, *J. Electroanal. Chem.* **103**, 11 (1979).

143. M. Guainazzi, G. Silvestri, and G. Serravalle, *J. Chem. Soc. Chem. Commun.* 200 (1975).

144. M. Faraday, *Philos. Trans. I*, **147**, 145 (1857).

145. M. Platt, R. A. W. Dryfe, and E. P. L. Roberts, *Electrochim. Acta*, **48**, 3037 (2003).

146. V. J. Cunnane, D. J. Schiffrin, C. Beltran, and G. Geblewicz, *J. Electroanal. Chem.* **247**, 203 (1988).

147. T. Solomon and A.J. Bard, *J. Phys. Chem.* **99**, 17487 (1995).

148. M. Tsionsky, A. J. Bard, and M. V. Mirkin, *J. Phys. Chem.* **100**, 17881 (1996).

149. V. J. Cunnane, G. Geblewicz, and D. J. Schiffrin, *Electrochim. Acta*, **40**, 3005 (1995).

150. B. Quinn, R. Lahtinen, L. Murtomäki, and K. Kontturi, *Electrochim. Acta*, **44**, 47 (1998).

151. R. A. Marcus, *J. Phys. Chem.* **94**, 4152 (1990).

152. R. A. Marcus, *J. Phys. Chem.* **99**, 5742 (1995).

153. R. A. Marcus, *J. Chem. Phys.* **43**, 679 (1965)

154. R. A. Marcus, *J. Phys. Chem.* **94**, 1050 (1990).

155. Z. F. Ding, B. M. Quinn, and A. J. Bard, *J. Phys. Chem. B* **105**, 6367 (2001).

156. N. Eugster, D. J. Fermín, and H. H. Girault, *J. Phys. Chem. B* **106**, 3428 (2002).

157. R. A. Marcus, *J. Phys. Chem.* **95**, 2010 (1991).

158. W. Schmickler, *J. Electroanal. Chem.* **428**, 123 (1997).

159. Y. F. Cheng and D. J. Schiffrin, *J. Chem. Soc. Farad. Trans.* **89**, 199 (1993).

160. Y. F. Cheng and D. J. Schiffrin, *J. Chem. Soc. Farad. Trans.* **90**, 2517 (1994).

161. M. Tsionsky, A. J. Bard, and M. V. Mirkin, *J. Am. Chem. Soc.* **119**, 10785 (1997).

162. C. E. D. Chidsey, *Science*, **251**, 919 (1991).

163. A. L. Barker, P. R. Unwin, S. Amemiya, J. F. Zhou, and A. J. Bard, *J. Phys. Chem. B* **103**, 7260 (1999).

164. A. L. Barker, P. R. Unwin, and J. Zhang, *Electrochem. Commun.* **3**, 372 (2001).

165. J. Zhang and P. R. Unwin, *Phys. Chem. Chem. Phys.* **4**, 3820 (2002).

166. Z. Q. Zhang, Y. Yuan, P. Sun, B. Su, J. D. Guo, Y. H. Shao, and H. H. Girault, *J. Phys. Chem. B* **106**, 6713 (2002).

167. Z. Samec, A. R. Brown, L. J. Yellowlees, and H. H. Girault, *J. Electroanal. Chem.* **288**, 245 (1990).

168. O. Dvorak, A. H. de Armond, and M. K. de Armond, *Langmuir* **8**, 955 (1992).

169. R. A. W. Dryfe, Z. F. Ding, R. G. Wellington, P. F. Brevet, A. M. Kuznetzov, and H. H. Girault, *J. Phys. Chem. A* **101**, 2519 (1997).

170. D. J. Fermín, H. D. Duong, Z. F. Ding, P. F. Brevet, and H. H. Girault, *J. Amer. Chem. Soc.* **121**, 10203 (1999).

171. N. Eugster, D. J. Fermín, and H. H. Girault, *J. Am. Chem. Soc.* **125**, 4862 (2003).

172. D. G. Georganopoulou, M. V. Mirkin, and R. W. Murray, *Nano Lett.* **4**, 1763 (2004).

173. B. Su, N. Eugster, and H. H. Girault, *J. Am. Chem. Soc.* **127**, 10760 (2005).

174. C. N. Shi and F. C. Anson, *J. Phys. Chem. B* **102**, 9850 (1998).

175. C. N. Shi and F. C. Anson, *J. Phys. Chem. B* **103**, 6283 (1999).

176. C. N. Shi and F. C. Anson, *J. Phys. Chem. B* **105**, 8963 (2001).

177. B. Liu and M. V. Mirkin, *J. Am. Chem. Soc.* **121**, 8352 (1999).

178. J. Zhang, A. L. Barker, and P. R. Unwin, *J. Electroanal. Chem.* **483**, 95 (2000).

179. C. X. Cai and M. V. Mirkin, *J. Am. Chem. Soc.* **128**, 171 (2006).

180. T. Kakiuchi, *Electrochim. Acta* **40**, 2999 (1995).

181. Y. B. Zu, F. R. F. Fan, and A. J. Bard, *J. Phys. Chem. B* **103**, 6272 (1999).

182. E. A. McArthur and K. B. Eisenthal, *J. Am. Chem. Soc.* **128**, 1068 (2006).

183. R. M. Penner, *J. Phys. Chem. B*, **106**, 3339 (2002).

184. M. Brust, M. Walker, D. J. Schiffrin, and R. Whyman, *J. Chem. Soc. Chem. Commun.* 801 (1994).

185. C. J. Murphy and N. R. Jana, *Adv. Mater.* **14**, 80 (2002).

186. C. N. R. Rao, G. U. Kulkarni, P. J. Thomas, V. V. Agrawal, and P. Saravanan, *J. Phys. Chem. B*, **107**, 7391 (2003).

187. J. K. Sakata, A. D. Dwoskin, J. L. Vigorita, and E. M. Spain, *J. Phys. Chem. B* **109**, 138 (2005).

188. Y. Cheng and D. J. Schiffrin, *J. Chem. Soc. Farad. Trans.* **92**, 3865 (1996).

189. V. J. Cunnane, and U. Evans, *Chem. Commun.* 2163 (1998).

190. U. Evans-Kennedy, J. Clohessy, and V. J. Cunnane, *Macromolecules* **37**, 3630 (2004).

191. C. Johans, R. Lahtinen, K. Kontturi, and D. J. Schiffrin, *J. Electroanal. Chem.* **488**, 99 (2000).

192. C. Johans, K. Kontturi, and D. J. Schiffrin, *J. Electroanal. Chem.* **526**, 29 (2002).

193. J. D. Guo, T. Tokimoto, R. Othman, and P. R. Unwin, *Electrochem. Commun.* **5**, 1005 (2003).

194. M. Platt, E. P. L. Roberts, and R. A. W. Dryfe, *Electrochim. Acta* **49**, 3937 (2004).

195. M. Platt and R. A. W. Dryfe, *Phys. Chem. Chem. Phys.* **7**, 1807 (2005).

196. M. Platt and R. A. W. Dryfe, *J. Electroanal. Chem.* **599**, 323 (2007).

197. A. Trojanek, J. Langmaier, and Z. Samec, *J. Electroanal. Chem.* **599**, 160 (2007).

198. R. Aveyard and J. H. Clint, *J. Chem. Soc. Faraday Trans.* **92**, 85 (1996)

199. C. Johans, P. Liljeroth, and K. Kontturi, *Phys. Chem. Chem. Phys.* **4**, 1067 (2002).

200. F. Reincke, S. G. Hickey, W. K. Kegel, and D. Vanmaekelbergh, *Angew Chem. Int. Ed.* **43**, 458 (2004).

201. D. Yogev and S. Efrima, *J. Phys. Chem.* **92**, 5754 (1988).

202. B. Su, J. P. Abid, D. J. Fermín, H. H. Girault, H. Hoffmannova, P. Krtil, and Z. Samec, *J. Am. Chem. Soc.* **126**, 915 (2004).

203. Y. Lin, H. Skaff, T. Emrick, A. D. Dinsmore, and T. P. Russell, *Science* **299**, 226 (2003).

204. M. G. Nikolaides, A. R. Bausch, M. F. Hsu, A. D. Dinsmore, M. P. Brenner, C. Gay, and D. A. Weitz, *Nature* **420**, 299 (2002).

205. B. P. Binks, *Curr. Opin. Coll. Int. Sci.* **7**, 21 (2002).

206. R. D. Webster, R. A. W. Dryfe, B. A. Coles, and R. G. Compton, *Anal. Chem.* **70**, 792 (1998).

207. R. M. Allen and D. E. Williams, *Farad. Disc.* **104**, 281 (1996).

208. F. O. Laforge, J. Carpino, S. A. Rotenberg, and M. V. Mirkin, *Proc. Natl. Acad. Sci. USA* **104**, 11895 (2007).

209. C. W. Monroe, L. I. Daikhin, M. Urbakh, and A. A. Kornyshev, *Phys. Rev. Lett.* **97**, 136102 (2006).

THE PHYSICS OF ULTRATHIN SOLID–FLUID–SOLID FILMS: FROM SURFACE INSTABILITIES TO ISOLATED POCKETS OF FLUID

GAVIN A. BUXTON AND NIGEL CLARKE

Department of Chemistry, University of Durham, Durham, DH1 3LE, United Kingdom

CONTENTS

I. INTRODUCTION

This chapter describes the physical processes that occur in free-standing ultrathin trilayer films consisting of a fluid layer sandwiched in between two solid layers. We have recently referred to these films as "Dutcher films"; these are named

Advances in Chemical Physics, Volume 141, edited by Stuart A. Rice

after the group of John Dutcher, whose experiments motivated these computational studies [1–3]. An interesting instability arises in these films if the total film thickness is thin enough that van der Waals forces acting across it are sufficiently strong to deform the solid layers. In other words, the van der Waals forces can favor the formation and growth of instabilities (resulting in the dewetting of fluid films on substrates or film breakup in free-standing fluid films). In contrast, the elasticity of the solid capping layers resist the bending deformations required for instability growth. However, for very thin films the van der Waals forces are sufficiently strong that the solid layer begins to deform. The surface instabilities grow with time because of the van der Waals forces, but are restricted by both the deformation energy of the solid layer and the transport kinetics of the fluid layer. In particular, small-wavelength deformations in the solid layer are more energetically unfavorable, while large-wavelength deformations would require the transport of fluid material over kinetically-prohibitive distances. Hence, the competition between energetics and kinetics selects a characteristic wavelength for undulation growth.

To further explain this process we refer to Fig. 1, which shows a schematic of instability growth. Some regions of the film are thicker than others, and this difference is further driven by the fact that the van der Waals forces become larger as the thickness decreases. The van der Waals forces (numbered 1) are depicted in Fig. 1 as the arrows acting from the outer solid surface toward the center of the film (here we are concerned with polymer fluid and polymer solids, where the properties are assumed to be similar). The van der Waals forces exert

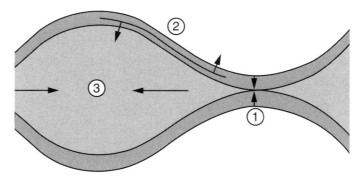

Figure 1. Schematic of ultrathin solid–fluid–solid film and undulation growth. (1) The van der Waals forces cause undulations to grow, locally minimizing the distance between the solid capping layers. (2) The elasticity of the solid capping layers resist bending and energetically oppose undulation growth (more for smaller wavelengths, which involve more bending). (3) The transport of fluid from regions where the film is thinning to regions where the film is thickening kinetically opposes undulation growth (more for larger wavelengths, which involve the transport of fluid over further distances). The competition between energetics and kinetics results in the emergence of a characteristic wavelength of undulation growth.

a pressure on the fluid and result in greater variations in the height of the film surface (bending the solid capping layers). The bending of the solid capping layers, in response, also results in a pressure being applied to the fluid because the elastic solid would energetically prefer to be in an undeformed state (numbered 2 in Fig. 1). The pressures from both the deformed elastic solid and van der Waals forces, or rather the gradients in pressure, result in fluid flow. This is indicated by the arrows near the number 3 in Fig. 1. Here we report on how we can computationally capture the physics of these films and, more interestingly, simulate the dynamics of the system as the top and bottom layers eventually meet. At this point, fluid is expelled from regions of the film and corraled into isolated pockets of fluid. In other words, the system prefers to increase the contact between the top and bottom solid layers and corral the fluid into smaller and smaller regions of the film.

In this chapter we show how we can model the complete evolution of these films, from the growth of initial instabilities to the final formation of isolated pockets of fluid. Furthermore, we discuss how we might be able to manipulate the instabilities or final morphology (through, for example, applying an external strain field or introducing elastic heterogeneity) and tailor the undulations to meet specific technological needs.

II. BACKGROUND

A. Ultrathin Solid–Fluid–Solid Films

The trilayer films of interest in this study [1–5] are free-standing ultrathin solid–fluid–solid films that are thin enough that van der Waals forces acting across the film can deform the solid layers. These films are typically on the order of 100 nm thick. Because these films are so thin, the van der Waals forces in these systems favor a reduction in film thickness. In other words, the energy decrease in making the film thinner in one region is greater than the energy increase as a result of the film becoming thicker in another region (note that mass is conserved and if the film becomes thinner in one region, then the film must become thicker in another region). Therefore, if the film has initial height variations (as a consequence of thermal noise, say), then these variations will grow and the film is inherently unstable. If the film was entirely fluid, then the fluctuations would progress until holes appeared in the film and eventually the entire film would break through the coalescence of these holes [6]. In the films of interest here, however, the fluid layer in the middle is capped on either side by solid layers, which maintain the mechanical integrity of the film. The fluid can still flow in the center of the film and, as the van der Waals forces favor reduced film thicknesses, the system will eventually evolve to the point of contact between the top and bottom solid layers.

As already mentioned, the elastic properties of the solid capping layers resist bending deformations and small-wavelength undulations (or wrinkles) are more energetically unfavorable than larger-wavelength undulations. This is simply because smaller-wavelength undulations involve more bending of the film than do higher-wavelength undulations. Furthermore, recall that the fluid is incompressible and, therefore, a reduction in film thickness in one region of the film must correspond with an increase in film thickness in neighboring regions. The transport of fluid from regions of decreasing thickness to regions of increasing thickness kinetically favors smaller-wavelength undulations. In other words, the fluid can travel shorter distances in less time and, therefore, it is faster for the system to grow smaller-wavelength undulations. This competition between energetics and kinetics results in the emergence of a characteristic length-scale for undulation growth. The characteristic wavelength for undulation growth is large enough that the driving force is high (van der Waals forces minus restoring elastic forces) and small enough that kinetically the fluid can flow (in response to the driving force) in a reasonable time.

The solid capping layers become increasingly deformed, as the amplitude of the surface undulations increase, until the top and bottom solid layers meet. The film cannot become any thinner because of the mechanical integrity offered by the solid layers. At this point the central fluid layer is corraled into isolated pockets separated by increasingly large areas of the film where the top and bottom layers are in contact. It is the evolution of these films, along with the manipulation of their structures to meet technological needs, which is the focus of this chapter. In particular, it would be desirable to know how the initial instabilities grow and how we might exhibit some control over these instabilities. Furthermore, the creation of isolated pockets of fluid (hundreds of nanometers in height and microns in diameter) are interesting enough, but it would also be desirable to manipulate these pockets of fluid and perhaps create nanoscale channels for fluid flow. Before we turn our attention to the fascinating physics (and potential importance) of Dutcher films, however, it is worth noting that surface undulations commonly occur in other thin-film structures.

B.   Substrate–Fluid–Solid Films

Closely related substrate–fluid–soid films can also exhibit undulation growth of the top solid layer. This is thought to occur due to the buckling of the top layer in response to a thermal expansion mismatch between solid and fluid layers [7] and van der Waals forces are not thought to play a significant role [8, 9]. This may seem curious to the reader, because we have just explained how van der Waals forces can play such a significant role in the trilayer films we are interested in. It should first be noted that in the trilayer films in which we are interested, the van der Waals forces are always attractive, due to the symmetry of the problem, while this is not necessarily the case for the substrate–fluid–solid films. However,

Sharp et al. [9] noted that the film under study in their work showed no signs of instability as the temperature was raised to a value greater than the glass-transition temperature of the solid (now fluid) capping layer. Because van der Waals forces are ever-present, the argument was made that they cannot be significant because the film did not dewet. It should be noted, however, that surface tension effects might offer a greater resistance to undulation growth than the elasticity of the solid layer and, therefore, this is not justification for presuming that van der Waals forces do not play a significant role. Surface tension can stop undulations from forming in thin films, if the wavelength of undulation growth approaches macroscopic dimensions or if undulation growth becomes kinetically restricted. Rather than the stresses, which arise due to the thermal expansion mismatch, dissipating at the expense of fluid flow (as one might expect), it has been argued that stress transfer between the viscoelastic fluid and the solid layer causes the buckling [7, 8, 10]. Furthermore, buckling does not occur *until* the temperature of the film is raised to above the glass-transition temperature of the fluid layer. Even if the rubbery viscoelastic fluid layer could transfer enough stress to buckle the top layer, however, the stresses would still be expected to dissipate with time and the undulations would gradually disappear [11]. This does not appear to be the case in the studies reported in the literature, and therefore the fundamental physics at work in these systems is still unclear. The compressive strains that cause the buckling instability might arise as a consequence of a thermal expansion mismatch between sample and holder. Raising the temperature above the glass-transition temperature of the fluid layer might be necessary because a fluid layer offers less resistance to deformation. Because of the compressive stresses on the sample from the holder, the film can buckle in its own right, without the need for van der Waals forces or thermal expansion mismatch between layers to play a part. That said, instabilities have been reported in substrate–fluid–solid films and described in terms of a thermal expansion mismatch between solid and fluid layers.

C. Substrate–Solid–Solid Films

A more common example of undulation growth, as a consequence of thermal expansion mismatch, occurs in substrate–solid–solid films. Here, the interface between the two layers can support stresses because both layers are solid. The thermal contraction of an underlying layer can induce buckling in an adjacent thin solid layer [12–14]. In other words, as the underlying layer contracts, it transfers stress to the top layer, and this effectively stretches the underlying layer and compresses the top layer. Stretching the lower layer does not lead to any interesting behavior, other than a slightly stretched substrate. However, compressing the top layer can lead to interesting deformations when the compressive stresses in the top layer are relieved through buckling instabilities. In other words, the film surface exhibits undulations or ripples as the top solid layer buckles (the top layers undergoes bending, warping, or crumpling as it tries to relieve the compressive stresses). The solid capping layer in these systems is typically metal while the underlying layer is a compliant elastomeric polymer [12], although any two solid layers with different thermal expansions should exhibit this effect. In a similar manner, buckling instabilities can occur in thin films deposited on prestressed substrates [15, 16]. If the substrate is stretched and a top layer is deposited on top of the stressed substrate, then the system consists of a stretched substrate and a unstretched top layer. Upon relaxing the substrate and removing the stresses that were stretching the substrate, the top layer becomes compressed. The top and bottom layers are attached at the interface; and as the substrate tries to regain its original undeformed shape, it tries to compress the top layer. To relieve the compressive stresses in the top layer of the film, the top layer buckles and these buckling instabilities result in surface undulations. In such systems the surface undulations can be relatively large (on the order of microns) while in the ultrathin trilayer films of interest here the surface undulations are restricted by the thickness of the fluid layer (and, therefore, typically on the order of 50 nm). This has important implications for the final applicability of these films. For example, the undulations in substrate–solid–solid films can significantly increase the surface area of the film (as opposed to the solid–fluid–solid films of interest in this study), and this can affect such things as the surface tension of the film [17] or the use of such surfaces for tissue growth [18].

D. Alternative Mechanisms for Driving Instabilities

Besides the mechanical forces just mentioned, along with the van der Waals forces present in very thin films, acoustic Casimir forces may also play a part in thin films. The acoustic Casimir effect arises due to the geometrical confinement of thermally excited acoustic waves [19, 20]. A thin layer of material, with reflective boundaries, can only support standing sound waves whose wavelength is less than twice the thickness of this layer [19]. In contrast, outside of the film phonon frequencies have no such limitations and this difference is believed to result in a net pressure acting on the surface of the film. It appears counterintuitive that sound waves (which consist of both compressions and rarefactions) would produce a net force, however, Steiner and co-workers [20] have shown that films where van der Waals forces are expected to be stabilizing can still develop instabilities, indicating the presence of an alternative destabilising force.

    Another source for a destabilizing force in thin films is electrostatic pressure. It has been found in substrate–fluid–fluid systems that if the central fluid layer is a dielectric and the substrate and top fluid layer are both conducting (with differing work functions), then an electric field across the dielectric can drive instabilities [21, 22]. In other words, the presence of the two different work functions (of the substrate and top layer) straddling the insulating layer will result in an internal electric field. These films have been reported to exhibit instabilities driven by this electric field. A similar effect could exist in the ultrathin solid–fluid–solid films of interest here, if the top and bottom layers are constructed from metals (or even conducting polymers) with differing work functions or if an external electric field is applied across the film. This could provide an alternative mechanism for driving instability growth in solid–fluid–solid films and allow these films to be created with thicknesses much greater than 100 nm (above this thickness, van der Waals forces are not necessarily enough to drive undulation growth and an alternative driving force may be required).

III. CONTROLLING SURFACE UNDULATIONS

While the development of surface undulations in thin films (and the fundamental physics responsible for these instabilities) are of academic interest, from a technological perspective, it may be desirable to exhibit some control over these surface undulations. Potential applications of surface undulations have ranged from optical properties (regular corrugated surfaces for tunable diffraction) to substrates for tissue growth (tissue is often grown on porous media, because this has a greater surface area, but systematically changing the surface area without otherwise changing the surface properties can be problematic). Furthermore, micropatterned substrates with adhesive islands have been used to control cell growth where geometry played a crucial role [23]. Textured surfaces can also improve the efficiency of LEDs by over 30% by reducing the total internal reflection at the semiconductor–air interface [24].

Once such method of controlling the surface undulations is through stress-guided self-assembly. Stress-driven self-assembly is a promising route to either induce ripples in a desired pattern [12] or guide undulation growth such that a specific corrugation is produced [4].

In the ultrathin solid–fluid–solid films of interest here, the growth of undulations in the tensile direction can be either suppressed (through stretching the film) or enhanced (through compressing the film) and the resultant anisotropic corrugations can exhibit long-range order [4]. In other words, the system would like instabilities to grow, but recall that these are governed by the competition between the elasticity of the solid capping layers, the van der Waals forces, and the kinetics of the fluid transport in the central layer. By applying external deformations to the thin film, we can make it easier (or harder) for the solid capping layers to bend in a given direction. This has the effect of encouraging (or suppressing) undulation growth in a given direction and results in anisotropic corrugations. In other words, well-ordered perfectly aligned stripes of alternating elevated and depressed regions can emerge on the film surface extending over macroscopic distances.

Herringbone, as well as striped wrinkles, can also arise in thin elastic films by applying a more complex biaxial stress to the system [25, 26]. Therefore, by controlling the nature of the applied deformation, we can exhibit some control over the buckling instability and, ultimately, *pattern* the surface with regular undulations. The amplitude of these surface deformations (whether herringbone or striped undulations) can be controlled by varying the degree of compression, giving a materials scientist a good degree of control over the final structures [15].

Another interesting method of controlling surface undulations is through the introduction of complex topography. For example, stress-driven self-assembly has also been used to create highly ordered surface structures in Ag-core–SiOx-shell spherical microsctructures [27]. Compressive stresses in the shell, due to the mismatch in thermal expansion coefficients between core and shell, in conjunction with the spherical geometry were found to result in triangular and Fibonacci number patterns. Not only is the shell undergoing a buckling instability, but it also has to contend with the geometrical constraints of being on a spherical surface. With the recent application of core–shell polymer nanoparticles in targeted drug delivery [28] (and the obvious interest in how the surface of these particles interact with tumour cells), creating patterned undulations on the surface of core–shell structures is of technological relevance.

Randomly oriented undulations can also be directed and *forced* to conform to a given pattern. In particular, the undulations in buckling thin metal films have recently been directed by placing an elastomeric mold, with a periodic pattern, on top of the surface [7, 29]. Highly ordered buckled structures were found to occur as a consequence of the confined geometry provided by the mold. The buckling layer will buckle in a way that minimizes the deformation of the elastomeric mold placed on top of the film. This trick, however, is limited to patterns that possess similar length-scale to the original buckling instability.

It has also been found that the wrinkles in buckling thin metal layers of substrate–solid–solid films form perpendicularly to the interface of elevated regions [12] and defects [13]. By creating simple patterns in the underlying polymer film (such as regular arrays of elevated square and circular regions), the wrinkles in the top metal film form ordered structures in the regions in between these features. Again, this provides an excellent way of controlling the surface undulations; and if the underlying features are close together, then the induced patterning can span the entire film.

IV. METHODOLOGY

We now return our attention to the evolution of ultrathin solid–fluid–solid films, which are the focus of this study. Using a computer model, which couples the fluid flow of the polymer with the elastic deformation of the capping layers, we can simulate the evolution of the these films. In particular, we evolve the ultrathin solid–fluid–solid films from the initial growth of surface undulations through to the formation of isolated pockets of fluid. The model, therefore, provides insights into the complete structural evolution of the films during film processing. Computer simulations not only can heip predict the structures in ultrathin solid–fluid–solid films, but also can help identify methods for optimising the surface structures for a given application. One of the main reasons for doing such computer simulations is to identify methods of controlling the undulations and, ultimately, help guide future experimental studies.

A. Theory

Ultrathin solid–fluid–solid films are considered to be on the order of 100 nm thick, and the polymer physics at this length-scale generally requires microscopic models. However, the wavelength of undulations is on the order of micrometers, which is inaccessible to microscopic models, and continuum modles are required. We, therefore, currently use continuum fluid and elasticity theories and, as such, assume that the film thickness is large enough for continuum approximations to be applicable [4, 5]. While this is strictly not the case (slip boundary condition and discrete ordering of polymers have been observed at this length-scale), a continuum model is still expected to capture the fundamental physics of these systems. Furthermore, such an approach is widely found in dewetting simulations (where the fluid layer dewets from a substrate) and has been found to give very good agreement with experimental studies [30].

We use a thin-film model where the film thickness is assumed to be much smaller than the characteristic wavelength of the undulations in the lateral plane. Under such a long-wave approximation the Navier–Stokes equations lead to the following boundary-layer equations [31–33]:

$$\mu \frac{\partial^2 v_x}{\partial z^2} = \frac{\partial P}{\partial x}, \qquad \mu \frac{\partial^2 v_y}{\partial z^2} = \frac{\partial P}{\partial y} \qquad (1)$$

where μ is the fluid viscosity, P is the pressure acting on the film surface, and $\mathbf{v}$ is the local velocity of the fluid. Under no-slip boundary conditions the above equations will provide the correct Poiseuille flow behavior (the laminar Poiseuille flow profile is parabolic, with a maximum velocity in the middle of the channel and zero velocity at the walls).

The mass-conserving kinematic equation for this system is of the form [32]

$$\frac{\partial h}{\partial t} + \frac{\partial}{\partial x} \left[\int_{-h}^{h} v_x \, dz \right] + \frac{\partial}{\partial y} \left[\int_{-h}^{h} v_y \, dz \right] = 0 \qquad (2)$$

where h is the height of the fluid film. In the current study we take h to be the half thickness, and therefore the total fluid layer extends from $z = -h$ to $z = h$.

Assuming such symmetry about the mid-plane, as well as no-slip boundary conditions at the fluid–solid interface, we obtain the following thin-film equation

$$\frac{\partial h}{\partial t} = \frac{2}{3\mu} \left(\frac{\partial}{\partial x} \left[h^3 \frac{\partial P}{\partial x} \right] + \frac{\partial}{\partial y} \left[h^3 \frac{\partial P}{\partial y} \right] \right) \tag{3}$$

The pressure acting on the fluid layer, from the solid capping layers, consists of four parts [5]. The first pressure contribution comes from the van der Waals interactions across the film [34],

$$P_{vdW} = \frac{1}{6\pi} \left[\frac{A_{232}}{(2h)^3} - \frac{A_{123}}{(2h+d)^3} + \frac{A_{121}}{(2h+2d)^3} \right] \tag{4}$$

where A_{ijk} is a nonretarded Hamaker constant describing the interaction between media i and k across media j, and d is the thickness of both the upper and lower solid capping layers. In the above equation, media 1, 2, and 3 correspond to the air, solid layer, and fluid layer, respectively. In the current study we assume that the polymer solid and polymer fluid are similar in properties and that the Hamaker constant between them is zero ($A_{232} = A_{123} = 0$). The van der Waals forces are assumed to arise from the solid–air interface ($A_{121} = 7.38 \times 10^{-20} J$).

The pressure on the fluid from the elastic deformation of the solid layers has both bending and stretching contributions. The pressure due to bending is given by [35]

$$P_{bend} = -\frac{Ed^3}{6(1-\sigma^2)} \left[\frac{\partial^4 h}{\partial x^4} + \frac{\partial^4 h}{\partial y^4} + \frac{\partial^4 h}{\partial x^2 \partial y^2} \right] \tag{5}$$

where E is the Young's modulus, and σ the Poisson's ratio, of the solid capping layers.

We introduce a pressure due to the contact mechanics between the upper and lower solid capping layers because we wish to simulate the local contact between top and bottom layers and the ejection of fluid from this region. We adopt an exponential repulsive term of the form [36]

$$P_{contact} = -\frac{\epsilon}{\xi} \exp\left(-\frac{h}{\xi} \right) \tag{6}$$

where ϵ dictates the strength of the interaction and $\xi = 1$ nm is a characteristic length-scale. This short-range interaction energetically penalizes contact, and potential overlap, between the two solid layers. The presence of large deformations in the elastic layers means that we must take into consideration the stretching of the solid layers. The free energy for this stretching is of the form [35]

$$H_s = \frac{Ed}{2(1+\sigma)} \int u_{xx}^2 + u_{yy}^2 + 2u_{xy}^2 + \frac{\sigma}{1-2\sigma}(u_{xx} + u_{yy})^2 \, d\Omega \tag{7}$$

where the integration extends over the system and u_{ij} is the strain tensor, including terms that are quadratic in the derivatives of h,

$$u_{ij} = \frac{1}{2}\left(\frac{\partial u_i}{\partial x_j} + \frac{\partial u_j}{\partial x_i} + \frac{\partial h}{\partial x_i}\frac{\partial h}{\partial x_j}\right) \tag{8}$$

Terms quadratic in the derivatives of u_i, are omitted because there are first-order terms present, while this is not the case for the derivatives of h. Using functional derivatives (see Appendix 1), we calculate the forces in the solid layers $(F_{u_i} \sim \delta H_s/\delta u_i)$, thereby resulting in the following equations for the forces in the x and y directions:

$$
\begin{aligned}
F_{ux} = & -\frac{Eh}{2(1+\sigma)}\left[2\frac{\partial^2 u_x}{\partial x^2} + 2\frac{\partial h}{\partial x}\frac{\partial^2 h}{\partial x^2} + \frac{\partial^2 u_x}{\partial y^2} + \frac{\partial^2 u_y}{\partial x \partial y} + \frac{\partial h}{\partial x}\frac{\partial^2 h}{\partial y^2} + \frac{\partial h}{\partial y}\frac{\partial^2 h}{\partial x \partial y}\right. \\
& \left. + \frac{\sigma}{1-2\sigma}\left(2\frac{\partial^2 u_x}{\partial x^2} + 2\frac{\partial h}{\partial x}\frac{\partial^2 h}{\partial x^2} + 2\frac{\partial^2 u_y}{\partial x \partial y} + 2\frac{\partial h}{\partial y}\frac{\partial^2 h}{\partial x \partial y}\right)\right]
\end{aligned}
\tag{9}
$$

and

$$
\begin{aligned}
F_{uy} = & -\frac{Eh}{2(1+\sigma)}\left[2\frac{\partial^2 u_y}{\partial y^2} + 2\frac{\partial h}{\partial y}\frac{\partial^2 h}{\partial y^2} + \frac{\partial^2 u_y}{\partial x^2} + \frac{\partial^2 u_x}{\partial x \partial y} + \frac{\partial h}{\partial y}\frac{\partial^2 h}{\partial x^2} + \frac{\partial h}{\partial x}\frac{\partial^2 h}{\partial x \partial y}\right. \\
& \left. + \frac{\sigma}{1-2\sigma}\left(2\frac{\partial^2 u_y}{\partial y^2} + 2\frac{\partial h}{\partial y}\frac{\partial^2 h}{\partial y^2} + 2\frac{\partial^2 u_x}{\partial x \partial y} + 2\frac{\partial h}{\partial x}\frac{\partial^2 h}{\partial x \partial y}\right)\right]
\end{aligned}
\tag{10}
$$

By putting these equations to zero, we can calculate the equilibrium lateral displacements. In other words, we can calculate the deformations of the elastic layers in the x and y directions, from the current local variations in film height. Furthermore, by taking the functional derivative of the stretching free energy with respect to the height of the film, we also obtain the pressure contribution, which acts on the fluid layer.

$$
\begin{aligned}
P_s = & -\frac{Eh}{2(1+\sigma)}\left[\left(1 + \frac{\sigma}{1-2\sigma}\right)\left(2\frac{\partial u_x}{\partial x}\frac{\partial^2 h}{\partial x^2} + 2\frac{\partial^2 u_x}{\partial x^2}\frac{\partial h}{\partial x} + 3\left(\frac{\partial h}{\partial x}\right)^2\frac{\partial^2 h}{\partial x^2}\right.\right. \\
& \left.\left. + 2\frac{\partial u_y}{\partial y}\frac{\partial^2 h}{\partial y^2} + \frac{\partial^2 u_y}{\partial y^2}\frac{\partial h}{\partial y} + 3\left(\frac{\partial h}{\partial y}\right)^2\frac{\partial^2 h}{\partial y^2}\right)\right.
\end{aligned}
$$

$$+2\left(\frac{\partial u_x}{\partial y}+\frac{\partial u_y}{\partial x}\right)\frac{\partial^2 h}{\partial x \partial y}+\frac{\partial h}{\partial x}\left(\frac{\partial^2 u_x}{\partial y^2}+\frac{\partial^2 u_y}{\partial x \partial y}\right)+\left(\frac{\partial h}{\partial x}\right)^2\frac{\partial^2 h}{\partial y^2}$$

$$+\frac{\partial h}{\partial y}\left(\frac{\partial^2 u_x}{\partial x \partial y}+\frac{\partial^2 u_y}{\partial x^2}\right)+\left(\frac{\partial h}{\partial y}\right)^2\frac{\partial^2 h}{\partial x^2}+4\frac{\partial h}{\partial x}\frac{\partial h}{\partial y}\frac{\partial^2 h}{\partial x \partial y} \tag{11}$$

$$+\frac{2\sigma}{1-2\sigma}\left(\frac{\partial u_y}{\partial y}\frac{\partial^2 h}{\partial x^2}+\frac{\partial h}{\partial x}\frac{\partial^2 u_y}{\partial x \partial y}+\frac{1}{2}\left(\frac{\partial h}{\partial y}\right)^2\frac{\partial^2 h}{\partial x^2}+2\frac{\partial h}{\partial x}\frac{\partial h}{\partial y}\frac{\partial^2 h}{\partial x \partial y}\right.$$

$$\left.+\frac{\partial u_x}{\partial x}\frac{\partial^2 h}{\partial y^2}+\frac{\partial h}{\partial y}\frac{\partial^2 u_x}{\partial x \partial y}+\frac{1}{2}\left(\frac{\partial h}{\partial x}\right)^2\frac{\partial^2 h}{\partial y^2}\right)\Bigg]$$

Therefore, given the deformations in the x and y directions of the capping solid layers, we can determine the pressure acting on the fluid due to this stretching or compression [5].

In summary, the different pressures are shown schematically in Fig. 2. The attractive van der Waals forces act across the film and favor thinner film thicknesses, the bending elastic forces depend on the fourth-order derivatives of the height variations and essentially penalizes curvature in the height, the contact pressure is included for numerical stability and ensures that the solid capping layers do not overlap, and finally the stretching elastic forces describe the pressure on the fluid due to in-plane displacements in the solid capping layers. The behavior of the model depends upon the interactions of all these pressures, as well as the kinetics of fluid transport, in response to these pressures.

We have described the model used in this study to describe the evolution of the ultrathin solid–fluid–solid films. In particular, this model can capture large-amplitude height variations and simulate the late stages of undulation growth. We now describe the computational implementation of this model.

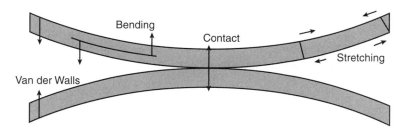

Figure 2. Schematic of various pressures accounted for in the model. We include the pressures due to the bending and stretching of the solid capping layers, the van der Waals forces across the film, and a pressure term that accounts for contact mechanics between the top and bottom solid layers.

B. Computational Details

The simulation progresses through the iterative update and calculation of the equations described in the Section IV.A. In particular, the first step is to obtain the lateral displacement field in the solid capping layers from the current height variations. The second step is to calculate the pressures acting on the fluid layer, given the current height variations and lateral displacement field. Finally, the third step is to update the height variations using Eq. (3) and the calculated pressures acting on the fluid layer. The updated height variations are fed into the next iteration and system evolves through the iterative solution of these three steps.

The first step of the above iterative cycle involves the solution of Eq. (9) and (10). If we assume that there are no externally applied forces, then Eq. (9) and (10) can both be set to zero. Using finite difference approximations, we can rewrite Eq. (9) as

$$S(h) = \left(2 + \frac{2\sigma}{1-2\sigma}\right) \frac{u_x(i+1,j) + u_x(i-1,j) - 2u_x(i,j)}{\Delta x^2}$$
$$+ \frac{u_x(i,j+1) + u_x(i,j-1) - 2u_x(i,j)}{\Delta x^2}$$
$$+ \left(1 + \frac{2\sigma}{1-2\sigma}\right)$$
$$\times \frac{u_y(i+1,j+1) + u_y(i-1,j-1) - u_y(i+1,j-1) - u_y(i-1,j+1)}{4\Delta x^2}$$

$$(12)$$

Similarly, the force in the y direction gives

$$T(h) = \left(2 + \frac{2\sigma}{1-2\sigma}\right) \frac{u_y(i+1,j) + u_y(i-1,j) - 2u_y(i,j)}{\Delta x^2}$$
$$+ \frac{u_y(i,j+1) + u_y(i,j-1) - 2u_y(i,j)}{\Delta x^2}$$
$$+ \left(1 + \frac{2\sigma}{1-2\sigma}\right)$$
$$\times \frac{u_x(i+1,j+1) + u_x(i-1,j-1) - u_x(i+1,j-1) - u_x(i-1,j+1)}{4\Delta x^2}$$

$$(13)$$

where the displacement field is now assumed to be on a regular grid (in x and y) and $u_x(i, j)$ and $u_y(i, j)$ represent the displacement in the x and y direction, respectively, at coordinates i and j. Likewise, the discrete form of the height variations is $h(i,j)$ at the node (i,j). The spatial discretization is Δx, and $S(h)$ and $T(h)$ are the relevant collections of derivatives in h (simply obtained by rearranging Eqs. (9) and (10), which are considered to be constant (during the solution of the above equations).

The two equations above are now defined at every single point on the grid ($2N^2$ equations, where N is the number of grid points in one direction). This may seem like a large number of equations to solve, but we are aided in the fact that only nodes that are close to each other appear in a single equation. In other words, we can write the very large set of equations as a sparse matrix and solve it using the conjugate gradient method [37]. Without any further explanation, we give the conjugate gradient algorithm here. Imagine the matrix is written in the form $\mathbf{A} \cdot \mathbf{x} = \mathbf{b}$ and we define $\mathbf{r}$ as the residual. The iteration of the following five steps will converge to an accurate solution (defined as the solution obtained when the sum of the residual squared has decreased to a significantly small value).

$$
\begin{aligned}
\mathbf{r}^k &= \mathbf{A} \cdot \mathbf{x}^k - \mathbf{b} \\
\beta_k &= \frac{(\mathbf{r}^{k+1})^T \cdot \mathbf{A} \cdot \mathbf{d}^k}{(\mathbf{d}^k)^T \cdot \mathbf{A} \cdot \mathbf{d}^k} \\
\mathbf{d}^{k+1} &= -\mathbf{r}^{k+1} + \beta_k \cdot \mathbf{d}^k \qquad (\mathbf{d}^0 = -\mathbf{r}^0) \\
\lambda_k &= -\frac{(\mathbf{d}^k)^T \cdot \mathbf{r}^k}{(\mathbf{d}^k)^T \cdot \mathbf{A} \cdot \mathbf{d}^k} \\
\mathbf{x}^{k+1} &= \mathbf{x}^k + \lambda_k \cdot \mathbf{d}^k
\end{aligned}
\tag{14}
$$

This is widely regarded as the quickest way in which to solve these equations. In this manner we can efficiently obtain the displacement field as a function of height variations and the elastic properties of the film.

The second step in an iteration, while simulating the evolution of the ultrathin trilayer solid–fluid–solid films, is to calculate the pressures acting on the fluid. The van der Waals pressure is obtained locally from the height of the film using Eq. (4). The elastic bending pressure is also obtained from the height of the film, but this time a finite-difference approximation is required to obtain the derivatives in discrete form. In particular,

$$
\frac{\partial^4 h}{\partial x^4} = h(i-2,j) - 4h(i-1,j) + 6h(i,j) - 4h(i+1,j) + h(i+2,j) \tag{15}
$$

and

$$
\begin{aligned}
\frac{\partial^4 h}{\partial x^2 \partial y^2} =\ & h(i+1,j+1) + h(i+1,j-1) + h(i-1,j+1) + h(i-1,j-1) \\
& - 2h(i,j+1) - 2h(i,j-1) - 2h(i+1,j) - 2h(i-1,j) + 4h(i,j)
\end{aligned}
\tag{16}
$$

The third pressure term, representing the contact mechanics between top and bottom solid layers, like the van der Waals pressure, simply depends on the height of the film. The final pressure term comes from Eq. (11) and represents the pressure exerted on the fluid due to the in-plane stretching or compression of the solid layers. This term involves derivatives of the height but also derivatives of the displacement field. Using finite differences, the derivatives can be obtained from the discrete data (recall in step one we obtained the discrete displacement field from the height variations). Now we can calculate the pressure acting on the fluid at every point in our discretized system (at all nodes (i,j)).

The third, and final, step in an iteration is simply updating the height using Eq. (3). Given that we now have both the pressure and previous height data at all points in the x and y directions we can simply solve Eq. (3) using a finite difference approximation. In this manner, the height at a given time can be obtained from the height, displacement, and pressure data from an earlier time (and assuming that the time step is kept sufficiently small, our discrete set of equations correctly captures the continuum equations outlined in the previous section).

We now present some results from these simulations and demonstrate the application of the model.

V. RESULTS

We can use the model, described in the previous section, to simulate the dynamics of ultrathin solid–fluid–solid films and, more importantly, gain insights into what controls the dynamics of the films and how to manipulate the final structures. We now present some results obtained from our model.

A. Evolution of Ultrathin Solid–Fluid–Solid Films

Recall that the early stages of the evolution of a ultrathin Solid–Fluid–Solid film consist of the growth of surface undulations. That is, the film is initially unperturbed with the exception of thermal noise (which we mimic by including small random fluctuations in the height). Small fluctuations, however, grow with time as the van der Waals forces drive the instability of the film. Eventually, surface undulations of a characteristic wavelength emerge. An example of these surface undulations is depicted in Fig. 3, which shows the height variations of the surface of the film. In the current study we investigate systems where the thickness of the fluid layer is 50 nm and the thicknesses of the solid layers are both 25 nm. The thickness of the entire film, therefore, is 100 nm. The system size is $256 \times 256\,\mu m^2$ and the height variations are presented in nanometers. Figure 3 shows a snapshot of this system after a time of 5 minutes. The surface undulations have grown to a height of roughly 3 nm and, driven by the van der Waals interactions across the film, the surface can be seen to be perturbed.

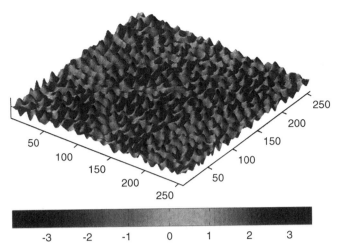

Figure 3. Plot of surface undulations in nanometers for a film of dimenions $256^2 \mu m^2$ after 5 minutes. The undulations have grown due to van der Waals driven instabilities, but, as yet, the top and bottom surfaces are still separated by the fluid layer.

In order to gain more insight into the mechanisms that drive these instabilities, we depict the forces that are driving the system (at the time depicted in Fig. 3) in Fig. 4. In particular, Fig. 4a shows the van der Waals forces that act across the film and encourage the growth of height fluctuations. Meanwhile, Fig. 4b and 4c show the elastic forces from the bending and stretching of the elastic solid layers, respectively, which suppress fluctuation growth. The van der Waals pressure in Fig. 4a is seen to fluctuate around a positive value of 4 Pa, because the van der Waals forces acting across the film always cause a positive pressure on the fluid layer. However, the fluctuations in film thickness cause variations in the van der Waals pressure which seek to increase these fluctuations. In other words, areas of the film where the height is reduced experience a higher van der Waals pressure whereas areas of the film where the height is increased experience lower van der Waals forces. As fluid in the central layer flows from regions of high pressure to regions of low pressure, this local variation in pressure drives undulation growth. In contrast, the bending elastic energy in Fig. 4b opposes undulation growth and regions of positive pressure in Fig. 4b correspond with regions of greater film thickness. Recall that the elasticity of the top and bottom solid layers would prefer that the solid layers were undeformed and that there were no height variations in the system. However, it is worth noting that the order of magnitude of these variations is roughly ± 0.3 Pa, as opposed to the magnitude of the variations in van der Waals pressure of ± 0.8 Pa. The variation in van der Waals forces are larger than the elastic forces, and recall from Eq. (3) that variations in pressure drive fluid flow

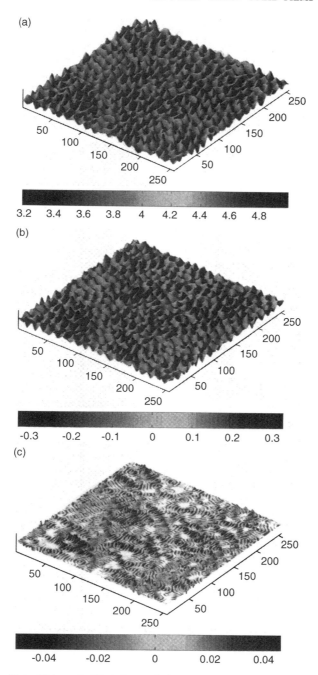

Figure 4. Plots of (a) van der Waals forces, (b) bending elastic forces, and (c) stretching elastic forces in Pascals for the system shown in Fig. 1 after 5 minutes. The variations in these forces are responsible for fluid flow and the evolution of the system.

and ultimately the evolution of the system. Therefore, the van der Waals forces, dominate and fluctuation growth occurs.

The final pressure comes from the stretching of the elastic solid layers and is depicted in Fig. 4C. As the undulations grow, not only are the elastic solid layers being bent but they also stretch slightly. This stretching, along with the higher-magnitude pressures that result from this stretching, appears to occur at the interfaces between regions of the film that are thicker and regions of the film that are thinner. In other words, gradients in film thickness cause the elastic solid layers to become stretched, and this produces localized pressures in the fluid which (like the bending elastic pressures) oppose undulations growth. The magnitude of the pressures from the stretching energy, however, are relatively small (compared to the other pressures in the system) and are not important at these early stages of undulation growth. As the height variations in the film become comparable to the thickness of the solid layers, the film will become significantly stretched and pressures due to the stretching of the solid layers will play a significant role in ultrathin solid–fluid–solid film evolution. For the time being, however, we consider only the initial growth of surface fluctuations.

While the elasticity of the solid capping layers opposes the growth of all surface fluctuations, undulations of lower wavelengths involve more bending of the solid layer and are more energetically unfavorable. In other words, if the solid layers have large wavelength undulations then, while this is still energetically unfavorable for the solid films, the curvature is relatively small and the energy penalty is relatively small. On the other hand, if the wavelength of the undulations are small, then the solid layers are subject to more deformation of greater curvature and the energy penalty is also greater. However, this system does not only depend on energetics, since the growth of large wavelength undulations is kinetically restricted because the fluid has to be transported from regions of the film which are thinning to regions of the film where the thickness is increasing. This competition between energetics and kinetics results in undulations of a characteristic wavelength growing faster than other wavelengths. Dalnoki-Veress et al. [1] theoretically determined this wavelength to be

$$\lambda = 4\pi \left[\frac{\pi E}{4A_{121}(1 - \sigma^2)} \right]^{1/4} d^{3/4}(h + d) \qquad (17)$$

where A_{121} is the Hamaker constant and the effects of varying the Young's modulus of the film are depicted in Fig. 5. In particular, we compare the theoretical predictions with the predictions from our simulations and find that the two results are in very good agreement. The simulated wavelength is obtained from the first moment of the radially averaged fast Fourier transform of the height variations. The data obtained from the fast Fourier transform of the height

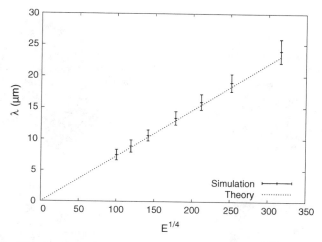

Figure 5. Wavelength of undulation growth as a function of Young's modulus, or, more precisely, Young's modulus to the power of one-fourth. Simulations results from the model described above are directly compared with theory of Dalnoki-Veress et al.. The agreement is found to be satisfactory and the wavelength increases with Young's modulus as bending the solid layers becomes increasingly energetically unfavorable.

variations is averaged over three independent simulations. The error bars reflect the standard deviation and are obtained from the second moment of the radially averaged fast Fourier transform. The simulated predictions for the characteristic wavelength are consistently higher than the theoretical predictions, and this might be due to the assumption of corrugated ID undulations in the theory. However, the theoretical line falls well within the error bars from our analysis. Note that both the theory and simulations assume similar continuum approximations; therefore, such an agreement should be expected.

As the Young's modulus is increased, the bending of the elastic solid layers becomes increasingly energetically unfavorable and the characteristic wavelength of the system increases. In other words, the elastic bending pressure (see Fig. 4B) and (more importantly) the variations in the elastic bending pressure are proportional to the Young's modulus, and increasing this Young's modulus increases the suppression of shorter wavelength undulations. Therefore, the balance between energetics and kinetics described earlier is shifted to increasingly longer wavelength undulations. By tuning the mechanical properties of the solid layers, we can exhibit some control over the wavelength of the undulations that grow in the early stages in the evolution of these films.

We now turn our attention to the latter stages in the evolution of these ultrathin solid–fluid–solid films. At some point during the growth of the surface undulations, the top and bottom solid layers meet and fluid is expelled from

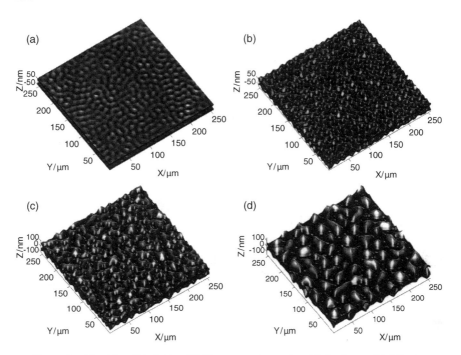

Figure 6. Ultrathin solid–fluid–solid film geometries at time (a) 5 minutes, (b) 7.5 minutes, (c) 10 minutes, and (d) 15 minutes for a film of dimensions $256^2 \mu m^2$.

between the solid layers. The van der Waals forces, which drive the system, have minimized the thickness of the film to the limit of contact between the top and bottom solid layers. The evolution of the structure of the film as the undulations initially grow, and beyond when the top and bottom layers are brought into contact, is depicted in Fig. 6. Both the top and bottom solid layers of the trilayer films are depicted in these plots. Note that, for clarity, the top surface is colored from black (thin regions) to white (thicker regions). The edges of the top and bottom solid layers are colored black. Figure 6a shows the system at a time of 5 minutes (and corresponds with the same structure seen in Fig. 3). In Fig. 6b the system is shown at a time of 7.5 minutes, when the top and bottom layers have met and the film, in these regions, cannot become any thinner. Therefore, in order to minimize the van der Waals energy, the area of contact between the top and bottom layers is maximized. Recall that the van der Waals interactions favored thinner regions of the film, at the expense of pushing fluid to other regions of the films which became thicker. Now the film thickness cannot decrease anymore; in the regions of contact between the two solid layers, the van der Waals energy can still be minimized by simply increasing the area of contact between the solid layers. At the same time both the bending and stretching elastic free energies favor larger wavelength deformations. The consequences of this are that the fluid is corraled into isolated pockets and this process can be seen in Fig. 6c. Half of the film consists of these isolated pockets of fluid, and the other half of the film consists of regions where the top and bottom layers of the film are in contact. The area of contact between the top and bottom solid layers continues to grow at the expense of the fluid being pushed into increasingly larger pockets of fluid. This is the situation depicted in Fig. 6d, which shows the system after a time of 15 minutes. The fluid has been corraled into isolated pockets separated by regions of the thin film where the top and bottom layers are in contact.

B. Manipulating the Undulations and Final Film Structure

It is desirable to control the evolution of ultrathin solid−fluid−solid films so that we might tailor the structure to meet the needs of a given application. One way to induce large-scale order in ultrathin solid−fliud−solid films is to apply a global deformation to the film—in other words, to either compress or stretch the film in an in-plane direction and encourage or suppress undulation growth in this direction. We have previously found this method to produce anisotropic corrugations [4], or ID undulations perfectly aligned over large distances. The evolution of an ultrathin solid−fluid−solid film subject to a global deformation, beyond the point when the top and bottom surfaces meet, is now considered in Fig. 7. The film is compressed in the x direction through the application of a global strain of $\partial u_x / \partial x = 1 \times 10^{-4}$; and the growth of the undulations are, in part, caused by this deformation. In other words, by applying a global compression to the system, not only are the undulations, driven by the van der Waals forces, encouraged to grow in a given direction, but the undulations could emerge entirely independent of van der Waals forces due to the buckling of the top and bottom solid layers in response to the compression. Figure 7a, shows the system at a relatively early time of 7.5 seconds (while the system is still experiencing undulation growth, and the top and bottom surfaces have not yet been brought into contact). The undulations grow anisotropically and form a regular corrugated pattern on the surface. Eventually the top and bottom surfaces meet; at this point, undulations begin to grow in both the x and y directions. As the surfaces meet, the van der Waals forces seek to maximize the contact between the top and bottom solid surfaces. This results in the formation of isolated pockets of fluid which are elongated slightly in the y direction, but do not form percolating channels of fluid. Furthermore, the anisotropy decreases with time as the strain field due to the height variations in the stretched solid layers (obtained using Eqs. (9) and (10)) become significantly higher than the imposed strain field that was driving the corrugation.

The wavelength of the initial undulations is modified by the application of a global deformation. Recall how varying the Young's modulus changed the

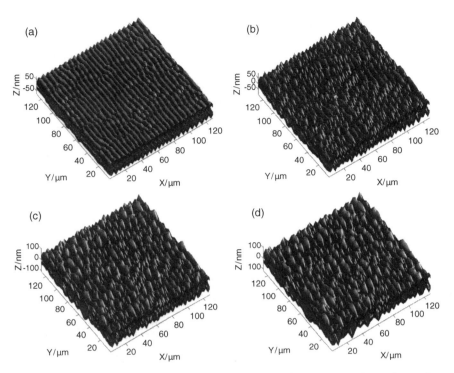

Figure 7. Ultrathin solid–fluid–solid film geometries, subject to compressive strains, at time (a) 7.5 seconds, (b) 15 seconds, (c) 30 seconds, and (d) 45 seconds for a film of dimensions $128^2 \mu m^2$. The compressive strains in the x direction cause the undulations to elongate, and the final pockets of fluid are elongated in the y direction.

energy of elastic deformation, which, in turn, changed the wavelength of undulation growth. If a compressive strain is applied to the film, then this encourages undulation growth in the tensile direction (and may even cause the film to buckle) at a lower wavelength. Alternatively, stretching the film suppresses the growth of undulation in the tensile direction, and the corrugations occur at a higher wavelength. In other words, in a manner similar to increasing the Young's modulus, stretching the film slightly makes the growth of undulations harder, and as a consequence the wavelength of undulation growth increases. This is quantified in Fig. 8, which shows the relative variation in wavelength as a function of the magnitude of the applied deformation (both compressive and stretching are contrasted). We plot the wavelength of undulation growth relative to the undeformed wavelength, λ_0, and plot the magnitude of the imposed strain field (which can be either compressing or stretching the film). The solid line represents the stretched system; as we expect,

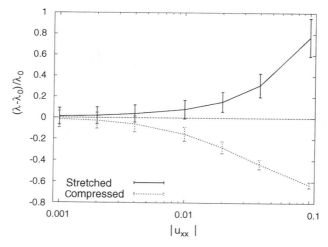

Figure 8.   Relative wavelength of surface corrugations as a function of the magnitude of the imposed strain field. Both compressed and stretched systems are contrasted, and the wavelength of undulation growth is seen to increase or decrease with stretching or compression, respectively.

the wavelength increases with increasingly large stretching of the film. The dashed line represents the compressed systems, and here shorter wavelengths are found to emerge from the deformed systems.

An alternative way in which we might impose long-range order on the undulation growth, and more interestingly on the isolated pockets of fluid, is through mechanical heterogeneity. That is, we can make regions of the solid layers either stiffer or softer and follow the evolution of the ultrathin solid–fluid–solid film in the presence of these heterogeneities. This could be achieved by having regions of the solid layer be of a different material to the rest of the solid layer (one way of achieving this might be to have the solid layers made of a phase-separating polymer blend). As an example, however, we imagine that the ultrathin solid–fluid–solid film consists of solid layers that are stiffer in striped regions through the center of the simulation. In particular, we can make the Young's modulus 10 times greater in this narrow strip (the strip is currently taken to be 10 μm wide and the system considered is 128^2 μm^2). The evolution of the film, containing this strip of stiffer material, is shown in Fig. 9. In the present simulation we consider the stripe to be in both the top and bottom surfaces, but a similar (but less pronounced) effect would be expected to occur if the stripe was only in one of the layers (see Appendix 2). Again we show both the lower and upper solid layers, and the height variations are depicted on the top surface.

The region of stiffer material is clearly evident in Fig. 9a, which shows the system at a time of 5 minutes. The stiffer material opposes undulation growth;

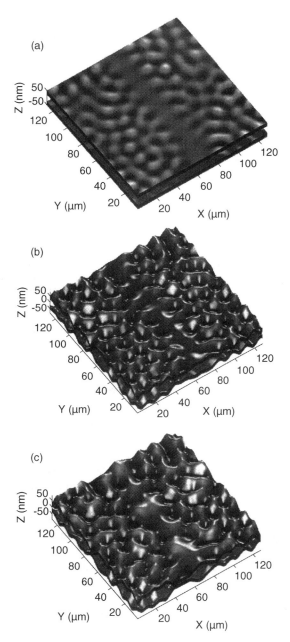

Figure 9.   Ultrathin solid–fluid–solid film geometries, in the presence of a stiff stripe, at time (a) 5 minutes, (b) 7.5 minutes, and (c) 9.25 minutes for a film of dimensions $128^2\,\mu m^2$. The stiff stripe is clearly seen as the region of reduced height variations at early times.

therefore, this region appears less perturbed than the neighboring regions. However, undulations are still growing in the stiffer region, albeit at a slower rate; and as the top and bottom layers meet, the undulations in the stiffer regions can be seen to emerge at this later time. Furthermore, the wavelength of the undulations in the stiffer region are much larger than in the rest of the film (as is expected as the characteristic wavelength varies with Young's modulus; see Fig. 5). After a time of 9.25 minutes (Fig. 9c), relatively larger isolated pockets of fluid exist in the stiffer region than in the rest of the film. However, the presence of the stripe of stiffer solid layer has not led to any long-range order in the system in the timescales simulated here. Further studies (over longer simulations times, or looking at equivalent experimental systems) are required to see if the energetic penalty of bending the stiff solid layers in the striped regions will eventually result in a single isolated pocket covering the entire stiff region.

We now turn our attention to the case where the mechanical heterogeneity consists of a stripe of softer material. The evolution of this ultrathin solid–fluid–solid film is seen in Fig. 10 to be quite different. The more compliant stripe is now much easier to deform than the rest of the solid layer; therefore, undulations grow faster in the striped region. Furthermore, the undulations are expected to be smaller in wavelength. Figure 10a shows the system after 5 minutes, when the top and bottom solid layers have met, but only in the more compliant region. In the remainder of the film the undulations are still growing but appear to be influenced by the undulation growth in the compliant stripe. In particular, the undulations in the film in the vicinity of the stripe appear to be elongated in the direction of the stripe and the system is anisotropic. As the top and bottom solid layers meet throughout the film (and not just in the more compliant striped region), the anisotropy gradually disappears and the isolated pockets of fluid are similar to what are seen in a homogeneous film (see Fig. 6). However, in the more compliant section in the center of the film there is a significant reduction in the number of isolated pockets of fluid forming. The van der Waals forces exert more influence in this region because the more compliant solid layers offer less resistance to deformation. In other words, it is easier for the solid layers to deform in these regions and allow the van der Waals forces to press the top and bottom solid layers together.

VI. SUMMARY AND CONCLUSIONS

We show how computer simulations can capture the physics of ultrathin solid–fluid–solid films from the initial growth of surface fluctuations to the emergence of isolated pockets of fluid. The benefit of using computer simulations is the ability to see the evolution of these films in their entirety, down to the local contributions of various phenomena (be it van der Waals interactions or elasticity) to the fluid pressures. Furthermore, it is possible in simulations to

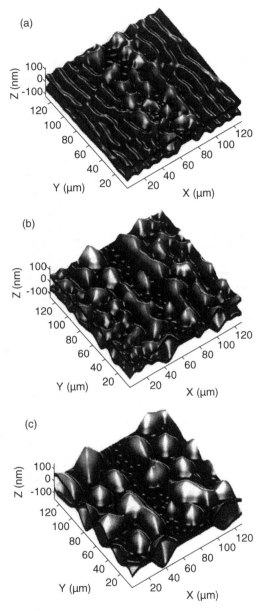

Figure 10. Ultrathin solid–fluid–solid film geometries, in the presence of a softer stripe, at time (a) 5 minutes, (b) 7.5 minutes, and (c) 12.5 minutes for a film of dimensions $128^2\,\mu m^2$.

imagine (in an experimentally unrealistic way) a hypothetical system where, for example, van der Waals forces can be turned off. This allows us to see the effects and consequences of different phenomena and gain a great deal of insight into the evolution of these films.

Initially the van der Waals forces across the film drive undulation growth in these systems and cause the surface instability to occur. The growth is resisted by the elastic deformation of the solid capping layers, which also maintains the films structural integrity. Eventually, the solid capping layers locally meet and the fluid is ejected from these regions of the film to form isolated pockets of fluid.

The orientation of the undulations can be directed by either compressing or stretching the thin films during undulation growth. This stretching or compression can either inhibit or encourage undulation growth and results in isolated corrugated structures on the surface of the film. We have also shown that the undulations, and resulting isolated pockets of fluid, can be manipulated to some degree by incorporating regions of mechanical heterogeneity. Further studies are required, however, to determine if this method can result in the control of these thin-film structures.

The method of manipulating these ultrathin trilayer solid–fluid–solid films is likely to depend on the final application of the film. These films could find application as a method of testing the mechanical properties of the thin solid layers [12, 38]. Furthermore, the creation of regular corrugated structures could lead to optical applications, such as diffraction grating or optical sensors [12]. However, the unique feature of these ultrathin solid–fluid–solid films (over other thin-film systems) is the final structures that consist of isolated pockets of fluid. It would, therefore, be highly desirable to exhibit some control over these pockets and guide the structures to form ordered channels of fluid.

One method for achieving a control of the distribution and orientation of the isolated pockets of fluid, besides the methods already considered in the current study, might be through the use of chemically heterogeneous solid layers. Chemically heterogeneous substrates have been found to guide the dewetting process in thin fluid films [39]. By making regions of the solid layer locally attractive to the fluid, the fluid layer should be preferentially corraled into isolated pockets that correspond to regions of the chemical heterogeneity. In effect, by patterning the solid layers, it should be possible to create isolated pockets of fluid which conform to this pattern.

An alternative, and interesting, possibility is to introduce a phase-separating blend as the fluid component (a schematic of this system is shown in Fig. 11). The phase-separating A–B polymer blend will evolve, and phase separate, at its own length- and timescales. However, to minimize the interface between the A and B domains of the polymer blend it may be desirable for the length-scale of phase separation to conform to the wavelength of undulation growth found in

Figure 11. Schematic of potential phase separation in these films. The interaction between the length-scale of phase separation and the length-scale of undulation growth could result in the emergence of novel structures.

the solid–fluid–solid films. Recently, Clarke [40] simulated the phase separation of a thin-film polymer blend during dewetting using a model compatible with the model described here, opening up the possibility of being able to simulate this system in the near future.

As the interest in nanotechnology and nanoscale structures increases these ultrathin solid–fluid–solid films could find a wide range of applications. Future studies will, therefore, seek to tailor the evolution of these structures to meet the needs of these applications.

APPENDIX 1

We use functional derivatives to calculate the forces from the elastic energy, and it is worth giving an example of this process. We, therefore, include a derivation of the forces for the bending of an elastic plate in one dimension. The extension of this to two dimensions, along with the derivation of stretching forces, is achieved in a similar manner. The bending free energy associated with a elastic plate in one dimension is obtained from Landau and Lifshitz [35] to be of the form

$$H = \frac{Ed^3}{24(1-\sigma^2)} \int_{-\infty}^{\infty} \left(\frac{\partial^2 h}{\partial x^2}\right)^2 \partial x \qquad (18)$$

where now h represents the height variation perpendicular to the plate surface. The force in this direction is defined as

$$F = -\frac{\delta H}{\delta h(x')} \qquad (19)$$

where, using functional derivatives, we can write

$$\frac{\delta H}{\delta h(x')} = \frac{H[h + \epsilon\delta(x - x')] - H[h]}{\epsilon} \qquad (20)$$

This is correct in the limit of $\epsilon \to 0$ (where ϵ represents a small change in h) and δ is the Dirac delta function. The energy in the system, subject to this small change, is of the form

$$
\begin{aligned}
H[h + \epsilon\delta(x - x')] &= \frac{Ed^3}{24(1 - \sigma^2)} \int_{-\infty}^{\infty} \left[\frac{\partial^2 (h + \epsilon\delta(x - x'))}{\partial x^2} \right]^2 \partial x \\
&= \frac{Ed^3}{24(1 - \sigma^2)} \int_{-\infty}^{\infty} \left(\frac{\partial^2 h}{\partial x^2} \right)^2 + 2\epsilon \frac{\partial^2 h}{\partial x^2} \frac{\partial^2 \delta(x - x')}{\partial x^2} + O(\varepsilon^2) \partial x
\end{aligned}
$$

$$(21)$$

Given that we take the limit $\epsilon \to 0$ terms of order ϵ^2 can, to a first approximation, be ignored. The derivative can now be expressed in the following form:

$$
\frac{\delta H}{\delta h(x')} = \frac{Ed^3}{12(1 - \sigma^2)} \int_{-\infty}^{\infty} \frac{\partial^2 h}{\partial x^2} \frac{\partial^2 \delta(x - x')}{\partial x^2} \partial x
\tag{22}
$$

Integration by parts once gives

$$
\frac{\delta H}{\delta h(x')} = \frac{Ed^3}{12(1 - \sigma^2)} \int_{-\infty}^{\infty} \frac{\partial^3 h}{\partial x^3} \frac{\partial \delta(x - x')}{\partial x} \partial x
\tag{23}
$$

and integrating by parts a second time yields the final result:

$$
F = \frac{Ed^3}{12(1 - \sigma^2)} \frac{\partial^4 h}{\partial x^4}
\tag{24}
$$

The films considered in this study consist of two solid layers; therefore, the pressures are doubled.

APPENDIX 2

In the current simulations we have assumed that the top and bottom solid surfaces are the same and that the system is completely symmetrical. It is worth pointing out, however, that the simulation is not limited to this case. We, therefore, include the evolution equation for a system that, rather than extending from $-h$ to h, extends from h_1 to h_2:

$$
\begin{aligned}
\frac{\partial(h_1 - h_2)}{\partial t} = \frac{1}{12\mu} \Bigg(& \frac{\partial}{\partial x} \left[(h_1^3 - h_2^3 + 3h_1 h_2^2 - 3h_1^2 h_2) \frac{\partial P}{\partial x} \right] \\
& + \frac{\partial}{\partial y} \left[(h_1^3 - h_2^3 + 3h_1 h_2^2 - 3h_1^2 h_2) \frac{\partial P}{\partial y} \right] \Bigg)
\end{aligned}
$$

$$(25)$$

The evolution equation dictates the evolution of the total film thickness, however, and it is necessary to assume that the stiffer elastic material will respond to the change in pressures less than the compliant material. For example, if the top layer was twice as stiff as the bottom layer, then the top height would be expected to respond half as much as the bottom height.

References

1. K. Dalnoki-Veress, B. G. Nickel, and J. R. Dutcher, Dispersion-driven morphology of mechanically confined polymer films. *Phys. Rev. Lett.* **82**, 1486 (1999).

2. J. R. Dutcher, K. Dalnoki-Veress, B. G. Nickel, and C. B. Roth, Instabilities in thin polymer films: From pattern formation to rupture. *Macromol. Symp.* **159**, 143 (2000).

3. C. A. Murray, S. W. Kamp, J. M. Thomas, and J. R. Dutcher, Onset and manipulation of self-assembled morphology in freely standing polymer trilayer films. *Phys. Rev. E* **69**, 061612 (2004).

4. G. A. Buxton and N. Clarke, Stress-guided self-assembly in Dutcher films. *Phys. Rev. E* **73**, 041801 (2006).

5. G. A. Buxton and N. Clarke, Structural evolution and control of Dutcher films. *Soft Matter* **2**, 678 (2006).

6. C. B. Roth and J. R. Dutcher, Hole growth as a microrheological probe to measure the viscosity of polymers confined to thin films. *J. Polym. Sci. B* **44**, 3011 (2006).

7. P. J. Yoo, K. Y. Suh, Y. Park, and H. H. Lee, Physical self-assembly of microstructures by anisotropic buckling. *Adv. Mater.* **14**, 1383 (2002).

8. P. J. Yoo, K. Y. Suh, H. Kang, and H. H. Lee, Polymer elasticity-driven wrinkling and coarsening in high temperature buckling of metal-capped polymer thin films. *Phys. Rev. Lett.* **93**, 034301 (2004).

9. J. S. Sharp, D. Vader, J. A. Forrest, M. I. Smith, M. Khomenko, and K. Dalnoki-Veress, Spinodal wrinkling in thin-film poly(ethylene oxide)/polystyrene bilayers. *Eur. Phys. J. E* **19**, 423 (2006).

10. A. Serghei, H. Huth, M. Schellenberger, C. Schick, and F. Kremer, Pattern formation in thin polystyrene films induced by an enhanced mobility in ambient air. *Phys. Rev. E* **71**, 061801 (2005).

11. J. A. Forrest, K. Dalnoki-Veress, J. R. Stevens, and J. R. Dutcher, Effect of free surfaces on the glass transition temperature of thin polymer films. *Phys. Rev. Lett.* **77**, 2002 (1996).

12. N. Bowden, S. Brittain, A. G. Evans, J. W. Hutchinson, and G. M. Whitesides, Spontaneous formation of ordered structures in thin films of metal supported on an elastomeric polymer. *Nature* **393**, 146 (1998).

13. M. Müller-Wiegand, G. Georgiev, E. Oestschulze, T. Fuhrmann, and J. Salbeck, Spinodal patterning in organic-inorganic hybrid layer systems. *Appl. Phys. Lett.* **81**, 4940 (2002).

14. S. J. Kwon and H. H. Lee, Theoretical analysis of two-dimensional buckling patterns of thin metal-polymer bilayer on the substrate. *J. Appl. Phys.* **98**, 063526 (2005).

15. C. Harrison, C. M. Stafford, W. Zhang, and A. Karim, Sinusoidal phase grating created by a tunably buckled surface. *Appl. Phys. Lett.* **85**, 4016 (2004).

16. J. S. Sharp, K. R. Thomas, and M. P. Weir, mechanically driven wrinkling instability in thin film polymer bilayers. *Phys. Rev. E* **75**, 011601 (2007).

17. G. McHale, N. J. ShirtCliffe, S. Aqil, C. C. Perry, and M. I. Newton, Topography driven spreading. *Phys. Rev. Lett.* **93**, 036102 (2004).

18. W.-J. Li, C. T. Laurencin, E. J. Caterson, R. S. Tuan, and F. K. Ko, Electrospun nanofibrous structure: A novel scaffold for tissue engineering. *J. Biomed. Mater. Res.* **60**, 613 (2002).

19. E. Schäffer and U. Steiner, Acoustic instabilities in thin polymer films. *Eur. Phys. J. E* **8**, 347 (2002).

20. M. D. Morariu, E. Schaffer, and U. Steiner, Molecular forces caused by the confinement of thermal noise. *Phys. Rev. Lett.* **92**, 156102 (2004).

21. S. Herminghaus, Dynamical instability of thin liquid films between conducting media. *Phys. Rev. Lett.* **83**, 2359 (1999).

22. N. E. Voicu, S. Harkema, and U. Steiner, Electric-field-induced pattern morphologies in thin liquid films. *Adv. Funct. Mat.* **16**, 926 (2006).

23. C. S. Chen, M. Mrksich, S. Huang, G. M. Whitesides, and D. E. Ingber, Geometric control of cell life and death. *Science* **276**, 1425 (1997).

24. R. Windisch, P. Heremans, A. Knobloch, P. Kiesel, G. H. Dohler, B. Dutta, and G. Borghs, Light-emitting diodes with 31% external quantum efficiency by outcoupling of lateral waveguide modes. *Appl. Phys. Lett.* **74**, 2256 (1999).

25. Z. Huang, W. Hong, and Z. Suo, Evolution of wrinkles in hard films on soft substrates. *Phys. Rev. E* **70**, 030601 (2004).

26. R. Huang, and S. H. Im, Dynamics of wrinkle growth and coarsening is stressed thin films. *Phys. Rev. E* **74**, 026214 (2006).

27. C. Li, X. Zhang, and Z. Cao, Triangular and Fibonacci number patterns driven by stress on core/shell microstructures. *Science* **309**, 909 (2005).

28. Y. Wang, S. Gao, W.-H. Ye, H. S. Yoon, and Y.-Y. Yang, Co-delivery of drugs and DNA from cationic core-shell nanoparticles self-assembled from a biodegradable copolymer. *Nat. Mat.* **5**, 791 (2006).

29. K. Y. Suh, S. M. Seo, P. J. Yoo, and H. H. Lee, Formation of regular nanoscale undulations on a thin polymer film imprinted by a soft mold. *J. Chem. Phys.* **124**, 024710 (2006).

30. J. Becker, G. Griin, R. Seemann, H. Mantz, K. Jacobs, K. R. Mecke, and R. Blossey, Complex dewetting scenarios captured by thin-film models. *Nat. Mat* **2**, 59 (2003).

31. M. B. Williams and S. H. Davis, Nonlinear theory of film rupture. *J. Colloid Interface Sci* **90**, 220 (1982).

32. A. Oron, S. H. Davis, and S. G. Bankoff, Long-scale evolution of thin liquid films. *Rev. Mod. Phys.* **69**, 931 (1997).

33. H. Schlichting, *Boundary-Layer Theory1*, McGraw-Hill, New-York, 1979.

34. J. Israelachvili, *Intermolecular and Surface Forces*, Academic Press, London, 1992.

35. L. D. Landau and E. M. Lifshitz, *Theory of Elasticity*, Pergamon Press, Oxford, 1970.

36. M. H. Muser, L. Wenning, and M. O. Robbins, Simple microscopic theory of Amontons's laws for static friction. *Phys. Rev. Lett*, **86**, 1295 (2001).

37. W. H. Press, B. P. Flannery, S. A. Teukolsky, and W. T. Vetterling, *Numerical Recipes: The Art of Scientific Computing*, Cambridge University Press, Cambridge, 1986.

38. C. M. Stafford, C. Harrison, K. L. Beers, A. Karim, EJ. Amis, M. R. Van Landingham, H.-C. Kim, W. Volksen, R. D. Miller, and E. E. Simony, A buckling-based metrology for measuring the elastic moduli of polymeric thin films. *Nature*, **3**, 545 (2004).

39. R. Konnur, K. Kargupta, and A. Sharma, Instability and morphology of thin liquid films on chemically heterogeneous substrates. *Phys. Rev. Lett.*, **84**, 931 (2000).

40. N. Clarke, Toward a model for pattern formation in ultrathin-film binary mixtures. *Macromolecules* **38**, 6775 (2005).

DYNAMICS OF THERMOTROPIC LIQUID CRYSTALS ACROSS THE ISOTROPIC–NEMATIC TRANSITION AND THEIR SIMILARITY WITH GLASSY RELAXATION IN SUPERCOOLED LIQUIDS

DWAIPAYAN CHAKRABARTI

Department of Chemistry, University of Cambridge, Cambridge CB2 1EW, United Kingdom

BIMAN BAGCHI

Solid State and Structural Chemistry Unit, Indian Institute of Science, Bangalore 560 012, India

CONTENTS

Advances in Chemical Physics, Volume 141, edited by Stuart A. Rice
Copyright © 2009 John Wiley & Sons, Inc.

I. INTRODUCTION

The study of molecular relaxation in soft condensed matter systems has unraveled many fascinating phenomena, leading to better understanding of the relationship between their structure and dynamics [1]. Molecular motion in these systems entails widely different length- and timescales, which are probed by a variety of experimental techniques. Owing to the great advancement in experimental techniques in recent years, one can now study electronic, vibrational, and rotational relaxation processes *directly in the time domain*. While electronic and vibrational relaxation processes are mostly coupled to intramolecular modes, rotational relaxation is more susceptible to evolution of the local environment, often involving intermediate length and timescales and thereby bridging the gap between the two extremes. In addition, a rather large number of experimental techniques, namely infrared spectroscopy, fluorescence depolarization, nuclear magnetic resonance spectroscopy, dielectric relaxation spectroscopy, and optical Kerr effect, provide access to rotational relaxation. Together with translational motion (which can be studied by neutron scattering), rotational relaxation has thus been widely used to probe the dynamics of condensed phases, in particular, of soft condensed matter systems, which are also sometimes referred to as complex systems.

The field of soft condensed matter/complex systems spans a seemingly diverse spectrum, including isotropic liquids with short-range order at the one end, and liquid crystals with long-range order in their anisotropic phases, often called mesophases, at the other [2–4]. Anisotropy in molecular shape plays a crucial role in the rich phase behavior that thermotropic liquid crystals exhibit upon cooling from the high-temperature isotropic phase. The simplest of the ordered phases is the nematic, which has long-range orientational order without the presence of any long-range translational order. Although the orientational dynamics across the isotropic–nematic (*I–N*) phase transition have been discussed in great detail, much of the dynamics seem to have been studied in the past at long times. The rotational relaxation slows down dramatically near the *I–N* phase boundary, often extending to the timescales of milliseconds.

A series of optical Kerr effect measurements with a number of thermotropic liquid crystals near the *I–N* phase transition have recently resulted in renewed interests—in particular, at short to intermediate timescales where power law decay has been observed surprisingly [5–7]. Another remarkable feature of these measurements is the revelation of a striking similarity in the orientational dynamics between the isotropic phase of thermotropic liquid crystals near the *I–N* phase boundary and supercooled molecular liquids [8]. The puzzles that dynamics of supercooled liquids pose have drawn intense research activities over several decades from experimental, theoretical, and computational researchers alike. One of the central issues in this area has been strongly nonexponential relaxation, discussed often in terms of power laws and stretched exponential function. Understanding the molecular mechanisms for complex slow dynamics in supercooled liquids is still elusive, though significant progress has been made in last few decades [9–11]. Computer simulation studies, which often provide a microscopic view, have proved to be indispensable to understand structure and dynamics of soft matter systems [12]. A number of theoretical and computational studies have recently been devoted to explore and understand the apparent analogy between thermotropic liquid crystals and supercooled molecular liquids in a variety of approaches.

In the present review we shall focus primarily on the orientational dynamics of thermotropic liquid crystals across the *I–N* transition. We shall also discuss the dynamics of supercooled liquids in connection with their apparently analogous relaxation behavior. The main aim of the present review is understanding of the power law decay in these systems and the origin of their similar dynamical behavior. To this end, we shall review experimental, theoretical, and computer simulation results together. In particular, we shall address the influence of the nature of the *I–N* phase transition in the dynamics that thermotropic liquid crystals exhibit in the vicinity of the *I–N* phase boundary. The issue assumes special relevance because the emergence of slow dynamics in supercooled liquids is often attributed to a dynamical transition. The weakly first-order nature of the *I–N* phase transition, however, suggests growth of significant static correlations as the *I–N* phase boundary is approached from the isotropic side. At a fundamental level, the potential energy surface of a system governs its structure, thermodynamics, and dynamics [13]. The review will also highlight the scope of energy landscape analysis in search of a plausible unified description of slow dynamics in soft condensed matter systems.

It is beyond the scope of this review to be exhaustive in the field of supercooled liquids that has drawn intense research activities over several decades. Reviews that are exclusive for this field and deal with specific topics in considerable detail are recommended for supplemental reading [9–11]. In view of the scope this chapter, the next section provides the readers with a brief introduction to the systems of interest and associated nomenclature. Section III sets up the background by reviewing experimental results on the dynamics of thermotropic liquid crystals across the I–N transition, then introducing the central issues in the dynamics of supercooled liquids, and finally comparing the dynamics of the two systems in the light of recent experiments. Section IV presents a summary of some of the well-known theoretical approaches to liquid crystals. Section V provides a detailed account of computational efforts. Finally, we conclude in Section VI with a list of problems for future work.

II. OVERVIEW OF THE SYSTEMS

A. Thermotropic Mesophases and Order Parameters

Liquid crystals are of considerable technological importance that arise in most cases from their ability to switch in response to external stimuli coupled with their optical, electrical, and mechanical properties [14]. They are broadly classified into *thermotropic* or *lyotropic*, depending upon the stimulations that drive their phase behavior. While thermotropic liquid crystals display rich phase behavior upon temperature variation, lyotropic mesomorphs are observed in multicomponent systems under the influence of a solvent. Lyotropic mesogens are typically amphiphilic in the sense that they are composed of both lyophilic (solvent-attracting) and lyophobic (solvent-repelling) moieties. On the other hand, anisotropy in molecular shape is the key for thermotropism. The general shape of mesogens serves the basis of further classification of thermotropic liquid crystals into *calamitic* or *discotic*. Calamitic liquid crystals consist of rod-shaped mesogens while disc-shaped mesogens comprise discotic liquid crystals.

The concepts of symmetry and order parameter play crucial roles in the classification of various mesophases that thermotropic liquid crystals exhibit. The isotropic phase, which has no long-range order, is the most symmetric one. In the *nematic* phase, there is a long-range, orientational order arising from a preferred alignment of the mesogens, on the average, with their symmetry axis parallel to a spatial direction. The preferred direction is represented by a unit vector $\hat{n}$, called the *director*. The nematic phase is devoid of any long-range translational order. As a result, there is only diffuse scattering in the X-ray diffraction spectra. The long-range orientational order is reflected in all macroscopic tensor properties.

For example, the nematic phase formed by calamitic systems is optically positive and uniaxial, with the optical axis along $\hat{\mathbf{n}}$. In the absence of an external field, the two directions, $\hat{\mathbf{n}}$ and $-\hat{\mathbf{n}}$ are, however, indistinguishable. The nematic phase, formed by the disc-shaped mesogens, is denoted by N_D to distinguish it from the one formed by the calamitic systems. Unlike the latter phase, the N_D phase is optically and diamagnetically negative [15].

The *cholesteric* mesophase is observed mostly in the calamitic systems either with a pure, optically active mesogen or even with a small amount of a non-mesomorphic substance added to a nematic phase. The cholesteric mesophase is orientationally ordered, but the preferred direction of local molecular alignment rotates following a helical pattern. The director is spatially periodic with the periodicity in the order of several thousands angstroms, comparable with the wavelength of visible light. Such periodicity results in Bragg scattering of light beams.

The appearance of long-range translational order in addition to long-range orientational order results in a *smectic* phase for calamitic liquid crystals. In a smectic phase, the mesogens are distributed in layers leading to a, spatially modulated density. There are several variants of the smectic phase depending on the degree of order within the layers and/or the presence of any tilt of the director with respect to the layer normal. For example, the director is parallel to the layer normal for the smectic-A *(SmA)* and smectic-B *(SmB)* phases, but is tilted with respect to the layer normal in the smectic-C *(SmC)* phase. While there is no long-range order within a layer in the *SmA* and *SmC* phases, the *SmB* phase has hexagonal order within each layer.

For discotic liquid crystals, superposition of long-range translational order on long-range orientational order results in a *columnar* phase rather than a smectic phase. The columnar phase in its simplest form consists of disc-shaped mesogens stacked one on top of another aperiodically to form columnar structures, which thus have liquid-like order in one dimension. These columnar structures constitute a two-dimensional lattice, which commonly has either hexagonal or rectangular symmetry. The spatial arrangement of the mesogens within a column may vary, resulting in an ordered or a tilted columnar phase.

The most fundamental character of thermotropic mesophases, from a microscopic point of view, is the presence of long-range orientational order while long-range translational order is limited or completely absent. In order to study a mesophase, it is often useful to define an order parameter, which provides a quantitative measure of the order present therein. The order parameters serve, in particular, to monitor phase transformations. To a good approximation, both rod-shaped and disc-shaped mesogens can be modeled as ellipsoids of revolution [16], as illustrated in Fig. 1. In the spirit of Fig. 1, we introduce the order parameters for a system of N ellipsoids of revolution, each having a single-site representation. We represent the ith ellipsoid of revolution in terms of the position of its center of mass $\mathbf{r}_i$ and a unit vector $\mathbf{e}_i$ along the principal symmetry axis.

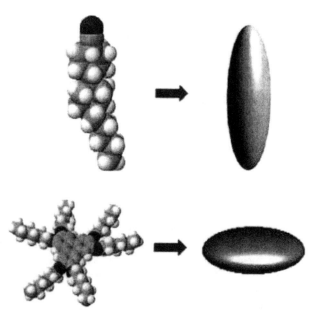

Figure 1. Atomistic and molecular level models for rod-shaped (top) and disc-shaped (bottom) mesogens. Examples shown: 4′-n-pentyl-4-cyanobiphenyl (5CB) (rod-shaped) and hexakis(n-hexyloxy)triphenylene (HAT6) (disc-shaped). (This figure is Reproduced from Ref. 16.)

For a measure of orientational order, one defines an orientational order tensor $\mathbf{S}$:

$$S_{\alpha\beta} = \frac{1}{N}\sum_{i=1}^{N}\frac{1}{2}\left(3e_{i\alpha}e_{i\beta} - \delta_{\alpha\beta}\right) \tag{1}$$

where $\alpha, \beta = x, y, z$ are the indices referring to the space fixed frame, $e_{i\alpha}$ is the α-component of the unit vector $\mathbf{e}_i$, and $\delta_{\alpha\beta}$ is the Kronecker symbol. $\mathbf{S}$ is a real, symmetric, and traceless tensor. For the more common uniaxial nematic phase, the orientational order tensor can be diagonalized into a form

$$\mathbf{S} = \begin{pmatrix} -\frac{1}{2}S + \xi & 0 & 0 \\ 0 & -\frac{1}{2}S - \xi & 0 \\ 0 & 0 & S \end{pmatrix} \tag{2}$$

where the biaxiality parameter ξ is identically equal to zero. The scalar orientational order parameter S is thus obtained as the largest eigenvalue of the orientational order tensor [2]. The corresponding eigenvector gives the director $\hat{\mathbf{n}}$.

For a measure of the one-dimensional translational order in a layer perpendicular to the director, the smectic order parameter ψ is defined as the magnitude of the Fourier component of the normalized density along the director [17]:

$$\psi = \frac{1}{N}\left|\sum_{j=1}^{N} \exp(ik_z z_j)\right| \tag{3}$$

where $i = \sqrt{-1}, k_z = 2\pi/d, d$ being the periodicity of the smectic layers, and z_j is the z-coordinate of the center of mass of the jth ellipsoid of revolution, the z-axis being taken to be along the director. The order parameter ψ vanishes in the nematic phase as $N \rightarrow \infty$ and assumes a finite value less than unity for the SmA and SmB phases.

In order to obtain a measure of hexagonal order within a smectic layer and thus to distinguish between the SmA and SmB phases, the sixfold bond orientational parameter Ψ_6 is defined [18]. To this end, we imagine a bond between the centers of mass of two neighbors as the straight line connecting them. Taking a bond between an ellipsoid of revolution and one of its nearest neighbors for reference, we define ϕ_{jk} as the angle between the bond connecting an ellipsoid of revolution k with one of its nearest neighbors j and the reference bond. The sixfold bond orientational order parameter Ψ_6 is then defined as

$$\Psi_6 = \frac{1}{N}\left|\sum_{k=1}^{N} \Psi_k\right| \tag{4}$$

where

$$\Psi_k = \frac{1}{6}\sum_{j} \exp\left(i6\phi_{jk}\right) \tag{5}$$

the sum going over all the nearest neighbors j of the ellipsoid of revolution k. In the absence of hexagonal order within a smectic layer, the order parameter Ψ_6 is averaged to zero as $N \rightarrow \infty$.

B.   Supercooled Liquids and the Glass Transition

A liquid can be supercooled at a temperature below its freezing temperature T_m by cooling sufficiently fast to circumvent crystallization [9–11]. The supercooled liquid regime has a lower bound at the glass transition temperature T_g. At T_g, which usually occurs around $2T_m/3$ [9, 10], the metastable liquid transforms to an amorphous solid, called glass, that lacks long-range spatial order. The glass

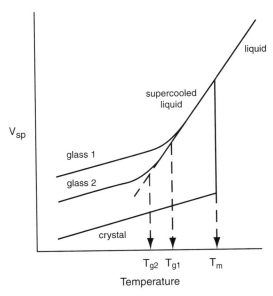

Figure 2. Schematic representation of the specific volume V_{sp} of a glass-forming liquid as a function of temperature T. The dependence of the glass transition temperature T_g on the cooling rate is shown. A slower cooling rate was applied for the formation of glass 2 as compared to that of glass 1. (This figure is reproduced from Ref. 11.)

transition is marked with an abrupt, though continuous, decrease in the rate of change of volume or enthalpy with respect to temperature to a value comparable to that of a crystalline solid. The intersection of the extrapolated portions of the curve on either side of the transformation range provides a way to define T_g, as illustrated in Fig. 2 with a schematic representation of the temperature dependence of the specific volume [9, 11]. In calorimetric experiments, one observes a rapid decrease in the isobaric heat capacity from a liquid-like value to a crystal-like one when a glass-forming liquid is subjected to isobaric cooling. Such a drop in the isobaric heat capacity is taken to be the signature of the glass transition. When subsequently reheated, there occurs an overshoot of the isobaric heat capacity. T_g is alternatively defined to be the temperature that marks the onset of the overshoot of the heat capacity on being reheated [19, 20]. Figure 3 schematically shows the typical temperature behavior of the enthalpy H and the heat capacity C_p of a glass-forming liquid as it undergoes isobaric cooling and subsequent reheating at a constant rate over a wide temperature range across the transition.

The avoidance of crystallization upon rapid cooling results in an excess of entropy of a supercooled liquid over that of the corresponding crystal. The excess entropy diminishes upon further lowering the temperature because the heat

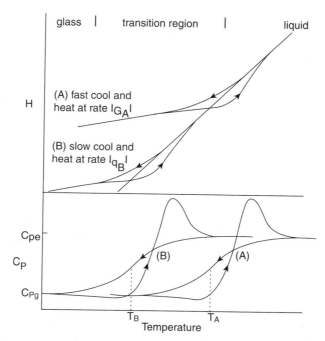

Figure 3. Schematic behavior of enthalpy and isobaric heat capacity of a glass-forming liquid subject to isobaric cooling and subsequent reheating at a constant rate through the glass transition region. The two temperatures T_A and T_B correspond to two different values obtained for the glass transition temperature T_g for different rates. (Reproduced from Ref. 19.)

capacity of a liquid is higher than that of the crystal. If this trend were to continue, the liquid entropy would become (unphysically) smaller than the crystal's entropy below a temperature, known as the Kauzmann temperature T_k [21]. Since the entropy of the crystal approaches zero as T tends to zero, the entropy of the liquid would eventually become negative in conflict with the third law of thermodynamics. In practice, the glass transition intervenes at a temperature $T_g > T_k$. The apparently paradoxical implication of the extrapolated heat capacity for a supercooled liquid into a regime that is beyond direct observation is a scenario what constitutes an entropy crisis [21–23].

In laboratory, T_g, defined in either way, is found to increase with increasing cooling rate [19]. However, the dependence of T_g on the cooling rate is weak. Typical cooling rates in laboratory experiments are $0.1-100\,\text{K min}^{-1}$ [11]. T_g changes by 3–5 K upon change in cooling rate by an order of magnitude [11]. The dependence of T_g on the cooling rate and the absence of a discontinuous change in any physical property together seem to rule out the involvement of a true phase transition. Now there appears a consensus that the laboratory glass transition is a kinetic phenomenon caused by a dramatic slowdown of molecular motion once the crystallization is avoided by sufficiently fast cooling. Upon cooling below T_m, the rearrangement of molecules becomes progressively slow. At low temperatures, molecules fail to adequately sample configurations in the time permitted by the cooling rate for equilibrium to be maintained. The structure of the material eventually appears *frozen* on the laboratory timescale, resulting in the glass transition across a narrow transformation range. T_g increases with increasing cooling rate because faster cooling allows less time for configurational sampling at each temperature, inducing falling out of the equilibrium at higher temperatures [19].

III. BACKGROUND

A. Dynamics across the Isotropic–Nematic Phase Transition

One of the most notable effects of the onset of long-range orientational order on crossing the *I–N* transition from the isotropic side is that the translational diffusion becomes anisotropic [24]. In the uniaxial nematic phase, the diffusion description invokes $D_\parallel$ and $D_\perp$, the principal components of the second-rank diffusion tensor, for translational motion parallel and perpendicular to the macroscopic director, respectively [2]. A variety of experimental techniques, namely radioactive tracer diffusion, quasi-elastic neutron scattering (QENS), nuclear magnetic resonance (NMR), forced Rayleigh scattering (FRS), and optical microscopy, probe the anisotropic translational diffusion in the nematic phase [24–29]. The most general feature that these experiments reveal for calamitic liquid crystals is that $D_\parallel > D_\perp$ throughout the nematic range. A monotonic decrease in $D_\perp$ is observed as temperature drops.

A surprising result on the behavior of $D_\parallel$ was reported by an early molecular dynamics simulation study of rod-like hard ellipsoids of revolution [30]. $D_\parallel$ was found to increase initially before it started falling as the density was increased across the nematic phase [30]. Such nonmonotonic behavior of $D_\parallel$ was attributed to the opposing effects of increasing molecular alignment and increasing collision rate. Similar observations were made subsequently in molecular dynamics simulations of a system of ellipsoids of revolution interacting through a soft pair potential and in Brownian dynamics simulations of hard spherocylinders [31, 32], However, the signature of such nonmonotonic behavior of $D_\parallel$ in the nematic phase of calamitic liquid crystals is yet to be conclusively found in experiments [27–29].

We now discuss the orientational dynamics of mesogens. The orientational dynamics in the isotropic phase of thermotropic liquid crystals near the *I–N* transition have drawn much attention over the years [5–7, 33–40]. The focus was initially on the verification of the Landau–de Gennes (LdG) theory, which predicts a long-time exponential decay with a strongly temperature-dependent time constant [33–37]. However, a series of optical heterodyne-detected optical Kerr effect (OHD–OKE) measurements by Fayer and co-workers [5–7] recently revealed a rather complex relaxation pattern in the isotropic phase of a number of calamitic systems near the *I–N* transition. In these time-domain experiments with molecular liquids, the observable is the time derivative of the correlation function of the total linear susceptibility (measured at zero wavevector) for times greater than ~1 ps. In this time domain, the linear susceptibility may be considered as the sum of molecular polarizability tensors for a liquid with rigid molecules. The OKE signal is thus essentially a measure of the time derivative of the polarizability–polarizability time correlation function and is sensitive mainly to the collective orientational dynamics at long times and to both orientational and translational dynamics at short-to-intermediate times. The translational dynamics affects OKE signal through interaction induced polarizability effects. However, these translational effects are expected to be restricted below 10–50 ps. It is safe to assume that above 100 ps, the main contribution to OKE comes from orientiational dynamics.

In a series of OHD–OKE experiments early in this decade, Gottke et al. [5] investigated the orientational dynamics of three calamitic systems in the isotropic phase at several temperatures down to $\sim T_{IN}$ for more than five decades of time (1 ps to 100 ns). The systems studied were 4'-pentyloxy-4-biphenylcarbonitrile (5-OCB), 4'-pentyl-4-biphenylcarbonitrile (5-CB), and 1-isothiocyanato-(4-propyl-cyclohexyl)benzene (3-CHBT), having aspect ratios between 3 and 4. The OKE signal was found to decay exponentially only at long times beyond several nanoseconds. The temperature dependence of the time constant $\tau(T)$ of this long-time exponential decay was well-described by the Landau–de Gennes theory, which predicts

$$\tau(T) = \frac{V_{eff}^*\eta(T)}{k_B(T - T^*)} \tag{6}$$

where $\eta(T)$ is the viscosity which has temperature dependence, V_{eff}^* is a parameter providing a measure of the effective molecular volume, and k_B is the Boltzmann constant. The striking feature of this study, however, concerned short to intermediate times where a power law decay ($\sim t^{-z}$) was observed in all three cases. The values of the power law exponent z, which appeared to be temperature independent, were found in the range between 0.63 and 0.76 for the systems studied. The long-time exponential decay was attributed to the randomization of the pseudonematic domains that developed within the isotropic phase near the *I–N* transition while the short-to-intermediate time power law decay was suggested to reflect the intradomain dynamics. A typical example of the complex relaxation pattern revealed in these experiments is illustrated in Fig. 4. In subsequent OHD–OKE experiments with the calamitic

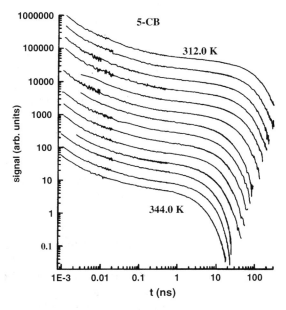

Figure 4. Time dependence of the OHD–OKE signal in a log–log plot at a series of temperatures in the isotropic phase for 5-CB over five decades of time. The data sets were offset for clarity of presentation. (Reproduced from Ref. 6.)

liquid crystal $4'$-octyl-4-biphenylcarbonitrile (8-CB) in the isotropic phase near the I–N transition, Fayer and co-workers [7] observed a similar complex relaxation pattern. The power law exponent z was found to decrease linearly with increasing aspect ratio of the mesogens [7].

More recently, Fayer and co-workers studied the orientational dynamics of a homeotropically aligned nematic liquid crystal, 5-CB, over more than six decades of time (500 fs to 2μs) using OHD–OKE experiments [41]. The OKE signal was found to follow a rather temperature-independent power law decay at long times spanning over more than two decades in contrast to the strongly temperature-dependent exponential decay that is commonly observed in the isotropic phase near the I–N transition. On shorter timescales (from ~ 3 ps to $\sim$1 ns), a temperature-dependent power law ($\sim t^{-z}$) was observed. The exponent of the temperature-dependent power law, called the intermediate power law to distinguish it from the final power law observed at long times, was found to increase linearly with the increase in the orientational order parameter.

B. Dynamics across the Supercooled Regime

We shall only briefly discuss below some of the well-known anomalies in this much studied and much reviewed field.

1. Temperature Dependence of the Shear Viscosity

The slowdown in dynamics is manifested by a phenomenal increase in the shear viscosity η (or equivalently the characteristic relaxation time τ) as a supercooled liquid is cooled toward T_g. The growth of the viscosity can be as much as 14 orders of magnitude [9–11]. In fact, T_g, in yet another convenient definition, is taken to be the temperature at which the shear viscosity reaches 10^{13} poise [9–11]. The Arrhenius plot, which shows the shear viscosity η in the logarithmic scale versus the inverse of T_g-scaled temperature as illustrated in Fig. 5, underlies Angell's empirical classification scheme for glass-forming liquids [42–44]. A linear behavior, describing the Arrhenius temperature dependence, is a characteristic of strong liquids, such as SiO_2 and GeO_2, which form network structures. The deviation from the Arrhenius behavior is taken as a signature of fragile liquids. The $\eta(T)$ data (also $\tau(T)$ data) for fragile liquids

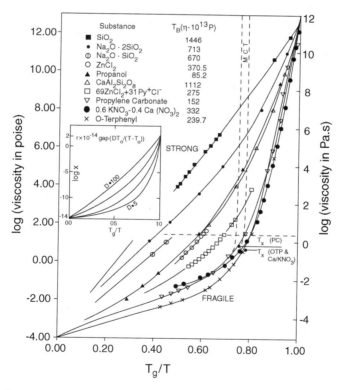

Figure 5. Arrhenius representation of liquid viscosity illustrating Angel's strong–fragile classification scheme. Here T_g is the glass transition temperature defined in terms of $\eta(T_g) = 10^{13}$ poise. (Reproduced from Ref. 44.)

are most frequently fitted to the so-called Vogel–Fulcher–Tammann (VFT) equation [45]

$$\eta(T) = A \, \exp[B/(T - T_0)] \qquad (7)$$

where A, B, and T_0 are temperature-independent parameters that are found to remain constant over a temperature range in which η varies by 2–4 orders of magnitude [9, 10]. T_0, known as the Vogel temperature, is the temperature of apparent divergence of the viscosity and is typically 30–50 K below T_g. The strong liquid limit is described by $T_0 = 0$.

2. Nonexponential Relaxation

A basic feature of the response of fragile liquids to various perturbations is the pronounced nonexponential relaxation behavior. The relaxation function $\phi(t)$ typically exhibits a two-step feature. The fast relaxation at short times is generally associated with vibrational degrees of freedom. The long-time decay of the relaxation function $\phi(t)$, which is governed by the structural relaxation, can often be described by the stretched exponential or the Kohlrausch–Williams–Watts (KWW) function

$$\phi(t) = C \, \exp[-(t/\tau)^{\beta_{KWW}}] \qquad (8)$$

where τ is the characteristic relaxation time, β_{KWW} is the stretch exponent $(0 < \beta_{KWW} < 1)$, and C is the renormalization constant. The limiting value of $\beta_{KWW} = 1$ describes an exponential relaxation that is a characteristic of strong liquids. For a typical fragile glass-former, β_{KWW} generally decreases from near 1 at high temperatures to below 0.5 close to T_g with a display of monotonic temperature dependence [10, 46]. Such a stretched exponential description is particularly suitable for long-time behavior of $\phi(t)$, which typically exhibits a two-step feature.

The origin of the nonexponential relaxation in supercooled liquids could be debated over two fundamentally different propositions [47]. The homogeneous description portrays a picture where each molecule undergoes a nearly identical relaxation in an intrinsically nonexponential fashion [47]. The other description involves the growth of spatially heterogeneous relaxing domains where each individual displays a nearly exponential relaxation with the relaxation time varying significantly among different domains [46, 48, 49]. The latter description has gained in strength with supporting evidence provided by recent experimental and computational studies [46, 48–65].

3. Dynamic Decoupling Phenomena

The dielectric loss spectra, have traditionally yielded useful information on relaxation in liquids [66, 67]. At sufficiently high temperatures, only one peak appears in the spectra, indicating a single relaxation mechanism. On approaching T_g upon cooling, this peak often gets split into two, one corresponding to the slow α relaxation and the other to the faster β relaxation [68]. This decoupling is known as the α–β bifurcation. The α peak, which is associated with the structural relaxation, exhibits a non-Arrhenius temperature dependence for the peak-frequency before disappearing at T_g [68]. The frequency of the β peak shows the Arrhenius temperature behavior, and the peak continues to exist even below T_g [68]. The extrapolation of the experimental data previously suggested the bifurcation temperature to be $T_B \simeq 1.2T_g$ [69], but recent broadband dielectric relaxation measurements have shown that α and β relaxations merge together well above this extrapolated temperature (possibly around T_m) [70].

The slowdown of molecular motion with the growth of the shear viscosity across the supercooled regime has been extensively investigated with a variety of experimental techniques. The Stokes–Einstein (SE) relation describes a coupling between the translational diffusion coefficient $D_t(T)$ and the shear viscosity $\eta(T)$ at a temperature T:

$$D_t(T) \propto \frac{T}{\eta(T)} \tag{9}$$

Similarly, the rotational diffusion coefficient at temperature $T, D_r(T)$, is predicted to be coupled to $\eta(T)$ by the Debye–Stokes–Einstein (DSE) relationship:

$$D_r(T) = \frac{1}{l(l+1)\tau_l(T)} \propto \frac{T}{\eta(T)} \tag{10}$$

where $\tau_l(T)$ is the lth rank orientational correlation time at T. While the SE relationship holds good for moderately supercooled liquids, it breaks down at $T \simeq 1.2T_g$ as has been found in a variety of experimental measurements since the early 1990s [53, 71–73]. The breakdown of the SE relationship was, however, claimed to be suggested much earlier in an analysis of crystal growth rate measurements [74], which revealed a stronger temperature dependence of $\eta(T)$ as compared to $D_t(T)$ [75]. The DSE relationship, on the contrary, continues to conform well to the experimental data of the rotational correlation time $\tau_c(= \tau_2)$, obtained from NMR as well as from time-resolved optical spectroscopic studies [50, 53, 71, 72]. The SE and DSE relationships together suggest a coupling between translational diffusion and rotational diffusion with the prediction of the temperature independence of the product $D_t(T)\tau_c(T)$ [53]. The breakdown of the SE relationship results in yet another decoupling, this time between translational diffusion and rotational diffusion. Near T_g, the translational diffusion coefficient can be more than two orders of magnitude larger than what is expected on the basis of the rotational correlation time [53, 76]. Several attempts have been made to interpret such decoupling [55, 77, 78]. It has been often argued that this decoupling is a consequence of spatially heterogeneous dynamics [46, 55]. In the heterogeneous picture, translational diffusion has dominant contribution from the regions with faster-than-average mobility, in contrast to the rotational correlation time that derives major contribution from the regions having slower-than-average mobility [46].

C. Similarity in Orientational Dynamics Between the Isotropic Phase of Liquid Crystals and Supercooled Liquids

Based on their OHD–OKE data for the isotropic phase of four calamitic liquid crystals and for five supercooled molecular liquids, Fayer and co-workers [8] recently drew a comparison between these two kinds of systems in their orientational relaxation behavior. The four calamitic systems considered were 8-CB, 5-OCB, 5-CB, and 3-CHBT, and the five supercooled molecular liquids investigated were benzophenone (BZP), 2-biphenylmethanol (BMP), ortho-terphenyl (OTP), salol, and dibutylphthalate (DBP). The liquid crystalline systems were investigated over a range of temperatures from well above T_{IN} down to $\sim T_{IN}$; the supercooled liquids were studied from above the melting point down to $\sim T_c$, the mode coupling theory critical temperature. The OHD–OKE signal was probed over a time window spanning from subpicoseconds to tens of nanoseconds. The most striking feature that emerged from this comparison was that an identical description conformed well to the OHD–OKE data for all nine liquids. The description involved a power law at short times, and an exponential decay at long times, with a second power law in the crossover region [8]. Figure 6 illustrates the similarity in the complex relaxation pattern with the OHD–OKE data shown for benzophenone (BZP) and $4'$-pentyloxy-4-biphenylcarbonitrile (5-OCB).

Over the wide range of temperatures investigated, the OHD–OKE data for each of the nine liquids were found to be well-described for times beyond ~ 2 ps by the same fitting function

$$F(t) = [pt^{-z} + dt^{b-1}] \exp(-t/\tau) \qquad (11)$$

where p and d are the amplitudes of the power law components [8]. The first term within the square brackets with the exponent $z \leq 1$ corresponds to the intermediate power law, so called as it appeared following the ultrafast dynamics and preceding the long-time slow dynamics. The second term is the von Schweidler law, which appeared in the crossover region with the values of b varying between 0.73 and 0.97. The exponential decay described the long-time structural relaxation in supercooled liquids and was consistent with the prediction of the LdG theory for liquid crystals. One of the key features was that the exponent of the intermediate power law was independent of temperature [8]. It was conjectured that similar underlying physical mechanisms were operative for

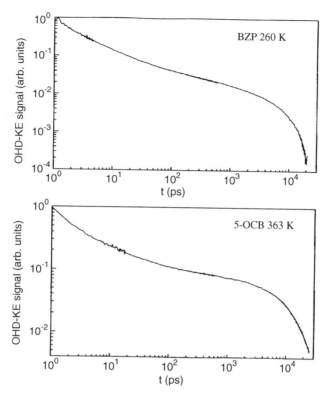

Figure 6. Time dependence of the OHD–OKE signal in a log–log plot for benzophenone (BZP) (top) and 4′-pentyloxy-4-biphenylcarbonitrile (5-OCB) (bottom) from 1 ps to more than 20 ns. (Reproduced from Ref. 8.)

orientational dynamics in these two kinds of molecular liquids even in the absence of a thermodynamic singularity at the glass transition [8].

The intermediate power law observed in time-domain OHD–OKE experiments has manifestation in frequency-domain measurements as well. When expressed in terms of frequency dependence, the intermediate power law in the time domain (corresponding to nearly logarithmic decay of the correlator) translates into a very weak frequency dependence of the imaginary part of the susceptibility, a phenomenon known as nearly constant loss (NCL) [8]. This has been observed for a number of glass-formers, especially for ionic systems [79–81]. In particular, the depolarized light-scattering (DLS) data of polyisobutylene suggested the existence of NCL over a rather broad frequency range extending beyond 10 GHz across the glass transition temperature [80]. However, the origin of NCL is also not well understood at present [82]. In a recent work, Brodin and Rössler discussed the OKE and DLS data of a number of molecular liquids on a comparative note [83].

These authors noted that the intermediate power law (i.e., $\sim t^{-1+\gamma}$, with a small positive γ) of the OKE data was formally equivalent to the excess wing $v^{-\gamma}$ in the frequency-dependent susceptibility, the latter discussed in the dielectric literature since 1951. Brodin and Rössler argued that the intermediate power law observed in the OKE data was in essence a manifestation of the excess wing of the corresponding frequency-domain data, known long since from broadband dielectric spectroscopy and anticipated from DLS studies of supercooled liquids [83]. More recently, these authors showed that the excess wing was an equally common feature of the DLS data and discussed the merits of the Mode coupling theory analysis of the time and frequency-domain data [84].

IV. THEORETICAL FRAMEWORK

The thermodynamics of the I–N phase transition has been extensively investigated for resolving the issue concerning the order of the transition. Following the Ehrenfest scheme, a phase transition is classified into a first-order transition or a second-order one, depending upon the observation of finite discontinuities in the first or the second derivatives of the relevant thermodynamic potential at the transition point. An experimental assessment of the order of the I–N transition has turned out to be not a simple task because of the presence of only small discontinuities in enthalpy and specific volume. It follows from high-resolution measurements that I–N transition is weakly first order in nature [85].

A. The Onsager Theory

Onsager's treatment of the density driven transition in a system of long hard rods in the 1940s is reckoned as the first statistical mechanical theory of phase transitions in liquid crystals [86, 87]. Hard rods can be modeled by sphero-cylinders with length L and diameter D [87]. The theory invokes the assumptions that (i) the rods cannot interpenetrate each other, (ii) the rods are very long ($L \gg D$), and (iii) the volume fraction $\Phi = \rho \frac{1}{4}\pi L D^2$, ρ being the number density, is much less than unity. With these assumptions, the rather involved approach of Onsager's theory is based on an expansion of the free energy in a power series in density involving virial coefficients. The approach takes into account only the first term of the expansion, associated with the second virial coefficient. Onsager's second virial approximation is shown to be valid only for very high values of L/D—that is, for very long and thin rods. Onsager's treatment results into the prediction of a first-order transition from the isotropic to the nematic phase. The volume fraction Φ occupied by the rods in the nematic phase just at the transition point is

$$\Phi_n^c = 4.5D/L \tag{12}$$

At the same point, the value of Φ for the isotropic phase in equilibrium with the nematic phase is

$$\Phi_i^c = 3.3D/L \tag{13}$$

The order parameter S for a system of rods can be defined by $S = 1/2 < 3\cos^2\theta - 1 >$, where θ is the angle the axis of the rod makes with the director $\hat{n}$ and the angular brackets denote an ensemble average. The value for S in the nematic phase at the transition point, as predicted by Onsager's theory, turns out to be quite high ($S_c \simeq 0.84$).

B. The Maier–Saupe Theory

The Maier–Saupe theory of nematic liquid crystals is founded on a mean field treatment of long-range contributions to the intermolecular potential and ignores the short-range forces [88, 89]. With the assumption of a cylindrically symmetrical distribution function for the description of orientation of the molecules and a nonpolar preferred axis of orientation, an appropriate order parameter for a system of cylindrically symmetrical molecules is

$$S = \frac{1}{2} < 3\cos^2\theta - 1 > = \frac{1}{2}\int (3\cos^2\theta - 1)f(\theta)d\Omega \tag{14}$$

where θ is the angle that the molecular symmetry axis makes with the preferred axis of orientation, the angular brackets denote an ensemble average, and $f(\theta)d\Omega$ gives the fraction of molecules, which are oriented at angles between θ and $\theta + d\Omega$, in the solid angle $d\Omega$. The theory relies on the enhancement of favorable attractive interactions arising from the van der Waals forces between the molecules being aligned to outweigh the concomitant loss of the orientational entropy. It is assumed that the position of the centers of mass of the molecules is not affected by the orientation dependent interaction. The treatment of Maier and Saupe invokes the phenomenological assumption that the orientation-dependent energetic interaction is simply a quadratic function of the order parameter S so that the change in the free energy per molecule on going from the isotropic to the nematic phase can be written as

$$\Delta F = -\frac{1}{2}US^2 + k_BT\int f(\theta)\ln[4\pi f(\theta)]d\Omega \tag{15}$$

where the parameter U is positive and the second term accounts for the decrease in entropy due to the anisotropy in the orientational distribution. In the original presentation of the theory, U is taken to be temperature-independent. The minimization of the free energy in the nematic phase subject to the constraint of the normalization condition for the orientational distribution function leads in a variation equation. The latter results in the normalized orientational distribution function

$$f = \frac{1}{4\pi Z} \exp(m\cos^2\theta), \tag{16}$$

where

$$m = \frac{3}{2}\frac{US}{k_BT} \tag{17}$$

with the normalization constant Z given by

$$Z = \int_0^1 dx \exp(mx^2) \tag{18}$$

The imposition of the self-consistency condition for S using Eq. (14) gives

$$S = -\frac{1}{2} + \frac{3}{2}\frac{\partial \ln Z}{\partial m} \tag{19}$$

One can obtain the free energy as a function of S for various values of k_BT/U from the solutions of Eqs. (19) and (17). For high values of k_BT/U, the minimum in the free energy is found for $S = 0$ corresponding to the isotropic phase. As the value of k_BT/U falls below 4.55, the minimum in the free energy is found for a nonzero value of S; that is, the nematic phase becomes stable. For this critical value of $k_BT/U = 4.55$, there is a discontinuous change in the order parameter from $S = 0$ to $S \simeq 0.44$. The Maier–Saupe theory thus predicts a first-order transition from the isotropic to the nematic phase.

C. The Landau–de Gennes Theory

The Landau–de Gennes (LdG) theory is the application of the Landau theory of phase transition to the context of I–N transition as developed elegantly by de Gennes [2]. In the framework of LdG theory, the free energy is expanded in a power series of a macroscopic tensor order parameter $\mathbf{Q}$ across the I–N transition. A typical difference found in the measurement of all macroscopic tensor properties on either side of the I–N transition underlies the following definition of $\mathbf{Q}$. The magnetic susceptibility tensor χ, which relates the magnetic moment $\mathbf{M}$ to the field $\mathbf{H}$, is chosen here for illustration as per convention. In the isotropic phase, one observes $\chi_{\alpha\beta} = \chi\delta_{\alpha\beta}$, where $\alpha, \beta = x, y, z$ are the indices referring to the space fixed frame, and in the uniaxial nematic phase we have

$$\chi = \begin{pmatrix} \chi_\perp & 0 & 0 \\ 0 & \chi_\perp & 0 \\ 0 & 0 & \chi_\| \end{pmatrix} \tag{20}$$

where the z-axis is taken to be parallel to the director. The anisotropic part of the magnetic susceptibility $\chi_{\alpha\beta}$ can be extracted to define the macroscopic tensor order parameter as

$$Q_{\alpha\beta} = G\left(\chi_{\alpha\beta} - \frac{1}{3}\delta_{\alpha\beta}\sum_{\gamma}\chi_{\gamma\gamma}\right) \tag{21}$$

The normalization constant G is often conveniently defined by setting Q_{zz} equal to unity in perfectly oriented system. By definition, $\mathbf{Q}$ is real, symmetric, and traceless. $\mathbf{Q}$ vanishes in the isotropic phase as per the requirement of its suitability for an order parameter to describe the I–N transition. The macroscopic tensor order parameter $\mathbf{Q}$ can always be diagonalized:

$$\mathbf{Q} = \begin{pmatrix} -\frac{q}{2} + \eta & 0 & 0 \\ 0 & -\frac{q}{2} - \eta & 0 \\ 0 & 0 & q \end{pmatrix} \tag{22}$$

where η is the biaxiality parameter, which is identically equal to zero in the case of a uniaxial nematic phase as mentioned before.

Taking into consideration the invariance of the free energy under rotation, one obtains the following Landau-type expansion correct up to the fourth order in $\mathbf{Q}$ in the absence of any external field

$$F = F_0 + \frac{1}{2}A(T)Tr\mathbf{Q}^2 - \frac{1}{3}B(T)Tr\mathbf{Q}^3 + \frac{1}{4}C(T)[Tr\mathbf{Q}^2]^2 \tag{23}$$

For the uniaxial case, one gets

$$F = F_0 + \frac{3}{4}A(T)q^2 - \frac{1}{4}B(T)q^3 + \frac{9}{16}C(T)q^4 \tag{24}$$

For the minimum in the free energy to correspond to the zero value of q at $T > T^*$, T^* being the stability limit of the isotropic phase, A must be positive, while for a nonzero value of q to correspond to a stable state at $T < T^*$, A has to be negative. Therefore, A must vanish at $T = T^*$. The simplest choice of $A(T)$ that satisfies these conditions is $A(T) = a(T - T^*)$. In practice, the coefficients of higher-order terms are assumed to be temperature-independent, as in the well-known Landau theory.

The presence of the cubic term, which is not precluded by the symmetry consideration, in the free energy expansion rules out the possibility of a continuous transition from the isotropic to the nematic phase. The otherwise continuous I–N transition at T^* is preempted by a first-order transition at T_{IN} given by

$$T_{IN} = T^* + \frac{B^2}{27aC} \tag{25}$$

At $T = T_{IN}$, the mesophase that is in coexistence with the isotropic phase has an order parameter $q_c = 2B/9C$. The parameter B is small for a weakly first-order I–N transition as observed experimentally.

There exist pre-transition effects in the isotropic phase heralding the I–N phase transition. Such pre-transition effects, which are consistent with the weakly first-order nature of the I–N transition, can be attributed to the development of short-range orientational order, which can be characterized by a position-dependent local orientational order parameter $Q(\mathbf{r})$, where all component indices have been omitted [2]. In the Landau approximation, the spatial correlation function $< Q(0)Q(\mathbf{r}) >$ has the Ornstein–Zernike form: $< Q(0)Q(\mathbf{r}) > \approx \exp(-r/\xi)/r$, where ξ is the coherence length or the second-rank orientational correlation length. The coherence length is temperature-dependent and the Landau-de Gennes theory predicts

$$\xi(T) = \xi_0 \left(\frac{T^*}{T - T^*} \right)^{1/2} \tag{26}$$

where ξ_0 is a molecular length. The coherence length defines the size of the so-called pseudonematic domains.

The LdG theory can be used to develop a macroscopic description of the dynamics of the pseudonematic domains in terms of the rate equations for the order parameters $Q_{\alpha\beta}$ [2]. An exponential relaxation for $Q_{\alpha\beta}$ is then obtained with a rate

$$\Gamma(T) = \frac{A(T)}{\nu} \tag{27}$$

The slowdown of the relaxation in the isotropic phase near T_{IN} is attributed to the smallness of $A(T)$.

D. Mode Coupling Theory Analysis

Gottke et al. [5] offered a theoretical treatment of collective motions of mesogens in the isotropic phase at short to intermediate time scales within the framework of the Mode coupling theory (MCT). The wavenumber-dependent collective orientational time correlation function $C_{lm}(k, t)$ is defined as

$$C_{lm}(k,t) = \left\langle \sum_{\alpha} \exp[-i\mathbf{k} \cdot \mathbf{r}_\alpha(0)]Y_{lm}^*(\omega_\alpha(0)) \times \sum_{\beta} \exp[i\mathbf{k} \cdot \mathbf{r}_\beta(t)]Y_{lm}^*(\omega_\beta(t)) \right\rangle \tag{28}$$

where $\omega_\alpha(t)$ represents the Euler angles giving the orientations of αth molecule at time t, Y_{lm} is the spherical harmonics of degree l and order m, and the angular brackets denote average over time origins [90]. The advantage of $C_{lm}(\mathbf{k}, t)$ is that it allows study of length dependence of correlation and also includes effects of translational diffusion on orientational relaxation. This time correlation function plays an important role in the study of solvation dynamics and dielectric relaxation in dipolar liquids for $l = 1$ [90]. The response in OKE experiments is governed by the $k = 0$ limit of $C_{20}(k, t)$. The long-time decay of $C_{20}(k, t)$ is given by the Landau–de Gennes theory, which predicts an exponential decay. The Landau–de Gennes time constant τ_{LdG} increases rapidly near the I–N transition. The physical origin of this slowdown in orientational relaxation could be interpreted from a mean-field theory, which gives

$$\tau_{LdG} = \frac{S_{220}(k = 0)}{6D_R} \tag{29}$$

where $S_{220}(k)$ is the (200) wavenumber-dependent orientational structure factor in the intermolecular frame with $k = |\mathbf{k}|$, the vector $\mathbf{k}$ being aligned along the z-axis in the intermolecular frame, and D_R is the rotational diffusion coefficient of mesogens in the isotropic phase [90]. As the I–N transition is approached from the isotropic side, $S_{220}(k = 0)$ undergoes pronounced growth, reflecting the appearance of long-range orientational correlation near the phase boundary. Note that $S_{220}(k = 0)$ does not actually diverge as the weakly first-order I–N phase transition is preempted. In the prescription of Gottke et al., the rotational friction Γ_R is split into a short-time part, Γ_0, and a singular part, Γ^{sing}:

$$\Gamma_R = \Gamma_0 + \Gamma^{sing} \tag{30}$$

where Γ_0 is determined by local, short-range interactions. Their treatment yields a typical MCT expression for Γ^{sing}, which under certain approximations gives

$$\Gamma^{sing}(t) \approx A_0/\sqrt{t} \tag{31}$$

where A_0 is a constant dependent on temperature. It was shown that the orientational time correlation function could then be given by

$$C_{20}(t) \approx \exp(a^2 t) erfc(a\sqrt{t}) \tag{32}$$

where $erfc$ is the complementary error function and the quantity a depends on orientational structure factor and rotational diffusion coefficient [5, 6]. Such a description is expected to be valid in a time window that is short compared to τ_{LDG}, but long compared to the ultrashort timescale of collisional dynamics. Equation (32) seemed to predict well the experimentally observed short-time decay of the OKE signal [5, 6]. Gottke et al. suggested that the weak time dependence on timescales short compared to τ_{LDG} was a direct consequence of pseudonematic domain formation near the I–N phase boundary [5, 6].

Li et al. subsequently developed a schematic mode coupling theory description of the short-time as well as long-time dynamics of mesogens in the isotropic phase near the I–N transition [91]. Their treatment started with a very general form [92] for the kinetic equation of the autocorrelation function $\phi_2(t)$ of the anisotropy of polarizability

$$\dot{\phi}_2(t) = -\int_0^t dt' M(t - t')\phi_2(t') \tag{33}$$

where $M(t)$ is the memory function associated with $\Phi(t)$. A schematic mode coupling model [93–95] expresses the memory function as

$$M_{MCT}(t) = \Omega_2^2 K_2(t) \tag{34}$$

Here Ω_2 is the characteristic frequency, $K_2(0) = 1$, and the time dependence of $K_2(t)$ is expressed in terms of its memory function $m_2(t)$ in the following way:

$$\dot{K}_2(t) = \mu_2 K_2(t) - \Omega_2^2 \int_0^t dt' m_2(t - t')K_2(t') \tag{35}$$

where μ_2 is the damping constant. The standard assumption about $m_2(t)$ in schematic mode coupling theory is

$$m_2(t) = k\phi_2(t)\phi_1(t) \tag{36}$$

where $\phi_1(t)$ is the solution of an F_{12} model for the density correlator [93, 94].

The LdG theory predicts exponential decay of the correlation function at long times

$$\dot{\phi}_2(t) = -\Gamma\phi_2(t) \tag{37}$$

where $\Gamma^{-1}(\Gamma^{-1} = \tau_{LdG})$ is the relaxation time. Equation (37) is a specific form of Eq. (33), where the memory function is of the form

$$M_{LdG} = \Gamma\delta(t) \tag{38}$$

As per the prescription of Li et al., a simple way to get a combined theory is to assume that the total memory function of the correlator of interest is the sum of the mode coupling memory function and the Landau–de Gennes memory function. The total memory function $M(t)$ can thus be written as

$$M(t) = M_{MCT}(t) + M_{LdG}(t) \tag{39}$$

This is a conjecture that seems to be reasonable because the two theories describe very different effects that dominate on different timescales [91].

Using Eqs. (33)–(39), Li et al. obtained the following second-order differential equations:

$$\ddot{\phi}_1(t) = -\Omega_1^2 \phi_1(t) - \mu_1 \dot{\phi}_1(t) - \Omega_1^2 \int_0^t dt' m_1(t - t') \dot{\phi}_1(t') \tag{40}$$

where $m_1(t) = v_1 \phi_1(t) + v_2 \phi_1^2(t)$, v_1 and v_2 being the coupling constants in the memory kernel for the density correlator, and the initial conditions are $\phi_1(0) = 1$ and $\dot{\phi}_1(0) = 0$.

$$\ddot{\phi}_2(t) = - (\Omega_2^2 + \mu_2\Gamma)\phi_2(t) - (\mu_2 + \Gamma)\dot{\phi}_2(t) - \Omega_2^2 \int_0^t dt' m_2(t - t') \dot{\phi}_2(t')$$
$$- \Omega_2^2\Gamma \int_0^t dt' m_2(t - t')\phi_2(t') \tag{41}$$

where the initial conditions are $\phi_2(0) = 1$ and $\dot{\phi}_2(0) = -\Gamma$.

Li et al. employed Eqs. (40) and (41) to fit the temperature-dependent OHD–OKE data on mesogens in the isotropic phase [91]. Equation (40) is identical to the one use in the analysis of the supercooled liquid data. The difference between the schematic model developed by Li et al. and the one applied to supercooled liquids is Eq. (41). The schematic mode coupling theory developed by Li et al. was found to be successful in reproducing the OHD–OKE data on three mesogens in the isotropic phase on all timescales and at all temperatures investigated [91].

V. COMPUTATIONAL APPROACH

Computer simulations of model liquid crystals have undergone an upsurge in recent times due to, at least in part, the continuous development of computer resources [96–98]. An integral part of this approach is molecular modeling that is rather involved so as to yield the rich phase behavior typical of liquid crystals [16]. While atomistic models could in principle serve the purpose of molecular models where mesogens are approximated with particles with well-defined anisotropic shape, as illustrated in Fig. 1, find their utility in obtaining a generalized view. A simple approach along this line involves consideration of purely repulsive models involving hard bodies [99]. This rather extreme choice is inspired by the idea that the equilibrium structure of a dense liquid is essentially determined by the repulsive forces that fix the molecular shape [100]. Such an approach is appealing for its simplicity [99]. However, temperature plays no direct role in purely repulsive models [99]. In thermotropic liquid crystals, on the contrary, the approach to increasingly ordered mesophases is temperature-driven. In reviewing studies of thermotropic liquid crystals, we thus primarily focus on simple molecular models with pairwise interaction potentials, which have both attractive and repulsive contributions.

A. Model Systems

Perhaps the most extensively used generic model that includes soft anisotropic interactions in simulations of thermotropic liquid crystals is the one suggested by Gay and Berne [101]. In this spirit, we carried out molecular dynamics simulation studies of model thermotropic liquid crystalline systems where mesogens, approximated with ellipsoids of revolution, interact with the Gay–Berne (GB) pair potential or one of its variants. The GB pair potential, where each ellipsoid of revolution has a single-site representation, is an elegant generalization of the extensively used isotropic Lennard-Jones potential to incorporate anisotropy in both the attractive and the repulsive parts of the interaction [101]. In the GB pair potential, the ith ellipsoid of revolution is represented by the position $\mathbf{r}_i$ of its center of mass and a unit vector $\mathbf{e}_i$ along the long axis in the case of a prolate. The GB interaction between ellipsoids of revolution i and j is given by

$$U_{ij}^{GB}(\mathbf{r}_{ij}, \mathbf{e}_i, \mathbf{e}_j) = 4\epsilon(\hat{\mathbf{r}}_{ij}, \mathbf{e}_i, \mathbf{e}_j)(\rho_{ij}^{-12} - \rho_{ij}^{-6}) \qquad (42)$$

where

$$\rho_{ij} = \frac{r_{ij} - \sigma(\hat{\mathbf{r}}_{ij}, \mathbf{e}_i, \mathbf{e}_j) + \sigma_{GB}}{\sigma_{GB}} \qquad (43)$$

Here σ_{GB} defines the cross-sectional diameter, r_{ij} is the distance between the centers of mass of the ellipsoids of revolution i and j, and $\hat{\mathbf{r}}_{ij} = \mathbf{r}_{ij}/r_{ij}$ is a unit vector along the intermolecular separation vector $\mathbf{r}_{ij}$. The molecular shape parameter σ and the energy parameter ϵ both depend on the unit vectors $\mathbf{e}_i$ and $\mathbf{e}_j$ as well as on $\hat{\mathbf{r}}_{ij}$ as given by the following set of equations:

$$\sigma(\hat{\mathbf{r}}_{ij}, \mathbf{e}_i, \mathbf{e}_j) = \sigma_{GB} \left[1 - \frac{\chi}{2} \left\{ \frac{(\mathbf{e}_i \cdot \hat{\mathbf{r}}_{ij} + \mathbf{e}_j \cdot \hat{\mathbf{r}}_{ij})^2}{1 + \chi(\mathbf{e}_i \cdot \mathbf{e}_j)} + \frac{(\mathbf{e}_i \cdot \hat{\mathbf{r}}_{ij} - \mathbf{e}_j \cdot \hat{\mathbf{r}}_{ij})^2}{1 - \chi(\mathbf{e}_i \cdot \mathbf{e}_j)} \right\} \right]^{-1/2} \qquad (44)$$

with $\chi = (\kappa^2 - 1)/(\kappa^2 + 1)$ and

$$\epsilon(\hat{\mathbf{r}}_{ij}, \mathbf{e}_i, \mathbf{e}_j) = \epsilon_{GB}[\epsilon_1(\mathbf{e}_i, \mathbf{e}_j)]^{\nu}[\epsilon_2(\hat{\mathbf{r}}_{ij}, \mathbf{e}_i, \mathbf{e}_j)]^{\mu} \qquad (45)$$

where the exponents μ and ν are adjustable, and

$$\epsilon_1(\mathbf{e}_i, \mathbf{e}_j) = [1 - \chi^2(\mathbf{e}_i \cdot \mathbf{e}_j)^2]^{-1/2} \tag{46}$$

and

$$\epsilon_2(\hat{\mathbf{r}}_{ij}, \mathbf{e}_i, \mathbf{e}_j) = 1 - \frac{\chi'}{2}\left[\frac{(\mathbf{e}_i \cdot \hat{\mathbf{r}}_{ij} + \mathbf{e}_j \cdot \hat{\mathbf{r}}_{ij})^2}{1 + \chi'(\mathbf{e}_i \cdot \mathbf{e}_j)} + \frac{(\mathbf{e}_i \cdot \hat{\mathbf{r}}_{ij} - \mathbf{e}_j \cdot \hat{\mathbf{r}}_{ij})^2}{1 - \chi'(\mathbf{e}_i \cdot \mathbf{e}_j)}\right] \tag{47}$$

with $\chi' = (\kappa'^{1/\mu} - 1)/(\kappa'^{1/\mu} + 1)$. Here $\kappa = \sigma_{ee}/\sigma_{ss}$ is the aspect ratio of the ellipsoid of revolution with σ_{ee} denoting the molecular length along the principal symmetry axis and $\sigma_{ss} = \sigma_{GB}$; $\kappa' = \epsilon_{ss}/\epsilon_{ee}$, where ϵ_{ss} is the depth of the minimum of the potential for a pair of ellipsoids of revolution aligned parallel in a side-by-side configuration; and ϵ_{ee} is the corresponding depth for the end-to-end alignment.

It follows that the GB pair potential defines a family of models, each member of which is characterized by a set of four parameters $(\kappa, \kappa', \mu, \nu)$. Following the suggestion of Bates and Luckhurst [102], each member is represented by $GB(\kappa, \kappa', \mu, \nu)$. Figure 7 shows U_{ij}^{GB} plotted as a function of interparticle separation for four different relative orientations of the particles i, j corresponding to three members of the GB family. The choice of a $GB(\kappa, \kappa', 0, 0)$ corresponds to a soft repulsive potential between ellipsoids of revolution and a $GB(0, 0, \mu, \nu)$ corresponds to a spherical Lennard-Jones potential [16]. The GB pair potential has been used extensively in the study of thermotropic liquid crystals with the original parameterization $(3, 5, 2, 1)$ for which the phase diagram is well known [103, 104]. In Fig. 8, the phase diagram of the $GB(3, 5, 2, 1)$ model is shown on the density-temperature plane. This will be instructive, since much of the work discussed here involved the study of the $GB(3, 5, 2, 1)$ system along isochors or along isotherms.

B. Dynamics of Thermotropic Liquid Crystals across the Isotropic–Nematic Transition

Despite extensive investigation of phase behavior of liquid crystals in computer simulation studies [97–99], the literature on computational studies of their dynamics is somewhat limited. The focal point of the latter studies has often been the single-particle and collective orientational correlation functions. The lth rank single-particle orientational time correlation function (OTCF) is defined by

$$C_l^s(t) = \frac{\langle \sum_i P_l(\mathbf{e}_i(0) \cdot \mathbf{e}_i(t)) \rangle}{\langle \sum_i P_l(\mathbf{e}_i(0) \cdot \mathbf{e}_i(0)) \rangle} \tag{48}$$

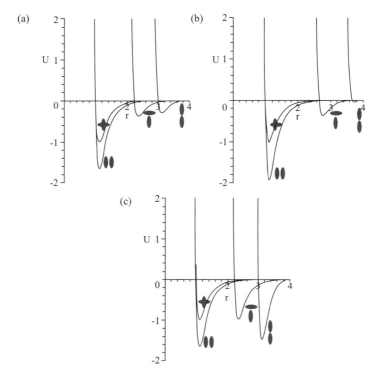

Figure 7. The Gay–Berne pair potential as a function of interparticle separation for side-by-side, cross, T-shaped, and end-to-end configurations of the pair corresponding to three members of the GB family: (a) GB(3, 5, 2, 1), (b) GB(3.6, 5, 2, 1), and (c) GB(3, 5, 1, 1). The interparticle separation is scaled by σ_{GB} and the potential by ϵ_{GB}. The potential is cut at $4\sigma_{GB}$ and shifted to zero at the cutoff. (Reproduced from Ref. 97.)

where P_l is the lth rank Legendre polynomial and the angular brackets stand for ensemble averaging. The corresponding collective OTCF is defined by

$$C_l^c(t) = \frac{\left\langle \sum_i \sum_j P_l(\mathbf{e}_i(0) \cdot \mathbf{e}_j(t)) \right\rangle}{\left\langle \sum_i \sum_j P_l(\mathbf{e}_i(0) \cdot \mathbf{e}_j(0)) \right\rangle} \tag{49}$$

The first-rank and second-rank OTCFs are mostly studied because of their relevance to experiments. One of the early computational studies of orientational dynamics in the isotropic phase near the I–N transition is due to Allen and Frenkel [105]. In their molecular dynamics simulation study of a system of $N = 144$ hard ellipsoids of revolution, the slowdown of orientational dynamics on approaching the I–N transition was captured—in particular, in the time evolution of the

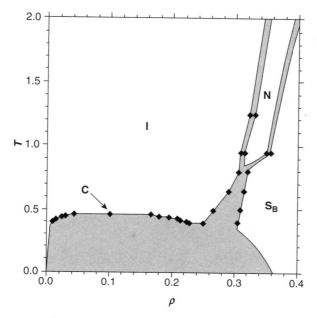

Figure 8. Phase diagram of the Gay–Berne model with the original and the most-studied parameterization $(\kappa = 3, \kappa' = 5, \mu = 2, \nu = 1)$ in the density-temperature plane as obtained from computer simulations. Filled diamonds mark simulation results; the phase boundaries away from these points are drawn as a guide only. The domains of the thermodynamic stability of the isotropic (I), nematic (N), and smectic (S_B) phases are shown. The liquid–vapor critical point is denoted by C. Two-phase regions are shaded. (Reproduced from Ref. 104.)

collective second-rank OTCF [105]. The emphasis was on long times where an exponential decay was observed. The time constants of the single-particle and collective second-rank OTCFs are related to the static second-rank Kirkwood factor g_2 by the expression

$$\tau_2^s/\tau_2^c = (1 + j_2)/(1 + g_2) \tag{50}$$

where

$$g_2 = \frac{1}{N}\sum_i \sum_{j\neq i} P_2(\mathbf{e}_i, \mathbf{e}_j) \tag{51}$$

and j_2 is a dynamical quantity that can be expressed in terms of the memory function of the angular velocity time correlation function. An examination of the above relationship suggested that the slowdown of orientational dynamics on approaching the I–N transition was largely due to the growth in static correlations, as measured by g_2[105].

The growth of orientational correlations and the slow down of collective orientational dynamics were subsequently investigated using a soft potential [106]. Allen and Warren (AW) studied a system consisting of $N = 8000$ particles of ellipsoids of revolution, interacting with a version of the Gay–Berne potential, GB (3, 5, 1, 3), originally proposed by Berardi et al. [107]. AW computed the direct correlation function, $c(1, 2)$, in the isotropic phase near the I–N transition. The direct correlation function is defined through the Ornstein–Zernike equation [108]

$$h(1,2) = c(1,2) + \frac{\rho}{4\pi} \int d\mathbf{r}_3 d\mathbf{e}_3 h(1,3) c(3,2) \qquad (52)$$

where ρ is the number density and the pair correlation function, $h(1,2) = g(1,2) - 1 = g(\mathbf{r}_1 - \mathbf{r}_2, \mathbf{e}_1, \mathbf{e}_2) - 1$, can be expanded in a set of rotationally invariant functions

$$h(1,2) = \sum_{mnl} h^{mnl}(r) \Phi^{mnl}(\mathbf{e}_1, \mathbf{e}_2, \hat{\mathbf{r}}) = 4\pi \sum_{mnx} h_{mnx}(r) Y_\chi^m(\mathbf{e}_1) Y_{-\chi}^m(\mathbf{e}_2) \qquad (53)$$

Here the first set of expansion coefficients refers to a space-fixed laboratory frame while the second set refers to a coordinate system based on the intermolecular frame. The two sets of coordinates are interconvertible. A similar expansion holds good for $c(1, 2)$, and it is possible to express the mechanical stability of the isotropic phase relative to the nematic phase in terms of these expansion coefficients and thus to obtain an estimation of T^* [109, 110]. In spite of the rapid growth of the second-rank orientational correlation length, ξ_2, on approaching the I–N transition from the isotropic side, the simulation results showed the associated component of the direct correlation function to remain short-ranged and showed its spatial integral to approach the mechanical instability limit of the isotropic phase [106].

In order to study the collective dynamics, AW defined the following dynamical variable:

$$Q_{\alpha\beta}(\mathbf{k}, t) = \sum_j [\frac{3}{2} e_{j\alpha}(t) e_{j\beta}(t) - \frac{1}{2} \delta_{\alpha\beta}] \exp[i\mathbf{k} \cdot \mathbf{k}_j(t)] \qquad (54)$$

where $\alpha, \beta = x, y, z$ and $\delta_{\alpha\beta}$ is the Kronecker delta. The time correlation function

$$C(k, t) = \sum_\alpha \sum_\beta \langle Q_{\alpha\beta}(-\mathbf{k}, 0) Q_{\alpha\beta}(\mathbf{k}, t) \rangle$$

$$\propto \sum_{ij} < P_2(\mathbf{e}_i(0) \cdot \mathbf{e}_j(t)) \exp[i\mathbf{k} \cdot (\mathbf{r}_i(0) - \mathbf{r}_j(t))] \qquad (55)$$

was measured. The simulation results suggested roughly exponential decay $C(k, t) = A(k)\exp[-t/\tau(k)]$ with $\tau(k) \sim k^{-2}$. At $k = 0$, it was found that $\tau^{-1} \propto \xi_2^{-2} \propto (T - T^*)$, in agreement with the description of Landau and de Gennes [106].

In a contemporary effort, Perera et al. [111] studied a system of mesogens, modeled by GB(3, 5, 2,1) in molecular dynamics simulations along an isotherm at the reduced temperature $T = 1.25$. In particular, the single-particle and collective orientational dynamics were investigated as the system was driven to the I–N phase boundary by increasing density. For both the first and second ranks, the long-time decay was found to be exponential and the slowdown of the decay of the second-rank collective OTCF was dramatic. Although the Debye model of rotational diffusion, which predicts the ratio τ_1^s/τ_2^s to be equal to 3, was observed to be obeyed in the isotropic phase far away from the I–N phase boundary, it was found to break down on approaching the transition. The long-time dynamics was analyzed in terms of the memory functions of the OTCFs. The collective orientation relaxation was found to approach the Markovian behavior near the I–N phase boundary [111].

Until recently, the focus of the computational studies of the orientational dynamics near the I–N transition was on the long-time decay. The revelation of intriguing dynamics at short to intermediate timescales has now changed the scenario. In search of temporal power laws, Jose and Bagchi [112] investigated a system of Gay–Berne mesogens, modeled by $GB(3, 5, 2, 1)$, in molecular dynamics simulations near the I–N phase boundary. In this case, the focus was on short-to-intermediate times. The study revealed the emergence of power law decay in the single-particle second-rank OTCF near the I–N phase transition, as shown in Fig. 9a. The linear regime in the log-log plot corresponds to a power law decay at the intermediate timescale. In agreement with recent OKE experiments, the negative of the time derivative of the collective second-rank OTCF, which provides a measure of the OKE signal derived from this system, also showed temporal power law in the intermediate time window near the I–N transition, as illustrated in Fig. 9b.

Jose and Bagchi further investigated the time dependence of the OKE signal derived frm the GB(3, 5, 2, 1) system at three state points across the nematic phase, as shown in Figure 10 [113]. It is computationally expensive to obtain the whole decay of the OKE signal with high resolution. Figure 11 shows the decay of the OKE signal at short times with high resolution. It is evident from Fig. 10 and 11 that the decay involves multiple power law regimes. Two distinct power law decay regimes are observed, one at short times and the other at long times; the intervening crossover region could also be described by a power law. No exponential decay was observed at long times in agreement with recent experimental observation [41]. The authors argued that the short-time power law might originate from the local orientational density fluctuations. The origin of the power law decay at long times

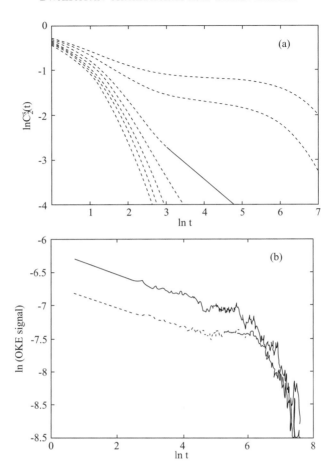

Figure 9. Orientational relaxation in the model liquid crystalline system GB(3, 5, 2, 1) ($N = 576$) at several densities across the $I-N$ transition along the isotherm at $T = 1$. (a) Time dependence of the single-particle second-rank orientational time correlation function in a log–log plot. From left to right, the density increases from $\rho = 0.285$ to $\rho = 0.315$ in steps of 0.005. The continuous line is a fit to the power law regime. (b) Time dependence of the OKE signal, measured by the negative of the time derivative of the collective second-rank orientational time correlation function $C_2^c(t)$ in a log–log plot at two densities. The dashed line corresponds to $\rho = 0.31$ and the continuous line to $\rho = 0.315$. (Reproduced from Ref. 112.)

is not clear, but could be, related to long-range fluctuations such as smectic like density fluctuations [113]. Bertolini et al. studied orientational dynamics of the isotropic phase of a model calamitic liquid crystal with a slightly longer aspect ratio [114]. Their model consists of an array of nine soft spheres arranged linearly so that the overall particle resembles a spherocylinder of $\kappa = 5 - 6$. The time derivatives

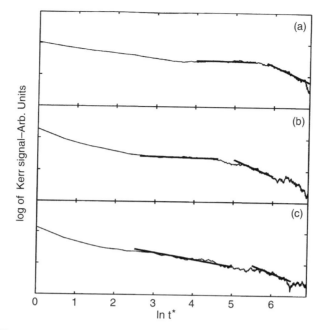

Figure 10. Time dependence of the OKE signal obtained from the system GB(3, 5, 2, 1) ($N = 576$) in a log–log plot at three state points across the nematic phase. (a) $\rho = 0.32$, $T = 1$; (b) $\rho = 0.34$, $T = 1.25$; (c) $\rho = 0.33$, $T = 1.0$. The thick straight lines correspond to linear fits to the data, showing the power law decay regimes. (Reproduced from Ref. 113.)

of both single-particle and collective second-rank orientational time correlation functions were found to have, in their respective log–log plots, a linear regime at short times, in qualitative agreement with recent experiments.

In the quest for a universal feature in the short-to-intermediate time orientational dynamics of thermotropic liquid crystals across the I–N transition, Chakrabarti et al. [115] investigated a model discotic system as well as a lattice system. As a representative discotic system, a system of oblate ellipsoids of revolution was chosen. These ellipsoids interact with each other via a modified form of the GB pair potential, GBDII, which was suggested for disc-like molecules by Bates and Luckhurst [116]. The parameterization, which was employed for the model discotic system, was $\kappa = 0.345$, $\kappa' = 0.2$, $\mu = 1$, and $\nu = 2$. For the lattice system, the well-known Lebwohl–Lasher (LL) model was chosen [117]. In this model, the particles are assumed to have uniaxial symmetry and represented by three-dimensional spins, located at the sites of a simple cubic lattice, interacting through a pair potential of the form

$$U_{ij} = -\epsilon_{ij}P_2(\cos\theta_{ij}) \tag{56}$$

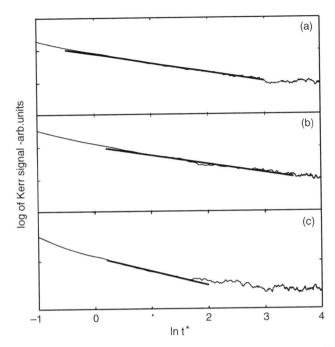

Figure 11. Time dependence of the OKE signal obtained from the system GB(3, 5, 2, 1) ($N = 576$) in a log–log plot at three state points across the nematic phase. (a) $\rho = 0.32$, $T = 1$; (b) $\rho = 0.34$, $T = 1.25$; (c) $\rho = 0.33$, $T = 1.0$. The short-to-intermediate time data are shown here at high resolution. The thick lines correspond to linear fits to the data, showing the power law decay regimes. (Reproduced from Ref. 113.)

Here ϵ_{ij} is a positive constant ϵ for nearest-neighbor spins i and j and zero otherwise, P_2 is the second rank Legendre polynomial and θ_{ij} is the angle between the spins i and j. The simplicity of the model allowed the authors to study a larger system size [115, 118]. Figures 12a and 12b show the time evolution of the single-particle second-rank OTCF and the OKE signal derived from the discotic system at several temperatures along an isobar. Figures 13a and 13b show the time dependence of the single-particle second-rank OTCF and the OKE signal derived from the lattice system at temperatures near the *I–N* phase boundary. The emergence of the power law decay regime in the isotropic phase near the *I–N* transition was observed again for both systems. The short-to-intermediate time power law decay thus seems to be a universal feature in the orientational relaxation of thermotropic liquid crystals near the *I–N* transition. We note that while a power law decay of the OKE signal was recently observed in experiments for calamitic liquid crystals near the *I–N* phase boundary and in the nematic phase [5–7, 41], this prediction on the discotic system is yet be tested in experiments.

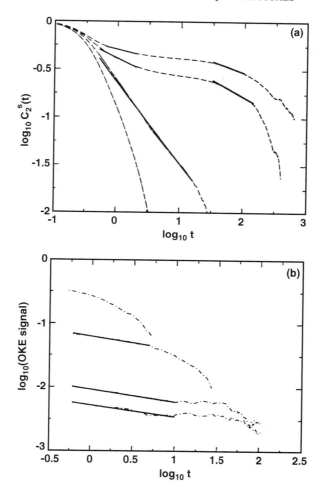

Figure 12. Orientational dynamics of the discotic system GBDII ($N = 500$) at several temperatures across the isotropic–nematic transition along the isobar at pressure $P = 25$. (a) Time evolution of the single-particle second-rank orientational time correlation function in a log–log plot. Temperature decreases from left to right. (b) Time dependence the OKE signal at short-to-intermediate times in a log–log plot. Temperature decreases from top to bottom on the left side of the plot: $T = 2.991, 2.693, 2.646$, and 2.594. The dashed lines are the simulation data and the continuous lines are the linear fits to the data, showing the power law decay regimes at temperatures. (Reproduced from Ref. 115.)

When a different isobar was chosen at a lower pressure for the discotic system, a direct transition from the isotropic phase to the columnar phase was observed [119]. In this case, decay of neither the single-particle second-rank OTCF nor the OKE signal showed any power law regime [119]. As compared

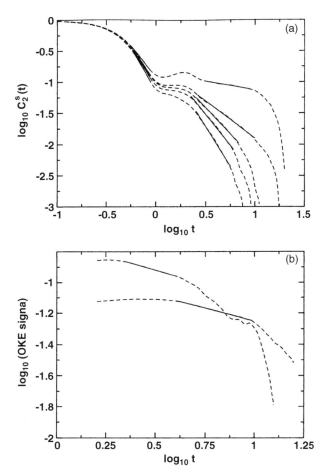

Figure 13. Orientational dynamics of the Lebwohl–Lasher lattice model $(N = 1000)$ at temperatures near the isotropic–nematic transition. (a) Time evolution of the single-particle second-rank orientational time correlation function in a log–log plot at temperatures $T = 1.213, 1.176$, $1.160, 1.149, 1.134$. Temperature decreases from left to right. (b) Decay of the OKE signal in a log–log plot at short-to-intermediate time window at temperatures $T = 1.176$, and 1.149. Temperature decreases from top to bottom on the left side of the plot. The dashed lines are the simulation data and the continuous lines are the linear fits to the data. The system undergoes a transition from the isotropic to the nematic phase at $T \simeq 1.14$. (Reproduced from Ref. 115.)

to the I–N transition, a much larger change in the density was observed across the isotropic–columnar $(I$–$C)$ transition as expected for a strongly first-order transition. The weakly first-order nature of the I–N transition thus appears to play a role in the origin of the short-to-intermediate time power law decay in orientation relaxation of thermotropic liquid crystals near the I–N phase boundary. However, the lack of power law decay in orientational relaxation of the discotic system in the isotropic phase near the I–C phase boundary is in contrast with the observation made in a more recent OHD–OKE experimental study [120]. We offer the following explanation for the disagreement between simulation and experiment.

The strongly first-order isotropic to columnar phase transition is not accompanied by any noticeable growth in orientational pair correlation. Therefore, the phase transition proceeds by nucleation, as in any first-order phase transition. The lack of growth of orientational pair correlation near the I–C phase transition has been confirmed by computer simulation [119]. So, the simulation study seems self-consistent. What then could be the reason for the appearance of power law in the experimental results of Fayer and co-workers? One possibility is that the experimental study was carried out close to the I–N–C columnar triple point so that the system experiences a considerable degree of nematic fluctuations to give rise to the power law decay. The other explanation, which appears to be favored by Fayer group, is the packing-induced slowdown of relaxation as in supercooled liquid and can thus be explained by mode coupling theory, even without imposing growth of orientational correlations. If the latter explanation is correct, it might mean that the intermediate timescale decay is essentially a density effect and not due to orientational correlation. The alternative explanations suggest need for further theoretical and experimental work. This remains an intriguing problem.

Jose and Bagchi also investigated the viscoelasticity near the I–N phase boundary in a study of the GB(3, 5, 2,1) system [121]. The viscoelasticity of the system is given by the frequency dependent shear viscosity $\eta_{\alpha\beta}(\omega)$ [122, 123]. The $\alpha\beta$ component of the shear viscosity in the isotropic phase is defined by

$$\eta_{\alpha\beta} = \frac{V}{\kappa_B T} \int_0^\infty dt \langle \sigma_{\alpha\beta}(t)\sigma_{\alpha\beta}(0)\rangle \tag{57}$$

where V is the volume of the system and $\sigma_{\alpha\beta}$ is the shear stress tensor [108], which in turn is defined by

$$\sigma_{\alpha\beta} = \frac{1}{V}\left(\sum_i \frac{p_{\alpha i} p_{\beta i}}{m} + \sum_i \sum_{j>i} r_{\alpha ij} f_{\beta ij}\right) \tag{58}$$

where $p_{\alpha i}$ is the α component of the momentum of ith particle and $\mathbf{f}_{ij}$ is the α component of the force between the pair of particles i and j. The viscoelasticity can be decomposed into real and imaginary parts

$$\eta_{\alpha\beta}(\omega) = \eta'_{\alpha\beta}(\omega) + i\eta''_{\alpha\beta}(\omega) \tag{59}$$

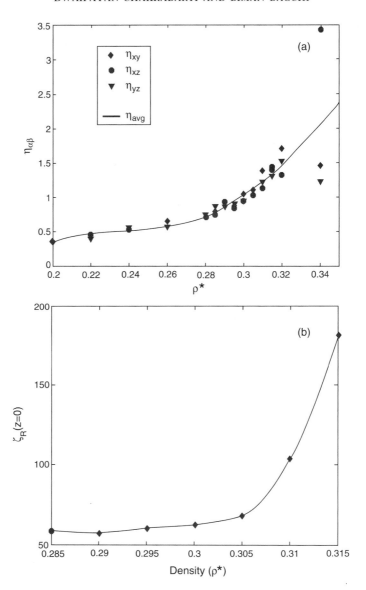

Figure 14. (a) Shear viscosity of the system GB(3, 5, 2, 1) ($N = 576$) versus density in the isotropic phase. When the I–N transition is approached, various components of the shear viscosity become anisotropic. (b) Zero-frequency rotational friction versus density for the system GB(3, 5, 2, 1) ($N = 576$) in the isotropic phase. (Reproduced from Ref. 121.)

where the real part is given by

$$\eta_{\alpha\beta}(\omega) = \frac{V}{\kappa_B T} \int_0^\infty dt \langle \sigma_{\alpha\beta}(t)\sigma_{\alpha\beta}(0)\rangle \cos(\omega t) \tag{60}$$

Although the shear viscosity was found to undergo somewhat rapid increase as the phase boundary was approached from the isotropic side, it did not diverge as evident in Fig. 14a. Figure 14b shows that the rotational friction, however, undergoes pronounced growth on approaching the phase boundary. The frequency dependence of the shear viscosity showed nonmonotonic behavior with the maximum observed at an intermediate frequency, as shown in Fig. 15. The lack of divergence or any rapid growth in the value of the shear viscosity is in agreement with known experimental results [121]. However, this result seems to be counter to the argument that the slow relaxation observed near the *I–N* transition is a purely packing-induced effect as in supercooled liquids.

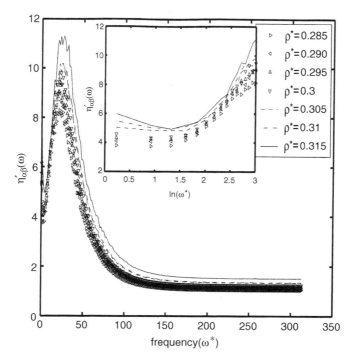

Figure 15. Frequency dependence of shear viscosity for the system GB(3, 5, 2, 1) ($N = 576$) at several densities along the isotherm at temperature $T = 1$. The inset shows the low-frequency data $\eta'_{\alpha\beta}(\omega)$ in a semilog plot. (Reproduced from Ref. 121.)

C. Supercooled Liquids and Thermotropic Liquid Crystals: Similarity in Dynamics

In this section of the review we restrict ourselves to those aspects of dynamics of supercooled liquids that appear to be analogous with dynamical features of thermotropic liquid crystals across the I–N transition. Power law relaxation in short-to-intermediate time scales is well known for supercooled liquids. We first review recent computational efforts that provided insights into power law relaxation in supercooled liquids. We then focus on a model system tailored specifically to study analogous dynamical features of the two seemingly different classes of soft matter systems.

1. Power Law Relaxation

The mode-coupling theory (MCT) makes detailed predictions about the time dependence of any correlator $\Phi(t) = <X(0)Y(t)>$, whose dynamic variables X and Y are coupled to density fluctuations ρ_q for any wavevector $\mathbf{q}$. Within the framework of the MCT, the β relaxation regime, which follows the short-time ballistic regime, encompasses a period during which the particles get trapped in cages formed by their neighbors, resulting in the development of a plateau. The MCT predicts the decay to the plateau to follow a power law $\Phi(t) = f + At^{-a}$, while it predicts the decay from the plateau to follow another power law $\Phi(t) = f - Bt^b$, where the exponents bear a relationship. The second power law is known as the von Schweidler law. Beyond the β relaxation regime, what follows is the α relaxation regime over which the correlator decays to zero for temperatures above T_c as the cages break up and reform. A great deal of computer simulation work has been devoted to test the predictions of the MCT. Many of these studies make use of the model of atomic liquids that is a 80:20 binary mixture of Lennard-Jones particles interacting with a specific set of length and energy parameters (and its minor versions, which differ in ways of truncating the potential beyond a certain distances) [124, 125]. In one of the early studies, Kob and Anderson showed that the self-intermediate-scattering function in the β relaxation regime follows a temporal power law consistent with the MCT prediction of the von Schweidler law [124].

In connection with the recently suggested analogy, we investigated the orientational dynamics across the supercooled regime [126]. The focus of molecular dynamics simulation studies on supercooled liquids has largely been on atomic systems involving only translational degrees of freedom (TDOF) even though most of the good glass-formers in reality are molecular systems [125, 127]. We considered a 50:50 mixture of spheres and ellipsoids of revolution as a model system. In this binary mixture, the spheres interact with each other via the Lennard-Jones pair potential and the interaction between two ellipsoids of revolution is modeled by the Gay–Berne pair potential, GB(2, 5, 2, 1). The interaction between a sphere and an ellipsoid of revolution (S–E) is given by the pair potential suggested by Cleaver and coworkers [128, 129]. For the binary mixture, all quantities are given in reduced units, defined in terms of the Lennard-Jones potential parameters σ_{LJ} and ϵ_{LJ}. The various energy and length parameters of the interaction potentials for the binary mixture were chosen as follows: $\epsilon_{LJ} = 1.0, \epsilon_{GB} = 0.5, \epsilon_{SE} = 1.5, \sigma_{LJ} = 1.0, \sigma_{GB} = 1.0$, and $\sigma_{SE} = 1.0$. For the interaction between a sphere and an ellipsoid of revolution, the ratio, ϵ_S/ϵ_E which controls the configurational side-to-end well-depth anisotropy, was set equal to 5 and the exponent, μ was taken to be 2. The choice of a smaller aspect ratio for the ellipsoids of revolution, the presence of spheres as the second component, and the chosen set of energy and length parameters ensured that there was neither any phase separation nor any appearance of a liquid crystalline phase with orientational order even at the lowest temperature studied at a high density within the length of our molecular dynamics simulations.

Figures 16a and 16b show the time evolution of the single-particle second-rank OTCF and the OKE signal derived from the system in log–log plots at several temperatures down to $\sim T_c$. The slowdown in the single-particle dynamics is evident on approaching T_c upon cooling. The single-particle second-rank OTCF shows a shoulder at intermediate timescales below a certain temperature. The shoulder develops into a plateau as the temperature is further lowered. It is evident in Figure 16b that a rather long power law decay regime ($\sim t^{-\alpha}$, with α falling between 0.45 and 0.495) is observed spanning over almost two decades of time in the intermediate time window. The exponent of this power law thus appears to be only weakly temperature-dependent, if at all, Although recent OHD–OKE data for supercooled molecular liquids suggested a temperatures-independent exponent for the intermediate power law [8], the values of the exponent obtained here do not seem to agree with the experimentally observed values that are close to -1. The finite system size might have contributed to this lack of agreement.

2. Heterogeneous Dynamics

A large number of experimental and computer simulation studies of supercooled liquids have revealed signatures of heterogeneous dynamics, which are now believed to underlie several features of complex dynamics observed in the supercooled regime [46, 48, 49]. Although the isotropic phase of a liquid crystalline system is macroscopically homogeneous, near the I–N transition a local nematic-like order, which is attributed to the weakly first-order nature of the transition, persists over a length scale that characterizes the so called pseudone-matic domains. The long-time exponential decay is ascribed to the randomization of these pseudonematic domains. It is intuitive that the appearance of pseudonematic domain in the isotropic phase near T_{IN} would result in heterogeneous dynamics. This led us to investigate heterogeneous dynamics

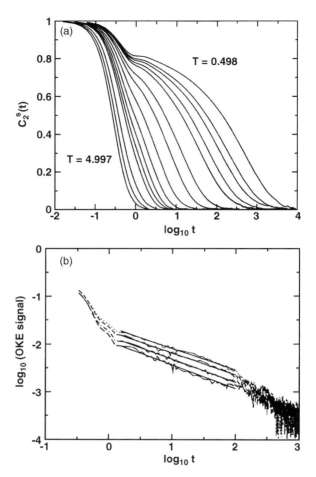

Figure 16. Orientational dynamics of the ellipsoids of revolution in the 50:50 binary mixture ($N = 256$) at several temperatures along the isochor at density $\rho = 0.8$. (a) Time evolution of the single-particle second-rank orientational time correlation function in a log–log plot. Temperature decreases from left to right ($4.997 \geq T \geq 0.498$). (b) Time dependence of the OKE signal in a log–log plot at temperatures $T = 0.574, 0.550, 0.529$, and 0.498. The continuous lines are linear fits, showing the power law decay. Temperature decreases from top to bottom across the linear regime. (Reproduced from Ref. 126.)

near the I–N transition and across the supercooled regime and compare our results [130, 131]. As a diagnostic of dynamical heterogeneity in ODOF, we defined the rotational non-Gaussian parameter (NGP) as

$$\alpha_2^R(t) = \frac{< \Delta\phi(t)^4 >}{2 < \Delta\phi(t)^2 >^2} - 1 \tag{61}$$

where $< \Delta\phi(t)^{2n} > = \frac{1}{N}\sum_{i=1}^{N} < |\phi_i(t) - \phi_i(0)|^{2n} >$. Here ϕ_i is the rotational analogue of the position $\mathbf{r}_i$ of the ith ellipsoid of revolution, the change of which is defined [132, 133] by $\Delta\phi_i(t) = \phi_i(t) - \phi_i(0) = \int_0^t dt' \omega_i(t')$, with ω_i being the corresponding angular velocity. Figure 17a shows the time dependence of the rotational NGP $\alpha_2^R(t)$ for the model liquid crystalline system GB(3, 5, 2, 1) at

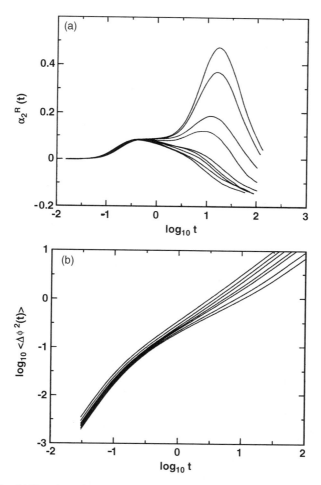

Figure 17. (a) Time dependence of the rotational non-Gaussian parameter $\alpha_2^R(t)$ in a semilog plot for the model calamitic system GB(3, 5, 2, 1) ($N = 500$) at several temperatures $2.008 \geq T \geq 1.102$ across the I–N transition along the isochor at $\rho = 0.32$. The long-time peak gets stronger as the temperature gradually falls. (b) Time dependence of the mean-square angular displacement for the same system at the same set of temperatures mentioned above. Temperature decreases from top to bottom. The curves corresponding to two temperatures, where the system is in the isotropic phase, are removed for the clarity of the figure. (Reproduced from Ref. 131.)

several temperatures across the I–N transition, A bimodal feature appears with the growth of a second peak at longer times on approaching the I–N transition from the isotropic side. On crossing the I–N phase boundary, the long-time peak becomes the dominant one with a shoulder at shorter times. As the temperature is lowered, the onset of the growth of the long-time peak in the rotational NGP is accompanied by a signature of a subdiffusive regime in the temporal evolution of the mean-square angular deviation as shown in Figure 17b. The timescale of the shoulder is also found to nearly coincide with that of the onset of the subdiffusive regime. The shoulder in the rotational NGP can therefore be ascribed to what may be called the rotational analogue of rattling within a cage. Subsequent to the shoulder, the dominant peak appears around a time t_{max}^r, which shifts rather slowly to higher values as the temperature falls. We note that t_{max}^r is comparable to the onset of the diffusive motion in ODOF. We further note that the timescale of the dominant peak also coincides with that of the plateau, which is observed in the time evolution of $C_2^s(t)$ as evident in Figure 9. It follows that the long-time peak appears with the growth of the pseudonematic domains that have random local directors and becomes pronounced as the orientational correlation grows. In Figure 18, we show the time evolution of the rotational NGP as well as the translational NGP for the ellipsoid of revolution in the binary mixture. The latter is defined by

$$\alpha_2^{Tr}(t) = \frac{3 < \Delta \mathbf{r}(t)^4 >}{5 < \Delta \mathbf{r}(t)^2 >^2} - 1 \qquad (62)$$

where $< \Delta \mathbf{r}(t)^{2n} > = \frac{1}{N} \sum_{i=1}^{N} < |\mathbf{r}_i(t) - \mathbf{r}_i(0)|^{2n} >$. It is evident in Figure 18a that the growth of the rotational NGP to its maximum value before it starts decaying is rather smooth at all the temperatures studied. The maximum is found to be reached on a timescale that characterizes the onset of the diffusive motion in ODOF. This time t_{max}^r gets lengthened with decreasing temperature. While the short-time shoulder is not observed in the case of the rotational NGP, it appears in the time evolution of the translational NGP as evident in Figure 18b . At high temperatures, starting from $\alpha_2^{Tr}(t = 0) = 0$, the translational NGP rises smoothly to a value on a timescale that characterizes the crossover from the ballistic to the diffusive motion in TDOF. It then starts falling off to reach the long time limit. As the temperature drops, a shoulder or a step-like feature appears in between the initial rise and subsequent growth to its maximum value. The shoulder is found to appear as a subdiffusive regime sets in between the ballistic and diffusive motion. It is further observed that the shoulder in the translational NGP appears on a timescale that coincides with that of the second maximum of the velocity autocorrelation function (data not shown). The velocity autocorrelation function shows oscillatory character at low temperatures. The present analysis suggests that the shoulder at small times in the translational NGP can be attributed to what

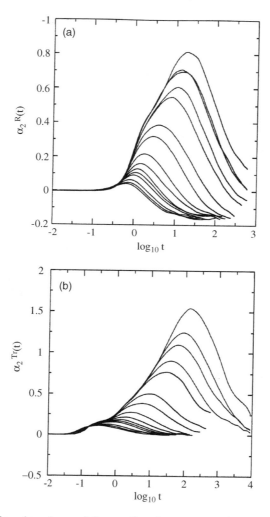

Figure 18. Time dependence of the non-Gaussian parameters in a semilog plot for the ellipsoids of revolution in the 50:50 binary mixture ($N = 256$) for all temperatures investigated. Temperature decreases from bottom to the top. (a) Time dependence of the rotational non-Gaussian parameter $\alpha_2^R(t)$; (b) time dependence of the translational non-Gaussian parameter $\alpha_2^R(t)$. (Reproduced from Ref. 130.)

is known as the rattling within a cage while the dominant peak appears on a timescale that characterizes the escape of a particle from the cage formed by its nearest neighbors. The absence of the step-like feature in the time dependence of the rotational NGP may be ascribed to the less pronounced freezing of the motion along ODOF.

In a recent work, the translational motion of 4-n-hexyl-4'-cyanobiphenyl (6CB) was studied in the isotropic phase by atomistic molecular dynamics simulation [134], The mean-square displacement showed evidence of sub-diffusive dynamics, with a plateau that became very apparent at the lowest temperatures. A three-time self-intermediate scattering function revealed that this plateau was connected with a homogeneous dynamics that, at longer times, became heterogeneous and finally exponential. These features, which are shared by, for example, a high-density system of hard spheres, support the universal character of the translational dynamics of liquids in their supercooled regime.

3.   Microscopic View

In Figure 19, we present a microscopic view of single-particle trajectories in the orientational space. Such a view clearly demonstrates the onset of localization of the orientational motion around a preferred alignment as the I–N transition is approached upon cooling. The single-particle trajectories in Fig. 19a provide evidence for the rotational symmetry breaking on crossing the I–N phase boundary from the isotropic phase. In Fig. 19b, we display typical single-particle trajectories in the orientational space for the ellipsoids in the binary mixture. While the dynamics are ergodic at high temperatures, the signature of nonergodicity is evident at the lowest temperature studied. A comparison between Figs. 19a and 19b is revealing. For the model liquid crystalline system, the seemingly "nonergodic" behavior is driven by the rotational anisotropy that emerges near the I–N transition [130].

4.   Fragility of Calamitic Liquid Crystals

The remarkable similarity in the spectrum of relaxation behavior exhibited by calamitic liquid crystals across the I–N phase boundary and supercooled liquids calls for a quantitative estimation of the extent of glassy dynamics in the former. For glass-forming liquids, the plot, which displays the shear viscosity (or the structural relaxation time, the inverse diffusivity, etc.) in a logarithmic scale as a function of the inverse temperature scaled by T_g [43, 44, 135], provides the basis of a quantitative measure in terms of the fragility index m. In a similar spirit, the single-particle second-rank orientational correlation time $\tau_2^s(T)$ was plotted in the logarithmic scale as a function of the inverse temperature along three isochors for three Gay–Berne systems GB(κ, 5, 2, 1), which differed only in the choice of the aspect ratio [136](Fig. 20a). In Fig. 20a, the temperature was scaled by the I–N transition temperature T_{I-N}, which was taken as the temperature at which the average orientational order parameter S of the system was 0.35. An estimate of $\tau_2^s(T)$ was obtained as the time taken for $C_2^s(t)$ to decay by 90%, that is, $C_2^s(t = \tau_2^s) = 0.1$ at respective temperatures. For all the three systems, two distinct features are found to be common: (i) In the isotropic phase far away from the I–N transition, the orientational correlation time exhibits Arrhenius

(a)

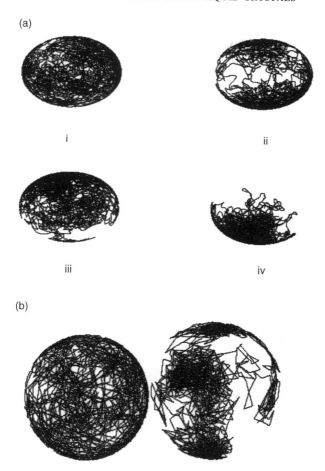

(b)

Figure 19. Typical trajectories of the unit orientation vector for a single ellipsoid revolution in two different systems, (a) Calamitic system GB(3,5,2,1) at four temperatures: (i) $T = 2.008$ (in the isotropic phase), (ii) $T = 1.396$ (close to the I–N transition), (iii) $T = 1.310$ (close to the I–N transition), and (ii) $T = 1.192$ (in the nematic phase), (b) Binary mixture at the highest temperature (left) and the lowest (right) temperature studied. (Reproduced from Ref. 131.)

temperature dependence, that is, $\tau_l^s(T) = \tau_0 \exp[E_0/(\kappa_B T)]$, where the activation energy

E_0 and the infinite temperature relaxation time τ_0 are independent of temperature, and (ii) in the isotropic phase near the I–N transition, the temperature dependence of $\tau_2^s(T)$ shows marked deviation from Arrhenius behavior and can be well-described by the Vogel–Fulcher–Tammann (VFT) equation $\tau_2^s(T) = \tau_{VFT} \exp[B/(T - T_{VFT})]$, where τ_{VFT}, B, and τ_{VFT} are constants, independent of temperature. Again these features bear remarkable similarity with

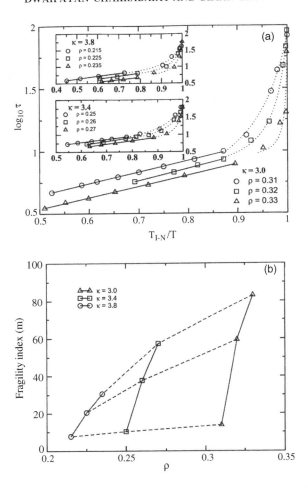

Figure 20. (a) Orientational correlation time τ in the logarithmic scale as function of the inverse of the scaled temperature, with the scaling being done by the isotropic to nematic transition temperature with T_{I-N}. For the insets, the horizontal and the vertical axis labels read the same as that of the main frame and are thus omitted for clarity. Along each isochor, the solid line is the Arrhenius fit to the subset of the high-temperature data and the dotted line corresponds to the fit to the data near the isotropic–nematic phase boundary with the VFT form, (b) Fragility index m as a function of density for different aspect ratios of model calamitic systems. The systems considered are GB(3, 5, 2, 1), GB(3.4, 5, 2, 1), and GB(3.8, 5, 2, 1). In each case, $N = 500$. (Reproduced from Ref. 136.)

those observed for fragile glass-forming liquid. A non-Arrhenius temperature behavior is taken to be the signature of fragile liquids. For fragile liquids, the temperature dependence of the shear viscosity follows Arrhenius behavior far above T_g and can be fitted to the VFT functional form in the deeply supercooled regime near T_g [43, 44, 135], In the same spirit that offers a quantitative estimation of the fragile behavior of supercooled liquids, we defined the fragility index of a liquid crystalline system as

$$m = \frac{dlog_{10}\tau(T)}{d(T_{I-N}/T)}\Big|_{T=T_{I-N}} \tag{63}$$

Figure 20b shows the density dependence of the fragility index for the three systems with different aspect ratios. For a given aspect ratio, the fragility index increases with increasing density. The numerical values of the fragility index are found to be comparable to those of supercooled liquids. The change in the fragility index for a given density difference $(\Delta\rho)$ increases with the decrease in the aspect ratio.

D.   Energy Landscapes

A useful framework for interpreting the thermodynamic and dynamic properties of condensed phase is provided by the energy landscape picture [13], which, in particular, has been extensively used for elucidation of dynamics of supercooled liquids [137–142]. By energy landscape, what is meant is the hyperspace generated upon representing the potential energy of an N-body system as a function of all configurational coordinates. The inception of the energy landscape picture was the classic contribution by Goldstein in 1969 when he formulated what he called "a potential energy barrier picture" of flow and relaxation in viscous liquids. Subsequently, Stillinger and co-workers, building upon the foundation laid down by Goldstein, proposed a computational approach to develop the energy landscape picture [143]. In this approach, the potential energy surface is partitioned into a large number of basins, each defined as the set of points in the multidimensional configuration space such that a local minimization of the potential energy via a steepest descent quench maps each of these points to the same local minimum. The configuration corresponding to a minimum is known as an inherent structure (IS) [143].

For an N-particle system confined to a fixed volume V, the energy landscape of the system is independent of temperature [9]. However, the manner in which the system samples its energy landscape is sensitive to temperature and has influence on its dynamics [139]. In an appealing landscape study of a binary Lennard-Jones mixture, certain changes in the dynamical behavior were found to occur around a crossover temperature, below which the dgpth of the potential energy minima explored by the system started growing, as shown in Fig. 21 [139]. The dynamics was said to be *landscape-influenced* below this crossover temperature, which marked the onset of the nonexponential structural relaxation and the breakdown of the Arrhenius behavior of the structural relaxation time [139]. The cooling rate was found to affect the depth of the potential energy minima explored by the

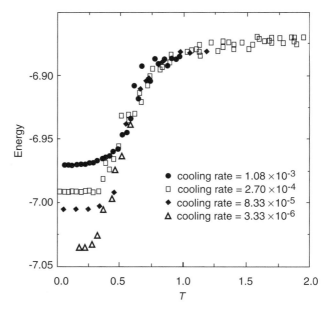

Figure 21. Average energy per particle of the local potential energy minima explored by a binary model system, well known as a good glass-former, as a function of temperature for different cooling rates. (Reproduced from Ref. 139.)

system on an average; the slower the cooling, the deeper were the potential energy minima explored.

1. Exploration of Energy Landscapes

The similarity in the dynamic behavior between calamitic liquid crystals near the I–N transition and supercooled liquids motivated Chakrabarti and Bagchi to investigate the energy landscape of calamitic liquid crystals. In a work probably first of its kind, they studied how a family of Gay–Berne models sample their underling potential energy surfaces by investigating the local potential energy minima [144]. Figure 22a displays the average inherent structure energy per particle as the change in the temperature drives the system GB(3, 5, 2, 1) across mesophases along three different isochors. The isochors were so chosen that the range of the nematic phase along these isochors varied considerably [104]. Apart from the isotropic phase, this system is known to exhibit the nematic and smectic-B phases, but no smectic-A phase [104]. Figure 22b shows the concomitant evolution of the average orientational order parameter for both the inherent structures and the corresponding prequenched configurations. It is evident that the average inherent structure energy remains fairly insensitive to temperature in the isotropic phase before it starts undergoing a steady fall below a certain temperature that

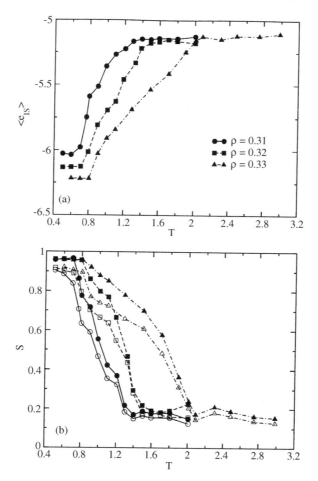

Figure 22. Potential energy landscape explored by the model calamitic system GB(3, 5, 2, 1) ($N = 256$) as the system makes a transit through mesophases upon cooling, (a) Temperature dependence of the average inherent structure energy per particle, $\langle e_{IS} \rangle$, along three isochors at densities $\rho = 0.31, 0.32$, and 0.33. (b) Evolution of the average second-rank orientational order parameter S with temperature both for the inherent structures (filled) and for the instantaneous configurations (opaque). (Reproduced from Ref. 144.)

corresponds to the onset of growth of the orientational order. As the orientational order grows through the nematic phase, the system continues to explore increasingly deeper potential energy minima. The transition from the nematic phase to the smectic phase in the parent system, signaled by the appearance of a one-dimensional density wave in the parallel radial distribution function computed from prequenched configurations [104], results again in an average inherent structure energy that is roughly independent of temperature.

The Gaussian form for the number density of the inherent structure energy predicts a linear variation of $<e_{IS}>$ with the inverse temperature:

$$< e_{IS}(T) >= e_0^{eff} - \sigma^2/(2/N^2 k_B T) \qquad (64)$$

where e_0 and σ are parameters independent of temperature. Figure 23a demonstrates that the prediction holds good over the temperature range through which $\langle e_{IS} \rangle$ decreases along all the three isochors studied. It then follows that a plot of $< e_{IS} > -e_0^{rff}$ versus $\sigma^2/(2N^2 k_B T)$ would result in a collapse of the data for all the densities onto a straight line with negative unit slope. This is indeed found to be true, as shown in Figure 23b, implying the validity of the Gaussian model in

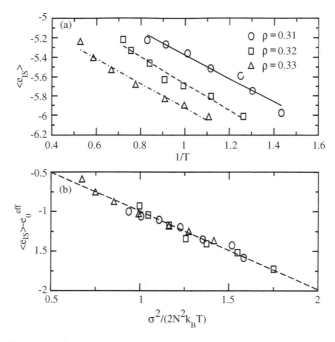

Figure 23. Validity of the Gaussian model for the number density of the inherent structure energy for the model calamitic system GB(3, 5, 2, 1) ($N = 256$). (a) Average inherent structure energy per particle as a function of the inverse temperature at different densities. The continuous line, dashed line, and dot–dashed line are the linear fits to the data at densities $\rho = 0.31$, 0.32, and 0.33, respectively. (b) Displaced average inherent structure energy per particle versus a scaled inverse temperature along the same three isochors. If the Gaussian model for the number density of the inherent structure energy is validated, a collapse of the data for all densities is expected onto a straight line with negative unit slope. The straight line is shown. Here data are shown over the temperature regimes through which the average inherent structure energy decreases. (Reproduced from Ref. 136.)

this case over a temperature range where the orientational order continues to grow upon cooling [136].

The manner in which the present system samples its potential energy landscape is remarkably similar to what was observed for a model system of glass-forming liquids, even though the latter avoided a phase transition [139, 142]. In an appealing landscape study of a binary mixture of atomic liquids, Sastry et al. [139] showed that a crossover temperature, below which the average depth of the potential energy minima explored by the system grows, marks the onset of the breakdown of Arrhenius behavior in the temperature dependence of the structural relaxation time. In search of a correlation between the exploration of the underlying energy landscape by the system and its dynamics, we show the Arrhenius representation of the temperature-dependent data of the single-particle orientational relaxation times in Fig. 24. In this case [144], a measure of the temperature dependent relaxation time $\tau_l^s(T)$ was taken as $C_l^s(t = \tau_l^s) = e^{-1}$. Figure 24 illustrates the dramatic slowdown of orientational dynamics with decreasing temperature near the I–N transition for $l = 1, 2$. It follows that in the isotropic phase far from the I–N transition region, $\tau_l^s(T)$ exhibits Arrhenius behavior, that is, $\tau_l^s(T) = \tau_{0,l} \exp[E_l/(k_B T)]$, where the activation energy E_l and

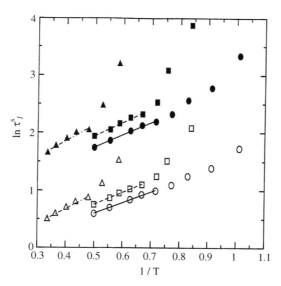

Figure 24. Breakdown of Arrhenius behavior of the single-particle orientational correlation times for the model calamitic system GB(3, 5, 2, 1) ($N = 256$). The inverse temperature dependence of the single-particle orientational correlation times $\tau_l^s(T), l = 1$ (filled) and $l = 2$ (opaque), in the logarithmic scale. The straight lines are the Arrhenius fits to the subsets of data points, with each set corresponding to an isochor: $\rho = 0.31$ (circle), $\rho = 0.32$ (square), and $\rho = 0.33$ (triangle). (Reproduced from Ref. 144.)

the infinite temperature relaxation time $\tau_{0,l}$ are independent of temperature. We find that Arrhenius behavior breaks down near the *I–N* transition, and this breakdown occurs at a temperature that marks the onset of growth of the average depth of the potential energy minima explored by the system [144].

For the binary mixture we investigated, the average depth of the potential energy minima sampled by the system was also found to grow with decreasing temperature below a certain onset temperature $T_o \simeq 1.0$ as shown in Fig. 25a. In this case [145], the low-temperature plateau, which is missing unlike in Ref. 139, seems to be beyond the temperature range of investigation. Figure 25b shows how

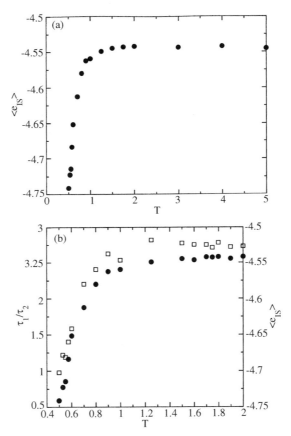

Figure 25. (a) Average inherent structure energy per particle, $\langle e_{IS} \rangle$, as a function of temperature for the 50:50 binary mixture of the Lennard-Jones spheres and the Gay–Berne ellipsoids of revolution ($N = 256$) over the whole temperature range studied. (b) Temperature dependence of τ_1/τ_2, the ratio of the single-particle first-rank orientational correlation time to the single-particle second-rank orientational correlation time (squares). On a different scale (appearing on the right), shown again is the temperature dependence of $\langle e_{IS} \rangle$ for the purpose of comparison (circles). (Reproduced from Ref. 145.)

the ratio of the first-rank orientational correlation time to the second rank orientational correlation time, τ_1/τ_2, evolves as temperature drops. In this case, the lth rank orientational correlation time τ_l is denned as

$$\tau_l = \int_0^\infty C_l^s(t) \, dt \qquad (65)$$

It is evident in Fig. 25b that the ratio has a value close to 3 at high temperatures ($T > 1.0$) and *declines steadily* below $T \simeq 1.0$ until it reaches a value nearly equal to unity at low temperatures. While the Debye model of rotational diffusion, which invokes small steps in orientational motion, predicts the ratio τ_1/τ_2 to be equal to 3, a value for this ratio close to 1 is taken to suggest the involvement of long angular jumps [146, 147]. The ratio was observed to deviate from the Debye limit at lower temperatures in a recent molecular dynamics simulation study as well [148]. The onset temperature was thus found to mark the breakdown of the Debye model of rotational diffusion [145]. Recently, the Debye model of rotational diffusion was also demonstrated to break down for calamitic liquid crystals near the I–N phase boundary due to the growth of the orientational correlation [149].

It is to be noted that while the emergence of the power law decay near the I–N transition is coincident with growing orientational correlation, the system explores deeper potential energy minima as the orientational order parameter grows through the nematic phase upon cooling. This suggests a plausible connection between the power law relaxation and the exploration of the underlying energy landscape. In fact, the exponent of the intermediate power law observed in the OKE experiments in the nematic phase has been found to be linearly dependent on the orientational order parameter [41]. It would therefore be instructive to explore the correspondence between the power law relaxation and the features of the energy landscape in further details.

2. Structural Characterization of Inherent Structures of Calamitic Liquid Crystals

The structural features of the inherent structures yielded important information on the interplay between the orientational and translational order in the calamitic mesophases [144]. It follows from Fig. 22b that quenching results in enhanced orientational order for inherent structures than the corresponding pre-quenched configurations. We now examine how the inherent structures evolve as revealed by the pair distribution functions that were obtained by averaging over the quenched configurations. An analysis through the parallel radial distribution function $g_\parallel(r_\parallel)$, which depends only on $r_\parallel$, the pair separation $\mathbf{r}$, parallel to the director $\hat{n}$, reveals a remarkable feature as illustrated in Fig. 26a. The onset of growth of the orientational order in the vicinity of the I–N transition induces a

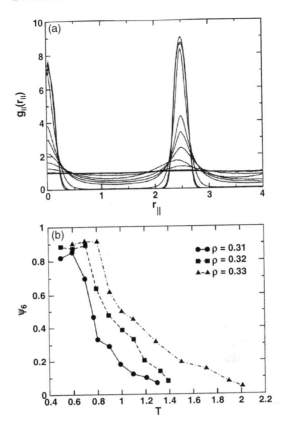

Figure 26. Characterization of the inherent structures for the model calamitic system GB(3, 5, 2, 1) ($N = 256$). (a) Parallel radial distribution function $g_\parallel(r_\parallel)$ for the inherent structures at all temperatures considered along the isochor at density $\rho = 0.32$. Note that the curves for the highest five temperatures are nearly superposed on each other. For others, the amplitude of the peaks gradually increases as the temperature drops. (b) Evolution of the 6-fold bond orientational order parameter Ψ_6 for the inherent structures with temperature at three densities. (Reproduced from Ref. 144.)

translational order in layers in the underlying quenched configurations. Such smectic-like layering is characterized by the one-dimensional density wave along the layer normal appearing in $g_\parallel(r_\parallel)$. The amplitude of the density wave in inherent structures grows stronger as the orientational order increases through the nematic phase of the parent system and tends to attain saturation as the smectic phase sets in. An exploration of the parameter space suggested that the inherent structures never sustain orientational order alone even when the parent phase is nematic if the nematic phase is sandwiched between the high-temperature isotropic phase and the low-temperature smectic phase [144]. Thus, the stability of the nematic phase appears to be due to a subtle balance between the energy and

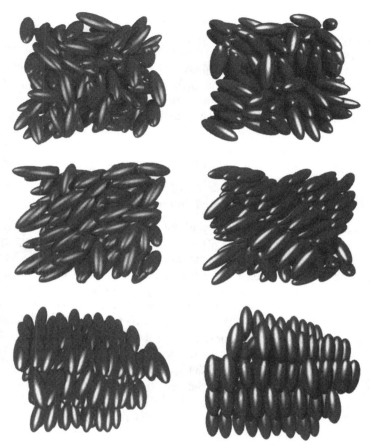

Figure 27. Snapshots of typical instantaneous configurations (left panel) and inherent structures (right panel) for the model calamitic system GB(3, 5, 2, 1) at various mesophases. From top to bottom: isotropic phase, nematic phase, and smectic phase. (Reproduced from Ref. 131.)

the entropy, with the former favoring it over the isotropic phase and the latter favoring it over the smectic phase. In Fig. 27, we show typical snapshots of the parent system in various phases obtained from molecular dynamics trajectories and the inherent structures. The induction of translational order in the inherent structures when the parent phase is nematic is evident.

In the evolution of the perpendicular radial distribution function for the inherent structures, the gradual enhancement of order is evident and the peaks that follow the dominant one are found to be split [144]. This is typical of quenched configurations and hence inconclusive for the inherent structures to be categorized as either smectic-A or smectic-B. We have, therefore, computed the

6-fold bond orientational order parameter Ψ_6 the inherent structures to characterize the smectic phase further on the basis of hexagonal symmetry [18]. As shown in Fig. 26b, we find that as the system passes through the nematic phase upon cooling, concomitant with the growth of the translational order in the inherent structures, Ψ_6 also grows until it attains saturation with the advent of the smectic-B phase in the system. Therefore, it seems reasonable to conclude that as the system makes a passage across the nematic phase from the high-temperature isotropic to the low-temperature smectic-B phase, the underlying potential energy minima evolve from the isotropic to the smectic-B phase through the smectic-A phase with no signature of the nematic phase.

3. Dynamical Signatures of the Interplay Between Orientational and Translational Order

Although a variety of experimental techniques probe anisotropic translational diffusion in the nematic phase [24, 27–29], a consensus regarding an appropriate dynamical model still lacks. In particular, the role of coupling between orientational and translational order parameters has been overlooked. On the contrary, the interplay between orientational and translational order has been extensively discussed in the context of the transition from the nematic to the Smectic-A phase [2, 150–158]. The well-known de Gennes–McMillan coupling refers to the occurrence of the one-dimensional translational (smectic) ordering being intrinsically coupled with an increase in orientational (nematic) ordering [2, 150, 151]. The de Gennes–McMillan coupling could drive, within a mean field approximation, an otherwise continuous nematic to smectic-A (NA) transition to first order for a narrow nematic range [150]. Halperin, Lubensky, and Ma later invoked the coupling between the smectic order parameter and the transverse director fluctuations in their theoretical treatment that predicted NA transition to be *at least weakly first order* [152, 153].

Intuitively, $D_\parallel$ appears to be well-placed to capture the dynamical signature of the coupling between orientational and translational order. In the energy landscape formalism the time-dependent position $\mathbf{r}_i(t)$ of a particle i can be resolved into two components: $\mathbf{r}_i(t) = \mathbf{R}_i(t) + \mathbf{S}_i(t)$, where $\mathbf{R}_i(t)$ is the spatial position of the particle i in the inherent structure for the basin inhabited at time t, and $\mathbf{S}_i(t)$ is the intrabasin displacement away from that inherent structure [159], It has been theoretically argued that the replacement of the real positions $\mathbf{r}_i(t)$ by the corresponding inherent structure positions in the Einstein relation yields an equivalent diffusion description [159, 160]. Such a proposition, which has been verified in simulations [159, 160], forms the foundation of the analysis presented here.

Figure 28 shows the temperature-dependent $D_\parallel$ and $D_\perp$ data in the Arrhenius plot for the 256-particle Gay–Berne system GB(3, 5, 2, 1) along two isochors [161]. $D_\parallel$ and $D_\perp$ are obtained from the slopes at long times of the respective

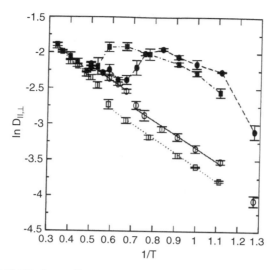

Figure 28. Self-diffusion coefficients $D_\parallel$ and $D_\perp$ in the logarithmic scale versus the inverse temperature along two isochors at densities $\rho = 0.32$ (circles) and 0.33 (squares). The dot-dashed and long-dashed lines lines are a guide to eye for the $D_\parallel$ data (filled symbols), and the solid lines and the dotted lines are the Arrhenius fits to the $D_\perp$ data (empty symbols) data for $\rho = 0.32$ and 0.33, respectively. $D_\perp$ data were considered separately across the isotropic phase and the nematic phase for the Arrhenius fits. (Reproduced from Ref. 161.)

mean-square displacements versus time plots [162]. For the finite size of the system, the average orientational order parameter S has a nonzero value even in the isotropic phase. This allows us to compute $D_\parallel$ and $D_\perp$ also in the isotropic phase. It is evident that both $D_\parallel$ and $D_\perp$, which have nearly identical values in the isotropic phase as expected, exhibit an Arrhenius temperature dependence in this phase. On crossing the isotropic–nematic (I–N) phase boundary as temperature drops, $D_\parallel$ *first increases and then decreases* while $D_\perp$ continues to undergo a monotonic decrease following an Arrhenius temperature behavior across the nematic phase. From the Arrhenius fits to the $D_\perp$ data, we find that the activation energy for the diffusive translational motion perpendicular to the director remains effectively unchanged on either side of the I–N transition.

A quantitative, *albeit indirect*, approach to capture the dynamical signature of the coupling between orientational and translational order is to compare the $D_\parallel$ and $D_\perp$ data obtained from our simulations with those predicted by the existing dynamical models, which ignore such coupling. In Fig. 29, we do so by considering two theoretical models [163, 164] that have been applied to trace experimental and molecular dynamics simulation data of anisotropic translational diffusion in the nematic phase of liquid crystalline systems [26–28, 163–165]: the Hess–Frenkel–Allen (HFA) model and the Chu and Moroi

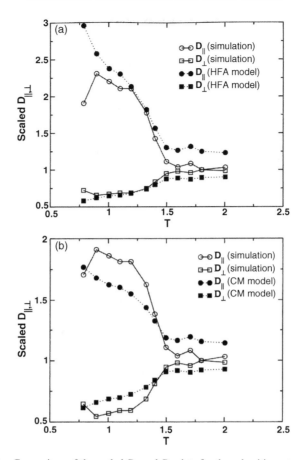

Figure 29. Comparison of the scaled $D_\parallel$ and $D_\perp$ data for the calamitic system GB(3, 5, 2, 1) ($N = 256$), obtained from simulations [161], with those predicted by two theoretical models, (a) The Hess–Frenkel–Allen (HFA) model; (b) the Chu and Moroi (CM) model. For the purpose of comparison, the scaling is done by $\langle D \rangle_g$ and $\langle D \rangle$, respectively. (Reproduced from Ref. 161.)

model. The latter gives relatively simple expressions for $D_\parallel$ and $D_\perp$ in terms of *only* the average orientational order parameter S and the shape factor $g = \pi/(4k)$:

$$D_\parallel = \langle D \rangle [1 + 2S(1 - g)/(2g + 1)] \tag{66}$$

and

$$D_\perp = \langle D \rangle [1 - S(1 - g)/(2g + 1)] \tag{67}$$

where the isotropic average is defined by $\langle D \rangle = (2D_\perp + D_{||})/3$. The HFA model invokes the concept of affine transformation from the space of isotropic hard spheres and yields the following expressions:

$$D_{||} = \langle D \rangle_g \alpha [\kappa^{4/3} - 2/3\kappa^{-2/3}(\kappa^2 - 1)(1 - S)] \qquad (68)$$

and

$$D_\perp = \langle D \rangle_g \alpha [\kappa^{-2/3} + 1/3\kappa^{-2/3}(\kappa^2 - 1)(1 - S)] \qquad (69)$$

where

$$\alpha = [1 + 2/3(\kappa^{-2} - 1)(1 - S)]^{-1/3}[1 + 1/3(\kappa^2 - 1)(1 - S)]^{-2/3} \qquad (70)$$

and the geometric average is defined by $\langle D \rangle_g = D_\perp^{2/3} D_{||}^{1/3}$. For the purpose of comparison, we plot scaled $D_{||}$ and $D_\perp$ data. It follows from Fig. 29 that neither model can capture the nonmonotonic temperature behavior of $D_{||}$. We next demonstrate *directly* by performing a landscape analysis that the nonmonotonic temperature behavior of $D_{||}$ could be due to the the coupling between orientational and translational order.

The smectic order parameter Ψ provides a quantitative measure of the one-dimensional translational order, which is a characteristic of the smectic phase. In Fig. 30, we show the evolution of the average smectic order parameter Ψ of the inherent structures with temperature. A steady increase in Ψ with the concomitant growth of S for the underlying inherent structures is apparent across the nematic phase.

The interplay between orientational order and translational order, shown in Fig. 30, is reminiscent of the de Gennes–McMillan coupling, which was originally conceived to be present near the nematic–smectic phase boundary in the parent system. Figure 31 confirms this with an explicit demonstration of the coupling between the smectic order parameter and the nematic order parameter for the pre-quenched configurations near the nematic–smectic transition region. While the fluctuation of the nematic order parameter is large in the nematic phase, it is the fluctuation of the smectic order parameter that is rather large in the smectic phase. A strong coupling between the two is evident at the nematic–smectic transition region where configurations with larger nematic order parameter values tend to have larger smectic order parameter values.

On scrutiny of Fig. 29 and 30, we find that the reversal in the temperature behavior of $D_{||}$ in the nematic phase occurs when the average smectic order parameter for the underlying inherent structures becomes significant (above 0.3) for the first time. The smectic order parameter is a measure of the translational order that appears in a layer perpendicular to the director. The induction of such

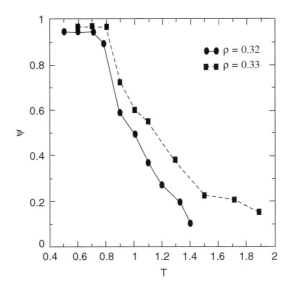

Figure 30. Evolution of the smectic order parameter Ψ for the inherent structures of the calamitic system GB(3, 5, 2, 1) ($N = 256$) with temperature at two densities.

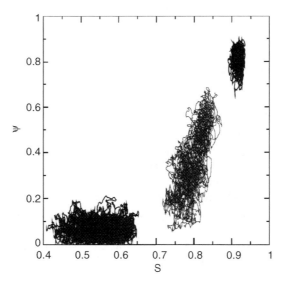

Figure 31. Coupling between the nematic order parameter S and the smectic order parameter Ψ for the calamitic system GB(3, 5, 2, 1) ($N = 256$) at three state points along the isochor at density $\rho = 0.32$. At the nematic phase ($T = 1.194$; bottom), at the smectic phase ($T = 0.502$; top), and at the nematic–smectic transition region ($T = 0.785$; middle). The order parameters are for instantaneous configurations. (Reproduced from Ref. 161.)

310

translational order makes the translational motion parallel to the director much difficult, resulting in a reducing effect on $D_{\parallel}$. From the viewpoint of the energy landscape analysis, translational order in the underlying inherent structures therefore appears to play a key role in inducing the nonmonotonic temperature behavior of $D_{\parallel}$. The latter can therefore be taken as a *dynamical signature of the de Gennes–McMillan coupling augmented in the potential energy landscape.*

This study throws light on the plausible role of the coupling between orientational and translational order in inducing a nonmonotonic temperature dependence of $D_{\parallel}$ in the nematic phase. While the competition between the alignment and thermal effects can also give rise to a nonmonotonic behavior, the importance of such a coupling cannot be ignored particularly when a low-temperature smectic phase exists. A comparison of the simulated $D_{\parallel}$ data with those predicted by two well-known theoretical models shows the inadequacy of these models to capture the observed non-monotonic temperature dependence of $D_{\parallel}$. The accompanying energy landscape analysis suggests the necessity of a theoretical treatment that includes the coupling between orientational and translational order, which has an augmented manifestation in the underlying energy landscape. Such a suggestion is likely to form the foundation of a theoretical framework to explain the features of anisotropic translational diffusion. It is worth noting here that in a recent study the effects of coupling between rotational and translational motion were elucidated by investigating the isolated ellipsoidal particles in water confined to two dimensions [166]. The results were interpreted by using the Langevin theory and numerical simulations [166].

VI. CONCLUSION

It is well known that the orientational relaxation in the isotropic phase of liquid crystals decays exponentially at long times near the isotropic–nematic phase boundary. The time constant of this decay grows rapidly as the *I–N* phase transition is approached upon cooling, and its temperature dependence can be explained quantitatively by using the mean-field theory of Landau and de Gennes. Much of these studies were restricted to times longer than hundreds of nanoseconds, and mostly in the time range from microseconds to milliseconds. Recent time domain OKE experiments, which could investigate a much wider time window ranging from subpicoseconds to tens of microseconds thanks to the advancement of optical and spectroscopic techniques, revealed rich dynamical behavior in the short-to-intermediate time window in the vicinity of the *I–N* phase boundary.

In this review, our focus has been largely on the orientational dynamics of calamitic liquid crystals across the *I–N* transition and their similarity with the dynamics of supercooled liquids. We have reviewed experimental, theoretical, and computer simulation studies together, with emphasis on the latter. Recent computational studies on model liquid crystals could capture their rich shortr-to-intermediate time orientational dynamics. Furthermore, these systems were found to exhibit an array of dynamical features, which are very similar to those of glassy dynamics, across the *I–N* transition. These include two-step decay of the single-particle second-rank orientational time correlation function, sub-diffusive regime in rotational motion, heterogeneous rotational dynamics, and the breakdown of the Debye model of rotational diffusion. Moreover, a quantitative estimate of glassy dynamics of calamitic liquid crystals near the *I–N* phase boundary in terms of a *fragility index* was found to have values comparable to those of glass-forming liquids.

The similarity between calamitic liquid crystals and glass-forming liquids in the exploration of the underlying potential energy landscape is indeed startling, in particular, in the absence of a thermodynamic singularity associated with the glass transition. The correspondence of the breakdown of Arrhenius temperature behavior of the relaxation times with the onset of exploration of deeper potential energy minima in these systems is remarkable. The striking resemblance might imply a unique underlying landscape mechanism for slow dynamics in soft condensed matter systems. However, the characterization of pathways between the potential energy minima and hence the connectivity of the potential energy surface would be critical in elucidation of the relaxation dynamics [13]. Another fascinating aspect of the energy landscape analysis for the calamitic systems is the observation that the inherent structures of the nematic phase do not retain its own morphology if the nematic phase is sandwiched between the high-temperature isotropic phase and the low-temperature smectic phase. This result emphasizes the role of entropy in the thermodynamic stability of the nematic phase. As the mesogens align, their excluded volume decreases and hence they are able to explore a larger fraction of the volume without being restricted by the excluded volume interactions. This causes the translational entropy to increase across the transition from the isotropic to the nematic phase, although the latter has lower orientational entropy than that of the isotropic phase. It is tempting to note here that the theory of Adam and Gibbs, which provides a connection between kinetics and thermodynamics for glass-forming systems, attributes the observed sluggishness near T_g to the decrease of configurational entropy [167].

We hope that the present review has been at least partly successful in bringing out the molecular aspects of dynamics in liquid crystals. This needs to be contrasted with the prevailing macroscopic theories based on long length-scale descriptions. While such a description is perhaps satisfactory for the description of slow long-time exponential decay, an understanding of short-to-intermediate timescale nonexponential decay requires microscopic considerations. Computer simulation studies have provided valuable information and insight in this regard.

A. Future Problems

The series of work reviewed here gives rise to several interesting problems for future work. Below we list them with a brief discussion on their relevance and importance.

1. The interplay between orientational density relaxation and isotropic density relaxation in the nematic phase needs to be explored and understood in greater detail than done till date. In an analytical theory, the coupling between translational and rotational motion often appears as a nonlinear term which makes the study of this aspect challenging.

2. The apparent disagreement between simulation results and experimental observations on the the existence of power law relaxation near the isotropic–columnar (I–C) phase boundary needs further attention. This appears to be associated with the existence of nematic-like fluctuations in the experimental system. This can be checked, on the experimental side, by studying the relaxation near the I–C phase boundary located away from the discotic–nematic phase. For simulation, a careful analysis should be done by changing the temperature and density along the phase coexistence line and studying the existence or lack of nematic-like orientational fluctuations.

3. A little work seems to have been carried out on the wavenumber-dependent orientational correlation functions $C_{lm}(k, t)$. These correlation functions can provide valuable insight into the details of microscopic dynamics of the system. A molecular level understanding of $C_{lm}(k, t)$ would first require the development of a molecular hydrodynamic theory that would have coupling between $C_{lm}(k, t)$ and the dynamic structure factor $S(k, t)$ of the liquid. A slowdown in $C_{lm}(k, t)$ may drive a slowdown in the dynamic structure factor. This would then give rise to a two-order parameter theory of the type develops by Sjögren in the context of the glass transition and applied to liquid crystals by Li et al. [91]. However, a detailed microscopic derivation of the hydrodynamic equations and their manifestations have not been addressed yet.

4. The role of smectic fluctuations in the transport properties of the nematic phase remains largely unexplored. We note that the importance of such fluctuations have been extensively discussed in the context of the nature of the transition from the nematic to the smectic-A phase. It has been shown recently that the anisotropic diffusion of mesogens in the nematic phase cannot be described by theories that employ only the nematic order parameter (S) (see Fig. 29) [161]. The smectic-like fluctuations seem to inhibit translational motion along the director. One can envisage of a free-energy based mean-field theory where the couplmg between the the nematic order parameter S and the smectic order parameter Ψ (see Fig. 31) can give rise to the localization. However, a detailed dynamical theory on this aspect is still lacking.

5. In a recent work, a fragility index was used to quantify the glassiness of orientational relaxation near the $I-N$ transition for calamitic liquid crystals [136]. It is likely that a correlation exists between the fragility index and the growth of orientational correlation. It is a worthwhile exercise to establish a correlation between this fragility index and the features of the potential energy landscape. Such investigation has been carried out with success on supercooled liquids [142]. In the context of liquid crystals, it would require calculation of the configurational entropy within the landscape paradigm. This is of course an interesting problem on its own right.

6. Understanding the organization of disc-shaped mesogens in columnar phases is of practical interest because of their useful optoelectronic applications [168–170]. It will be instructive to explore the role of fluctuations in influencing transport properties of thermotropic discotic liquid crystals—in particular, in the columnar phase.

7. Phase transitions and dynamics of mixtures of different mesogens have not been adequately studied in theoretical (and even in experimental) studies, despite the importance of these multicomponent systems in technology, especially in tuning the range of the nematic phase. These systems are worthy of future studies.

8. The study of the mesogens with limiting shapes can serve as useful models to further our understanding of liquid crystalline systems at the molecular level [171]. Although there have been some studies of diffusion for needles and flat discs [172, 173], a lot needs to be done. An advantage with the limiting cases is their tractability within theoretical framework.

9. An interesting area of research is the dynamics of nematic glass [174–176]. If the nematic phase is cooled sufficiently fast, then one can form nematic glass. Nematic glass is a problem of much current interest, although there exist only a few theoretical studies in the literature. The energy landscape study presented here can be of relevance to the study of nematic glass. In addition, study of the transport properties of this interesting system can be of great value.

10. It will be highly interesting to study molecular dynamics of cholesterol liquid crystals. The effects of asymmetry on the microscopic orientational dynamics may reveal exciting phenomenon.

The above list of future problems is by no means exhaustive, but provides a glimpse of several potentially exciting studies in this area of research. If theory, simulations, and experiments together address these problems, we can look forward to many fascinating results in near future.

Acknowledgments

We thank Dr, Prasanth P. Jose, Mr. Suman Chakrabarty, Mr. Biman Jana, Dr. S. Ravichandran, and Ms. R. Vasanthi for collaboration at various stages and helpful discussions. We thank Professor M. D. Fayer and Dr. David J. Wales for many stimulating discussions. We also thank Suman Chakrabarty and Biman Jana for critical reading of the manuscript. This research was supported in parts by grants from the Department of Science and Technology (DST), India. DC acknowledges the Marie Curie Incoming International Research Fellowship within the 6th European Community Framework Program. BB thanks JC Bose Fellowship for support.

REFERENCES

1. Y. P. Kalmykov, W. T. Coffey, and S. A. Rice, eds, *Advances in Chemical Physics*, Vols. **133A** and **133B**, John Wiley & Sons, New York, 2006.

2. P. G. de Gennes and J. Prost, *The Physics of Liquid Crystals*, Clarendon Press, Oxford, 1993.

3. S. Chandrasekhar, *Liquid Crystals*, Cambridge University Press, Cambridge, 1992.

4. P. M. Chaikin and T. C. Lubensky, *Principles of Condensed Matter Physics*, Cambridge University Press, Cambridge, 1998.

5. S. D. Gottke, D. D. Brace, H. Cang, B. Bagchi, and M. D. Fayer, *J. Chem. Phys.* **116**, 360 (2002).

6. S. D. Gottke, H. Cang, B. Bagchi, and M. D. Fayer, *J. Chem. Phys.* **116**, 6339 (2002).

7. H. Cang, J. Li, M. D. Fayer, *Chem. Phys. Lett.* **366**, 82 (2002).

8. H. Cang, J. Li, V. N. Novikov, and M. D. Fayer, *J. Chem. Phys.* **118**, 9303 (2003).

9. P. G. Debenedetti and F. H. Stillinger, *Nature* **410**, 259 (2001).

10. C. A. Angell, K. L. Ngai, G. B. McKenna, P. F. McMillan, and S. W. Martin, *J. Appl. Phys.* **88**, 3113 (2000).

11. M. D. Ediger, C. A, Angell, and S. R. Nagel, *J. Phys. Chem.* **100**, 13200 (1996).

12. C. Holm and K. Kremer, eds., *Advanced Computer Simulation Approaches for Soft Matter Sciences*, Volumes I & II, Springer-Verlag, Berlin Heidelberg, 2005.

13. D. J. Wales, *Energy Landscapes*, Cambridge University Press, Cambridge, 2003.

14. J. K. Viz, ed, *Advances in Chemical Physics*, Vol. **113**, John Wiley & Sons, New York, 2000.

15. S. Chandrasekhar and G. S. Ranganath, *Rep. Prog. Phys.* **53**, 57 (1990).

16. C. Zannoni, *J. Mater. Chem.* **11**, 2637 (2001).

17. J. M. Polson and D. Frenkel, *Phys. Rev. E* **56**, R6260 (1997).

18. K. J. Strandburg, ed., *Bond-Orientational Order in Condensed Matter Physics*, Springer-Verlag, New York, 1992.

19. C. T. Moynihan, A. J. Easieal, J. Wilder, and J. Tucker, *J. Phys. Chem.* **78**, 2673 (1974).

20. C. A. Angell, *Science* **267**, 1924 (1995).

21. W. Kauzmann, *Chem. Rev.* **43**, 219 (1931).

22. F. Simon, Z. Anorg. *Allg. Chem.* **203**, 219 (1931).

23. P. G. Wolynes, *J. Res. Natl. Inst. Standars Technol.* **102**, 187 (1997).

24. G. J. Krüger, *Phys. Rep.* **22**, 229 (1982).

25. D. R. Spiegel, A. L. Thompson, and W. C. Campbell, *J. Chem. Phys.* **114**, 3842 (2001).

26. A. J. Leadbetter, F. P. Temme, A. Heidemann, W. S. Howells, *Chem. Phys. Lett.* **34**, 363 (1975).

27. S. V. Dvinskikh and I. Furó, *J. Chem. Phys.* **115**, 1946 (2001).

28. S. V. Dvinskikh, I. Furó, H. Zimmermann, and A. Maliniak, *Phys. Rev. E* **65**, 61701 (2002).

29. M. P. Lettinga, E. Barry, and Z. Dogic, *Europhys. Lett.* **71**, 692 (2005).

30. M. P. Allen, *Phys. Rev. Lett.* **65**, 2881 (1990).

31. E. de Miguel, L. F. Rull, and K. E. Gubbins, *Phys. Rev. A* **45**, 3813 (1992).

32. H. Löwen, *Phys. Rev. E* **59**, 1989 (1999).

33. J. D. Litster and T. W. Stinson III, *J. Appl. Phys.* **41**, 996 (1970).

34. J. J. Stankus et al., *Chem. Phys. Lett.* **194**, 213 (1992).

35. J. J. Stankus, R. Torre, and M. D. Fayer, *J. Phys. Chem.* **97**, 9478 (1993).

36. J. Prost and J. R. Lalanne, *Phtys. Rev. A* **8**, 2090 (1973).

37. E. G. Hanson, Y. R. Shen, and G. K. L. Wong, *Phys. Rev. A* **14**, 1281 (1976).

38. J. C. Fillippini and Y. Poggi, *Phys. Lett.* **65A**, 30 (1978).

39. R. Torre, F. Tempestini, P. Bartolini, and R. Righini, *Philos. Mag. B* **77**, 645 (1998).

40. A. Drozd-Rzoska and S. J. Rzoska, *Phys. Rev. E* **65**, 041071 (2002).

41. J. Li, I. Wang and M. D. Fayer, *J. Phys. Chem. B* **109**, 6514 (2005).

42. C. A. Angell, in *Relaxation in Complex Systems*, K. Ngai and G. B. Wright, eds., National Technical Information Service, US department of Commerce, Springfield, VA, 1985, p. 1.

43. C. A. Angell, *J. Phys. Chem. Solids* **49**, 863 (1988).

44. C. A. Angell, *J. Non-Cryst. Solids* **131–133**, 13 (1991).

45. H. Vogel, *Phys. Zeit.* **22**, 645 (1921); G. S. Fulcher, *J. Am. Ceram. Soc.* **8**, 339 (1925); G. Tammann and W. Hesse, *Z. Anorg. Allg. Chem.* **156**, 245 (1926).

46. M. D. Ediger, *Annu. Rev. Phys. Chem.* **51**, 99 (2000).

47. R. Richert, *Chem. Phys. Lett.* **216**, 233 (1993).

48. H. Sillescu, *J. Non-Cryst. Solids* **243**, 81 (1999).

49. R. Richert, *J. Phys.: Condens. Matter* **14**, R703 (2002).

50. M. T. Cicerone, F. R. Blackburn, and M. D. Ediger, *J. Chem. Phys.* **102**, 471 (1995).

51. M. T. Cicerone and M. D. Ediger, *J. Chem. Phys.* **103**, 5684 (1995).

52. M. T. Cicerone, F. R. Blackburn, and M. D. Ediger, *Macromolecules* **28**, 8224 (1995).

53. M. T. Cicerone and M. D. Ediger, *J. Chem. Phys.* **104**, 7210 (1996).

54. R. Böhmer, G. Hinze, G. Diezemann, B. Geil, and H. Sillescu, *Europhys. Lett.* **36**, 55 (1996).

55. I. Chang and H. Sillescu, *J. Phys. Chem. B* **101**, 8794 (1997).

56. R. Richert, *J. Phys. Chem. B* **101**, 6323 (1997).

57. E. V. Russell, N. E. Israeloff, L. E. Walther, and H. A. Gomariz, *Phys. Rev. Lett.* **81**, 1468 (1998).

58. U. Tracht, M. Wilhelm, A. Heuer, H. Feng, K. Schmidt-Rohr, and H. W. Spiess, *Phys. Rev. Lett.* **81**, 2727 (1998).

59. C.-Y. Wang and M. D. Ediger, *J. Phys. Chem. B* **103**, 4177 (1999).

60. B. Doliwa and A. Heuer, *Phys. Rev. Lett.* **80**, 4915 (1998).

61. R. Yamamoto and A Onuki, *Phys. Rev. Lett.* **81**, 4915 (1998).

62. R. Yamamoto and A. Onuki, *Phys. Rev. E* **58**, 3515 (1998).

63. D. N. Perera and P. Harrowell, *Phys. Rev. E* **54**, 1652 (1999).

64. D. N. Perera and P. Harrowell, *J. Chem. Phys.* **111**, 5441 (1999).

65. C. Donati, S. C. Glotzer, P. H. Poole, W. Kob, and S. J. Plimpton, *Phys. Rev. E* **60**, 3107 (1999).

66. G. P. Johari and M. Goldstein, *J. Chem. Phys.* **53**, 2372 (1970).

67. G. P. Johari and M. Goldstein, *J. Chem. Phys.* **55**, 4245 (1971).

68. G. P. Johari, Ann. N.Y. *Acad. Sci.* **279**, 117 (1976).

69. E. Rössler, *Phys. Rev. Lett.* **65**, 1595 (1990).

70. T. Fujima, H. Frusawa, and K. Ito, *Phys. Rev. E* **66**, 31503 (2002).

71. F. Fujara, B. Geil, H. Sillescu, and G. Fleischer, Z. *Phys. B Condense Matter* **88**, 195 (1992).

72. I. Chang, F. Fujara, B. Geil, G. Heuberger, T. Mangel, and H. Sillescu, *J. Non-Cryst. Solids* **172–174**, 248 (1994).

73. S. F. Swallen, P. A. Bonvallet, R. J. McMahon, and M. D. Ediger, *Phys. Rev. Lett.* **90**, 015901 (2003).

74. J. H. Magill and D. J. Plazek, *Nature* **209**, 70 (1966); *J. Chem. Phys.* **46**, 3757 (1967).

75. K. L. Ngai, J. H. Magill, and D. J. Plazek, *J. Chem. Phys.* **112**, 1887 (2000).

76. F. R. Blackburne, C.-Y. Wang, and M. D. Ediger, *J. Phys. Chem.* **100**, 18249 (1996).

77. F. H. Stillinger and J. A. Hodgdon, *Phys. Rev. E* **50**, 2064 (1994); **53**, 2995 (1996).

78. G. Tarjus and D. Kivelson, *J. Chem. Phys.* **103**, 3071 (1995).

79. P. Lunkenheimer, A. Pimenov, and A. Loidl, *Phys. Rev. Lett.* **78**, 2995 (1997).

80. A. P. Sokolov, A. Kisliuk, V. N. Novikov, and K. L. Ngai, *Phys. Rev. B* **63**, 172204 (2001).

81. K. L. Ngai, J. Habasaki, Y. Hiwatari, and C. Leon, *J. Phys.: Condens. Matter.* **15**, S1607 (2003).

82. J. Habasaki and K. L. Ngai, *J. Non-Cryst. Solids* **352**, 5170 (2006).

83. A. Brodin and E. A. Rössler, *J. Chem. Phys.* **125**, 114502 (2006).

84. A. Brodin and E. A. Rössler, *J. Chem. Phys.* **126**, 244508 (2006).

85. B. V. Roie, J. Leys, K. Denolf, C. Glorieux, G. Pitsi, and J. Thoen, *Phys. Rev. E* **72**, 041702 (2005).

86. L. Onsager, *Phys. Rev.* **62**, 558 (1942); *Ann. N.Y. Acad. Sci.* **51**, 627 (1949).

87. G. J. Vroege and H. N. W. Lekkerkerker, *Rep. Prog. Phys.* **55**, 1241 (1992).

88. W. Maier and A. Saupe, Z. *Naturforsch.* **13a**, 564 (1959).

89. W. Maier and A. Saupe, Z. *Naturforsch.* **15a**, 287 (1960).

90. B. Bagchi and A. Chandra, *Adv. Chem. Phys.* **109**, 1 (1991).

91. J. Li, H. Cang, H. C. Andersen, and M. D. Payer, *J. Chem. Phys.* **124**, 014902 (2006).

92. W. Götze, in *Liquid Freezing and Glass Transition*, J. P. H. D. Levesque and J. Zinn-Justin, eds., North Holland, Amsterdam, 1990.

93. L. Sjögren, *Phys. Rev. A* **33**, 1254 (1986).

94. T. Franosch, W. Götze, M. R. Mayr, and A. P. Singh, *Phys. Rev. E* **55**, 3183 (1997).

95. W. Götze and M. Sperl, *Phys. Rev. Lett.* **92**, 105701 (2004).

96. P. Pasini and C. Zannoni, eds., *Advances in the Computer Simulations of Liquid Crystals*, Kluwer Academic Publishers, Dordrecht, 2000.

97. M. R. Wilson, *Int. Rev. Phys. Chem.* **24**, 421 (2005).

98. C. M. Care and D. J. Cleaver, *Rep. Prog. Phys.* **68**, 2665 (2005).

99. M. P. Allen, G. T. Evans, D. Prenkel, and B. M. Mulder, *Adv. Chem. Phys.* **86**, 1 (1993).

100. H. C Andersen, D. Chandler, and J. D. Weeks, *Adv. Chem. Phys.* **34**, 105 (1976).

101. J. G. Gay and B. J. Berne, *J. Chem. Phys.* **74**, 3316 (1981).

102. M. A. Bates and G. R. Luckhurst, *J. Chem. Phys.* **110**, 7087 (1999).

103. E. de Miguel and C. Vega, *J. Chem. Phys.* **117**, 6313 (2002).

104. J. T. Brown, M. P. Allen, E. M. del Ro, and E. de Miguel, *Phys. Rev. E* **57**, 6685 (1998).

105. M. P. Allen and D. Frenkel, *Phys. Rev. Lett.* **58**, 1748 (1987).

106. M. P. Allen and M. A. Warren, *Phys. Rev. Lett.* **78**, 1291 (1997).

107. R. Berardi, A. P. J. Emerson, and C. Zannoni, *J. Chem. Soc. Faraday Trans.* **89**, 4069 (1993).

108. J.-P. Hansen and I. R. McDonald, *Theory of Simple Liquids*, 2nd ed., Academic Press, London, 1986.

109. J. Stecki and A. Kloczkowski, *J. Phys. Paris* **C3**, 40 (1979); *Mol. Phys.* **51**, 42 (1981).

110. A. Perera, G. N. Patey, and J. J. Weis, *J. Chem. Phys.* **89**, 6941 (1988).

111. A. Perera, S. Ravichandran, M. Moreau, and B. Bagchi, *J. Chem. Phys.* **106**, 1280 (1997).

112. P. P. Jose and B. Bagchi, *J. Chem. Phys.* **120**, 11256 (2004).

113. P. P. Jose and B. Bagchi, *J. Chem. Phys.* **125**, 184901 (2006).

114. D. Bertolini, G. Cinacchi, L. De Gaetani, and A. Tani, *J. Phys. Chem. B* **109**, 24480 (2005).

115. D. Chakrabarti, P. P. Jose, S. Chakrabarty, and B. Bagchi, *Phys. Rev. Lett.* **95**, 197801 (2005).

116. M. A. Bates and G. R. Luckhurst, *J. Chem. Phys.* **104**, 6696 (1996).

117. P. A. Lebwohl and G. Lasher, *Phys. Rev. A* **6**, 426 (1972).

118. S. Chakrabarty, D. Chakrabarti, and B. Bagchi, *Phys. Rev. E* **73**, 061706 (2006).

119. D. Chakrabarti, B. Jana, and B. Bagchi, *Phys. Rev. E* **75**, 061703 (2007).

120. J. Li, K. Fruchey, and M. D. Fayer, *J. Chem. Phys.* **125**, 194901 (2006).

121. P. P. Jose and B. Bagchi, *J. Chem. Phys.* **121**, 6978 (2004).

122. W. G. Hoover, D. J. Evans, R. B. Hickman, A. J. C. Lad, W. T. Ashurst, and B. Moran, *Phys. Rev. A* **22**, 1690 (1980).

123. S. Tang, G. T. Evans, C. P. Mason, and M. P. Allen, *J. Chem. Phys.* **102**, 3794 (1995).

124. W. Kob and H. C. Andersen, *Phys. Rev. Lett.* **73**, 1376 (1994).

125. H. C. Andersen, *Proc. Natl. Acad. Sci. USA* **102**, 6686 (2005).

126. D. Chakrabarti and B, Bagchi, *J. Chem. Phys.* **126**, 204906 (2007).

127. W. Kob, *J. Phys.: Condens. Matter* **11**, R85 (1999).

128. D. J. Cleaver, C. M. Care, M. P. Allen, and M. P. Neal, *Phys. Rev. E* **54**, 559 (1996).

129. D. Antypov and D. J. Cleaver, *J. Chem. Phys.* **120**, 10307 (2004).

130. P. P. Jose, D. Chakrabarti, and B. Bagchi, *Phys. Rev. E* **71**, 030701 (R) (2005).

131. D. Chakrabarti and B. Bagchi, *J. Phys. Chem. B* **111**, 11646 (2007).

132. S. Kämmerer, W. Kob, and R. Schilling, *Phys. Rev. E* **56**, 5450 (1997).

133. C. De Michele and D. Leporini, *Phys. Rev. E* **63**, 036702 (2001).

134. L. De Gaetani, G. Prampolini, and A. Tani, *J. Phys. Chem. B* **111**, 7473 (2007).

135. R. Böhmer, K. L. Ngai, C. A. Angell, D. J. Plazek, J. Chem. Phys. **99**, 4201 (1993).

136. B. Jana, D. Chakrabarti, and B. Bagchi, *Phys. Rev. E* **76**, 011712 (2007).

137. M. Goldstein, *J. Chem. Phys.* **51**, 3728 (1968).

138. A. Heuer, *Phys. Rev. Lett.* **78**, 4051 (1997).

139. S. Sastry, P. G. Debenedetti, and F. H. Stillinger, *Nature (London)* **393**, 554 (1998).

140. D. J. Wales and J. P. K. Doye, *Phys. Rev. B* **63**, 214204 (2001).

141. T. F. Middelton and D. J. Wales, *Phys. Rev. B* 64, 024205 (2001).

142. S. Sastry, *Nature (London)* **409**, 164 (2001).

143. F. H. Stillinger and T. A. Weber, **28**, 2408 (1983).

144. D. Chakrabarti and B. Bagchi, *Proc. Natl. Acad. Sci. USA* **103**, 7217 (2006).

145. D. Chakrabarti and B. Bagchi, *Phys. Rev. Lett.* **96**, 187801 (2006).

146. D. Kivelson and S. A. Kivelson, *J. Chem. Phys.* **90**, 4464 (1989).

147. D. Kivelson, *J. Chem. Phys.* 95, 709 (1991).

148. J. Kim and T. Keyes, *J. Chem. Phys.* **121**, 4237 (2004).

149. P. P. Jose, D. Chakrabarti, and B. Bagchi, *Phys. Rev. E* **73**, 031705 (2006).

150. W. L. McMillan, *Phys. Rev. A* **4**, 1238 (1971).
151. P. G. de Gennes, *Solid State Commun.* **10**, 753 (1972).
152. B. I. Halperin, T. C. Lubensky, and S.-K. Ma, *Phys. Rev. Lett.* **32**, 292 (1974).
153. B. I. Halperin and T. C. Lubensky, *Solid State Commun.* **14**, 997 (1974).
154. C. Dasgupta, *Phys. Rev. Lett.* **55**, 1771 (1985).
155. P. E. Cladis et al., *Phys. Rev. Lett.* **62**, 1764 (1989).
156. M. A. Anisimov et al., *Phys. Rev. A* **41**, 6749 (1990).
157. A. Yethiraj and J. Bechhoefer, *Phys, Rev. Lett.* **84**, 3642 (2000).
158. I. Lelidis, *Phys. Rev. Lett.* **86**, 1287 (2001).
159. M. S. Shell, P. G. Debenedetti, and F. H. Stillinger, *J. Phys. Chem. B* **108**, 6772 (2004).
160. T. Keyes and J. Chowdhary, *Phys. Rev. E* **65**, 41106 (2002).
161. D. Chakrabarti and B. Bagchi, *Phys. Rev. E* **74**, 041704 (2006).
162. M. A. Bates and G. R. Luckhurst, *J. Chem. Phys.* **120**, 394 (2004).
163. K.-S. Chu and D. S. Moroi, *J. Phys. (Paris), Colloq.* **36**, C1-99 (1975).
164. S. Hess, D. Prenkel, and M. P. Allen, *Mol. Phys.* **74**, 765 (1991).
165. W. Urbach, H. Hervet, and F. Rondelez, *J. Chem. Phys.* **83**, 1877 (1985).
166. Y. Han, A. M. Alsayed, M. Nobili, J. Zhang, T. C. Lubensky, and A. G. Yodh, *Science* **314**, 626 (2006).
167. G. Adam and J. H. Gibbs, *J. Chem. Phys.* **43**, 139 (1965).
168. D. Adam et al., *Nature (London)* **371**, 141 (1994).
169. L. Schmidt-Mende, A. Fechtenötter, K. Mullen, E. Moons, R. H. Friend, and J. D. MacKenzie, *Science* **293**, 1119 (2001).
170. J. Kirkpatrick, V. Marcon, J. Nelson, K. Kremer, and D. Andrienko, *Phys. Rev. Lett.* **98**, 227402 (2007).
171. M. A. Bates and D. Frenkel, *Phys. Rev. E* **57**, 4824 (1998).
172. R. Vasanthi, S. Ravichandran, and B. Bagchi, *J. Chem. Phys.* **114**, 7989 (2001).
173. R. Vasanthi, S. Bhattacharyya, and B. Bagchi, *J. Chem. Phys.* **114**, 7989 (2001).
174. A. R. Brás, M. Diomísio, H. Huth, Ch. Schick, and A. Schönhals, *Phys. Rev. E* **75**, 061708 (2007).
175. R. Elschner, R. Macdonald, H. J. Eichler, S. Hess, and A. M. Sonnet, *Phys. Rev. E* **60**, 1792 (1999).
176. V. K. Dolganov, R. Fouret, C. Gors, and M. More, *Phys. Rev. E* **49**, 5230 (1994).

COMPLEX PERMITTIVITY OF ICE Ih AND OF LIQUID WATER IN FAR INFRARED: UNIFIED ANALYTICAL THEORY

VLADIMIR I. GAIDUK[†]

Institute of Radio Engineering and Electronics,
Russian Academy of Sciences, Fryazino, 141120, Moscow Region, Russia

DERRICK S. F. CROTHERS

Queen's University Belfast, Belfast, BT7 1NN, Northern Ireland

CONTENTS

[†]Deceased February 4, 2008

Advances in Chemical Physics, Volume 141, edited by Stuart A. Rice
Copyright © 2009 John Wiley & Sons, Inc.

---

[1]In this section we follow in general outline the review article GT2.

SYMBOLS

a.c.	Alternating current (time-varying field)	
ACF	Autocorrelation function	
GT1	Gaiduk and Tseitlin (1994) [9]	
GT2	Gaiduk and Tseitlin (2003) [10]	
H-bond	Hydrogen bond	
SF	Spectral function	
VIG	Gaiduk (1999) [11]	

Symbol	Value	Example
i	$\sqrt{-1}$	
$*$	Complex conjugation	$\chi^* = \chi' + i\chi''$
$\hat{E}$	Complex amplitude	
$\langle f \rangle$	Ensemble average of quantity $f(\Gamma)$	$\langle f \rangle \equiv \int_\Gamma f W \, d\Gamma$
$\in$	"is element of" sign	$x \in [a, b]$
$\propto$	Proportionality sign	$G \propto N$

$a = r/l$	Ratio of covalent bond over H-bond lengths in dimer
A	Action, Eq. (A7)
A	See Eq. (VI 104)
B	See Eq. (VI 104)
$b(t)$	Transverse shift of oxygen atom (see Fig. 1)
$c = 2.9979 \times$ $10^{10} \, \text{cm s}^{-1}, c_{\text{lt}}$	Speed of light
c_{rot}	Dimensionless rotary force constant
$C(t), \tilde{C}(\omega)$	Single-particle dipolar ACF and its spectrum
d	Distance between charges of the dimer (see Fig. 39b)
$d\Gamma$	Element of phase-space volume
$e = 4.803242$ units of CGSE	Charge of electron
$E = \text{Re}[\hat{E}e^{i\omega t}]$	A.c. electric field

$\hat{E}, E_m$	Complex, real amplitude of radiation field
E	Factor (33) [24]
$F(\gamma)$	Induced distribution
$\{F_B, F_G\}$	Boltzmann, Gross induced distributions
f	Form factor of the hat model
$G = \mu^2 N/(3k_B T)$	Normalized concentration of molecules
G_q	Dimensionless factor (21) [24]
G_μ	Dimensionless factor (21) [24]
g	Kirkwood correlation factor
$g_\perp$	Intensity factor (18d) introduced for transverse vibration
H	Steady-state energy (Hamiltonian) of a dipole
$h = H/(k_B T)$	Dimensionless energy of a dipole
H	Planck constant, see Appendix III.
I	Moment of inertia of a molecule
$I_{vib} = 3m_H r^2$	Moment of inertia pertaining to vibration of H-bonded molecule
k	Longitudinal force constant
$k_\perp$	Transverse force constant
$k_B = 1.3807\times 10^{-16} \text{erg} \cdot \text{K}^{-1}$	Boltzmann constant
k_μ	Factor correcting μ value in a liquid
$L(z)$	Spectral function
L_q	Spectral function (33) for longitudinal vibration
L_μ	Spectral function (33) for reorientation of H-bonded molecules
$L_\perp$	Spectral function (22) for transverse vibration
L	Projection of the normalized angular momentum on the symmetry axis of the hat potential
l	Length of the hydrogen bond (see Fig. 1)
M	Molecular mass (gram·mole)
M	Mass of a molecule
$m_H = 1.6726\times 10^{-24}\text{ g}$	Mass of the proton
$\bar{m} \approx \frac{1}{2}m$	Reduced mass of the water molecule
m_{or}	Number of librations of a dipole during lifetime of the hat potential
$m_\perp$	Number of vibrations during lifetime of transverse vibration; see Eq. (23)
N	Concentration
$N_A = 6.0220\times 10^{23}(\text{g}\cdot\text{mole})^{-1}$	Avogadro constant

N	Refraction index (real quantity)
$\hat{n} = \mathrm{Re}[\sqrt{\varepsilon}]$	Complex refraction index
$= n + i\kappa$	
$n_\infty\big\|_{\omega\to\infty}$	Optical refraction index
n_∞^2	Optical permittivity
$p_\perp$	Effective frequency factor for transverse vibration, Eq. (18b)
q	Charge of a nonrigid dipole (see Fig. 1)
$Q = \mu_E/\mu$	Normalized dipole moment (direction cosine)
r	Covalent-bond length (see Fig. 1)
r_{vib}	Proportion of vibrating particles
r_{OO}	Distance between oxygen atoms in dimer
s	Dimensionless longitudinal or transverse vibration deflection
s_m	Maximum transverse displacement (187)
T	Temperature
t, t_v	Time, lifetime between strong collisions
$U_{\mathrm{HB}} = 2.6$ kcal mol^{-1}	H-bond energy
u_{hm}	Reduced H-bond energy (14) [24]
U_0	Well depth
$u = U_0/(k_{\mathrm{B}}T)$	Dimensionless welldepth
$W(\mathrm{H}) = C\exp[-H/(k_{\mathrm{B}}T)])$	Boltzmann distribution function over energy
$x = \omega\eta = 2\pi c v \eta$	Normalized frequency of a.c. field
w	Factor (18c) introduced for transverse vibration
$y = \eta\tau^{-1}$	Normalized frequency corresponding to mean time
$= \tau^{-1}\sqrt{I(2k_{\mathrm{B}}T)^{-1}}$	between strong collisions
$z = x + iy$	Normalized complex frequency
z_{vib}	Complex frequency for harmonic vibration, Eq. (17)
$\alpha = \dfrac{\omega\varepsilon''}{cn} = 4\pi v\,\mathrm{Im}(\sqrt{\varepsilon})$	Absorption coefficient (in cm^{-1})
β	Libration amplitude of a dipole in the hat well
$\chi = \chi' + i\chi''$	Complex susceptibility
δ_∞	Optical permittivity in the case of ice
$\Delta l(t)$	Variable part of H-bond length (see Fig. 1)
$\Delta\varepsilon$	Contribution to complex permittivity due to vibrating molecules
$\Delta\varepsilon_q$	Contribution to ε due to longitudinal vibration
$\Delta\varepsilon_\mu$	Contribution to ε due to reorientation of vibrating molecule

$\Delta\varepsilon_\perp$	Contribution to ε due to transverse vibration
$\varepsilon = \varepsilon' + i\varepsilon''$, ε_s	Complex, static dielectric permittivity (dimensionless quantity)
ε_{or}	Contribution to complex permittivity due to librating dipoles
Φ	Dimensionless time (for various processes)
Φ	Dimensionless period (217) for libration of the permanent dipole in hat well
Φ	Dimensionless period of the $Q(t) = \mu_E(t)/\mu$ function
Φ	Dimensionless period (189) of transverse vibration
γ	Dimensionless coefficient correcting distance between charges of a dimer (see Fig. 39b)
Γ	Phase space
$\eta = \sqrt{I/(2k_B T)}$	Normalizing time parameter
η_{vib}	Normalizing time constant, Eq. (32) [24]
$\eta_\perp$	Normalizing time constant for transverse vibration, Eq. (18b)
κ	Imaginary part of refraction index
λ	Wavelength of electromagnetic radiation in vacuum
$\mu = \mu_0 k_\mu (n_\infty^2 + 2)/3$	Dipole moment of a molecule, librating in the hat well
μ	Dipole-moment vector of vibrating H-bonded molecule
$\bar{\mu}$	Elastic dipole moment at equilibrium, Eq. (163)
$v = \frac{\omega}{2\pi c} = \frac{1}{\lambda}$	Wavenumber (in cm^{-1}), termed "frequency"
v_q	Probe frequency of longitudinal vibration (about) 200 cm^{-1})
$\Pi = \int_{-\infty}^{\infty} \omega\chi''(\omega)\,d\omega$	Integrated absorption
θ	Angular deflection in the hat well (polar angle), Eq. (202)
θ_+ and θ_-	Angular deflections in dimer, respectively, of the covalent bond and H-bond (Fig. 39b)
Θ	Angle between direction of radiation field and symmetry axis (see Fig. 39c)
P	Density
τ_{or}	Mean lifetime (for reorientation)
τ_{vib}	Mean lifetime (for vibration)
τ_D	Debye relaxation time
τ_q	Lifetime for vibration along H-bond
τ_μ	Lifetime for vibration about H-bond
$\tau_\perp$	Lifetime for transverse vibration
$\Omega_q = \sqrt{k/\bar{m}}$	Probe frequency of longitudinal vibration (in rad s^{-1})
$\omega_D = \tau_D^{-1}$	Angular frequency of Debye loss maximum
ψ	Current precession shift of a dipole moment

Ψ	The angle between projections $\mu_\perp$ and $E_\perp$ on plane S (see Fig. 39c)
$\omega = 2\pi cv$	Angular frequency of radiation
$\zeta = \sqrt{g_\perp}$	Association factor (24)

I. INTRODUCTION

The work contains two parts. The first part (Sections II through V) concerns a brief description of the employed models, contains (without derivation) the formulas used in the calculation of the water/ice spectra, and includes a discussion of the results obtained for these fluids. This part, where the main results of our work are presented, could be studied independently of the second part.

The second part (Sections VI through X) is devoted to the derivations of the formulas used in calculations of the complex permittivity and absorption coefficient. Here the basics of the employed analytical approach are also briefly discussed.

A. Approach to Calculation of Spectra Based on Analytical Theory

The development of the theory, taking into account a leading role of the hydrogen bond, is the key problem in the molecular description of aqueous spectra—in particular, of spectra of such important fluids as ice and water [1, 2]. Describing water/ice spectra and the method of their calculation represents the objective of this work.

For water the broad frequency range $v \in 0$–$1000\,\mathrm{cm}^{-1}$, in which dielectric relaxation is stipulated by reorientation of polar molecules and by translation of charged ones, can *conventionally* be divided into four ranges (bands):

(i) the nonresonance low-frequency (Debye) range 0–$10\,\mathrm{cm}^{-1}$,

(ii) the nonresonance submillimetric range 10–$100\,\mathrm{cm}^{-1}$, where only several partial quantities experience resonance,

(iii) the resonance 100–$300\,\mathrm{cm}^{-1}$ band, often termed "translational" (T-band), and

(iv) the resonance librational band 300–$1000\,\mathrm{cm}^{-1}$.

As is accepted in spectroscopy, we term "frequency" the wavenumber $v = \lambda^{-1}$, where λ is radiation wavelength in free space.

As far as *ice* is concerned, we are interested mostly in the resonance far infrared (FIR) range 50–$1000\,\mathrm{cm}^{-1}$. However, the low-frequency nonresonance ice spectra will also be touched in passing.

It is often assumed that various ice/water properties are described most accurately by the numerical molecular dynamics (MD) simulations, in which pair interactions in ensembles, consisting of hundreds of molecules, are considered. However, the MD method is very complicated and not transparent.

For water it gives only a qualitative agreement between the theoretical and experimental FIR spectra [3–5]. Moreover, as will be demonstrated in Section X.A, at present the MD simulations usually cannot describe the T-band at v about $200\,cm^{-1}$.

As far as we know, application of the MD method to *ice* is based on the lattice-vibration concept. The theoretical spectra represent chaotic collections of peaks [6–8], while the experimental FIR spectra of permittivity $\varepsilon(v)$ and absorption $\alpha(v)$ are similar to such spectra in liquid water and therefore are rather smooth.

In this work we apply rather simple *analytical* formulas for $\varepsilon(v)$ and $\alpha(v)$ spectra. These formulas are based on calculation, in terms of classical theory, of the dipole autocorrelation functions (ACFs). Such theory gives a better description of water/ice spectra than does the MD method. Unlike the MD simulations, the linear dielectric response of an ensemble of rigid or nonrigid dipoles is studied using the *one-particle* approximation. Each dipole is assumed to move *without friction* in a *given* intermolecular potential well during a certain lifetime τ. The latter actually plays the role of the friction coefficient of the medium: The smaller the τ, the wider the linewidths of the calculated spectrum. The profile of the potential well used is chosen on the basis of intuitive ideas and experience. Due to such choice of the potential, the employed analytical theory generally turns out to be *less* fundamental than the MD approach. As a result, for the system under consideration the latter is capable of *predicting* the spectra using only the NVT *experimental data*. On the contrary, the analytical theory is generally parameterized *ad hoc*—that is, by taking into account the *experimental spectra*. However, this deficiency is fully compensated, since, as will be demonstrated in this chapter, the analytically obtained spectra, unlike those obtained by MD simulations, agree very well with experiment (in most cases, quantitatively).

The *basics* of the adopted analytical theory are described at great length in the review articles by Gaiduk and Tseitlin [9, 10] and in the monograph by Gaiduk [11]. These publications are, respectively, cited throughout as GT1, GT2, and VIG. An outline of our approach is briefly listed in Section IX.

Our objective is the calculation of the complex susceptibility $\chi(v) = \chi'(v) + i\chi''(v)$ and permittivity $\varepsilon(v) = \varepsilon'(v) + i\varepsilon''(v)$ (the complex-conjugation symbol is omitted) in terms of a molecular theory.

The theory was first elaborated for a rigid dipole librating (in other words, exerting restricted rotation) in a *given* rectangular potential well. Such an approach was applied for strongly absorbing "simple" nonassociated liquids (e.g., CH_3F).

Next, for describing the dielectric spectra of water and ice, it turned out to be optimal to extend the above theory to *two*[2] fractions, librational and vibrational—LIB and VIB fractions, in short.

In water/ice the LIB fraction describes the librational band, centered at 700–$800\,\mathrm{cm}^{-1}$ and located near the border with the IR range. In the case of water the LIB fraction explains also the nonresonance low-frequency relaxation range.

The second (VIB) fraction describes the *specific* spectrum observed in a rather narrow range at v about $200\,\mathrm{cm}^{-1}$ (the term "specific" coming from Walrafen [2]).

We remark that in "simple" liquids only one band arises, usually centered at several hundredths of cm^{-1}. On the contrary, the water spectrum, which covers up to $\approx 1000\,\mathrm{cm}^{-1}$, is two-humped. Its high-frequency part, bordering with the IR range, is to a certain extent similar to the spectrum of simple polar liquids, while a "specific" low-frequency part, with the *absorption* maximum near $200\,\mathrm{cm}^{-1}$, is typical of water/ice. The latter maximum, arising undoubtedly due to the existence of hydrogen bonds (HB), is lacking in simple liquids.

Analytical modeling of the 200-cm^{-1} band is generally met with considerable difficulty (as we mentioned above, the latter has not yet been overcome in the MD simulation method). Note, the development of our analytical theory took a long time.

B. Evolution of Molecular Modeling

In Fig. 1 we give illustrations for one- and two-fraction models (calculation schemes). A few stages of *previously* employed model complications are shown here.

The first stage concerns a one-fraction model regarding libration of a permanent (rigid) dipole in the potential well of the rectangular or hat-shaped profiles.

In the **variant 1a**, hydrogen bonds are lacking. This variant was applied to describing the dielectric spectra in nonassociated simple liquids.

In the **variant 1b** (which was successfully but rather *artificially* applied to water) a certain modification of the variant 1a was involved. In 1b the dipole-moment vector is represented as a sum of two components. The absolute moment $\bar{\mu}$ of the first component is set constant, as in the variant 1a, while the moment $\tilde{\mu}(t)$ of the second component is assumed to *change harmonically with time*. This allows a *formal* description of the T-band in ordinary and heavy water. A drawback of this approach concerns the lack of *explicit* consideration of HB motions and therefore of the T-band spectrum, induced by these motions.

---

[2]Application of the analytical theory to water *solutions* demands employment of *more fractions*—consequently, more potential wells. In each potential the dipole motion is regarded as independent of motion in the other wells (see, e.g., GT2 and VIG).

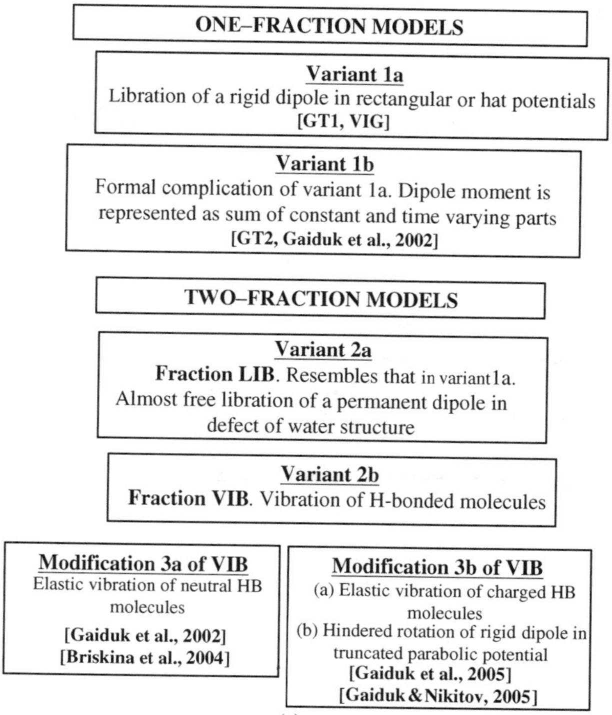

(a)

Figure 1a. The scheme pertaining to the evolution of molecular models preceding those employed in the present work.

This drawback is avoided in the **variant 2a + 2b**, which concerns the second stage of our modeling. An explicit but very simplified consideration of the HB-related effects is now given in terms of a *two-fraction* model. Here the *LIB fraction* is involved similarly with the variant 1a. In the case of water, the variant 2a explains the low-frequency relaxation (Debye) and the libration bands. In the case of ice only the latter band is considered.

Two realizations, **3a and 3b**, of the representation 2b, employed for the *VIB fraction*, are shown in Fig. 1.

The realization 3a concerns a rough calculation of the T-band based on consideration of elastic interactions in an ensemble of *neutral* water molecules. Restricted rotation of two polar water molecules, comprising a dimer, arises due to stretching and bending of hydrogen bonds. The longitudinal force constant k is parameterized to give the proper center frequency ν_{vib} of the T-band. In the experiment, $\nu_{vib}(D_2O) \approx \nu_{vib}(H_2O)$. To obtain such a result theoretically, it

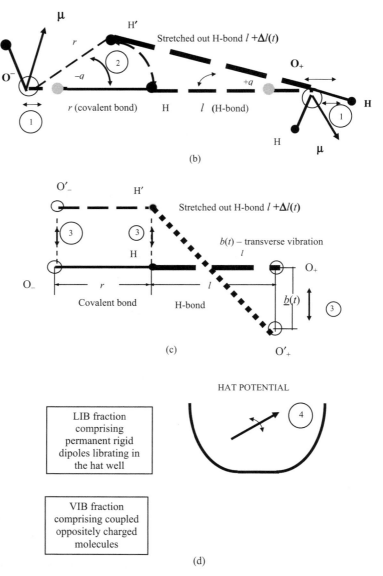

Figure 1 (b), (c): The schemes of the dimer performing (b) harmonic translations/reorientations and (c) non-harmonic transverse vibrations, (d): The scheme pertaining to the mixed model of water/ ice. Numbers 1–4 confined in the circles refer to the following types of motion: 1 – Harmonic longitudinal vibration of the oxygen atoms generates the translational band. 2 – Harmonic reorientation of the permanent dipole generates the V-band located nearby the translational band. 3 – Non-harmonic transverse vibration of the oxygen atoms generates the low-frequency wing of the V-band in the THz region. 4 – Nonharmonic libration of permanent dipole in the hat potential determines the far-infrared librational and the microwave relaxational bands; with the decrease of temperature the potential profile becomes close to the parabolic one.

turned out to be necessary to employ substantially different force constants in the cases of ordinary and heavy water: $k(D_2O) > k(H_2O)$.

The second realization 3b *cardinally* differs from 3a. Thus vibration of two oppositely *charged* water molecules is regarded in terms of the harmonic oscillator model. These molecules constitute a *nonrigid dipole* vibrating along the H-bond. The stretch of the H-bonded *charges* produces the translational band (since the system under consideration is *optically active*) placed about the frequency $\approx 200\,cm^{-1}$. The proper T-band frequency ν_{vib} is now determined by small differing molecular masses of heavy and light water. The difference of the fitted force constants $k(D_2O)$ and $k(H_2O)$ is noticeably less than such a difference in the realization 3a. Our calculations show that the vibrating *charges* exist in water and in ice during a very short period of the order of 0.1 ps.

The other mechanism, involved in the realization of 3b, is related to hindered rotation of a *neutral* molecule (viz., of a *rigid* dipole) in the truncated parabolic potential. Such a rotation was shown to produce the band centered at $\nu \approx 150\,cm^{-1}$. This band overlaps with the $200\text{-}cm^{-1}$ water band but appears in ice as a separate band.

Hence, the two-fraction combination of the LIB system (variant 2a), pertaining to a rigid dipole librating in the hat well, with the VIB system (variant 3b), pertaining to the vibrating HB molecules, allowed the description, for the first time, of the resonance *ice* spectrum in the far IR range $50\text{-}1000\,cm^{-1}$. Such an ice spectrum comprises two split peaks in the $200\,cm^{-1}$ band and the ibrational peak at $\approx 800\,cm^{-1}$. *For water* the 2a, 3b realization gives a reasonable description of the $\varepsilon(\nu)$ and $\alpha(\nu)$ spectra in a very broad range from 0 to $1000\,cm^{-1}$. The part of this range, placed at $\approx 200\,cm^{-1}$, unlike the variant 3a, originates quite *naturally* due to molecules, oscillating along the H-bond. Such a simple calculation was not made before.[3] Note that this mechanism was mentioned, for example, by Walrafen [2].

C. Proposed Model for Calculation of Water/Ice Spectra

Stating the latter result, we describe in this work a new two-fraction model based on a few papers of 2005 and 2006. The main points of this new model are as follows:

- A dimer, involved in the VIB state and formed by H_2O or D_2O molecules, is specified as compared with the variant 3b. The dimer represents a rough structure of the hydrogen bond. Dielectric relaxation arises, first, due to

---

[3]An exception concerns the previously proposed one-fraction calculation, made by Gaiduk and Vij [12] and Gaiduk and Tseitlin [10] in terms of the harmonic-oscillator model. This calculation explains the T-band in water.

longitudinal (along the H-bond) elastic vibration of charges, characterized by the HB force constant k. A small isotopic shift of the translational-band center is attributed to a small dependence of the H-bond rigidity on the type of isotope. Second, taking into account the H-bond rotation, the response to the reorientation of rigid dipoles is found. Using this mechanism allowed us to reject consideration of motion of a rigid dipole in the truncated parabolic potential, as was made in variant 3b.

- Unlike variant 3b, translation and reorientation are now *coupled*. A pair of opposite charges appears near two neighboring oxygen atoms. The lifetime τ_{vib} of these charges comprises only several hundredths of picoseconds. The dielectric relaxation of the VIB fraction is regarded in terms of the harmonic oscillator model as being induced by vibrations of charges along the H-bond. An essential distinction from the interpretation of the small isotopic shift, given in variant 3a, is now traditionally explained by the small difference between the isotopic masses and not by the difference of the longitudinal force constants $k(H_2O)$ and $k(D_2O)$, as was assumed in variant 3a. Hence, the translational vibrations of hydrogen-bonded molecules, forming the dimer, are now coupled with the fast rotational reorientations of HB molecules. Namely, it is assumed that the two oxygen atoms forming the dimer vibrate along the hydrogen bond and that four hydrogen atoms covalently bonded to them experience rotational vibrations.

- The derived system of equations of motion describes simultaneously a periodic variation of the H-bond length and rotation of this bond. We obtain dielectric response both to small translational oscillations of charges and to rigid-dipole reorientations. *Each* response (for charges and for dipoles) is characterized by *two* Lorentz lines.

- An important simplification of the VIB state is proposed, which allows consideration of the translational vibration of charges and reorientation of HB rigid dipoles as *uncoupled* processes. Then the VIB state is determined by two Lorentz lines.

- A new dielectric-relaxation mechanism, important in the THz region, is proposed, relevant to vibration of a nonrigid dipole in a direction perpendicular to that of the H-bond.

- The librational fraction is discussed in the context of the concepts of the water structure: The hat potential models the "defects" of the water (ice) structure and rigid polar molecules reorient relatively freely in these defects. In the case of water the lifetime τ_{or} of this fraction (on the order of 10^{-13} s) is several times greater than that of the H-bond.

- Our model yields an analytical description of the broadband water spectrum and FIR ice spectrum, agreeing with the experimental data.

II. MAIN MOLECULAR MECHANISMS
OF DIELECTRIC RELAXATION

In Section II, *four* molecular mechanisms—a, b, c, and d—are briefly observed. These mechanisms allow us to model analytically the water/ice spectra in terms of classical theory based on calculation of autocorrelation function (ACF). Our description is restricted by the *results* of this theory. A more detailed consideration is given in the second part of the review.

The first mechanism (a) refers to dielectric relaxation pertinent to a permanent dipole influenced by a rather narrow hat intermolecular potential; the next two (b, c) refer to the complex permittivity generated by two elastically vibrating hydrogen-bonded (HB) molecules. The last mechanism (d) refers to a nonrigid dipole vibrating in direction perpendicular to that of the undisturbed H-bond.

Our consideration concerns the wideband spectrum of liquid *water* (H_2O and D_2O) in the range 0–$1000 \, cm^{-1}$ and the spectrum of *ice* H_2O in the far-infrared range 10–$1000 \, cm^{-1}$. We apply the two-fraction (mixed) model, shortly described in Section A. In this model, one fraction refers to the librational (LIB) and the other refer to the vibrational (VIB) modes of molecular motion. In Sections B, C, and D we illustrate the loss spectra relevant to each of the four specific mechanisms (we reserve the term "loss" for the imaginary part of ε). In Section E we write down the formulas for the total permittivity/absorption spectra by taking into account all the above-mentioned mechanisms.

In Sections III–VI of this chapter we shall demonstrate that the main features of the complex permittivity $\varepsilon(v)$; and the absorption coefficient $\alpha(v)$, pertinent to water and ice, may be properly described if these four mechanisms are used.

A. LIB and VIB States

We consider a two-fraction (mixed) model comprising the librational (LIB) and vibrational (VIB) states illustrated by Fig. 1. We shall show that consideration of the LIB and VIB states enables the calculation of the water spectrum, as well as that of ice, irrespective of the nature of these states.

1. LIB State

This state describes (i) dielectric response arising from libration of a permanent dipole μ in a hat-like potential well [the relevant librational band is located near the border of the infrared region (at $\approx 700 \, cm^{-1}$)] and (ii) the nonresonance relaxation band, whose loss peak is located at microwaves. The lifetime τ_{or} of the LIB state is much less than a picosecond.

It appears (see Refs. 13 and 14) that during *cooperative* fast reorientations of hydrogen atoms, occurring inside a rather stable tetrahedral-like set of oxygen atoms, some "favorable" configuration may arise, in which a proton attached to a water molecule jumps to form another H-bond. From this viewpoint the lifetime

τ_{or} of the LIB state actually represents the waiting time before the proton jumps. This molecular picture is supported of the experimental studies by Teixeira et al. [15] in high-quality quasi-elastic incoherent neutron scattering in water. The jump diffusion of the proton across the tetrahedral angle is rather temperature independent. A very useful interpretation of this experiment made recently by Teixeira [16] is reproduced in Appendix I.

A convincing but indirect proof indicating a close connection between the motions of hydrogen atoms and the dielectric properties of aqueous fluids follows from the comparison of the static dielectric permittivity ε_s of ice I_h and ice II. In the former, where the proton disorder is emphasized, $\varepsilon_s \approx 100$, while in the latter, where such a disorder is lacking, $\varepsilon_s = 3.66$ [17].

2. VIB State

In terms of our classical representation the VIB state is formed by two vibrating charged hydrogen-bonded (HB) molecules constituting a dimer and participating in the b, c, and d mechanisms. In view of Figs. 1a, and 1b, these mechanisms concern the following:

- The b mechanism concerns the nonrigid $O^+–H \cdots O^-$ dipole that harmonically vibrates *along* the equilibrium H-bond.
- The c mechanism concerns two *rigid* dipoles μ—formed by the left and right water molecules—that reorient harmonically about this bond.
- The d mechanism concerns the same nonrigid $O^+–H \cdots O^-$ dipole performing a nonharmonic transverse vibration with respect to the equilibrium HB direction.

This system of charges is spectroscopically active. It contributes to the complex permittivity and absorption in the THz region.[4] In water the relevant absorption peak is located at $\approx 200\,cm^{-1}$. Our estimate shows that in the VIB state the concentration N_{vib} of water molecules is commensurable[5] with their total concentration N (estimation gives $N_{vib} = Nr_{vib}$ with $r_{vib} \approx 35$–45%). The b mechanism is responsible for the translational band (T-band) located in the vicinity of $180\,cm^{-1}$ and the c mechanism is responsible for the band that we term the V-band, with the center placed in the vicinity of $\approx 150\,cm^{-1}$.

The b and c vibrations, generated by the harmonic potentials, reveal themselves as the Lorentz lines, respectively, in T- and V-bands. These are

---

[4]Conditionally, the THz region is determined as that in which the frequency changes from 100 to 10^4 GHz—that is, with the range of frequencies v extending from ≈ 3.3 to $333\,cm^{-1}$. Conditionally, the far-IR range could be ascribed as the range 10–$1000\,cm^{-1}$ of wavenumbers v (as usual, v is termed "frequency").

[5]This contrasts with the concentration of the so-called excess protons in water, which determines the electric conductivity of ice and water and is very low with respect to N (see also Appendix I).

resolved in ice but overlap in water. In water the lifetime τ_q of harmonic potential is on the order 0.1 ps and is longer in ice. As for the d mechanism, the transverse displacement b, described by a nonlinear equation of motion, is less than the covalent-bond length $r \approx 1$ Å.

B. Nonharmonic Libration of a Permanent Dipole in a Deep Well

This motion, ascribed to the mechanism a, refers to the *LIB state* and concerns the quasi-resonance dielectric response pertinent to the librational band in the far-IR-range (around $\nu_{or} \approx 700$ and $\approx 900 \, cm^{-1}$, respectively, in water and ice). A similar mechanism determines also the far-IR band of a typical strongly absorbing liquid, such as CH_3F, where, however, libration of dipoles is less free (see details in VIG and GT2).

In terms of our approach, the far-IR band arises from a rather free reorientation of a permanent dipole μ in the hat-like potential well typical of broken or strongly bent hydrogen bonds.

For water, in the low-frequency limit, the mechanism a *formally* (after a proper parameterization) describes the nonresonance Debye relaxation band, whose loss peak is located at microwaves. A useful molecular interpretation of this band was given, for example, by Agmon [18]. Liebe et al. [19] obtained a convenient *empirical* formula for $\varepsilon(\nu)$.

For ice a similar nonresonance band is located at extremely low frequencies (in kHz region). An empirical description of $\varepsilon(\nu)$ in this region is known (see Hufford [20] and references therein). It appears, however, that a relevant molecular interpretation of this band is lacking.

In water the lifetime τ_{or} of the LIB state is commensurable with a part of a picosecond and is shorter in ice. An important feature of this state is a fast reorientation (sometimes termed *disorder*) of hydrogen atoms inside a rather stable tetrahedral-like network of oxygen atoms. As noted above and in the Appendix I, this reorientation is *cooperative*, since a rotating proton, attached to a water molecule, jumps in a certain "favorable" configuration to form another H-bond.

The following free parameters are employed in the hat model:

$$I_{or}, \tau_{or}, k_\mu, , f \tag{1}$$

The moment of inertia $I_{or}[I_{or}(H_2O) = 1.483 \, 10^{-40}, I_{or}(D_2O) = 2.765 \, 10^{-40}$ g $cm^2]$ is estimated for a symmetric top molecule freely rotating about the principal axis normal to the symmetry axis. This moment I_{or} determines the frequency/ timescales relevant to reorientation of dipoles in the hat well. The timescale is determined by the parameter

$$\eta_{or} = \sqrt{I_{or}(2k_B T)^{-1}} \tag{2a}$$

where k_B is the Boltzmann constant and T is the absolute temperature. Parameterization of the model determines the dimensionless strong-collisions frequency y_{or}. Its value $(y < 1)$ allows us to estimate the lifetime τ_{or} (usually being a part of a picosecond) of the hat potential:

$$\tau_{or} = \eta_{or}/y_{or} \tag{2b}$$

Libration of dipoles in the hat well is perhaps more chaotic in ice than in water, since $\tau_{or}(\text{ice}) \ll \tau_{or}(\text{water})$.

The permanent moment μ_{or} of a librating dipole is expressed through that (μ_0) of a free molecule and through the fitted dipole-moment factor k_μ, which is rather close to unity:

$$\mu_{or} = k_\mu \mu_0 (n_\infty^2 + 2)/3 \tag{3}$$

with $n_\infty^2 \approx 1.7$ being the optical (measured near frequency $1000 \, \text{cm}^{-1}$) permittivity.

The frequency-dependent complex permittivity $\varepsilon_{or}(v)$, generated by librating dipoles, reorienting in the hat well during mean time τ_{or}, is determined by such a dynamic quantity as the complex susceptibility $\chi_{or}(v)$:

$$\varepsilon_{or}(v) = (1/4)) \left\{ 12\pi\chi_{or}(v) + n_\infty^2 + \sqrt{[12\pi\chi_{or}(v) + n_\infty^2]^2 + 8n_\infty^4} \right\} \tag{4}$$

We employ the Gross collision model for which this susceptibility $\chi_{or}[z(v)]$ is related to the correlator ("spectral function") L_{or} of the hat model as

$$\chi_{or}(z_{or}) = \frac{g_{or}G_{or}z_{or}L_{or}(z_{or})}{g_{or}x_{or} + iy_{or}L_{or}(z_{or})} \tag{5}$$

where the complex frequency is given by $z_{or} = x_{or} + iy_{or}$; also, $x_{or} = \omega\eta_{or} = 2\pi c\eta_{or}v$, c is the speed of light,

$$G_{or} = \frac{\mu_{or}^2(1 - r_{vib})N}{3k_BT} = (1 - r_{vib})G; \qquad G \equiv \frac{\mu_{or}^2 N}{3k_BT} \tag{6}$$

$$g_{or} = \frac{(\varepsilon_s - \Delta\varepsilon_s - n_\infty^2)\left[2(\varepsilon_s - \Delta\varepsilon_s) + n_\infty^2\right]}{12\pi G_{or}(\varepsilon_s - \Delta\varepsilon_s)} \tag{7}$$

G_{or} is proportional to the librators' concentration $(1 - r_{vib})N$, g_{or} is the Kirkwood correction factor,

$$N = N_{A\rho}/M \tag{7a}$$

N_A is the Avogadro number, ρ is the density, M is the molecular mass, and ε_s is the static permittivity.

The contribution of the mechanism a to the static permittivity is $\varepsilon_{or,s} = \varepsilon_s - \Delta\varepsilon_s$. The symbol Δ in $\Delta\varepsilon_s$ refers to the contribution provided by the vibrating rigid and nonrigid dipoles pertinent to the VIB state. This symbol Δ is employed, since the optical constant n_∞^2 is included in the permittivity component $\varepsilon_{or,s}$. The value $\Delta\varepsilon_s$ will be estimated in Section E.

We use the relationships (4)–(7) to calculate the water spectra. In such a calculation for ice the static permittivity, ε_s(ice) is not involved. For molecules reorienting in the hat well the high-frequency approximation is employed. The complex permittivity of the LIB state is represented, instead of Eq. (4), as

$$\varepsilon_{or}(v) \approx \Delta\varepsilon_{or}(v) + \delta_\infty; \Delta\varepsilon_{or}(v) = 6\pi L_{or}[z_{or}(v)] \tag{7b}$$

so that the total permittivity is calculated from the expression

$$\varepsilon(v) \approx \varepsilon_{or}(v) + \Delta\varepsilon_{vib}(v) = \Delta\varepsilon_{or}(v) + \Delta\varepsilon_{vib}(v) + \delta_\infty \tag{7c}$$

The constant δ_∞ is found from the condition that the total dielectric constant $\varepsilon' = \mathrm{Re}[\varepsilon(v)]$, obtained from (7c), at its minimum (near the end of the far-IR region) should slightly exceed unity.[6] The expression for $\Delta\varepsilon_{vib}(v)$, namely the contribution of vibrating molecules to the total permittivity $\varepsilon(v)$, will be given below.

The spectral function (SF) L_{or} of the hat model is calculated from the integral

$$L_{or}(z) = \left(\frac{2}{\pi}\right)^5 \frac{4u(\beta f)^4}{D} \int_0^u S(h,z)[1 + \sigma(h)]^7 \exp(-h)h\,dh \tag{8}$$

over the reduced energy h of librators, where h is distributed from zero to some barrier energy u and where

$$S(h,z) = \sum_{n=1}^{\infty} \frac{\sin^2\left[\frac{2n-1}{2}\frac{\pi}{1+\sigma}\right]}{(2n-1)^2[(2n-1)^2\sigma^2 - (1+\sigma)^2]^2[(2n-1)^2 h - (2z\beta f/\pi)^2(1+\sigma)^2]}$$

$$\sigma \equiv \sigma(h) = \frac{\pi}{2}\frac{1-f}{f}\sqrt{\frac{h}{u}} \tag{9}$$

---

[6]In our calculation, δ_∞ is rather close to the optical permittivity n_∞^2 : $\delta_\infty = 1.9$, while $n_\infty^2 = 1.7$.

$$D = \frac{4\beta^2 fud}{\pi} \int_0^u \left[f + \frac{\pi}{2}(1-f)\sqrt{\frac{h}{u}} \right] \exp(-h) dh \quad (d=0.7) \qquad (10)$$

The hat model is parameterized to obtain an agreement between the calculated and experimental spectra in the region $400-1000\,\mathrm{cm}^{-1}$ and at low frequencies. To parameterize the hat model with respect to the relaxation band, in which the Debye relaxation time τ_D is regarded to be known, we set $x_{\mathrm{or}} = 0$ in the SF argument and vary y_{or} in order to satisfy the equation

$$\tau_\mathrm{D} \approx g_{\mathrm{or}} G_{\mathrm{or}} (1 - r_{\mathrm{vib}})[y_{\mathrm{or}} L_{\mathrm{or}}(iy_{\mathrm{or}})]^{-1} \qquad (11)$$

Note that τ_D determines the frequency $\nu_\mathrm{D} = \omega_\mathrm{D}/(2\pi c) = (2\pi c \tau_\mathrm{D})^{-1}$ of the microwave loss maximum, where ω_D is the angular frequency of this maximum.

The static permittivity ε_s and the relaxation time τ_D of *ordinary water* could be estimated in accord with Liebe et. al. [19], involving the temperature T in terms of $\theta(T) = 1 - 300T^{-1}$.

$$\varepsilon_s^{\mathrm{H_2O}}(T) = 77.16 - 103.3\,\theta(T) \qquad (12)$$

$$\tau_D^{\mathrm{H_2O}}(T) = \{2\pi\gamma_1[\theta(T)]\}^{-1}, \qquad h(\theta) = (20.2 + 146.4\theta + 316\theta^2)10^9 \qquad (13)$$

For heavy water we employ the formulas given in Gaiduk and Tseitlin [10, p. 198]:

$$\varepsilon_s^{\mathrm{D_2O}}(T_C) = 78.25\left[1 - 4.617\frac{T_C-25}{1000} + 12.2\left(\frac{T_C-25}{1000}\right)^2 - 27\left(\frac{T_C-25}{1000}\right)^3\right] \qquad (14a)$$

$$\tau_D^{\mathrm{D_2O}}(T) = \tau_D^{\mathrm{H_2O}}(T - 7.2) = \{2\pi\gamma_1[\theta(T-7.2)]\}^{-1} \qquad (14b)$$

Here the temperature T_C is expressed in $^\circ\mathrm{C}$ and T in K, $T_C \equiv T - 273.16$.

The experimental values of the imaginary (κ) and real (n) parts of the complex refraction index $\hat{n} = n + i\kappa = \sqrt{\varepsilon' + i\varepsilon''} = \sqrt{\varepsilon}$ are recorded for several temperatures in (Zelsmann [21] and Downing and Williams [22]). The real and imaginary parts of $\varepsilon = (n + i\kappa)^2$ and the absorption coefficient α are connected with n and κ by the relationships

$$\varepsilon' = n^2 - \kappa^2, \qquad \varepsilon'' = 2n\kappa, \qquad \alpha = 4\pi\nu\kappa \qquad (14c)$$

The profile of the hat well is characterized by the libration amplitude β, the normalized (divided by $k_\mathrm{B}T$) well depth u, and the form factor f determined as the ratio

$$f = (\text{potential width near the bottom})/(\text{potential width near the edge of the well}) \qquad (14d)$$

The hat well of a parameterized model has parabolic walls and a flat bottom (see Fig. 2c and 2d). Such a well is rather deep (u is about 8), the well width 2β comprises about 45°. Unlike the well width 2β and the well depth $U_{or} \equiv uk_BT$, the form factor f strongly depends on the structure of the fluid and temperature. For instance, in the case of water the well profile is at high temperature almost rectangular ($f \approx 0.9$ at 81.4°C), while at low temperature it becomes more rounded $f \approx 0.65$ at -5.6°C). In the case of ice the well becomes almost parabolic ($f \approx 0.15$).

The form factor f accounts implicitly for an influence on the spectrum of collisions of librating dipoles with the surrounding medium. The stronger the intermolecular interactions in a liquid, the smaller the fitted form factor f—that is, the more the potential profile declines from the rectangular one.

Unspecific interactions, characteristic for nonassociated liquids, generate frequency dependences typical for liquid water. If one is to apply the same hat

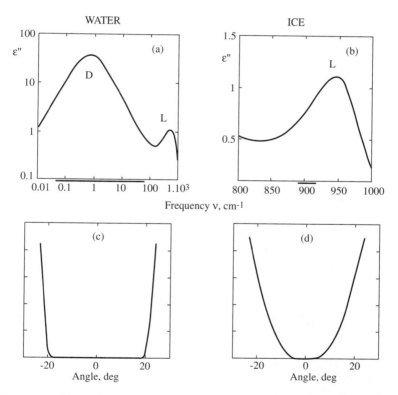

Figure 2 (a,c) The dielectric-loss spectrum pertinent to libration of a permanent dipole in the hat well (a) and profile of such a well (c), calculated for water at 27°C. (b, d) The loss spectrum (b) and the profile of the hat well (d), calculated for ice at -7°C. Symbols D and L refer, respectively, to the Debye relaxationai and the librational bands.

model for calculation of the absorption spectrum, then the main distinctions between nonassociated and associated (water) fluids are as follows: In the first medium (e.g., CH_3F) the number of reorientation cycles performed during the lifetime of the hat potential is much shorter than in the case of water, thus indicating that molecular rotation is more damped and chaotic than in associated liquids; there is also an absence of the translational band arising due to H-bonding of the molecules, and there is a greater form factor: $f(CH_3F) \approx 0.96$, while $f(H_2O) \approx 0.8$ and $f(D_2O) \approx 0.7$.

It was suggested therefore that the parameter f could be employed as a measure of localization of intermolecular interactions—that is, as a *spread of short-range forces* in a liquid. For lower f, these forces are revealed at larger distances.

Quantitatively, this spread could be estimated as a length of the curved part $(1 - f)\beta$ of the well:

$$\text{spread} = (1 - f)\beta r_{mol} \tag{15}$$

where r_{mol} is the mean radius of a reorienting molecule. For H_2O at room temperature, one can take $\beta = 21.5°, f = 0.7, r_{mol} \approx 1.5$ Å; then the spread is ≈ 0.12 Å. For D_2O an analogous estimate gives ≈ 0.2 Å, while for ice it gives as much as ≈ 0.5 Å. Hence, the short-range interactions become more enhanced in the following sequence: light water $\rightarrow$ heavy water $\rightarrow$ ice. Note in contrast, that in such a simple nonassociated liquid, as CH_3F, we have (Gaiduk and Tseitlin [10], p. 177) $f = 0.96, \beta = 22.9°, r_{mol} \approx= 3.9$ Å, so that an almost rectangular intermolecular well corresponds to these parameters characterized by a very small spread (≈ 0.03 Å).

The equation of motion of a reorienting dipole, governed by the hat potential, is nonlinear, so the law of periodic libration is rich in high harmonics, especially at high temperature. Due to a nonlinear equation of motion, the form of the librational band is far from being Lorentzian.

In Fig. 2a we depict the wideband loss spectrum of water calculated in terms of the hat model for room temperature (27°C). For ice at −7°C a similar calculation (but in a much narrower band) is presented in Fig. 2b. The fitted molecular parameters are presented in Table I (some of these parameters will be determined below). During the lifetime τ_{or} a dipole performs in water and ice about two librations and about six librations in supercooled water ($m_{or} = 5.6$).

An important feature of the LIB state (it is confirmed by experimental data) presents a strong isotopic shift of the loss-peak frequency ν_{or}. Since in the chosen (hat) potential a polar molecule librates almost freely, ν_{or} is determined by the moment of inertia $I_{or}(H_2O)$, which comprises about half of $I_{or}(D_2O)$. Therefore, $\nu_{or}(H_2O) \approx \sqrt{2}\nu_{or}(D_2O)$.

TABLE I
Molecular Constants. Fitted and Estimated Parameters of the Molecular Model

Water at 1.4°C

$\varepsilon_s = 61.8$	$\rho = 0.971\,\mathrm{g\,cm^{-3}}$	$\tau_D = 3.17$ ps	$n_\infty^2 = 1.7$
$\beta = 23.8°$	$f = 0.9$	$\tau_{or} = 0.08$ ps	$u = 6.8$
$r_{or} = 65\%$	$k_\mu = 1.19$	$\mu_{or} = 2.7$ D	$m_{or} = 1.5$
$k = 14{,}560$ dyne cm^{-1}	$\tau_q = 0.076$ ps	$v_q = 165\,\mathrm{cm^{-1}}$	$\mu_q = 5.3$ D
$\mu_\mu = 7$ D	$c_{rot} = 0.12$	$k_\perp = 4370$ dyne cm^{-1}	$\tau_\perp = 0.064$ ps
$p_\perp = 0.4$	$g_\perp = 8.4$	$m_\perp = 0.083$	

Water at 27°C

$\varepsilon_s = 77.6$	$\rho = 0.996$ g cm^{-3}	$\tau_D = 7.85$ ps	$n_\infty^2 = 1.7$
$\beta = 23°$	$f = 0.8$	$\tau_{or} = 0.12$ ps	$u = 8.0$
$r_{or} = 65\%$	$k_\mu = 1.08$	$\mu_{or} = 2.45$ D	$m_{or} = 2.3$
$k = 14{,}560$ dyne cm^{-1}	$\tau_q = 0.076$ ps	$v_q = 165\,\mathrm{cm^{-1}}$	$\mu_q = 7.0$ D
$\mu_\mu = 6.35$ D	$c_{rot} = 0.12$	$k_\perp = 4370$ dyne cm^{-1}	$\tau_\perp = 0.1$ ps
$p_\perp = 0.65$	$g_\perp = 3.9$		

Super-cooled water at -5.6°C

$\varepsilon_s = 90.2$	$\rho = 0.999$ g cm^{-3}	$\tau_D = 22.4$ ps	$n_\infty^2 = 1.7$
$m_\perp = 0.071$			
$\beta = 21.5°$	$f = 0.7$	$\tau_{or} = 0.28$ ps	$u = 8.5$
$r_{or} = 65\%$	$k_\mu = 1.08$	$\mu_{or} = 2.45$ D	$m_{or} = 5.6$
$k = 16{,}380$ dyne cm^{-1}	$\tau_q = 0.08$ ps	$v_q = 175\,\mathrm{cm^{-1}}$	$\mu_q = 6.15$ D
$\mu_\mu = 6.0$ D	$c_{rot} = 0.12$	$k_\perp = 6740$ dyne cm^{-1}	$\tau_\perp = 0.2$ ps
$p_\perp = 4.5$	$g_\perp = 154$	$m_\perp = 0.028$	

Ice at -7°C

$\varepsilon_s = 94$	$\rho = 0.92$ g cm^{-3}	$\delta_\infty = 1.9$	
$\beta = 23.5°$	$f = 0.15$	$\tau_{or} = 0.06$ ps	$u = 8.5$
$r_{or} = 70\%$	$k_\mu = 1.06$	$\mu_{or} = 1.9$ D	$k = 22{,}480$ dyne cm^{-1}
$\tau_q = 0.23$ ps	$v_q = 205\,\mathrm{cm^{-1}}$	$\mu_q = 7.0$ D	$\mu_\mu = 4.4$ D
$c_{rot} = 0.16$	$m_{or} = 2.1$	$k_\perp = 6740$ dyne cm^{-1}	$\tau_\perp = 0.03$ ps
$p_\perp = 0.4$	$g_\perp = 7$	$m_\perp = 1.31$	

Note. The employed molecular constants are: $M = 18$, $I_{or} = 1.483 \times 10^{-40}\,\mathrm{g\,cm^2}$, $I_{vib} = 3.38 \times 10^{-40}$ g cm^2, $\mu_0 = 1.84$ D, $s_{lim} = 0.2$.

C. Harmonic Vibration of Dipoles along and about a Hydrogen Bond

The VIB state concerns mechanisms b, c, and d, which are illustrated respectively by the vibrational modes 1, 2, and 3 in Figs. 1b and 1d.

We assume that two water molecules form a nonrigid dipole O^+–$H \cdots O^-$ or O^+–$D \cdots O^-$ (other molecular groups could also be attached to the oxygen atoms). The concentration N_{vib} of charged molecules is $N_{vib} = r_{vib}N$, where r_{vib} is the same as in Eq. (6), $r_{vib} \approx 1/3$. A longitudinal harmonic vibration of this dipole, giving rise to mechanism b, is governed by the elastic force constant k fitted in such a way that the relevant peak of the loss curve $\varepsilon_q''(v)$ should be located near the translational-band center. The constant k parameterizes the strained-state potential energy in the time-varying H-bond stretch $\Delta l(t)$ via $(1/2)k[\Delta l(t)]^2$, where l is the mean length of an undisturbed hydrogen bond. We take $r_{OO} = 2.85$ Å as the distance between oxygen atoms and $r = 1.02$ Å as the covalent-bond length, the H-bond length being $l = r_{OO} - r = 1.83$ Å. We also include the constant $a = r/l = 0.557$.

Since the equation of motion pertinent to longitudinal vibration is linear, the profile of the loss curve $\varepsilon_q''(v)$ is Lorentzian. The lifetime τ_q is very short (about 0.1 ps).

Mechanism c concerns the dielectric response of a rigid dipole reorienting about the hydrogen bond, the moment μ_μ of this dipole is rather close to μ_q. In a first approximation, mechanism c yields, just as in mechanism b, one Lorentz line (we term it the V-band). Its center frequency, v_μ is slightly less that v_q. In terms of our approach the frequency v_μ is determined by the dimensionless rotary force constant c_{rot} related to the rotational part of the strained-state time-varying part of the potential energy as $(1/2)c_{rot}kl^2\theta_+(t)$. Here $\theta_+(t)$ is the existing H-bond deflection at an instant t with respect to its equilibrium position.

The T- and V-bands overlap in water (the resulting band is often termed "translational") but in ice they reveal as separate bands.

Our study shows that the rotary-vibration lifetime τ_μ is approximately equal to the longitudinal-vibration lifetime τ_q. However, better agreement with experiment is found for $\tau_\mu > \tau_q$.

Below, unlike the work by Tseitlin, Gaiduk, Nikitov, 2005, denoted TGN, we shall *not* take into account a coupling of two harmonic motions: longitudinal vibration and reorientation. This coupling is considered in the second part of the review.

We consider collective motion of *pairs* of water molecules. Let the unit volume of the medium comprise $N_{vib}/2$ of such pairs, with N_{vib} being the concentration of molecules suffering elastic vibration. Using the high-frequency approximation, we calculate the complex susceptibility $\chi_{vib} = \chi_{vib}' + i\chi_{vib}''$ of the medium pertinent to harmonic vibration of the HB particles (we omit the complex-conjugation symbol). We assume that for an instant just after a strong collision, the velocities and position coordinates of the particles have Boltzmann distributions. Then the elastic-vibration complex susceptibility χ_{vib} and permittivity $\Delta\varepsilon_{vib}$ in view of TGN are determined by the formulas

$$\chi_{vib}(z) = G_q L_q(z_q) + G_\mu L_\mu(z_q) \tag{16a}$$

$$\Delta\varepsilon_{vib}(v) \approx 6\pi\chi_{vib}(v) = \Delta\varepsilon_q + \Delta\varepsilon_\mu \tag{16b}$$

$$\Delta\varepsilon_q = 6\pi\chi_q(z_q), \qquad \Delta\varepsilon_\mu = 6\pi\chi_\mu(z_q) \tag{16c}$$

where the complex frequency and constants G_q and G_μ are given by

$$z_{\text{vib}} = x_q + iy_q, \quad x_q = \omega\eta_{\text{vib}}, \quad y_q = \eta_{\text{vib}}/\tau_q, \quad \eta_{\text{vib}} = \sqrt{I_{\text{vib}}(2k_B T)^{-1}} \quad (17)$$

The calculated loss curves $\varepsilon_q''(v)$ are depicted in Fig. 3a for water at 27°C and in Fig. 3b for ice at -7°C. The fitted model parameters are presented in Table I. We see that the bandwidth of the loss curve $\varepsilon_q''(v)$ is several times wider in water than

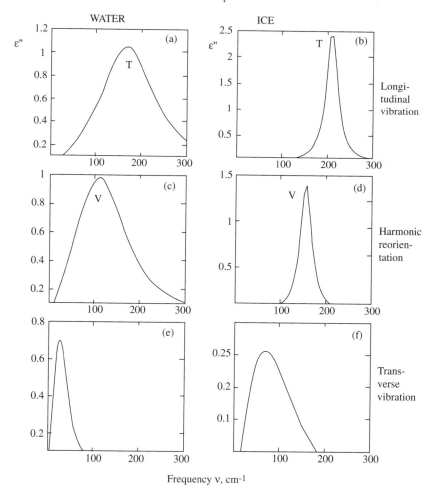

Figure 3 Contributions ε_j'' to the loss factor of water at 27°C (a, c, e) and of ice at -7°C (b, d, f) due to: longitudinal harmonic vibration of a nonrigid dipole (a, b); harmonic reorientation of a permanent dipole (c, d); and nonharmonic transverse vibration of a nonrigid dipole (e, f). Symbols T and V refer, respectively, to the T- and V-bands.

in ice. The probe frequency v_q and therefore the HB elastic constant k are greater in ice than in water but have the same order of magnitude.

The lifetime τ_q is very short in the case of water: It is near 0.08 ps. In ice this time is several times longer. It appears that the longitudinal vibration of H-bonded molecules is less chaotic in ice than in water. On the other hand, reorientation of permanent dipoles in defects of ice structure is more chaotic than above-mentioned longitudinal vibration. Thus,

$$\tau_q(\text{ice}) \gg \tau_q(\text{water}) \quad \text{and} \quad \tau_q(\text{ice}) \gg \tau_{\text{or}}(\text{ice})$$

It appears that the H-bond is polarized. In view of our estimate the mean moment μ_q of a nonrigid dipole is about 3 times larger than the moment μ_{or} of a dipole librating in the hat well.

D. Nonharmonic Transverse Vibration of a Nonrigid Dipole

This mechanism originates from an influence of the transverse force constant $k_\perp$ on displacement of two oppositely charged water molecules in the direction transverse to the equilibrium HB direction (see Fig. 1c). Parameterization of our model shows that $k_\perp$ comprises about one-third of the longitudinal constant k. The mean TV-vibration amplitude b_m is rather small: It comprises about 1/3 of the covalent-bond length r.

As is shown in the second part of the review, in spite of the smallness of b, the equation of motion, governing the $b(t)$ time dependence, is nonlinear. In terms of the dimensionless variables s and φ, it takes the form

$$\ddot{s} = -2ws^3 \tag{18a}$$

Here $s \equiv 2b(l+r)^{-1}$ is the transverse displacement of a water molecule from the equilibrium position, $\varphi = t/\eta_\perp$, so that $\dot{s} \equiv ds/d\varphi$; the normalizing constant $\eta_\perp$, having dimension of time, is given by

$$\eta_\perp \equiv p_\perp(l+r)\sqrt{\frac{m}{4k_BT}} \tag{18b}$$

where m is the mass of the vibrating water molecules. The factor

$$w \equiv \frac{k_\perp l^2(1+a)^4}{8k_BT} \tag{18c}$$

relates the dimensionless TV potential energy $u_\perp$ to the deflection s as $u_\perp = U_\perp/(k_BT) = ws^4$. The *effective* frequency $(p_\perp)$ and intensity $(g_\perp)$ factors allow variation, respectively, of the position $v_\perp$ and of the TV loss maximum $\varepsilon''_{\perp m}$.

The latter factor $(g_\perp)$ connects the transverse $(\mu_\perp)$ and longitudinal (μ_q) HB dipole moments by

$$\mu_\perp^2 = g_\perp \mu_q^2 \tag{18d}$$

To find the contribution $\Delta\varepsilon_\perp$ of the TV mechanism to the total complex permittivity ε, we introduce the complex frequency $z_\perp$:

$$z_\perp \equiv \hat{\omega}\eta_\perp = x_\perp + iy_\perp, \qquad x_\perp = \omega\eta_\perp, \qquad y_\perp = \eta_\perp/\tau_\perp \tag{19}$$

where $\tau_\perp$ is the TV lifetime. Then, as follows from Gaiduk and Crothers [23] (see also derivation in the part two) we have

$$\Delta\varepsilon_\perp = G_\perp L_\perp(z_\perp) \tag{20}$$

where

$$G_\perp = (\pi/k_B T)\mu_\perp^2 N_{\mathrm{vib}}, \qquad N_{\mathrm{vib}} = r_{\mathrm{vib}}N \tag{21}$$

r_{vib} is the proportion of the HB vibrating particles.

The spectral function (SF) $L_\perp(z_\perp)$ is found via integration over s:

$$L_\perp(z_\perp) = 3w \int_0^{s_{\mathrm{lim}}} \frac{(18Dws^2 - z_\perp^2)s^6 \exp(-ws^4)\,ds}{12w^2 s^4 + z_\perp^4} \bigg/ \int_0^{s_{\mathrm{lim}}} s^2 \exp(-ws^4)\,ds \tag{22}$$

s_{lim} being the limiting value of s, which is found from comparison of our theory with experiment ($s_{\mathrm{lim}} = 0.2$). It appears that s_{lim} is in principle determined by the molecular structure of the fluid. The factor $D \equiv [2\mathbf{E}(1/\sqrt{2})/\mathbf{K}(1/\sqrt{2})] - 1 \approx 0.627$, where $\mathbf{K}(\cdot)$ and $\mathbf{E}(\cdot)$ are the full elliptic integrals, respectively, of the first and second kind.

Calculation gives the following approximate formula for the number $m_\perp$ of the vibration cycles performed by a nonrigid dipole during the lifetime $\tau_\perp \approx \eta_\perp/y_\perp$:

$$m_\perp \approx [1 - \exp(-u_{\mathrm{lim}})]\left[8\sqrt{2w}\mathbf{K}(1/\sqrt{2})y_\perp \int_0^{s_{\mathrm{lim}}} s^2 \exp(-ws^4)\,ds\right]^{-1} \tag{23}$$

The loss frequency dependence $\varepsilon_\perp''(v)$, described by Eqs. (20) and (22), is non-Lorentzian. In view of Fig. 3e and 3f calculated, respectively, for water and ice. For reasonable dimer parameters, this dependence is located in the THz region. Transverse vibrations are very damped, especially at low temperatures, since the parameter $m_\perp$ (23) is very small. As is shown in Table I, in the case of water, $m_\perp$

varies from 0.03 to 0.08, when the temperature rises from -5.6 to $81.4\,^{\circ}$C. So low $m_{\perp}$ values could be ascribed to collective motions performed by associated nonrigid dipoles. This conclusion presents a certain generalization of our model, since according to Fig. 1c *formally* only two molecules participate in transverse vibration. Assuming that an *individual* dipole moment $\mu_{\perp}^{\text{ind}}$ of the transversally vibrating molecule is about that (μ_q) for the longitudinally vibrating one, we arrive at a rough estimate for an association factor ζ:

$$\zeta = \sqrt{g_{\perp}} \tag{24}$$

In supercooled (SC) water the factor ζ surprisingly increases up to ≈ 12, whereas in water at room temperature and in ice, ζ is much less:

$$\zeta(\text{water at } 300\,\text{K}) \approx 2, \qquad \zeta(\text{SC water}) \approx 12.4, \qquad \zeta(\text{ice}) \approx 2.8 \tag{25}$$

It is very interesting that the number $m_{\perp}$ strongly *increases* in *ice*:

$$m_{\perp}(\text{water at } 300\,\text{K}) \approx 0.07, \qquad m_{\perp}(\text{SC water}) \approx 0.03, \qquad m_{\perp}(\text{ice}) \approx 1.31 \tag{26}$$

So sharp an increase in the ratio $m_{\perp}(\text{ice})/m_{\perp}(\text{water})$, comprising about 20, undoubtedly corresponds to the phase transition water $\rightarrow$ ice. This transition is accompanied by an enhanced narrowing of the librational, translational, and V-bands (cf., respectively, Fig. 2b with 2a, Fig. 3b with 3a, and Fig. 3d with 3c). Surprisingly, the loss bandwidth pertinent to transverse vibrations broadens in ice as compared with that in water (cf. Figs. 3f and 3e).

E.   Overall Permittivity of Liquid Water and Ice

The total complex permittivity is found as a sum of the contributions from molecules, reorienting in the hat well, and from harmonically and non-harmonically vibrating HB molecules. Summing the relationships (4), (16) and (20), we write down the expression for the overall vibrational contribution $\Delta\varepsilon$ to the complex permittivity ε:

$$\Delta\varepsilon = \Delta\varepsilon_q + \Delta\varepsilon_\mu + \Delta\varepsilon_\perp = 6\pi[\chi_q(z_{\text{vib}}) + \chi_\mu(z_{\text{vib}}) + \chi_\perp(z_\perp)] \tag{27a}$$

$$= 6\pi[G_q L_q(z_{\text{vib}}) + G_\mu L_\mu(z_{\text{vib}})] + G_\perp L_\perp(z_\perp) \tag{27b}$$

The total permittivity is then

$$\varepsilon(v) = \varepsilon_{\text{or}}(v) + \Delta\varepsilon(v) \tag{28}$$

Note, the optical permittivity n_∞^2 is included in the first term $\varepsilon_{or}(v)$, so

$$\begin{aligned}
\text{Re}[\varepsilon(0)] &= \text{Re}[\varepsilon_{or}(0)] = \varepsilon_s, \qquad \text{Re}[\Delta\varepsilon(0)] = 0, \\
\text{Re}[\varepsilon(\infty)] &= \text{Re}[\varepsilon_{or}(\infty)] = n_\infty^2, \qquad \text{Re}[\Delta\varepsilon(\infty)] = 0
\end{aligned} \tag{29}$$

Then the overall *vibrational* contribution $\Delta\varepsilon_s$ to the static permittivity ε_s is

$$\Delta\varepsilon_s = \Delta\varepsilon_{qs} + \Delta\varepsilon_{\mu s} + \Delta\varepsilon_{\perp s} \tag{30}$$

so that the *librational* contribution $\varepsilon_{s,or}$ to the static permittivity ε_s is

$$\varepsilon_{or,s} = \varepsilon_s - \Delta\varepsilon_s \tag{31}$$

where

$$\Delta\varepsilon_{qs} = \frac{3\pi N_{vib}}{2k_B T} G_q L_q(iy_{vib}), \qquad \Delta\varepsilon_{\mu s} = \frac{2\pi N_{vib}}{k_B T} G_\mu L_\mu(iy_{vib}),$$

$$\Delta\varepsilon_{\perp s} = \frac{\pi N_{vib}}{k_B T} \mu_q^2 [L_\perp(iy_\perp)] \tag{32}$$

$$L_q(iy_{vib}) = \frac{E}{B} \frac{f_q^2}{f_q^2 + y_{vib}^2}, \quad L_\mu(iy_{vib}) = \frac{E}{A} \frac{p_q^2}{p_q^2 + y_{vib}^2}, \quad N_{vib} = \frac{r_{vib}\rho N_A}{M} \tag{33}$$

The imaginary part of the second term in (28) is

$$\text{Im}[\Delta\varepsilon(v)] = \Delta\varepsilon''(v) = \varepsilon_q''(v) + \varepsilon_\mu''(v) + \varepsilon_\perp''(v) \tag{34}$$

Hence, the total loss frequency dependence is given by

$$\varepsilon''(v) = \varepsilon_{or}''(v) + \varepsilon_q''(v) + \varepsilon_\mu''(v) + \varepsilon_\perp''(v) \tag{35}$$

Earlier in Section II we have calculated and depicted *each* term in (35). Below, in the first part of the review, we shall find the overall loss frequency dependence (35) and the total absorption coefficient

$$\alpha(v) = \frac{2\pi v \varepsilon''(v)}{n(v)} \tag{36a}$$

where the refraction index

$$n = \text{Re}(\sqrt{\varepsilon}) \tag{36b}$$

We shall also consider the partial absorption coefficients

$$\alpha_{or}(v) = \frac{2\pi v \varepsilon''_{or}(v)}{n(v)}, \quad \alpha_q(v) = \frac{2\pi v \varepsilon''_q(v)}{n(v)}, \quad \alpha_\mu(v) = \frac{2\pi v \varepsilon''_\mu(v)}{n(v)}, \quad \alpha_\perp(v) = \frac{2\pi v \varepsilon''_\perp(v)}{n(v)}$$

$$(36c)$$

determined by the relevant molecular mechanisms (a), (b), (c), and (d).

F. Additional Condition to Overall Permittivity Provided by Low-Frequency Raman Spectrum

The dielectric spectrum (DS), pertaining to the T-band region, arises also in a certain specific form in the low-frequency Raman spectrum (RS). Comparison of both spectra allows us to improve the parameterization of our molecular model. Namely, in view of recent works (Gaiduk et al. [24], Gaiduk [25], we may write down the following relationships for the so-called $R(v)$ and Bose–Einstein (BE) representations of the RS, denoted, respectively, $I_R(v)$ and $I_{BE}(v)$:

$$I_R(v) \approx v\text{Im}\{6\pi[\mu_q^2 G_q L_q(z_q) + \mu_\mu^2 G_\mu L_\mu(z_q)] + \mu_\perp^2 G_\perp L_\perp(z_\perp)\} \quad (37a)$$

$$I_{BE}(v) \approx \text{Im}\{6\pi[\mu_q^2 G_q L_q(z_q) + \mu_\mu^2 G_\mu L_\mu(z_q)] + \mu_\perp^2 G_\perp L_\perp(z_\perp)\} \quad (37b)$$

The first relationship, (37a), correlates with the reduced absorption $v\varepsilon''$ and the second one, (37b), with the dielectric loss ε''. It is convenient to relate both expressions (37) to the dielectric loss spectra setting the dipole moments in D. Then (Gaiduk [25]) we have

$$I_R(v) = v[(\mu_q 10^{18})^2 \varepsilon''_q + (\mu_\mu 10^{18})^2 \varepsilon''_\mu + (\mu_\perp 10^{18})^2 \varepsilon''_\perp] \quad (38a)$$

$$I_{BE}(v) = (\mu_q 10^{18})^2 \varepsilon''_q + (\mu_\mu 10^{18})^2 \varepsilon''_\mu + (\mu_\perp 10^{18})^2 \varepsilon''_\perp \quad (38b)$$

Therefore, the dimensionless RS are described by the same parameters as the DS, originated from the motion of the HB molecules pertaining to the VIB fraction.

It is important that the *total* complex permittivity (28) comprises also the term $\varepsilon_{or}(v)$, originated in the LIB fraction from motion of a permanent dipole in the hat well. Hence, the total loss (35) of water comprises an additional term $\varepsilon''_{or}(v)$, while in the RS, given by Eqs. (38), the corresponding term is lacking. This situation has a place, since we assume that rotation/libration of a polar molecule *as a whole*, which generates the librational band, does not contribute to the RS (in RS the latter band is very weak). Owing to this (viz., to absence of the term $\varepsilon''_{or}(v)$ in the formula for RS), the Raman spectra differ principally from the dielectric spectra. Another reason for such a difference arises due to the terms $(\mu_q)^2, (\mu_\mu)^2$ and

$(\mu_\perp)^2$ in (38), which in contrast with DS yields a different dependence on the dipole moments of vibrating HB molecules.

Hence the *calculated loss* $\varepsilon''(v)$ *and Raman* $I_{BE}(v)$ *spectra should differ.* Particularly, in the experimental Raman spectra the 50-cm^{-1} band is emphasized (it is seen in Fig. 4), while in the experimental dielectric spectra this band is missing (as will be demonstrated below) or is at least very weak. Due to this important circumstance, the parameterization of the molecular model should be such that the 50-cm^{-1} RS band, calculated from Eqs. (38), should appear in the *Raman* spectra, unlike such a band in the calculated and experimental *dielectric* spectra.

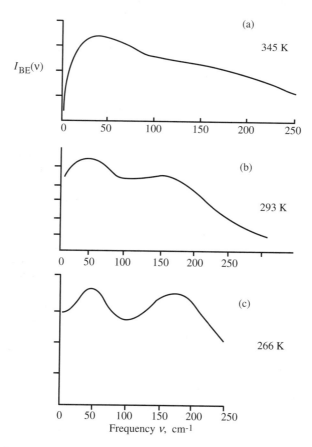

Figure 4 Experimental Raman scattering spectra of water for the Bose–Einstein representation at high (a), room (b), and low (c) temperatures. Parts (a) and (c) are reconstructed from Amo and Tominaga [26]; part (b) is reconstructed from Nielsen et al. [27].

APPENDIX I. TEIXEIRA'S INTERPRETATION
OF EXPERIMENTAL DATA

Below we present extracts from the letters to V. G. from J. Teixeira [16], who kindly agrees to publication.

My views (as experimentalist) are rather qualitative mostly based simply in many experimental results and in evaluations of the "orders of magnitude" of several physical properties. In the paper of 1985 (Teixeira, Bellissent-Funel, Chen, Dianoux) we deduced from the analysis of the quasi-elastic neutron scattering by pure water, that self diffusion of water molecules is mainly due to rotations of the molecule. In my view, at least at low temperatures, when, instantaneously, most of the pairs of molecules are bonded, the molecule can rotate around one bond if the other possible bonds are broken simultaneously.

In a perfect tetrahedral geometry, the length of the proton jump is equal $2r_{OH}$ sin $(104°/2) = 1.6$ Å, because the rotation is around the oxygen atom, the distance r_{OH} is 1 Å and the angle HOH is around $104°$. So, this "jump" is, more exactly, a hindered rotation of the whole molecule, where the moving H atom remains attached to the same molecule. This is what I call a molecular diffusion mechanism due to rotational jumps. Within this picture, the permanent dipole oscillates with librational motions and has a different orientation at each jump, but the molecule remains neutral.

Because the attractive potential is only around well defined directions, a molecule that forms an hydrogen bond (HB) with a next neighbour can rotate in a very short time breaking the bond and forming a new bond with another neighbouring molecule. Such picture is probably better for low temperature water for which many "intact bonds" are present.

HB are very directional and have relatively high energy as compared to $k_B T$ but, essentially because of librations, a proton goes frequently out of the region of high attractive potential and the bond "breaks". Because every molecule in liquid water is surrounded by 4 to 5 neighbours, the attraction by another lone pair is important what yields a rapid rotational jump. Instead, I remind that such jumps are totally independent of the jump (less frequent) between the two equilibrium positions along the bond H—O $\cdots$ O ("hopping", related to electric properties of water)." There are of course charged groups. But: a) their number is very small (as shown by the value of pH); b) the size of the groups can be large and the proton jumps take place inside them by the so-called Grotthuss mechanism.

There is a large difference between the diffusional jumps that I refer to corresponding to molecular diffusion and the hopping processes. First, the jumps that I mention take place "everywhere" in the liquid: the HB lifetime is of the order of 1 ps and the "residence time", i.e. the time that a molecule stays attached to its neighbours is only several ps. Instead, hopping takes place only in a few and isolated sites, probably (according to N. Agmon) inside regions where there is a large density of bonds (one can identify regions with groups $H_9O_4^+$, for example).

Even if a proton jump along a bond is very rapid, the conductivity is explained by the Grotthuss mechanism (it is not the same proton that jumps). Please notice that this jump is not a rotation; it takes place along the bond from one minimum of the potential to another, creating, for short time, an ion H_3O^+ and a correspondent OH^-. But, I insist, their concentration is very small. It is out of question to see such species, for example, in scattering experiments. Their impact on the transport properties (self-diffusion, molecular rotations) is totally negligible.

Thus, electric conductivity in water has another origin than self-diffusion in water.

In ice, the self-diffusion and rotations[7] is practically nonexistent (like for most solids). Instead, the proton jumps along the bonds remain possible. The situation is, consequently, almost the opposite of that of the liquid: in ice, the molecular motions are blocked, the HB lifetime is almost infinite but the proton can jump along bonds and generate a non-zero electrical conductivity.

These last years, theoretical work in water has been performed, almost exclusively, by computer simulations with efficient potentials that create exciting situations (for example, secondary critical points) which, in my opinion, are artifacts of the potentials and do not correspond to real water. This is why, in a sense, our knowledge of liquid water remains always qualitative and blocked since several years.

III. WIDEBAND WATER SPECTRUM AT ROOM TEMPERATURE

A. Introduction

In this section we calculate the water spectrum in the range $0-1000 \text{ cm}^{-1}$. This calculation is based on an analytical theory elaborated in 2005–2006 with the addition of a new criterion (38), related to the 50-cm^{-1} band in the low-frequency Raman spectrum. The calculation scheme was briefly described in Section II. One of our goals is to compare the spectra of liquid H_2O and D_2O and the relevant parameters of the model. Particularly, we consider the isotopic shift of the complex permittivity/absorption spectra and the terahertz (THz) spectra of both fluids. Additionally, in Appendix II we take into account the coupling of two modes, pertinent to elastically vibrating HB molecules.

The mixed model of water comprising the librational (LIB) and vibrational (VIB) fractions will be used. We employ four molecular mechanisms: a, b, c, and d. The first one, a, refers to the LIB state, in which a permanent dipole *reorients* in the hat potential formed by torn or weak hydrogen bonds (HB). The last three specific mechanisms (b, c, d), governed by *vibrating* hydrogen bonds, refer to the VIB state.

---

[7]The authors of this chapter consider that libration of a polar molecule exists also in ice, but it is more damped than in water. It is this motion that determines the intense librational band placed in ice around frequency 800 cm^{-1}. We also consider the HB lifetimes to be commensurable in ice and water.

The concentration N_{vib} of vibrating water molecules is commensurable with their total concentration N (an estimate gives $N_{vib} = Nr_{vib}$ with r_{vib} about 1/3).

For definiteness we choose as the room temperature $T_C = 20.2°C$. In the far-IR region the experimental complex-permittivity spectra of $\varepsilon = \varepsilon' + i\varepsilon''$ are given at this T_C by Zelsmann [21] for both fluids.

In the *low-frequency* region we parameterize our theory as follows. For ordinary water (OW) we use the empirical formula by Liebe et al. [19]. This formula will be used also for heavy water (HW) after a certain modification, based on the work by Rønne, Åstrand, and Keiding [28], denoted below RAK.

We calculate the complex permittivity ε and absorption coefficient α using the general formulas given in Section II.E. A good agreement between the *wideband* theoretical and experimental water spectra will be demonstrated.

B. Mixed Model of Water and LDL and HDL Concepts of Water

In this subsection we follow the RAK work (see also references therein), where the dielectric (0.1–2 THz) response of liquid H_2O and D_2O and the low-density-liquid (LDL) and high-density-liquid (HDL) concepts were considered with respect to isotopic dependence (ID) of the water spectra. Large ID of certain properties of water is well known, while other properties (e.g., the static dielectric constant ε_s) do not show ID. Such a behavior of water is still far from being understood at a molecular level.

In RAK a usage of the empirical double-Debye formula

$$\varepsilon(\omega) = \varepsilon' + i\varepsilon'' = \varepsilon_\infty + \frac{\varepsilon_s - \varepsilon_1}{1 - i\omega\tau_D} + \frac{\varepsilon_1 - \varepsilon_\infty}{1 - i\omega\tau_2} \tag{39}$$

for the complex permittivity of water was considered. This formula is characterized by the slow decay (relaxation) time τ_D and by the fast decay time τ_2. It was stated in RAK that a two-state temperature-dependent equilibrium exists between a LDL and HDL. The first (LDL) is declared to correspond to a *high local tetrahedral ordering*, while the molecular nature of second liquid (HDL) was not specified in RAK.

In (39) the slow time τ_D behaves identically for the two fluids, ordinary water (OW) and heavy water (HW), by *shifting* the temperature scale by

$$\delta T = 7.2 \, \text{K} \tag{40}$$

so that

$$\tau_D^{D_2O}(T) = \tau_D^{H_2O}(T - \delta T) \tag{41}$$

Hence, the low-frequency loss spectrum of heavy water is determined by an "effective temperature" $T_{eff} = T - 7.2 \, \text{K}$, which is lower than the temperature T "acting" on molecules of ordinary water.

On the other hand, the static permittivity ε_s and the fast decay time τ_2, approximately do not depend on a type of the isotope:

$$\varepsilon_s(H_2O) \approx \varepsilon_s(D_2O) \tag{42}$$

$$\tau_2(H_2O) \approx \tau_2(D_2O) \tag{43}$$

It appears that the two-component model of water, proposed in RAK, is somehow related to our two-fractional model, described in Section II.

Namely, the LDL is in some respect similar to our LIB fraction regarding its connection with the HB network and dominance in contribution to the low-frequency spectrum (described by the first term in (39)), in particular, to the static permittivity ε_s. However, in contrast to RAK, it is hardly reasonable to bring this fraction into correlation with the HB network itself due to the almost free libration of the dipoles in an intermolecular (hat) potential. On the other hand, it is reasonable to assign the VIB fraction to the HB network, which in our simplified calculation scheme is modeled by a dimer of oppositely charged water molecules connected by a hydrogen bond. Thus, in our opinion

The LDL–HDL equilibrium becomes apparent as an equilibrium between the "defect" of HB network (with weak or strongly bent bonds) and a "regular" part of this network (44)

The property (42) holds, since the *structures* of heavy and ordinary water are very close. However, the *dynamic* properties are isotopic-dependent in water. For instance, the relaxation is slower in D_2O, where the moment of inertia is larger than in H_2O.

We shall prove below that the isotopic dependence of the vibrational contribution $\Delta\varepsilon(v)$ on the permittivity $\varepsilon(v)$ is *small*, unlike the ID of the reorientation contribution $\varepsilon_{or}(v)$. It appears that the relaxation time τ_D differs in HW from that in OW, since τ_D strongly depends not only on the *structure* of liquid water but also on the strength of an *individual* hydrogen bond (a detailed analysis of dependence of τ_D on water structure is given by Agmon [18].

The property (43), connected with the representation (39), will be checked and analyzed further in the context of our mixed water model.

C. Spectra of Ordinary and Heavy Water

1. Isotopic Independence of Static Permittivity

Calculation from Eqs. (12) and (14a) gives almost the same static permittivity of H_2O and D_2O:

$$\varepsilon_s(OW) \approx 80 \tag{45}$$

$$\varepsilon_s(\text{HW}) \approx 80.2 \tag{46}$$

while estimation from (13) and (14b) yields the relaxation time for D_2O slower than for H_2O:

$$\tau_D(\text{OW}) \approx 9.34\,\text{ps} \quad \text{and} \quad \tau_D(\text{HW}) \approx 11.5\,\text{ps} \tag{47}$$

The result (45), (46) agrees with the property (42).

As we shall see, the second term $\Delta\varepsilon_s$ in Eq. (31) is much less than the first one:

$$\varepsilon_s \gg \Delta\varepsilon_s \tag{48}$$

Then the static permittivity ε_s of water in a good approximation is determined by the contribution $\varepsilon_{or,s}$ of permanent dipoles librating in the hat well:

$$\varepsilon_{or,s} \approx \varepsilon_s \tag{49a}$$

The same holds for the static susceptibility

$$\chi_{or,s} \approx \chi_s \tag{49b}$$

A general connection between the static susceptibility χ_s and permittivity ε_s, which accounts for the internal field correction is given by

$$\chi_s = \frac{\varepsilon_s - n_\infty^2}{4\pi} \frac{2\varepsilon_s + n_\infty^2}{3\varepsilon_s} \tag{50a}$$

Then, in view of Eqs. (49) we can write instead of (50a)

$$\chi_{or,s} = \frac{\varepsilon_{or,s} - n_\infty^2}{4\pi} \frac{2\varepsilon_{or,s} + n_\infty^2}{3\varepsilon_{or,s}} \tag{50b}$$

Here the left-hand part is equal to

$$\chi_{or,s} = gGL_{or}(iy_{or}) \tag{51}$$

$$G = \frac{\mu_{or}^2 N(1 - r_{vib})}{3k_B T} \tag{52}$$

where g is the Kirkwood correlation factor, L_{or} is the spectral function (SF) of the hat model, and $L_{or}(iy_{or})$ is the zero frequency value of this SF, and $y_{or} = \eta_{ot}/\tau_{or}$ is the dimensionless strong-collision frequency.

The property (42) leads to the conclusion about an isotropic independence of the right-hand side of Eq. (50a) and then, in view of Eq. (52), leads to the statement that

$$gL_{or}(iy_{or})\mu_{or}^2 N(1 - r_{vib}) \approx \text{isotopically independent} \qquad (53a)$$

where the concentration N of water molecules is given by

$$N = N_A \rho M^{-1} \qquad (53b)$$

2. Results of Calculation

Choosing room temperature as 20.2°C, we depict in Fig. 5a the wideband absorption frequency dependence $\alpha(v)$ of water H_2O and in Fig. 6a we depict that of water D_2O. The fitted parameters of the model are presented in Table II. The total loss spectrum $\varepsilon''(v)$ is shown in Figs. 5b and 6b, respectively, for OW and HW. The solid lines in Figs. 5a,b and 6a,b mark the results of our calculations.

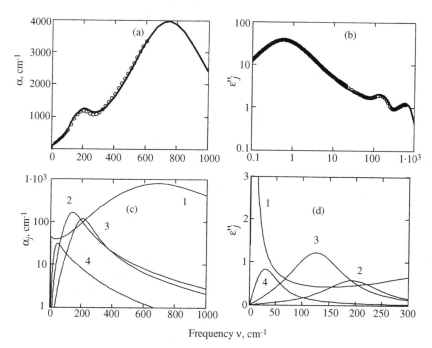

Frequency v, cm⁻¹

Figure 5 For H_2O: (a) wideband absorption frequency dependence; (b) total loss spectrum; (c) partial absorption contributions (1–4); (d) partial loss contributions (1–4); (e) is similar to (c) but for a reduced, domain, and on a linear ordinate scale; (f) is similar to (d) but on a reduced ordinate scale. (g) comprises the partial contributions to the Raman spectra in the $R(v)$ representation; (h) in the Bose–Einstein representation.

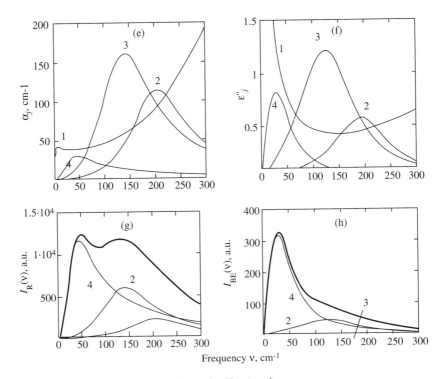

Figure 5 (*Continued*)

Open circles mark the results of measurements made by Zelsmann [21] and the estimations by Liebe et al. [19].

The partial absorption contributions $\alpha_j(\nu)$ are depicted by lines 1–4 in Figs. 5c and 5e for OW and in Figs. 6c and 6e for HW. The partial loss dependence $\varepsilon_j(\nu)$ is depicted in Figs. 5d and 5f for H_2O and in Figs. 6d and 6f for D_2O.

In Fig. 5g we depict by solid lines the Raman spectra (RS) for H_2O for the $R(\nu)$ representation, and solid lines in Fig. 5h depict the spectra for the Bose–Einstein representation. In Figs. 6g and 6h, similar spectra are shown for D_2O.

In Figs. 5 and 6, curves 1–4 refer, respectively, to mechanisms a–c—that is, to libration of rigid dipoles in the hat well (1), elastic reorientation of such dipoles (2), elastic translation of nonrigid dipoles (3), and their elastic transverse vibration (4).

As we see, the theoretical and experimental dielectric spectra (DS) agree *quantitatively*, while such *Raman spectra*, depicted in Figs. 4b and 5h, agree only *qualitatively*. Note also that the Raman and dielectric spectra *differ substantially*, since:

(i) In our calculation pertaining to Figs. 5a and 5b the peak at $\approx 50\,\mathrm{cm}^{-1}$ does *not* appear in contrast to the RS in Figs. 5g and 5h. In the latter figures, this peak is generated by transverse vibration ascribed by curves 4.

(ii) The *absorption* generated by reorienting librators is rather high at v about $300 \, cm^{-1}$ (curve 1 in Fig. 5e) and at $v < 25 \, cm^{-1}$ (curve 1 in Fig. 5e). As regards the RS, we deliberately omitted the contribution of this sort; see Figs. 5g and 5h, in which curves 1 are lacking.

(iii) The absorption and loss peaks, arising, respectively, in Figs. 5a and 5b in the region $150-200 \, cm^{-1}$ are provided by elastically reorienting rigid dipoles (curves 3) and by longitudinally vibrating charged dimer molecules (curves 2). These motions, ascribed to curves 2 and 3 in Figs. 5e and 5f, generate *also* the main peak of the $R(v)$ representation at $v \approx 170 \, cm^{-1}$ in Fig 5g and the shoulder of the BE representation at $v \approx 130 \, cm$ in Fig. 5h.

Thus, our theory explains the main peculiarities of the dielectric and Raman spectra for a particular case of room temperature. The theory allows also interpretation of the distinction between these spectra.

In this subsection we employ the *same* parameterization for the RS as for the DS. As shown in Section IV, better agreement with experiment of the RS is

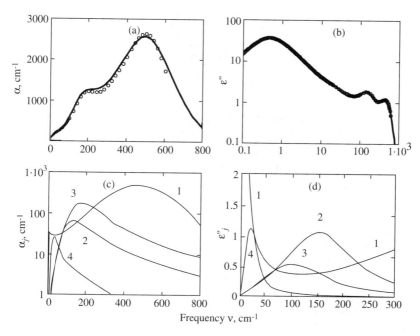

Figure 6 (a–h) Same as Fig. 5a–5h but for D_2O at 20.2°C.

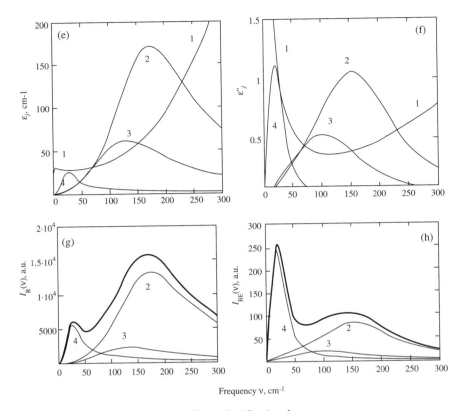

Figure 6   (*Continued*)

obtained, if some fitted model parameters now differ from those found from the condition of the best agreement of the theoretical and experimental DS. In this subsection this difference is neglected.

In view of the importance of Zelsmann's experimental data, we reproduce them in Table III.

Figures 5a–d and 6a–d refer to wideband dielectric spectra, Figs. (5e,f and 6e,f) refer to dielectric spectra in the range from 0 to $300\,\mathrm{cm}^{-1}$, and Figs. 5g,h and 6 refer to Raman spectra in the latter range. Open circles depict the experimental data by Zelsmann [21] and Liebe et al. [19]. Figure 5 is for H_2O at the temperature 20.2°C, and Fig. 6 is for D_2O at 20.2°C.

Thin lines in Figs. 5c–h and 6c–h refer to specific contributions due to: nonharmonic reorientation of a permanent dipole in the hat potential (1), harmonic longitudinal vibration of HB nonrigid dipole (2), harmonic reorientation of a permanent HB dipole (3), and nonharmonic transverse vibration of a nonrigid HB dipole (4).

TABLE II
Molecular Constants. Fitted and Estimated Parameters
of the Model (Temperature 20.2°C)

Water H_2O

$\varepsilon_s = 80$	$\rho = 0.996\,\mathrm{g\,cm^{-3}}$	$\tau_D = 9.34\,\mathrm{ps}$	$n_\infty^2 = 1.7$
spread $= 0.11$ Å	$\mu_0 = 1.84$ D	$M = 18$	$I_{or} = 1.483 \times 10^{-40}\,\mathrm{g\,cm^2}$
$\beta = 20.2°$	$f = 0.8$	$\tau_{or} = 0.15\,\mathrm{ps}$	$u = 8.2$
$r_{or} = 45\%$	$k_\mu = 1.35$	$\mu_{or} = 3.1$ D	$m_{or} = 3.2$
$y_{or} = 0.29$			
$k = 19,300\,\mathrm{dyne\,cm^{-1}}$	$\tau_q = 0.08\,\mathrm{ps}$	$v_q = 190\,\mathrm{cm^{-1}}$	$\mu_q = 4.4$ D
$\mu_\mu = 6.1$ D	$c_{rot} = 0.12$	$\Delta\varepsilon_s = 2.0$	$y_q = 0.8$
$k_\perp = 5792\,\mathrm{dyne\,cm^{-1}}$	$\tau_\perp = 0.078\,\mathrm{ps}$	$p_\perp = 0.8$	$g_\perp = 20$
$m_\perp = 0.07$	$S_{lim} = 0.2$	$\Delta\varepsilon_\perp = 1.8$	$y_\perp = 4$

Water D_2O

$\varepsilon_s = 80.2$	$\rho = 1.1\,\mathrm{g\,cm^{-3}}$	$\tau_D = 11.5\,\mathrm{ps}$	$n_\infty^2 = 1.7$
spread $= 0.25$Å	$\mu_0 = 1.84$ D	$M = 20$	$I_{or} = 2.765 \times 10^{-40}\,\mathrm{g\,cm^2}$
$\beta = 24°$	$f = 0.65$	$\tau_{or} = 0.16\,\mathrm{ps}$	$u = 8.2$
$r_{or} = 45\%$	$k_\mu = 1.1$	$\mu_{or} = 2.5$ D	$m_{or} = 2.2$
$y_{or} = 0.29$			
$k = 23,770\,\mathrm{dyne\,cm^{-1}}$	$\tau_q = 0.062\,\mathrm{ps}$	$v_q = 200\,\mathrm{cm^{-1}}$	$\mu_q = 8.8$ D
$\mu_\mu = 6.2$ D	$c_{rot} = 0.12$	$\Delta\varepsilon_s = 3.4$	$y_q = 1.5$
$k_\perp = 7130\,\mathrm{dyne\,cm^{-1}}$	$\tau_\perp = 0.12\,\mathrm{ps}$	$p_\perp = 0.8$	$g_\perp = 2.8$
$m_\perp = 0.08$	$S_{lim} = 0.2$	$\Delta\varepsilon_\perp = 1.65$	$y_\perp = 3.3$

Note. $I_{vib}(H_2O) = 3.38 \times 10^{-40}\,\mathrm{g\,cm^2}$, $I_{vib}(D_2O) = 7 \times 10^{-40\,\mathrm{g\,cm^2}}$

Thick lines in Figs. 5g,h and 6g,h refer to calculation of Raman spectra expressed in arbitrary dimensionless units for $R(v)$ (g) and Bose–Einstein (h) representations.

In Figs. 5a,b and 6a,b we have a good coincidence of the calculated and experimental absorption and loss spectra. In the far-IR range the Debye relaxation band is not revealed. Only three mechanisms (a–c) contribute to dielectric relaxation in this region. Mechanism d, depicted by the curves 4, plays a noticeable role only in the lower part of the THz region.

The *isotopic dependence* (ID) of the spectra is revealed as follows. In the librationai band the absorption-peak frequency $v_L(D_2O)$ is lower than $v_L(D_2O)$ (cf. Figs. 5a and 6a). Comparing peak positions v_L of curves 1 depicted in Figs. 5c and 6c, we may see that difference $v_L(OW) - v_L(HW) \approx 215\,\mathrm{cm^{-1}}$ is great, so that $v_L(OW)/v_L(HW) \approx 1.453 \approx \sqrt{2}$, as it should be, since the moment of inertia of D_2O is greater than that of H_2O: $I_{or}(HW) \approx 2I_{or}(OW)$. Correspondingly, for the Debye-relaxation time we have $\tau_D(D_2O) > \tau_D(H_2O)$. (see Eq. (47)).

On the other hand, a rather *isotopic independence* concerns the static permittivity of H_2O and D_2O (see Eqs. (45) and (46)) and to a lesser degree the fast times τ_{or} and τ_q fitted as the lifetimes of the hat and parabolic potentials (see Table II). It is important that the difference between the peak positions v_q of the

TABLE III
Real (n) and Imaginary (κ) Parts of the Complex Refraction Index n^* and Absorption $\alpha[\text{cm}^{-1}]$.
Liquid Water H_2O and D_2O at 20.2°C

v	n	κ	α	v	n	κ	α
10	2.600	1.090	137	460	1.504	0.404	2347
20	2.225	0.718	192	470	1.496	0.408	2423
30	2.150	0.527	210	480	1.488	0.411	2494
40	2.110	0.460	240	490	1.480	0.415	2565
50	2.070	0.438	290	500	1.470	0.418	2638
60	2.040	0.444	360	510	1.462	0.421	2709
70	2.020	0.450	429	520	1.451	0.423	2779
80	2.010	0.466	509	530	1.441	0.425	2842
90	2.000	0.487	594	540	1.431	0.426	2903
100	1.997	0.507	678	550	1.419	0.427	2964
110	1.982	0.532	773	560	1.407	0.427	3022
120	1.960	0.557	872	570	1.396	0.428	3077
130	1.929	0.577	967	580	1.385	0.427	3126
140	1.890	0.593	1065	590	1.372	0.425	3167
150	1.848	0.608	1165	600	1.361	0.423	3203
160	1.801	0.622	1266	610	1.348	0.420	3234
170	1.746	0.629	1358	620	1.335	0.417	3259
180	1.689	0.618	1412	630	1.324	0.412	3276
190	1.640	0.597	1437	640	1.313	0.408	3291
200	1.600	0.571	1445	650	1.303	0.403	3301
210	1.657	0.539	1434	660	1.289	0.397	3307
220	1.542	0.505	1407	670	1.277	0.392	3308
230	1.528	0.469	1364	680	1.264	0.386	3307
240	1.525	0.436	1323	690	1.249	0.379	3298
250	1.529	0.414	1310	700	1.236	0.373	3287
260	1.532	0.398	1311	710	1.223	0.365	3263
270	1.534	0.385	1317	720	1.213	0.356	3231
280	1.537	0.375	1331	730	1.201	0.347	3192
290	1.539	0.368	1351	740	1.189	0.338	3150
300	1.541	0.361	1374	750	1.182	0.328	3100
310	1.543	0.357	1401	760	1.171	0.317	3040
320	1.546	0.353	1432	770	1.157	0.305	2969
330	1.550	0.352	1472	780	1.142	0.292	2883
340	1.552	0.356	1532	790	1.138	0.277	2760
350	1.552	0.359	1593	800	1.134	0.260	2618
360	1.552	0.363	1658	810	1.130	0.243	2467
370	1.549	0.368	1724	820	1.130	0.226	2309
380	1.545	0.372	1793	830	1.132	0.208	2143
390	1.541	0.377	1862	840	1.131	0.192	1987
400	1.537	0.382	1933	850	1.132	0.176	1833
410	1.532	0.386	2004	860	1.132	0.159	1692
420	1.527	0.390	2072	870	1.135	0.144	1533
430	1.521	0.394	2143	880	1.139	0.130	1396
440	1.515	0.397	2210	890	1.143	0.118	1270
450	1.510	0.401	2280	900	1.149	0.107	1165

TABLE III
(*Continued*)

v	n	κ	α	v	n	κ	α
910	1.156	0.0973	1064	960	1.189	0.0661	770
920	1.162	0.0898	993	970	1.194	0.0622	733
930	1.168	0.0828	927	980	1.202	0.0589	702
940	1.174	0.0764	866	990	1.208	0.0557	673
950	1.181	0.0707	817	1000	1.214	0.0534	651

Source: Zelsmann [21].

specific loss curves 2, estimated from Figs. 5f and 6f, be rather small for mechanism b: $v_q(\text{OW}) - v_q(\text{HW}) \approx 27\,\text{cm}^{-1} \ll v_L(\text{OW}) - v_L(\text{HW})$.

A weak isotopic dependence of τ_{or}, τ_q, $\tau_{\perp}$, agrees with the property (43). The left-hand side of Eq. (53a) equals 2.26×10^{-14} and 2.51×10^{-14}, respectively, for H_2O and D_2O. Hence, the property Eq. (53a) approximately holds. Its validity is provided by the small magnitude of $\Delta\varepsilon_s$ as compared with ε_s. Indeed, in Table II we see that $\Delta\varepsilon_s(H_2O) \approx 2.12$ and $\Delta\varepsilon_s(D_2O) \approx 3.6$, while ε_s is given by the values in Eqs. (45) and (46).

As for item (iv) (I.A), the property $\tau_q(H_2O) \approx \tau_q(D_2O)$, pertinent to the translational band, is undoubtedly similar to the property (43) pertaining to the RAK work.

To elucidate the latter assertion, let us compare the water spectra, calculated in the THz region for our model and for the double-Debye representation (39). We substitute in Eq. (39) the following parameters.

(a) We take from RAK the static susceptibility as $\varepsilon_s = 87.91$ $\exp(-0.00458T_C)$ for H_2O and ε_s from Eq. (14a) for D_2O, T_C being temperature in °C.

(b) We take the principal relaxation time τ_d of H_2O from formula (13) by Liebe et al. [19] while for D_2O the modification (14b) of this formula is employed.

(c) At $T \approx 300\,\text{K}$ we take from Fig. 2b of RAK the fast time τ_2 equal to 0.1 ps.

(d) We rewrite the formula (39) for complex permittivity as the following function of wavenumber v:

$$\varepsilon(v) = \varepsilon'(v) + i\varepsilon''(v) = n_\infty^2 + \frac{\varepsilon_s - \varepsilon_1 - n_\infty^2}{1 - 2\pi civ\tau_D} + \frac{\varepsilon_1}{1 - 2\pi civ\tau\tau_2} \qquad (54)$$

with $\varepsilon_1 \equiv \Delta\varepsilon_s$ taken from Table II, namely with $\varepsilon_1(H_2O) = 2.12$, $\varepsilon_1(D_2O) = 3.6$. As above, $n_\infty^2 = 1.7$. This formula gives the correct values of the static and optical permittivity. In the THz region the first and second terms refer to the principal Debye process and the last term refers to the non-Debye process. The latter we ascribe mainly to fast vibration of HB molecules.

Note that a "pure" Debye process is described as

$$\varepsilon_D(v) = \varepsilon_1 + \frac{\varepsilon_s - \varepsilon_1}{1 - 2\pi c i v \tau_D} \qquad (55)$$

The water spectra, calculated from Eq. (54), are depicted by dash–dotted lines in Figs. 6a–d for H_2O and 6c–h for D_2O. The principal Debye process (55) is marked in Figs. 6a,d and 6g,h by open circles. Calculation for our mixed model is depicted by thick solid lines. To emphasize the contribution to ε of transverse vibrations, we show by dashed lines the permittivity components generated by the a + b + c mechanisms. Therefore, the values of s, marked by dashed curves, do not account for the $\varepsilon_\perp$ component.

The permittivity components ε' and ε'', calculated for our model (solid lines), agree rather well with those (dash–dotted lines) obtained from Eq. (54). However, these two approaches somewhat differ in the frequency region between 10 and $100 \, \text{cm}^{-1}$. Note, as declared in RAK, for the frequency $v > 70 \, \text{cm}^{-1}$ (marked by the vertical lines in Figs. 6c,d and 6g,h), Eqs. (44) and (54) are not applicable, unlike our composite (mixed) model, which holds up to $v \approx 1000 \, \text{cm}^{-1}$.

Since distinction between the solid and dashed lines is generally noticeable, in our model the *contribution to spectra of transverse vibrations is significant in all the region depicted in Fig. 6*. Still greater distinction is obtained in Figs. 6c,d and 6g,h between solid lines and curves depicted by open circles (the latter are calculated for the principal Debye representation (55)). Consequently, the latter does not hold for the frequency exceeding $10 \, \text{cm}^{-1}$.

The distinction between the spectra, calculated/measured for ordinary and heavy water, is rather small (cf. Figs. 6a–d and 6e–h). This distinction arises mainly due to a rather small difference of relaxation times (see Eq. (47)).

It should be noted that an emphasized distinction concerns at room temperature the elastic dipole moment μ_q, fitted for heavy and ordinary water. In view of Table II, μ_q is in HW three times as much as μ_q in OW. Moreover, the association factor (24) is noticeably greater in OW than in HW: $\zeta(\text{OW}) \approx \sqrt{20} \approx 4.5$, while $\zeta(\text{HW}) \approx \sqrt{2.8} \approx 1.3$. It is also interesting that the spread[8] S is about twofold

---

[8]In view of Eq. (15), S characterizes the profile of the rounded potential well.

greater for H_2O than for D_2O. Indeed, in view of Table II, $S(H_2O) \approx 0.11$ Å, while $S(D_2O) \approx 0.2$ Å. Hence, for D_2O the curvature of the hat potential is greater than for H_2O. Perhaps, these distinctions relate certain biological properties of two water isotopes.

Thus, the water spectra, pertinent to the LIB and VIB states, drastically differ. For instance, the corresponding peak-loss factors ε''_{max} strongly differ. At room temperature the ε''_{max} value, determined by the LIB state, comprises in the Debye region about 40 (see Figs. 7b and 7f), while the ε''_{max} value, generated by fast vibrations of HB molecules at v near $200 \, cm^{-1}$, reaches in the far-IR region only 1–2 (see Figs. 5b and 5f). We demonstrate these loss maxima (and compare them with the experimental data) in the Cole–Cole diagram $\varepsilon''[\varepsilon'(v)]$ (See Fig. 8).

Thus, in the case of heavy water the librational-band maximum is shifted to low frequencies due to increase of the moment of inertia I_{or}, while the translational-band maximum is located at approximately the same frequency $v_q \approx 200 \, cm^{-1}$ though doubling of the moment of inertia I_{vib} (see the note in Table II). The calculated microwave/THz spectra of OW and HW are also rather close (cf. Figs. 4 and 5, Figs. 6 and 7). This result corresponds to the experiment by Zelsmann [21].

In view of Table II the *main* difference of the parameters, fitted for HW, from those, fitted for OW, concerns (i) some increase of the libration amplitude β, (ii) decrease of the form factor f, (iii) decrease of the frequency v_q (the center frequency of the T-band) and increase of the moment μ_q, responsible for this band, and (iv) decrease of the intensity factor $g_\perp$, which strongly influences the THz band. Comparison of curves 3 in Figs. 4h and 5h shows that the partial dielectric loss peak $\varepsilon''_{q,max}$ of HW, located at v near $150 \, cm^{-1}$ and stipulated by harmonic longitudinal vibration of HB molecules, substantially exceeds such a peak of OW, since the "elastic" dipole moment $\mu_q(D_2O) \approx 8.8$ D exceeds the moment $\mu_q(H_2O) \approx 3.5$ D.

Finally, we estimate the difference of the partial absorption coefficients α_j in the THz region. Let us take the frequency $v = 3.16 \, cm^{-1}$, then we shall find the following α_j values (in cm^{-1}):

For OW: $\alpha_L = 76.2$, $\alpha_\mu = 0.2$, $\alpha_q = 0.015$, $\alpha_\perp = 1.87$
For HW: $\alpha_L = 67.8$, $\alpha_\mu = 0.13$, $\alpha_q = 0.11$, $\alpha_\perp = 1.94$

By inspection we see that only the values $\alpha_q(OW)$ and $\alpha_q(HW)$ differ substantially at this low frequency. They both are very small as compared with the absorption contribution α_L due to the rigid dipoles reorienting in the hat well. The inelastic transverse-vibration absorption $\alpha_\perp$, being noticeably greater, is nevertheless much smaller than α_L.

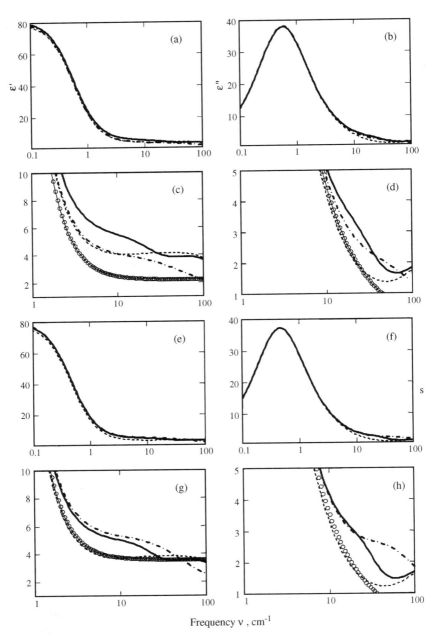

Frequency ν , cm^{-1}

Figure 7   Low-frequency and THz spectra of the dielectric constant ε' (a, c, e, g) and of the loss factor ε'' (b, d, e, f). Water H_2O (a–d) and D_2O (e–f), at the temperature 20.2°C. Other explanations are given in the text.

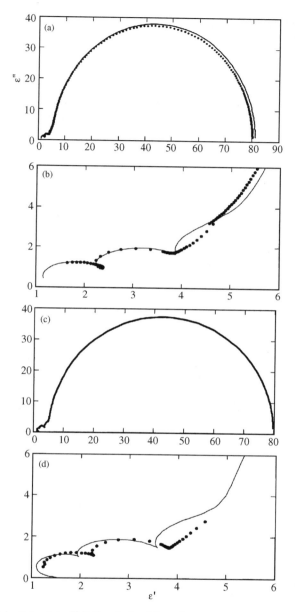

Figure 8 Cole–Cole plots $\varepsilon''[\varepsilon'(v)]$, calculated from the composite model (solid lines). Water H_2O (a, b) and water D_2O (c, d) at the temperature 20.2°C. Dots represent the experimental data by Zelsmann [21], Liebe et al. [19], and Rønne et al. [28]. The vertical lines mark the maximal radiation frequency v, for which the double Debye representation (39) and (54) holds.

APPENDIX II. EFFECT OF COUPLING BETWEEN LONGITUDINAL AND ROTATIONAL VIBRATIONS OF HB MOLECULES

We shall apply here the theory of harmonic vibrations, given in TGN and also in Part Two of this review, for calculation of dielectric response. For that we should replace the simplified spectral functions (19) by more rigorous ones, namely by

$$L_\mu(z) = \frac{E}{A} \frac{p_q^2(f_q^2 - z^2) - g_q w_q}{(p_q^2 - z^2)(f_q^2 - z^2) - g_q w_q}, \qquad L_q(z) = \frac{E}{B} \frac{f_q^2(p_q^2 - z^2) - g_q w_q}{(f_q^2 - z^2)(p_q^2 - z^2) - g_q w_q}$$

(A1)

Here $z \equiv z_q$; A, B, p_q, f_q are determined by (20) and (21), while other spectral parameters are given by

$$g_q = \frac{2Aa(1 - a^2)}{71(1 + a^2) - 2a}, \qquad w_q = \frac{2Ba(1 - a^2)}{71(1 + a^2) - 2a} \qquad \text{(for OW)}, \quad \text{(A2a)}$$

$$g_q = \frac{2Aa(1 - a^2)}{39(1 + a^2) - 2a}, \qquad w_q = \frac{2Ba(1 - a^2)}{39(1 + a^2) - 2a} \qquad \text{(for HW)} \quad \text{(A2b)}$$

In view of Section II.E the total permittivity is represented as a sum of two contributions

$$\varepsilon(v) = \varepsilon_{or}(v) + \Delta\varepsilon(v) \tag{A3}$$

generated by the LIB and VIB fractions. The *vibration* contribution $\Delta\varepsilon(v)$ is given by

$$\begin{aligned} \Delta\varepsilon &= \Delta\varepsilon_q + \Delta\varepsilon_\mu + \Delta\varepsilon_\perp, & \Delta\varepsilon_q &= 6\pi\, G_q L_q(z_q)\Delta\varepsilon_\mu \\ &= 6\pi G_\mu L_\mu(z_q), & \Delta\varepsilon_\perp &= g_\perp G_\perp L_\perp(z_\perp) \end{aligned} \tag{A4}$$

The total (α) and specific (α_{or}, α_μ, α_q, $\alpha_\perp$) absorption coefficients are

$$\alpha = \frac{2\pi v\varepsilon''}{n}, \qquad \alpha_{or} = \frac{2\pi v\varepsilon''_{or}}{n}, \qquad \alpha_\mu = \frac{2\pi v\varepsilon''_\mu}{n}, \qquad \alpha_q = \frac{2\pi v\varepsilon''_q}{n}, \qquad \alpha_\perp = \frac{2\pi v\Delta\varepsilon''_\perp}{n}$$

(A5a)

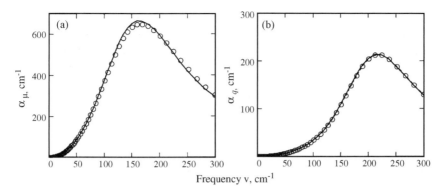

Figure 9 The partial absorption coefficients stipulated by harmonic reorientation (a) and longitudinal vibration (b) of HB molecules. Calculation for water H_2O at the temperature 20.2°C with account of coupling of these vibration modes (solid lines) and with neglect of such coupling (open circles).

where the refractive index

$$n(v) = \text{Re}(\sqrt{\varepsilon(v)}, \qquad \varepsilon_\mu'' \equiv \text{Im}(\Delta\varepsilon_\mu), \qquad \varepsilon_q'' \equiv \text{Im}(\Delta\varepsilon_q), \qquad \varepsilon_\perp'' \equiv \text{Im}(\Delta\varepsilon_\perp)$$

(A5b)

In Figure 9 we depict the frequency dependences of the partial absorption coefficients $\alpha_q(v)$ and $\alpha_\mu(v)$ pertinent to two harmonic-vibration modes. These frequency dependences are calculated from formulas (A6), (21) [24], (25), (28), and (29). When the above-mentioned coupling is accounted for (solid lines in Fig. 9), the spectral functions are taken from Eq. (A1). On the other hand, when the coupling is neglected (open circles in Fig. 9), then L_q and L_μ are found from Eq. (19). We see from Fig. 9a that for both cases the calculated partial absorption $\alpha_\mu(v)$ practically coincide. The same assertion is valid also for the partial absorption $\alpha_q(v)$ depicted in Fig. 8b. Hence, there is no practical need to account for the coupling between the harmonic reorientation and vibration of HB molecules for calculation of spectra in liquid water. However, the effect of such coupling becomes noticeable (being, however, a rather small) in the case of ice, where the absorption lines are much narrower.

APPENDIX III. APPLICABILITY OF CLASSICAL THEORY FOR HARMONIC LONGITUDINAL VIBRATION

In our classical theory the intuitively introduced intermolecular potentials allow calculation of the correlators (spectral functions) using the Newtonian dynamics. A multiparticle system of dipoles is reduced to a single-particle one. Application of force constants, which determine the dynamics of rigid and nonrigid dipoles, substantially simplify the problem, however, without exposing the meaning of these parameters on a more deep level. We recognize that the quantum theory—for instance, such as was used recently by Sharma, Resta, and Car [29]—could in principle yield a more adequate description of multiparticle interactions.

Taking into account the quantum principle of uncertainty, it is worthwhile to employ an *applicability criterion A* of classical theory:

$$A > h \tag{A6}$$

where h is Planck's constant. With respect to our spectral problem, we formulate this criterion as

$$A \equiv action \text{ of a specific dynamical system} > h \tag{A7}$$

Regarding the longitudinal harmonic vibration of H-bonded charged molecules, we determine A from *classical* theory as a quantity, which concerns *breaking of hydrogen bonds*:

$$A \equiv A_{\text{break}} = (\text{H-bond energy } U_{\text{HB}}) \times (\text{lifetime } \tau_q) \tag{A8}$$

For the criterion (A7) we distinguish the following three cases:

$$A_{\text{break}}/h \gg 1, \text{the classical theory of longitudinal vibration is well justified} \tag{A9a}$$

$$A_{\text{break}}/h \text{ is commensurable with} 1, \text{the quantum description} \\ \text{becomes important} \tag{A9b}$$

$$A_{\text{break}}/h < 1, \text{the classical description fails} \tag{A9c}$$

Note that the criterion (A9b) with A, given by (A8), occurs because the longitudinal vibration, being damped due to a very short lifetime τ_q, is nevertheless characterized by (i) an emphasized quasi-resonance frequency dependence $\alpha_q(v)$ and (ii) a small temperature dependence of the translational-band center frequency v_q. This is shown in the work by Gaiduk and Crothers [30] (see also Section IV). These results suggest that the chosen vibration is *essentially quantum*. Noting that $U_{\text{HB}} \approx 2.6 \, \text{kcal mol}^{-1} = 2.6 \times 6.95 \times 10^{-14} \text{erg s}^{-1}$ and $h \approx 6.7 \times 10^{-14} \text{erg s}$, we have from Eq. (A8)

$$A_{\text{break}}(\text{water}) \approx 1.6h \tag{A10a}$$

Hence, we have gotten the case (A9b), since A_{break}/h is commensurable with unity. Thus *the longitudinal elastic HB vibrations, arising in water, indeed have a feature of a quantum phenomenon.*

Regarding transition (A9b) $\rightarrow$ (A9c), it appears that *the mean lifetime* τ_q(water) *could hardly become much shorter than 0.05 ps*, since the quantity (A8) now falls near its lower limit equal to h. In other words, the bandwidth Δv of the absorption frequency dependence, generated by such vibrations, hardly could become much wider than the bandwidth depicted by curves 3 in Figs. 4h and 5h, respectively, for H_2O and D_2O.

In view of the work by Gaiduk and Crothers [31] (see also Section V), the translational bandwidth in ice is narrower than that in water. At the temperature $-7°C$ we may set for ice the lifetime $\tau_q \approx 0.23$ ps. Then we find from Eq. (A8)

$$A_{break}(ice) \approx 6.23h \qquad (A10b)$$

Thus, action(ice) is about six times the limiting action (equal to h). Therefore, ice behaves more as a "classical fluid" than does water, since for water Eq. (A10a) holds instead of Eq. (A10b). Correspondingly, in the translational band the absorption curve $\alpha(v)$ becomes less damped for ice than for water.

Let us touch now the opposite case of a rather *narrow* translational band. By a physical reasoning, the absorption bandwidth Δv *cannot* become extremely narrow, whatever the lifetime τ_q. We relate with v the period T_{rad} of electromagnetic radiation: $T_{rad} = (cv)^{-1}$, where c is the speed of light in vacuum. We have $\Delta T_{rad} \approx \Delta v(c\bar{v}^2)^{-1}$, where min($\Delta T_{rad}$) is meant to be positive and $\bar{v}$ is an average of $\bar{v}$ value in the frequency interval under investigation.

To estimate the minimal possible value Δv, denoted Δv_{min}, starting from an argument of a dimension, we replace in (A8) the action A_{break} by another one, comprising the quantity min(ΔT_{rad}):

$$A_{break} \rightarrow U_{HB} \times min(\Delta T_{rad}) = U_{HB} \times \Delta v_{min}(c\bar{v}^2)^{-1} \qquad (A11)$$

For the translational-band center we set here $v \approx 200\,cm^{-1}$. Then we obtain from (A11) an estimate

$$\Delta v_{min}(water) \approx 3 \times 10^{10} \times 4 \times 10^4\ 6.67 \qquad (A12)$$
$$\times 10^{-27}(2.6 \times 6.95 \times 10^{-14})^{-1} \approx 44.3\,cm^{-1}$$

Hence, the minimal bandwidth comprises at least several tens of cm^{-1}. Comparison with Fig. 4b shows this result to be reasonable. Since in ice the hydrogen-bond energy is approximately the same as in water and since Eq. (A11) has the only temperature-dependent parameter (U_{HB}), the following statement is true:

In ice and in water, perhaps the limiting bandwidths (A12)coincide   (A13)

It would be interesting to check this assertion by considering a wide range of temperatures.

IV. INFLUENCE OF TEMPERATURE ON WATER SPECTRUM

A.   Introduction to the Problem

In Section III we calculated the wideband spectra of liquid H_2O and D_2O at room temperature. In this section the results of analytical calculation of the complex permittivity ε and absorption coefficient α are presented for *ordinary* water for the temperature range extending from $-5.6°C$ (for supercooled water) to $81.4°C$. As in Section III, the dielectric response is modeled by two water fractions, LIB and VIB, with their lifetimes distributed in wide interval, viz. from about 0.01 to 0.25 ps. Motion of water molecules is governed in these fractions by four mechanisms a–d. The LIB fraction, pertaining to the mechanism a, consists of rigid permanent dipoles librating in a narrow and deep hat-like potential well. The other (VIB) fraction consists of H-bonded charged molecules performing: (b) fast elastic vibration along the hydrogen bond; (c) elastic reorientation around this bond; and (d) nonharmonic bending vibration perpendicular to the H-bond. The latter vibration is characterized by a noticeable association of water molecules. Our calculation reveals also the break of discontinuity in several fitted model parameters occurring in water at the temperature ≈ 300 K.

Studies of dielectric spectra of water in a range of temperatures present a fundamental physical problem that has also important practical applications. *Experimental* investigation of these spectra has a rich history. We refer here only to a few works. In Downing and Williams [22] and Zelsmann [21] tables for optical constants of water presented for the temperature $T = 300$ K and for a wide T-range, respectively. In recent publications by Vij et al. [32] and Zasetsky et al. [33] in addition to original investigations the results of many other works are also discussed. In work by Liebe et. al. [19] a useful empirical double Debye–double Lorentz formula for the complex permittivity $\varepsilon(v, T)$ is suggested.

The key theoretical problem consists in the difficulty of an adequate interpretation of the influence of temperature on the permittivity $\varepsilon(v)$ and absorption coefficient $\alpha(v)$. Recently, intense theoretical studies of water spectra were undertaken using the following:

- The molecular dynamics (MD) simulation, based on consideration of interactions in a large assembly of water molecules; see, for example, the paper by Zasetsky et al. [33] and references therein and the first-principles quantum study by Sharma et al. [29].

- Our approach discussed in Section II and with more details in the second part of the review.

The first (MD) approach actually comprises a computational experiment. Being in principle fundamental, the MD simulation however usually gives, up to now, only incomplete description of the experimental spectrum, since it cannot model the 200-cm^{-1} band in water. This simulation is complex and characterized by a high cost of computation. This drawback is avoided in an analytical one-particle approach, which, however, is less fundamental than the MD simulation, since the effective potential, in which a rigid or nonrigid dipole moves, is introduced on an *intuitive* basis.

Below, the analytical one-particle theory will be used. It is simple and yields good agreement between the calculated and experimental spectra.

B. Calculated Spectra

The calculations are made for seven temperatures: $T_1 = 268\,K, T_2 = 274\,K, T_3 = 293\,K, T_4 = 300\,K, T_5 = 312\,K, T_6 = 330\,K$, and $T_7 = 355\,K$—that is, for $T_{C_1} = -5.6°C, T_{C_2} = 0.4°C, T_{C_3} = 20.2°C, T_{C_4} = 27°C, T_{C_5} = 38.7°C, T_{C_6} = 57.2°C$, and $T_{C_7} = 81.4°C$.

In Fig. 10 the solid lines show, respectively in the right and left columns, the calculated loss spectra $\varepsilon''(v)$ and absorption coefficient $\alpha(v)$. These frequency dependences are compared with the experimental ones, depicted by open circles. For T_4 the experimental data from Downing and Williams [22] are used, while for other temperatures the data from Zelsmann [21] are used.

The adopted molecular constants, along with fitted and estimated parameters, are presented in Tables IV–VI. The absorption frequency dependences are depicted in Figs. 10e, 10c, and 10a, respectively, for the lowest (T_1), room[9](T_4), and highest (T_7) temperatures. The loss spectra are shown for the same temperatures in Figs. 10f, 10d, and 10b. The dash–dotted lines depict the contribution to loss spectra of transverse vibration (TV), and the dashed lines depict such spectra calculated without account of this vibration. We see that transverse vibrations play an important role in the THz region.

The loss peak revealed in Figs. 9b, 9d, and 9f at $\approx 200\,cm^{-1}$ originates from harmonic elastic *vibrations* (translations and reorientations) of HB molecules. These vibrations are ascribed, respectively, to mechanisms b and c.

The loss and absorption peaks at $v \approx 700\,cm^{-1}$, located near the border of the IR region, arise due to mechanism a—that is, due to *reorientation* of a rigid (permanent) dipole in the hat well. This mechanism is also responsible for the microwave loss peak located between the frequencies 0.1 and $1\,cm^{-1}$. The complex permittivity ε of the corresponding relaxation band is actually governed by Debye theory, which is involved formally in our calculation scheme.

The loss, relevant to mechanism d, is located in the region comprising several tens of cm^{-1} where the peak of the dash–dotted curve arises (see Figs. 10b, 10d, and 10f) but the overall loss curve $\varepsilon''(v)$ does not exhibit a relevant separate maximum.

The numbers $1, 2, \ldots, 7$ refer to the temperatures $T_1, T_2, \ldots T_7$. The solid and dashed lines in the right column of Fig. 11a show the contributions to the loss

---

[9]The term "room" is conditionally ascribed to 27°C, while in Section III it refers to 20.2°C.

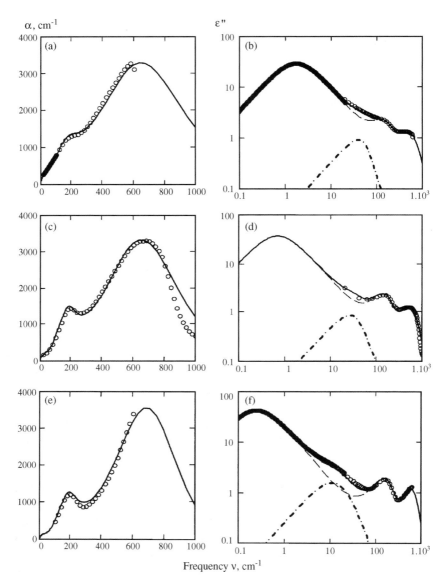

Figure 10a–f Calculated (solid lines) and experimental (open circles) absorption (a), (c), (e) and loss (b), (d), (f) spectra of water. The upper, middle, and lower rows correspond, respectively, to the high temperature (81.4°C), room temperature (27°C), and supercooled water (−5.6°C). Dash–dotted lines represent the contributions to dielectric loss pertinent to transverse vibration of H-bonded molecules. Dashed lines represent the loss spectra calculated without account of the latter mechanism.

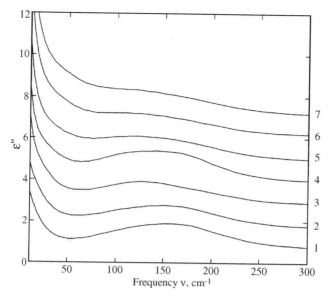

Figure 10g Evolution of the total calculated dielectric loss with the rise of temperature (from bottom to top). Curves 1, 2, ..., 7 correspond to the temperatures $T_1, T_2, \ldots, T_7$.

spectra due to, respectively, longitudinal vibration of nonrigid dipoles and elastic reorientation of rigid (permanent) dipoles.

The evolution with the temperature of the calculated overall loss spectrum $\varepsilon''(v)$ is illustrated for seven temperatures by Figure 10g and by the left column in Figure 11a. In the right column of Figure 11a the partial loss curves, governed by the mechanisms (b) and (c), are depicted for each temperature. In Fig. 11b the dielectric loss in the T-band region is compared with the low-frequency Raman spectrum calculated from the formula (38b).

TABLE IV
Static Parameters of the Model for Water H_2O

T	T (K)	T_C (°C)	ε_s	ρ (g cm^{-3})	τ_D (ps)	τ_{D2} (ps)	$\Delta\varepsilon_s$	k_μ
T_1	268	−5.6	90.2	0.999	22.4	0.564	4.33	1
T_2	274	0.4	87.6	1	17.7	0.444	3.96	1
T_3	293	20.2	80.0	0.998	9.34	0.235	1.72	1.35
T_4	300	27	77.6	0.996	7.85	0.197	2.74	1.08
T_5	312	38.7	73.7	0.992	6.07	0.153	2.12	1.18
T_6	330	57.2	68.2	0.984	4.38	0.110	1.99	1.16
T_7	355	81.4	61.8	0.971	3.17	0.080	2.05	1.19

Molecular constants: $\mu_0 = 1.84$ D, $n_\infty^2 = 1.7$, $I = 1.483 \times 10^{-40}$ g cm^2; $M = 18$

TABLE V
Parameters of the Hat Potential

T	u	β (deg)	f	τ_{or} (ps)	r_{or} (%)	μ_{or} (D)	m_{or}	S (Å)
T_1	8.98	23	0.65	0.25	65	2.27	4.9	0.21
T_2	8.78	22.7	0.68	0.22	70	2.27	4.3	0.19
T_3	8.19	20	0.8	0.15	55	3.06	3.2	0.10
T_4	8	23.2	0.8	0.13	65	2.45	2.4	0.12
T_5	7.7	22	0.8	0.11	65	2.68	2.2	0.16
T_6	7.27	23.6	0.9	0.095	65	2.63	1.8	0.086
T_7	6.77	23.8	0.9	0.079	65	2.69	1.5	0.062

Figures 10a–10f and the left column of Fig. 11a show that the theoretical dielectric spectra agree with the experimental spectra in a very wide range 0–$1000\,\text{cm}^{-1}$. These figures also show that:

(i) The loss maximum at $vu \approx 170\,\text{cm}^{-1}$ (see left column of Figure 11a) becomes shallow, when the temperature increases. At high temperature this maximum disappears, so that the loss curve $\varepsilon''(v)$ descends monotonically.

(ii) For a number of the fitted model parameters a monotonic dependence on T experiences a break at room temperatures. For instance, the loss maximum, produced by harmonic longitudinal translation, substantially lowers at T_3. Below (Section IV.D.5) we present a more detailed description/interpretation of such spectral peculiarity.

(iii) Raman spectrum, shown for three temperatures by thick lines in the right column of Fig. 11b, agrees with the experimental Raman spectra depicted in Fig. 4 for nearly the same temperatures. The RS substantially differs from the dielectric-loss spectrum calculated in the same T-band.

TABLE VI
Parameters of Harmonic Elastic Vibrations

T	Longitudinal Vibration of Charges						Elastic Reorientation
	τ_q (ps)	μ_q (D)	v_q, (cm^{-1})	$\Delta\varepsilon_{qs}$	K (dyn cm^{-1})	μ_μ, (D)	$\Delta\varepsilon_{\mu s}$
T_1	0.076	7.0	165	0.92	14,560	5.2	1.0
T_2	0.075	6.9	165	0.76	14,560	5.7	0.83
T_3	0.073	3.5	190	0.22	19,310	6.7	0.25
T_4	0.072	6.4	165	0.73	14,560	5.9	0.87
T_5	0.071	5.7	165	0.57	14,560	6.4	0.61
T_6	0.069	5.2	165	0.46	14,560	6.6	0.48
T_7	0.066	5.3	165	0.45	14,560	7.0	0.47

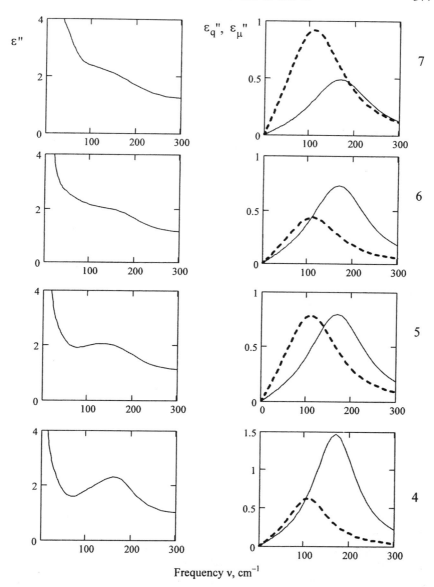

Figure 11a Evolution of the THz dielectric loss spectra stipulated by the rise of temperature (from bottom to top). Curves 1, 2, ..., 7 correspond to the temperatures $T_1, T_2, \ldots, T_7$. The left column represents the same $\varepsilon''(\nu)$ loss spectra as shown in Fig. 10. The right column represents the loss contributions ε_q'' and ε_μ'' pertinent to elastic longitudinal vibration (solid lines) and to elastic reorientation (dashed lines) of the H-bonded molecules.

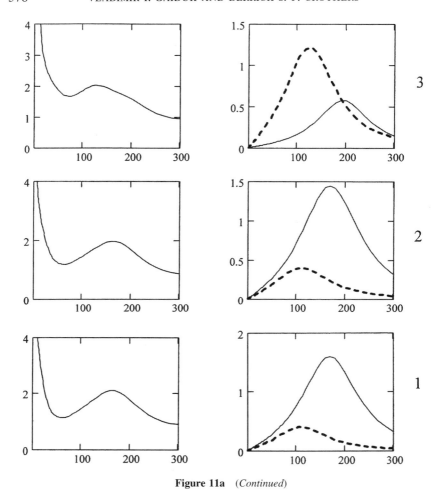

Figure 11a   (*Continued*)

Explanation of such a difference is given in Sections II.F and III.C.2. It should be emphasized that some of the model parameters, fitted for DS, differ (especially at low temperature) from those fitted for the best agreement with the DS. The latter difference, illustrated by Table VII, could be explained by an imperfection of our molecular model applied in describing of quite different physical phenomena (of dielectric and Raman spectra).

It seems that due to our intention to reach a *quantitative agreement* (demonstrated in Figs. 10a–10f) of the dielectric and experimental spectra, we "drive off" some model parameter in a rather unreasonable region. However, such parameters perhaps do *not* yield a correct description of the Raman spectra. This is especially true in an important case of supercooled (SC) water. For instance, in view of Table VII the best fit for the RS is obtained for the

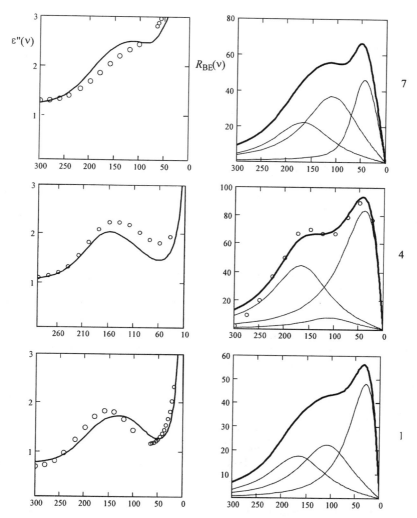

Figure 11b Evolution of the dielectric loss spectra (left column) and the Raman spectra for Bose-Einstein representation (right column) stipulated by the rise of temperature (from bottom to top). Thin lines in Figures of the right column show the contributions to Raman spectra due to (from left to right): longitudinal translation, elastic reorientation and transverse vibration of the HB molecules, Numbers 1, 4, and 7 refer, just as in Figure 11a, to the temperatures −5.6°C, 27°C, and 81.4°C. The parameters of the model are fitted for the best agreement with the experimental Raman spectra (see Table VII). The open circles refer to the experimental data.

intensity parameter $g_\perp$ equal to 32, while the best fit of the DS is obtained if $g_\perp \approx 112$! [We recall that $g_\perp$ is involved in Eq. (18d)]. The value $g_\perp = 32$ seems to be more reasonable than $g_\perp = 112$. We shall continue this discussion in Section C.3.

TABLE VII
Parameters Fitted with Account of Dielectric Spectra (A) and Raman Spectra (B)

	Fitted Parameters					Estimated Parameters	
T	$g_\perp$	$p_\perp$	$Y_\perp$	τ, ps	Z	τ_{vib}, ps	$g_\perp$, ps
			A. Best Fit for Raman Spectra				
T_1	111.8	4.50	7.5	0.25	10.6	0.0076	0.24
T_4	6.9	0.65	2.8	0.13	2.62	0.0072	0.0089
T_7	8.38	0.40	2.2	0.79	1.67	0.0066	0.0064
			B. Best Fit for Raman Spectra				
T_1	32	3.50	14.5	0.14	5.66	0.0076	0.0098
T_4	18	1.80	9.8	0.13	4.60	0.0075	0.0070
T_7	3	0.35	2.0	0.079	1.73	0.0075	0.0062

We assume that the above-indicated drawback of the present model can be avoided (or at least reduced) if a *new paradigm* [mentioned below in Section X.B.4(ii)] of the molecular model will be constructed. In our opinion, this drawback of the present model is stipulated by the following. In view of Eq. (11) the libration lifetime τ_{or} is determined by the experimental Debye relaxation time τ_D, so variation of τ_{or} cannot be used for other corrections of the calculated spectra. In the proposed new paradigm it is desirable to use τ_{or} for the latter purpose, while a correct describing of the low-frequency Debye spectrum is assumed to be reached by variation of additional parameter(s).

We see from Fig. 11b that the Raman spectrum changes rather weakly in the employed temperature range (see, e.g., the experimental RS in Fig. 4 and the calculated RS in the right column of Fig. 11b). In view of Table VII, such behavior of the RS is obtained if the fitted model parameters $g_\perp$, $p_\perp$, and $y_\perp$, pertaining to the transverse-vibration mechanism (d), exhibit a substantial change in the temperature interval of our interest. Due to the demonstrated *steepness* of the RS with respect to temperature, the agreement with experiment of the employed molecular model is attained for rather definite values of these parameters. We conclude that simultaneous application of the dielectric and Raman spectroscopy allows us to increase *reliability* of the employed molecular model.

C.   Fitted Parameters

1.   Steady State and LIB Subensemble

In the low-frequency region and in that adjacent with the IR region, the complex-permittivity and absorption spectra are described by the hat model. The formulas for $\varepsilon_{or}(v)$ and $\alpha_{or}(v)$ are given in Section II, Eqs. (4) and (38).

The hat model is parameterized to obtain an agreement between the complex permittivity ε and absorption α with experiment in the FIR range 400–1000 cm^{-1} and in the low-frequency region. To achieve correct low-frequency spectra, we set $x_{or} = 0$ in the argument of the spectral function L_{or} and vary the lifetime τ_{or} (and thus the collision frequency $y_{or} = \eta_{or}/\tau_{or}$) in order to satisfy the equation:

$$\tau_D \approx g_{or} G_{or} (1 - r_{vib}) [y_{or} L_{or}(iy_{or})]^{-1} \tag{56}$$

This procedure allows us to position correctly the loss maximum ε_D'' just at the frequency $v_D = \omega_D(2\pi c)^{-1} = (2\pi c \tau_D)^{-1}$ known from the experiment—for example, given below by the estimate (60).

In (58) c is the speed of light. The parameters of essentially the hat model are:

- the normalized (divided by $k_B T$) well depth u;
- the angular well width 2β, β being the libration amplitude;
- the form factor f, namely, the ratio of the well width at the bottom (where the potential is zero) to that at the top;
- the lifetime τ_{or} of the hat potential (namely, of the librational state).

The mean number of librations performed by a dipole during the lifetime τ_{or} is given by the approximate formula

$$m_{or} \approx \sqrt{3/2} \left\{ 4 y_{or} \beta \left[f + (\pi/2) \sqrt{\frac{3}{2u}} \right] \right\}^{-1} \tag{57}$$

To compare our theory with low-frequency experimental data, we estimate the static permittivity ε_s and the Debye relaxation time τ_D using for $\varepsilon(v)$ the empirical double Debye–double Lorentz formula by Liebe et al. [19], where the temperature T is involved in terms of $\theta(T) = 1 - 300T^{-1}$:

$$\varepsilon(v) = \varepsilon_s - \left(\frac{\varepsilon_s - \varepsilon_1}{v + i\gamma_1/c} + \frac{\varepsilon_1 - \varepsilon_\infty}{v + i\gamma_2/c} \right) v + \sum_{j=1,2} \left(\frac{A_j/c^2}{v_j^2 - v^2 - i\Gamma_j c^{-1} v} - \frac{A_j/c^2}{v_j^2} \right) \tag{58}$$

Here the first two terms comprise the double Debye representation,[10] where the temperature dependent parameters are

$$\varepsilon_s = 77.16 - 103.3\theta, \quad \varepsilon_1 = 0.0671\varepsilon_s, \quad \varepsilon_2 = 3.52 + 7.52\theta$$
$$\gamma_1 = (20.2 + 146.4\theta + 316\theta^2 \cdot) \times 10^9, \quad \gamma_2 = 39.8\gamma_1 \tag{59}$$

[10]In Section III.B the empirical formula (39), which is approximately equivalent to this representation, is applied for liquid H_2O and D_2O.

The principal Debye relaxation time τ_D and the corresponding frequency ν_D are then

$$\tau_D(\theta) = [2\pi\gamma_1(\theta)]^{-1}, \qquad \nu_D(\theta) = \gamma_1(\theta)c^{-1} \qquad (60)$$

Analogously the relevant quantities, pertinent to the second Debye region, are:

$$\tau_{D2}(\theta) = [2\pi\gamma_2(\theta)]^{-1}, \qquad \nu_{D2}(\theta) = \gamma_2(\theta)c^{-1} \qquad (61)$$

On the other hand, the sum over j in Eq. (58), whose parameters are temperature-independent, comprises the double-Lorentz representation of the FIR resonance spectrum:

$$\begin{aligned} \nu_1 &= 5.11 \times 10^{12}c^{-1}, & \Gamma_1 &= 4.46 \times 10^{12}, & A_1 &= 25.03 \times 10^{24} \\ \nu_2 &= 18.2 \times 10^{12}c^{-1}, & \Gamma_2 &= 15.4 \times 10^{12}, & A_2 &= 282.4 \times 10^{24} \end{aligned} \qquad (62)$$

Our analysis shows that the formula (68) gives reasonable spectra, only if the temperature is not very high, for example, $T < 300$ K. It seems, however, that in the work by Liebe et al. [19] the principal Debye region is also described correctly also in a greater temperature interval. However, for $T > 300$ K, application of formula (58) demands additional grounds, especially concerning the second Debye relaxation time τ_{D2}.

In Table IV we list the temperature-dependent steady-state parameters: static permittivity ε_s; density ρ; Debye relaxation time τ_D; second-Debye relaxation time τ_{D2}; the contribution $\Delta\varepsilon_s$ to ε_s due to fast vibrating HB molecules; and the dipole-moment factor k_μ involved in Eq. (3).

The quantity $\Delta\varepsilon_s$ is calculated in Section II.C, Eqs. (30)–(33), while ε_s, τ_D, and τ_{D2} are found from the empirical formulas by Liebe et al. [19]; see, respectively, Eqs. (12), (13), and (60). We also employ the molecular mass M, the moment of inertia I_{or}, and the dipole moment μ_0 of an isolated water molecule.

The moment of inertia I_{or} (its value is given in Table IV) is estimated for a freely rotating symmetric top molecule (rotation about the principal axis normal to the symmetry axis). The moment I_{or} determines the frequency scale relevant to reorientation of dipoles in the hat well. Thus, I_{or} determines the temperature dependence of the fitted lifetime τ_{or} and the strong-collision frequency y_{or} estimated from Eqs. (2a) and (2b) with account of lifetime τ_{or}. The dipole moment μ_{or} of a librating dipole is expressed from Eq. (3) through that (μ_0) of a free water molecule and through the fitted dipole-moment factor k_μ, with n_∞^2 being the optical (measured near the frequency 1000 cm^{-1}) permittivity ($n_\infty^2 \approx 1.7$). The factor k_μ is rather close to unity.

Assuming that the potential well depth U_{or} does not depend on T, we write

$$u(T) k_B T = U_{or} \qquad (63)$$

We set $u(T_4) = 8$ and then calculate $u(T)$ from (63) as a decreasing function of T. The libration amplitude β comprises about $23°$. However, at some notable point $T_3 \approx 20.2°C$ the fitted β value falls to $20°$. We shall return to this point below.

In view of Fig. 12d the form factor f decreases with the increase of T. At low temperature the hat well is almost rectangular ($f = 0.9$), while at a high temperature the potential well bottom becomes essentially rounded ($f \approx 0.6$). We

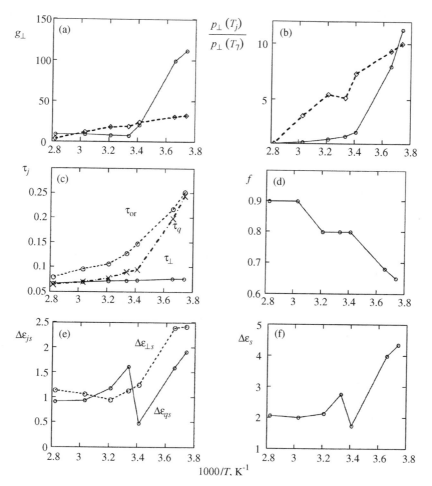

Figure 12 Fitted parameters of the molecular model: (a, b) Effective transverse intensity (a) and ratio of transverse frequency factors (b); solid curves refer to optimization o dielectric and dashed lines of Raman spectra. (c) Lifetimes of the hat potential, of transverse vibration, and of elastic vibration (from top to bottom). (d) form factor of the hat potential.(e) contributions of transverse (dashed line) and of elastic (solid line) vibrations to the static permittivity. (f) total vibration contribution to the static permittivity.

regard the *spread* S, namely the angle on the bottom, where the potential is inhomogeneous, as a measure of this rounding, see Eq. (15). At T_7 the spread comprises about one-third of the value at T_1. The ratio $\tau_{or}(T_7)/\tau_{or}(T_1)$ of lifetimes is also $\approx 1/3$. Note that the decrease of Debye relaxation times τ_D in the interval $[T_1, T_7]$, equal to about a factor of 7, is much greater. The proportion $r_{or} = 1 - r_{vib}$ of librating molecules comprises about 65%, except at the point T_3, where $r_{or} \approx 55\%$.

The fitted dipole moment μ_{or} of reorienting molecules increases with T from 2.3 to 2.7 D. However, at T_3 the fitted $\mu_{or} \approx 3.1$ D.

The mean number m_{or} of librations, performed by a rigid dipole during the lifetime τ_{or}, is estimated from Eq. (11). This number decreases with T from 5 to 1.5 (see Table V).

2. Harmonically Vibrating HB Molecules

The relationships concerning two elastically vibrating subensembles are presented in Section II.C. We note an approximate constancy of the fitted vibration frequency v_q, which represents the center of the loss curve $\varepsilon''_q(v)$, see Table VI and Figure 11a (right column, solid curve). This constancy presents a remarkable property of the hydrogen bond itself.

The fitted frequency $v_q \approx 165$ cm^{-1} is given in Table VI. At the break point T_3 the fitted frequency v_q is higher and the estimated loss intensity $\Delta\varepsilon''_q$ is lower. Also, in view of Fig. 12e, the contribution $\Delta\varepsilon_{qs}$ of longitudinal translation to the static permittivity ε_s drops at T_3. The same assertion holds also for the total vibration contribution $\Delta\varepsilon_s$ to ε_s (see Fig. 12f). Such an outcome for the model parameters could be ascribed to a structural transition occurring at room temperature. We shall return to this point in Section IV.D.5.

The dependence of other parameters on T could be slowly varying—for instance, the decrease, with T of the vibration lifetime τ_q (see Fig. 12c). This lifetime is proportional to $T^{-1/2}$, since $\tau_q(T) = \eta_{vib}(T)y_q - 1$, where the fitted $y_q \approx 0.9$ and $\eta_{vib}(T)$ is given by Eq. (17). The time τ_q, interpreted as the hydrogen-bond lifetime, is very short (less than 0.1 ps). Comparing the solid and dashed curves in Fig. 12c, we see that the reorientation time τ_{or} in water is longer than τ_q and its temperature dependence is stronger. The harmonic vibrations are damped, since the number of vibration cycles m_q performed during the lifetime τ_q and estimated as $\tau_q c v_q$, is only $\approx 1/3$. In spite of this, the loss frequency dependence (see solid curves in the right-hand column of Fig. 11) conserves its resonance character. We relate this property to the quantum origin of this molecular mechanism.

It appears that the fitted dipole moments μ_q and μ_μ, pertinent to elastic vibrations, substantially exceed the moment μ_{or} of a dipole reorienting in defects of water structure (see Tables V and VI). The excess of μ_q and μ_μ over μ_{or} arises possibly due to additional H-bond polarization produced by charged molecules.

The harmonic-vibration contributions $\Delta\varepsilon_{qs}$ and $\Delta\varepsilon_{\mu s}$ to the static permittivity ε_s and the fitted $\Delta\varepsilon_q$ value itself are presented in Figs. 12e and 12f.

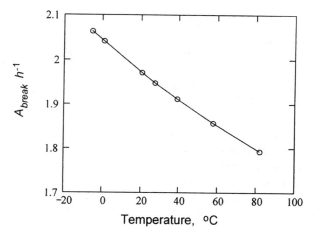

Figure 13 Temperature dependence of the quantum-break factor, estimated for H_2O.

Unraveling the concept of applicability of the classical theory, manifested in Appendix III, where the case of room temperature was considered, we present in Fig. 13 the temperature dependence of the quantum-break factor (A8a). As we see, this q-factor is commensurable with unity, decreases monotonically with the increase in T, and, just as in Appendix III, remains noticeably greater than the possible minimal value equal to 1.

Such a behavior appears to be reasonable. We shall return to this point in Section V, devoted to ice.

3. Transversely Vibrating HB Molecules

The formulas relevant to transverse vibration (TV) are presented in Section II.D. The results of calculations are illustrated by Table VIII and Figs. 8, 12c, 12e, and 14. Note that this table is obtained for the best fit of dielectric spectra (see also Table VIII and the relevant comments).

TABLE VIII
Parameters of Nonharmonic Transverse Vibrations ($j = 1, 2,...,7$) (the Best Fit for the Dielectric Spectra; Compare with Table VII)

T	$\tau_\perp$ (ps)	$g_\perp$	p_1/p_j	$\Delta\varepsilon_{\perp s}$	$m_\perp$	$v_\perp$ (cm^{-1})
T_1	0.244	112	1	2.42	0.028	10.6
T_2	0.198	99.4	1.41	2.39	0.032	12.6
T_3	0.094	19.4	5.63	1.25	0.070	20.0
T_4	0.089	6.87	6.92	1.13	0.071	26.6
T_5	0.077	7.22	8.18	0.94	0.072	29.9
T_6	0.069	8.43	10	1.06	0.079	35.5
T_7	0.064	8.38	11.25	1.13	0.083	39.8

We set the maximal dimensionless transverse displacement s_{lim} equal to 0.2. So, the validity condition of the theory (smallness of s) is satisfied. The corresponding TV amplitude b_m comprises about 1/3 of the covalent-bond length: $b_m \approx 0.285$ Å. A satisfactory description of spectra is obtained, if the transverse force constant $k_\perp$ comprises about 30% of the longitudinal force constant k (the latter is given in Table VI).

In the variant of the parameterization, in which we intend to achieve the best agreement of calculated and the experimental spectra, we see a sharp rise, for $T < 293K$, in the effective factors $g_\perp$ (see solid curve in Fig. 12a). However, in the alternative variant, pertaining to the best fit of the Raman spectra the $g_\perp(T^{-1})$ dependence is less emphasized (dashed curve in Fig. 12a). This behaviour seems to be more preferable. The association factor ξ defined by Eq. (24) rises with the decrease of T (see Fig. 12b), but in different fashion in these two variants.

The HB transverse vibrations are very damped. In view of Fig. 14a the number $m_\perp$ of vibration cycles, performed during the lifetime $\tau_\perp$, is small ($m_\perp < 0.1$). This number becomes as low as 0.03 in the case of supercooled water. One can interpret so small a $m_\perp$ value as evidence that the curve $\varepsilon_\perp''(v)$ and the parameter $m_\perp$ characterize transverse *collective* motions of the H-bonded molecules.

The transverse-vibration lifetime $\tau_\perp$ is commensurable with the reorientation lifetime τ_{or} (Fig. 12c). This lifetime strongly increases with decrease of temperature (see the dash–dotted curve in Fig. 12c).

In Fig. 14b we compare the center frequency $v_\perp$ of the TV loss dependence $\varepsilon_\perp''(v)$ (solid curve) with the frequency v_{D2} (dashed curve) pertinent to the so-called second-Debye region. The frequency v_{D2} was calculated from empirical formulas (61). Our estimated THz loss spectrum agrees with the approach by Liebe et al. [19] only for rather low temperatures ($T < 293$ K), when both curves in Fig. 14d are close to each other.

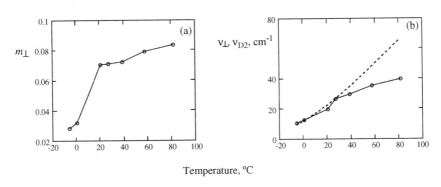

Temperature, °C

Figure 14 Estimated temperature dependences: mean number of transverse vibration performed during the lifetime $\tau_\perp$ (a); center $v_\perp$ of the partial-loss dependence $\varepsilon_\perp(v)$ (solid curve) and the second-Debye region frequency v_{D2} (dashed curve) (b).

In view of Fig. 12e we also remark that the contribution $\Delta\varepsilon_{\perp s}$ of nonharmonic transverse vibration to the static permittivity ε_s increases with decrease of T, as does the contribution $\Delta\varepsilon_q''$ of harmonic vibration (see also Tables VI and VII).

D. Discussion

1. On Two-Fractional Model of Water Spectrum

A variety of mixed models of water is widely available. We use a variant of such models for calculation of wideband spectra in liquid water. In terms of our analytical approach important arguments concerning the mixed model are provided by the different origins of the librational and translational bands. As was demonstrated in Section III, the center of the first is sensitive to the change of H_2O or D_2O, while the center of the second is insensitive.

We employ a classical description of the dynamical consequences of such a quantum object as the hydrogen bond. This concerns, for instance, the vibration of HB molecules. The price we pay for such an approach is that several fitted model parameters (e.g., force constants) are not related explicitly to the molecular structure of our object. Note that in the MD simulation method, based on application of various effective potentials, the classical theory is also often used [33–35]. A very detailed analysis of the problems pertinent to the two-fractional (mixed) models of water is given in the latter work (review) with respect to various (mostly steady-state) properties of water. In the context of our work, the use of a classical mixed model is justified by a possibility of considering a simplified picture of two-state molecular motion allowing a relatively simple analytical calculation of the complex permittivity $\varepsilon(\nu)$ given in Section II.

An interesting general question concerns the physical sense of the lifetimes τ_{or} and τ_q, which are tightly determined in our model by the experimental linewidths of the quasi-resonance spectrum.

The first time, τ_{or}, has the sense of a mean time, during which a permanent dipole librates rather freely due to a broken or strongly bent hydrogen bond. At high temperature the fitted τ_{or} comprises one-third of the τ_{or} value found for supercooled water (see Table V).

The other time, τ_q, pertinent to an elastic nonrigid dipole and comprising about 0.07 ps, decreases insignificantly with T in a wide temperature interval (see Table VI). This result may be interpreted as an evidence of its "quantum nature". The fitted τ_q(water) is less than the mean HB lifetime[11] τ_{HB} mentioned by Teixeira [16]. We should remark that our lifetime τ_q actually means a period, during which longitudinal vibration remains coherent.

It is especially interesting that in the case of ice the time τ_{HB} is generally accepted to be very long. Contrary to this, our *fitted* lifetime τ_q(ice) is only several

---

[11]A simple and generally accepted criterion for this notion is lacking [34].

times longer than τ_q(water) correspondingly with a narrowing of the linewidths in ice. A difference of these concepts may give rise to a new paradigm, which could be used for calculation of ice and water spectra (see Section 10).

One may consider rotation (libration) of a rather free permanent dipole as a motion typical for the LIB fraction, and one may regard a fast vibration of H-bonded molecules as that typical for the VIB fraction. In view of the data, presented in Sections II and V, where water and ice spectra are compared, we deduce the following:

(i) The fitted libration lifetime τ_{or} is in the case of ice several times shorter (see Table I) than the vibration lifetime τ_q. The converse applies in liquid water: $\tau_{or} > \tau_q$. The number m_{or} of librations suffered in water during the lifetime τ_{or} is much greater than the number m_q of vibrations performed during the lifetime τ_q (m_{or} decreases from 5 to 1.5 with rise of the temperature (see Table 3), while $m_q \approx 1/3$.

(ii) The number of librations is in the case of ice less than that of vibrations ($m_q = 2.1$, $m_{or} = 1.3$). Therefore, unlike water, vibrations in ice are less damped than librations.

2. Polarization of Hydrogen Bond

Our conclusion about existence of additional polarization of HB molecules follows from the relationship $\mu_q > \mu_{or}$ for the fitted dipole moments pertinent, respectively, to VIB and LIB subensembles (see Tables V and VI). A similar conclusion is made in many works. Thus, the works Ruocco and Sette [36] and Sampoli et al. [37] refer to the electrostatic long-range interactions; in Krasnoholovets [38] the spectrum of water around $200 \, cm^{-1}$ is ascribed to a strong ionic polarizability, since the pair potential of a water molecule is simulated in the form of an ionic crystal potential. We remind the reader that in our approach we postulate existence of oppositely charged molecules at both ends of the dimer shown in Fig. 1b.

3. Connection of Transverse Vibration (TV) with Water Spectrum

We demonstrate in Table VII and Figs. 8b and 8d, a considerable effect of TV on wideband THz spectra of water, especially at low temperatures. This result quantitatively confirms a well-known suggestion by Walrafen et al. [39] that "the $170 \, cm^{-1}$ Raman band arises from the vibration of an oxygen atom relative to its nearest-neighbor oxygen when a linear or at most weakly bent HB, O–H$\cdots$O, is present between them. This motion is often described alternatively as a restricted translation of H_2O molecules. The $60 \, cm^{-1}$ band might also be thought of as a restricted translation, but one that is perpendicular to the hydrogen-bond linear O–H$\cdots$O direction, i.e., as a bending of the hydrogen bond."

We should remark that the experimental (Downing and Williams [22], Zelsmann [21]) and our calculated absorption and loss THz spectra do *not* reveal a separate $\alpha(v)$ and $\varepsilon(v)$ band at 60 cm^{-1}, unlike the *partial* $\varepsilon''_{\perp}(v)$ dependence pertinent to transverse vibrations. The estimated center of this band is, however, located at the frequency $v_{\perp}$, which is less than 60 cm^{-1} (see Table VII and Figs. 8b and 8d, (dashed curves)).

An important role of bending motion of molecules is emphasized in the numerical MD simulation of liquid water by Ruocco and Sette [36].

A strong temperature dependence of the fitted effective parameters $g_{\perp}$ and $p_{\perp}$, depicted in Figs. 12a and 12b indirectly confirms the results of an experimental investigation of micron-sized water droplets in a cryogenic flow tube and the results of the MD simulation by Zasetsky et al. [33]. An increase of low-density domains (LDD) on cooling between 300 and 240 K was discovered in this work. It was suggested there that supercooled water contains, apart from "normal" water, some additive spectroscopically identifiable component. Increase of that LDD fraction with decrease of T correlates also with the neutron-scattering experiments (Belissent-Funel [40]). It follows from our calculations that this increase correlates with the contribution of transverse vibrations. On the contrary, in view of Fig. 11 (solid curves in the right-hand column) the decrease of T *reduces* the intensity of the longitudinal oscillations near the mode at 100 cm^{-1}. This result agrees with that found in Zasetsky et al. [33].

4. Power Absorption Coefficient

Our calculation of the absorption coefficient α in the translational-band region (Fig. 15a) agrees with recent experimental data by Vij et al. [32] depicted in Fig. 15b. The absorption maximum α_{max} is reached in both cases at $\approx 30°C$. It is interesting that at higher temperature the α_{max} value substantially decreases.

5. Temperature Dependences of the Model Parameters: Break of Continuity

In Fig. 12 we already met sharp turns in several such dependences. To continue studies of this interesting phenomenon, we also show in the range of temperatures:

(i) The *maximum* $\alpha_{L,m}$ of absorption frequency dependence $\alpha(v)$ at $v \approx 700$ cm^{-1} and location v_L of this maximum (see Figs. 16a and 16c). These data, found for permanent dipoles librating in the hat well, were obtained from Figs. 8a, 8c, and 8d and from others, similar to them, but not shown above.

(ii) The *maxima* of partial loss frequency dependences $\varepsilon''_q(v)$ and $\varepsilon''_{\mu}(v)$ (see Figs. 16b, and 16d) calculated for harmonically vibrating HB dipoles. These data were obtained from Fig. 11 (right column).

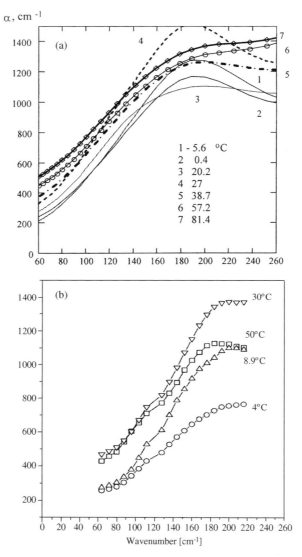

Figure 15 Theoretical (a) and experimental (b) power absorption coefficient. (Fig. 13b is reproduced from Vij et al. [32]).

A distinct break of continuity is manifested in Fig. 16 at a certain room temperature T_{break} (specific curves could be shifted a little with respect to the other). It is reasonable to explain such a break by a change of water structure occurring over a narrow temperature interval near T_{break}.

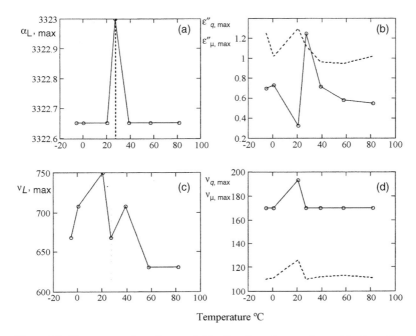

Figure 16 Temperature dependences of: libration-absorption maximum (a); absorption-peak frequency (c); maximum of partial loss, pertinent to elastic longitudinal (solid curve) and elastic rotational (dashed curve) vibration (b); frequencies of the corresponding loss peaks (d). Vertical lines in parts a and c refer to the temperature 27.6°C.

Independent and reliable evidence that the temperature T_{break} near 300 K is a singular point for water is given by the method of nuclear magnetic resonance (NMR) relaxation spectroscopy.

Thus, Fig. 17a reproduces the temperature dependence of the transverse relaxation time T_2 measured by Gaiduk and Emetz [41]. In this experiment the proton magnetic resonance was studied in water at the frequency 15.9 MHz. The measured dependence $T_2(T)$ was approximated by two straight lines, which intersect each other at a point with the abscissa 300 K.

A similar specific feature was also observed in the temperature dependence of the spin-lattice relaxation time T_1. Such a dependence, described in the monograph by Vashman and Pronin [42], is reproduced in Fig. 17b. The passage from one approximating straight line to the other takes place in a rather narrow range. It is stated in the above-mentioned reference that describing the temperature dependence in terms of the function

$$logT_1 = f\left(\frac{1000}{T(K)}\right)$$ (64)

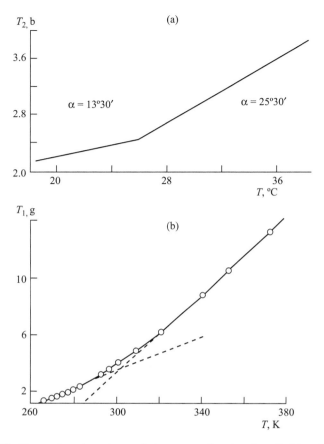

Figure 17 Temperature dependence of: transverse relaxation time T_2 (a) and spin-lattice relaxation time T_1 (b). Measurements in liquid water by NMR spectroscopy. (Reproduced from Fig. 6 of the work by Gaiduk and Nikitov [44].)

is *impossible* if *one* activation energy E is employed in the required temperature range,[12]. namely,

(i) one T_1 value concerns the interval 278–298 K, and the other interval 313–363 K; or

(ii) the interval 273–303 K and the other interval 318–333 K.

---

[12]Thus, in the work by Dass and Varshoeya [43] the following empirical dependence $T_1 = C \exp(E/R(T - T_0)$ is proposed with T_0 corresponding to Debye temperature ($T_0 = 150K$) found for the quasi-crystal model of water.

In accord with Vashman and Pronin [42] such a change of water structure in many observations concerns the interval 303–308 K, where the change of the slope of dependence (64) occurs. This interval is rather close to that illustrated by Figs. 17a and 17b.

Hence, the conclusions provided by NMR observations agree with our results (Figs. 14 and 16) concerning the continuity break of various parameters.

It is also important to note that the solid curve in Fig. 14d, which represents the left and right wings of the transverse-frequency dependence $v_\perp(T)$, resembles two straight lines with different slopes, depicted in Fig. 17. Hence, behavior of transverse vibrations, aggravated by a molecule association, is perhaps related to the break-point temperature T_{break} (near 300 K).

E. Concluding Remarks

The object and purpose of our calculations were mainly the same as were described in the previous review GT2. However, the model, used in this review, was substantially modified. Our description effects a good agreement of calculated and experimental spectra obtained over a range of temperatures.

Such an agreement was not reached in GT2, and in several subsequent works (e.g., Gaiduk and Nikitov [44], no agreement was reached for supercooled water or for temperatures near the boiling point. Also important is the change of the fitted parameters. For instance, the proportion r_{vib} of vibrating molecules, found in terms of our mixed model theory, is about 30% or 45%, while in Gaiduk and Nikitov [44] r_{vib} is one-quarter of that.

A very good classical description range of the complex-permittivity spectra, obtained in a wide T range, is based on *four basic* molecular mechanisms a–d, which allow coherent interpretation of these spectra.

(a) *In the LIB state* a rather free libration of a rigid permanent dipole occurs in a narrow ($\approx 48°$) and deep (7–9 k_BT) potential well. The concentration of molecules is about 55–70%. The libration lifetime is 0.3 ps for $T_1 = 268$ K and 0.08 ps for $T_7 = 355$ K. The employed hat potential, flat near the bottom, has parabolic walls. The flat part is small at T_1 and is rather wide at T_7.

The lifetime τ_{or} of the LIB state is interpreted as an average lifetime of a broken or strongly bent hydrogen bond. This lifetime decreases from 0.3 ps at T_1 to 0.08 ps at T_7. The striking property of the LIB state is a chaotic but collective motion of protons.

Mechanisms b, c, and d refer in our model to charged water molecules. We assume that the VIB state arises almost instantaneously in reality for a short period due to the LIB → VIB transition and comes to an end due to the inverse VIB → LIB transition. The latter transition lats during a short period (about 1 ps or less) when a hydrogen-bonded molecule exists.

(b) Elastic harmonic vibration of molecules along the hydrogen bond. In view of Fig. 13, an application of the classical theory, being valid (since $A_{break} > h$) is rather conditional, since A_{break} is commensurable with h.

(c) Elastic harmonic reorientation of permanent dipoles about the hydrogen bond. In view of Fig, 16b, the loss maximum pertinent to this molecular mechanism (solid curve) is close to that for mechanism (b) (see dashed curve), but the frequency relevant to this maximum is noticeably less (cf. solid and dashed curves in Fig. 16d).

(d) Nonharmonic bending vibration of molecules in a direction perpendicular to the H-bond. In view of Fig. 14b (solid curve) the change, at room temperature, of the slope of the curve $v_{\perp}(T)$, relevant to transverse vibration, manifests the change of water structure. The same conclusion follows from the NMR data (Fig. 17). Similarly the continuity-breaks of various fitted and estimated parameters arise as demonstrated in Section IV.D.5.

The three latter processes (b, c, d) are characteristic of *the VIB state*. In this state the lifetime τ_q of harmonic motion is about 0.07 ps. The lifetime $\tau_{\perp}$ of a nonharmonic vibration is 0.2 ps for T_1 and 0.06 ps for T_7. In supercooled water, this vibration is accompanied by a noticeable association of water molecules.

The proposed theory facilitates an uncovering of the intricate interrelations between microscopic dynamics of molecules in water and the spectra of water, observed over the range of temperatures.

Since water and ice are kindred fluids, it is interesting to continue this investigation in the case of ice (see Section V).

V. ICE SPECTRUM

A. Introduction to the Problem

In this section we apply to ice Ih, the two-fractional (mixed) molecular model that was briefly described in Section II and applied to water H_2O in Sections III and IV. For convenience we give in Appendix IV the list of formulas used in this Section V. We calculate analytically the complex dielectric permittivity $\varepsilon(v)$ and absorption $\alpha(v)$ of ice in the far-infrared and submillimeter wavelength regions.

The model comprises the librational (LIB) and vibrational (VIB) fractions. A rigid permanent dipole, constituting the first (LIB) fraction, performs nonharmonic reorientation in the part of the structure formed by weak or torn hydrogen bonds (HB). This fraction stipulates the librational band of ice located at $\approx 800\ cm^{-1}$. The second (VIB) fraction, constituted by two elastically vibrating oppositely charged HB molecules, describes T- and V-bands, located in the far-IR region 100–300 cm^{-1}, and the nonresonant dielectric-loss background at submillimeter wavelengths.

We present in detail the loss and absorption ice spectra at the temperatures $-7°C$ and 100 K and also calculate more superficially the influence of T on the ice spectra at intermediate temperatures. The ice spectra will also be compared with water spectra.

It should be noted that in the literature (e.g., in the works by Walrafen [45], Mikhailov et al. [46], Gurikov [47,48], Eisenberg and Kauzman [1] the mixed model of water usually concerns the description of *thermodynamic and other steady-state properties* of the fluid in terms of *structures differing in the way the H_2O molecules are packed*. In our analytical approach to the modeling of spectra, the different *types of molecular motion* are accentuated rather than those of water structures. Our approach, being rather artificial, allows modeling, in terms of a simplified classical description, of several ice bands differing by intensity and bandwidth.

Simulation of the ice spectrum using the numerical molecular-dynamics (MD) method in terms of lattice vibration usually represents (Nielson et al. [7], Marchi [8] this spectrum as a set of chaotic peaks whereas the experimental far-IR ice spectrum $\varepsilon(v)$, $\alpha(v)$ is rather smooth, similar to the spectrum of liquid water: compare Figs. 20a and 20c and also 20b and right side of Fig. 20d.

The MD simulation by Marchi [8] of the librational band (Fig. 18) is rather poor in this example. A better simulation is for the translational band (solid line in Fig. 19). The dashed curve represents here the 'energy loss function' determined by the dielectric constant ε' and calculated in the cited work by Marchi. The shift between the solid and dashed curves represents a specific parameter of the fluid

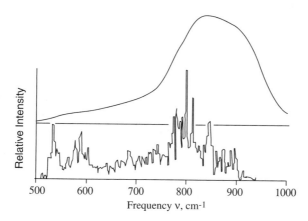

Figure 18 Comparison of the observed (above) and calculated by the MD method (below) absorption ice spectrum in the libration-mode region. (Reproduced from Marchi [8].

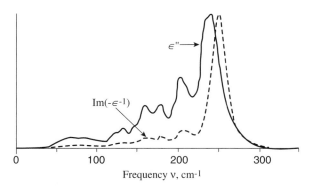

Figure 19 Frequency dependence of the loss factor ε'' measured for ice in the translational band at $T = 100\,\mathrm{K}$ (solid curves). Dashed curve represents the energy loss function. (Reproduced from Marchi [8]).

and also of the model describing the relevant spectrum. We shall return to this question below (see Fig. 24).

B.   Resonance Spectrum at Rather High Temperature

1.   Results of Calculations

For the frequency range $10\text{--}1000\,\mathrm{cm}^{-1}$ we present in Fig. 20a, b by solid lines the far-IR *ice spectra* of the absorption coefficient $\alpha(\nu)$ and of dielectric loss $\varepsilon''(\nu)$ calculated for the temperature $-7°\mathrm{C}$. The symbols V, T, L refer, respectively, to the V-, translational, and librational bands. The open circles mark the experimental data by Warren [49]; these data are reproduced in Table IX. The fitted model and molecular parameters, used in this calculation, are given in Table X.

In Figs. 20c and 20d for comparison the solid curves represent analogous spectra for liquid water calculated at the temperature $27°\mathrm{C}$ using the same molecular model; the employed parameters are also given in Table X. The experimental data obtained by Downing and Williams [22] and Liebe et al. [19] are marked by open circles. The symbol D in Fig. 20d marks the relaxation band of water, in which the loss peak is placed at the microwave region. The profiles of the hat well, estimated for ice and water, are depicted, respectively, in Figs. 21a and 21b.

In the range $0\text{--}800\,\mathrm{cm}^{-1}$ our composite model of water gives a quantitative agreement between the measured and calculated spectra, while in the region $800\text{--}1000\,\mathrm{cm}^{-1}$ the calculated absorption (Fig. 20c) exhibits too slow an approach to the transparency region, failing to decrease sufficiently rapidly near the HF edge of the dielectric-relaxation region. In Fig. 20c we see a two-humped absorption

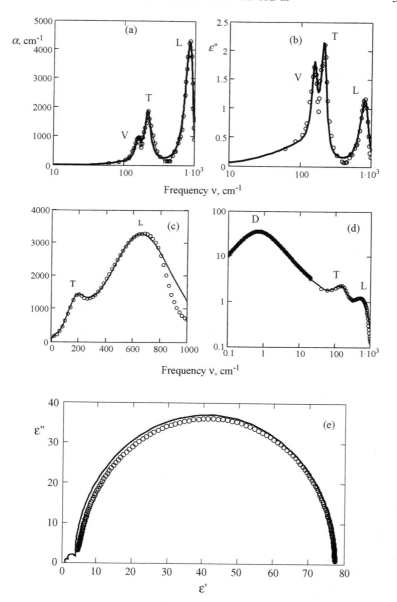

Figure 20 Frequency dependences of absorption coefficient (a, c) and of dielectric loss (c, d) and Cole–Cole plot for water (e). Solid lines represent calculation for ice at $-7°C$ (a, b) and for liquid water at $27°C$ (c, d). Open circles refer to the experimental data obtained for ice by Warren [49] and for water by Downing and Williams [22] and Liebe et al. [19].

TABLE IX

Optical Constants of Ice I_h at $T = 266.16$ K, viz. at $T_C = -7°C$

v (cm^{-1})	λ (micron)	n	κ	v (cm^{-1})	λ (micron)	n	κ
5.988E+01	1.670E+02	1.8296	8.300E-02	3.226E+02	3.100E+01	1.3104	7.550E-02
6.667E+01	1.500E+02	1.8227	9.620E-02	3.500E+02	2.857E+01	1.3489	5.300E-02
8.333E+01	1.200E+02	1.8191	1.200E-01	3.846E+02	2.600E+01	1.3895	3.400E-02
1.000E+02	1.000E+02	1.8322	1.390E-01	4.000E+02	2.500E+01	1.4068	3.000E-02
1.051E+02	9.517E+01	1.8425	1.490E-01	4.065E+02	2.460E+01	1.4138	2.890E-02
1.101E+02	9.080E+01	1.8536	1.620E-01	4.237E+02	2.360E+01	1.4316	2.730E-02
1.152E+02	8.680E+01	1.8670	1.771E-01	4.338E+02	2.305E+01	1.4422	2.700E-02
1.176E+02	8.500E+01	1.8750	1.865E-01	4.425E+02	2.260E+01	1.4518	2.750E-02
1.205E+02	8.297E+01	1.8874	1.981E-01	4.500E+02	2.222E+01	1.4607	2.900E-02
1.250E+02	8.000E+01	1.9033	2.350E-01	4.801E+02	2.083E+01	1.4905	4.500E-02
1.311E+02	7.629E+01	1.9043	2.935E-01	4.900E+02	2.041E+01	1.4986	5.500E-02
1.333E+02	7.500E+01	1.8992	3.118E-01	5.000E+02	2.000E+01	1.5015	6.700E-02
1.370E+02	7.300E+01	1.8911	3.433E-01	5.099E+02	1.961E+01	1.4993	7.500E-02
1.418E+02	7.053E+01	1.8741	3.880E-01	5.200E+02	1.923E+01	1.4968	7.600E-02
1.449E+02	6.900E+01	1.8568	4.193E-01	5.299E+02	1.887E+01	1.5010	7.200E-02
1.479E+02	6.760E+01	1.8330	4.508E-01	5.501E+02	1.818E+01	1.5171	7.900E-02
1.503E+02	6.655E+01	1.8034	4.766E-01	5.800E+02	1.724E+01	1.5329	1.070E-01
1.525E+02	6.558E+01	1.7686	4.873E-01	5.900E+02	1.695E+01	1.5334	1.160E-01
1.546E+02	6.467E+01	1.7371	4.899E-01	5.999E+02	1.667E+01	1.5322	1.230E-01
1.568E+02	6.378E+01	1.7066	4.890E-01	6.101E+02	1.639E+01	1.5330	1.250E-01
1.600E+02	6.250E+01	1.6662	4.779E-01	6.200E+02	1.613E+01	1.5381	1.280E-01
1.633E+02	6.125E+01	1.6141	4.671E-01	6.398E+02	1.563E+01	1.5508	1.420E-01
1.639E+02	6.100E+01	1.5992	4.509E-01	6.502E+02	1.538E+01	1.5587	1.520E-01
1.667E+02	6.000E+01	1.5851	3.902E-01	6.601E+02	1.515E+01	1.5667	1.640E-01
1.687E+02	5.929E+01	1.5960	3.517E-01	6.798E+02	1.471E+01	1.5792	1.980E-01
1.712E+02	5.840E+01	1.6240	3.170E-01	6.998E+02	1.429E+01	1.5803	2.460E-01
1.740E+02	5.746E+01	1.6849	2.835E-01	7.102E+02	1.408E+01	1.5732	2.710E-01
1.754E+02	5.700E+01	1.7233	3.056E-01	7.199E+02	1.389E+01	1.5628	2.940E-01
1.773E+02	5.640E+01	1.7501	3.376E-01	7.299E+02	1.370E+01	1.5490	3.150E-01
1.794E+02	5.574E+01	1.7648	3.771E-01	7.402E+02	1.351E+01	1.5333	3.350E-01
1.818E+02	5.500E+01	1.7674	4.203E-01	7.502E+02	1.333E+01	1.5165	3.540E-01
1.844E+02	5.424E+01	1.7486	4.707E-01	7.599E+02	1.316E+01	1.4962	3.740E-01
1.869E+02	5.350E+01	1.7206	4.883E-01	7.698E+02	1.299E+01	1.4717	3.890E-01
1.896E+02	5.275E+01	1.6981	5.070E-01	7.800E+02	1.282E+01	1.4448	4.030E-01
1.950E+02	5.128E+01	1.6571	5.292E-01	8.000E+02	1.250E+01	1.3857	4.220E-01
1.999E+02	5.003E+01	1.6296	5.433E-01	8.197E+02	1.220E+01	1.3231	4.220E-01
2.050E+02	4.878E+01	1.6188	5.654E-01	8.403E+02	1.190E+01	1.2582	4.090E-01
2.083E+02	4.800E+01	1.6038	6.250E-01	8.598E+02	1.163E+01	1.2020	3.790E-01
2.111E+02	4.736E+01	1.5532	6.790E-01	8.803E+02	1.136E+01	1.1478	3.410E-01
2.141E+02	4.671E+01	1.4649	7.091E-01	9.001E+02	1.111E+01	1.1065	2.800E-01
2.167E+02	4.615E+01	1.3866	6.839E-01	9.091E+02	1.100E+01	1.0925	2.480E-01
2.198E+02	4.550E+01	1.3278	6.362E-01	9.200E+02	1.087E+01	1.0873	2.040E-01
2.243E+02	4.458E+01	1.2817	5.731E-01	9.302E+02	1.075E+01	1.0908	1.680E-01
2.295E+02	4.358E+01	1.2417	5.247E-01	9.398E+02	1.064E+01	1.1013	1.340E-01
2.339E+02	4.276E+01	1.2217	4.642E-01	9.497E+02	1.053E+01	1.1181	1.080E-01
2.431E+02	4.114E+01	1.2086	3.906E-01	9.597E+02	1.042E+01	1.1370	8.800E-02
2.525E+02			3.960E+01	9.699E+02	1.031E+01	1.1553	7.500E-02
2.615E+02	3.824E+01	1.1964	2.746E-01	9.804E+02	1.020E+01	1.1715	6.500E-02
2.703E+02	3.700E+01	1.2023	2.229E-01	1.000E+03	1.000E+01	1.1991	5.100E-02
2.806E+02	3.564E+01	1.2166	1.761E-01				
2.900E+02	3.448E+01	1.2366	1.350E-01				
3.000E+02	3.333E+01	1.2642	1.060E-01				

$$n + i\kappa = \sqrt{\varepsilon} = \sqrt{\varepsilon' + i\varepsilon''}$$

$$\varepsilon' = n^2 - \kappa^2; \quad \varepsilon'' = 2n\kappa; \quad \alpha = 4\pi v\kappa$$

Source: Reproduced from Warren [49].

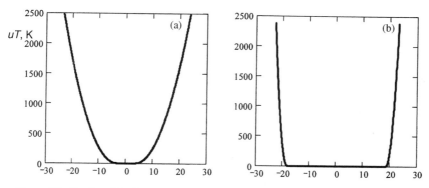

Figure 21 Profile of the hat potential expressed in degrees of K. Calculation for ice at the temperature −7°C (a) and for water at 27°C (b).

curve typical for water, which exhibits maxima near 200 and 700 cm^{-1}. Correspondingly, the loss-factor spectrum $\varepsilon''(v)$ comprises two maxima in the right side of Fig. 20d. On the left side the low-frequency Debye maximum arises. The Cole–Cole plot $\varepsilon''(\varepsilon')$ is shown for water in Fig. 20e, where the Debye semicircle dominates. In Fig. 21d the high-frequency part of the Cole-Cole plot is depicted in more detail.

Since the structures of ice and water are similar, the ice spectrum (Fig. 20a,d) resembles in main features the water spectrum (Figs. 19c,d) but differs from it in the following:

(a) The librational band is substantially narrower, its center is located at $v_{\mathrm{L}} \approx 830\ \mathrm{cm}^{-1}$ instead of $\approx 700\ \mathrm{cm}^{-1}$ in the case of water.

(b) An additional low-intensity band (which we term the V-band) arises with the center at $\approx 150\ \mathrm{cm}^{-1}$ located near the T-band with the center at $\approx 220\ \mathrm{cm}^{-1}$. Note that in the case of liquid water the V-band is not resolved, since it merges with the translational band.

(c) As shown in Section D (see Fig. 26), at lower frequencies (in the submillimeter wavelength region) the ice spectrum becomes nonresonant: The loss factor ε'' monotonically diminishes when the frequency v decreases, with the dielectric constant ε' being frequency-independent ($\varepsilon' \approx 3.15$).

(d) With further decrease of frequency the $\varepsilon''(v)$ dependence goes through a minimum located in the interval of wavelengths λ from 10 to 100 cm.

(e) At still lower frequencies (namely, in the kHz range), the loss factor ε'' exhibits the Debye maximum.

First we consider in detail only the above-mentioned dispersion ranges (a)–(c), while the range (d) comprising the minimum-loss point and the Debye-dispersion

TABLE X
Parameters of the Molecular Model Fitted/Estimated for Ice at $-7°C$ and for Water at $27°C$

Water at $27°C$

$\varepsilon_s = 77.6$	$\rho = 0.996$ g cm^{-3}	$\tau_D = 7.85$ ps	$n_\infty^2 = 1.7$
$k_\mu = 1.08$	$\mu_0 = 1.84$ D	$\beta = 23°$	$f = 0.8$
$\tau_{or} = 0.12$ ps	$u = 8.0$	$r_{or} = 65\%$	$\mu_{or} = 2.45$ D
$m_{or} = 2.3$	$k = 14{,}560$ dyne cm^{-1}	$\tau_q = \tau_\mu = 0.076$ ps	$v_q = 165$ cm^{-1}
$\mu_q = 7.0$ D	$\mu_\mu = 6.35$ D	$c_{rot} = 0.12$	$k_\perp = 4370$ dyne cm^{-1}
$\tau_\perp = 0.1$ ps	$p_\perp = 0.65$	$g_\perp = 3.9$	

Ice at $-7°C$

$\varepsilon_s = 94$	$\rho = 0.92$ g cm^{-3}	$\delta_\infty = 1.9$	$k_\mu = 1.06$
$\mu_0 = 1.45$ D,	$\beta = 23.5°$	$f = 0.15$	$\tau_{or} = 0.06$ ps
$u = 9.3$	$r_{or} = 70\%$	$\mu_{or} = 1.52$ D	$\theta_\tau = 9.9°$
$k = 22{,}480$ dyne cm^{-1}	$\tau_q = 0.23$ ps	$v_q = 205$ cm^{-1}	$\mu_q = 6.2$ D
$\mu_\mu = 4.2$ D	$c_{rot} = 0.16$	$\tau_\mu = 0.34$ ps	$k_\perp = 6740$ dyne cm^{-1}
$\tau_\perp = 0.024$ ps	$p_\perp = 0.3$	$g_\perp = 8$	

Note: The following values of molecular constants are used in calculations: $M = 18$, $I_{or} = 1.483 \times 10^{-40}$ g $\times$ cm^2, $I_{vib} = 3.38 \times 10^{-40}$ g $\times$ cm^2, $s_{lim} = 0.2$.

range (e) will be mentioned only in passing. We shall return to a very wide range of frequencies, where the loss $\varepsilon''(v)$ goes through a minimum, in Section F.

We shall also not consider the spectra pertinent to ice modifications differing from ice I$_h$ (the former exist at lower temperatures and/or at higher pressures).

It follows from comparison of Figs. 20a,b with Figs. 20c,d and also from Table X that transition from water to ice leads to:

(i) narrowing of the librational and translational bands and their shift to higher frequencies;

(ii) increase of the longitudinal (k), transverse ($k_\perp$), and rotational (c_{rot}) force constants;

(iii) quadrupling of the vibration lifetimes τ_q and τ_μ, and diminution of the transverse-vibration lifetime $\tau_\perp$;

(iv) rapprochement of the hat-potential profile and the parabolic one, and correspondingly to decrease of the form factor f.

It is interesting to estimate the mean angular turn θ_τ of a rigid dipole, occurring in the hat potential during the lifetime time τ_{or}. This turn is characterized by the number m of librations

$$\theta_\tau = m\langle\theta_m\rangle \tag{65}$$

θ_m is the libration amplitude in the absence of collisions and the angle bracket denotes ensemble averaging:

$$m = \tau_{or}/(\eta_{or}\Phi) = (2\Phi y_{or})^{-1} \tag{66}$$

where Φ is the libration period. Since the formfactor f is rather small (in the case of ice the fitted $f \approx 0.15$ as compared with $f = 0.8$ in the case of water), roughly one may replace the hat potential by a parabolic one with the same reduced well depth u. Assuming rotation of a dipole to be one-dimensional, let us write down the law of motion of a rigid dipole using formula (3.129) in VIG:

$$\theta(\varphi) = \left(\sqrt{h}/p_{or}\right)\sin(p_{or}\varphi) \tag{67}$$

where $\varphi = t/\eta$ is the dimensionless lifetime and $p_{or} = \sqrt{u}/\beta$ is the steepness of the hat potential. Then we find that the dimensionless libration period Φ and oscillation amplitude θ_m are, respectively,

$$\Phi = 2\pi/p_{or} \quad \text{and} \quad \theta_m = \sqrt{h}/p_{or} \tag{68}$$

In view of (65) and (66) the turn θ_τ is determined by the well width 2β and by the normalized strong-collision frequency y_{or} as

$$\theta_\tau = \left\langle \sqrt{h}/p_{or} \right\rangle \frac{p_{or}}{2\pi y_{or}} = \left\langle \sqrt{h} \right\rangle (2\pi y_{or})^{-1} \tag{69}$$

To find the mean value $\left\langle \sqrt{h} \right\rangle$ in the interval $h \in [0, u]$ of reduced (divided by $k_B T$) energies h, we take into account the Boltzmann distribution of dipoles over h:

$$\left\langle \sqrt{h} \right\rangle = \int_0^u \sqrt{h}\exp(-h)\,dh \bigg/ \int_0^u \exp(-h)\,dh$$

Calculating this integral, we derive the following expression after substitution into Eq. (69):

$$\theta_\tau \approx \left[\sqrt{\pi}\mathrm{erf}\left(\sqrt{u}\right) - 2\sqrt{u}\exp(-u)\right]\{4\pi y_{or}[1 - \exp(-u)]\}^{-1} \tag{70}$$

Estimation shows that the angular turn θ_τ of a rigid dipole during the lifetime τ_{or} in the parabolic (in our case) potential is *less* than the angular half-width β of this potential. Indeed, in accord with (70) $\theta_\tau(\text{ice}) \approx 9.9°$. Therefore this quantity is less than $<2\beta \approx 23.5°$.

We remark that the potential widths 2β are almost equal in the cases of water and ice. A rigid dipole, rotating in the hat potential, lives in water longer than in ice, since during the lifetime τ_{or}(water) such a dipole performs about two librations ($m_{or} \approx 2$), while during τ_{or}(ice) it performs about a half of one libration.

We do *not* consider the low-frequency spectra for ice, since the contribution to complex permittivity of rigid reorienting dipoles is calculated from the simplified expression (A29), which is applicable only in the high-frequency approximation. Indeed, the ice permittivity is found for $v > 0.1 \text{ cm}^{-1}$ (see Figs. 20a,b and 24a), while for liquid water Eq. (4) is used, applicable also in the relaxation region.

Other distinctions from water of the employed calculation scheme, described in Appendix IV, consist of the following.

(i) We introduce for ice *two* vibration lifetimes τ_q and τ_μ relevant, respectively, to motion of a nonrigid dipole along the H-bond and to reorientation of a rigid dipole about it ($\tau_q < \tau_\mu$; see Table IX). On the other hand, noting that T- and V-bands overlap in water, we take τ_q(water) $= \tau_\mu$(water).

(ii) Since in the case of ice we employ the high-frequency approximation for a librating permanent dipole, we involve in Eq. (A28) the constant δ_∞, found from the condition that the dielectric constant ε' slightly exceeds unity the minimum point of the $\varepsilon'(v)$ curve (see Fig. 22a). In our example ($T_C = -7°\text{C}$) the constant δ_∞ is close to the optical permittivity n_∞^2. Indeed, we see in Table IX that $n_\infty^2 = 1.7$ and $\delta_\infty = 1.9$.

We demonstrate for ice and water the frequency dependence $\varepsilon'(v)$ of the dielectric constant calculated in the far-IR region (Fig. 22a). In Fig 22b the Cole–Cole diagram $\varepsilon''[\varepsilon'(v)]$ is also shown in this region. The theoretical dependences agrees reasonably with the experimental ones. The latter, shown by open circles, are obtained from data of Warren [49] and Downing and Williams [22].

2. Discussion

First let us compare the librational mode found for ice by Marchi [8] using lattice mode spectroscopy (see Fig. 18) with our smooth "water-like" absorption dependence illustrated by the right-hand part of Fig. 20a. This spectrum evidently agree much better with the experimental data than the noise-like pattern calculated by the MD method and depicted in the lower part of Fig. 18. It is important that in the work by Marchi [8] "the dipole-induced-dipole terms provide the dominant contribution to the total intensities." However, in view of a recently MD modeling of water spectra by Sharma et al. [29], the polarization effects depend on the environment in a strong *intermolecular* way, which possibly was not taken into account in the cited treatment by Marchi.

In the frequency range $400–1000 \text{ cm}^{-1}$ the $\alpha(v)$ and $\varepsilon''(v)$ ice spectra are determined mostly by reorientation of rigid dipoles in the hat potential whose

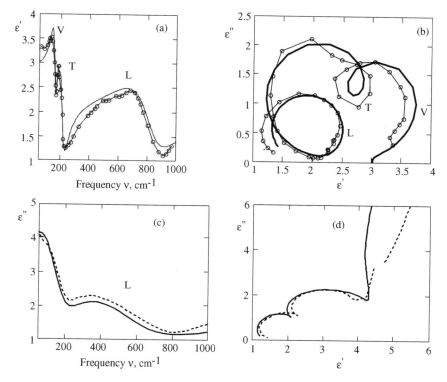

Figure 22 Frequency dependence of dielectric constant ε' in the far-IR region, calculated for ice at $-7°C$ (a) and for water at $27°C$ (c) and the relevant Cole–Cole diagrams pertinent to ice (b) and water (d). Open circles refer to the experimental data by Warren [49] and Downing and Williams [22]. Notation V, T, L refer, respectively to the V-, T-, and L-bands.

profile is depicted in Fig. 21a. Our successful description in ice of the librational band (shown in the right-hand parts of Figs. 20a and 20b), as well as in water (see parts L of Figs. 20c and 20d) represents, perhaps, the main achievement of the proposed modeling.

Molecules, reorienting in the hat well, belong to the main fraction of the fluids under consideration. This fraction comprises, in both water and ice, about 70% of all molecules. An underlying molecular mechanism, which governs the librational band, can be interpreted as follows. During the mean lifetime τ_{or}, a dipole performs in the hat well on average about two (in water) and a half (in ice) libration cycles. In the case of *water* this motion is rather free, since a large part of the hat well constitutes a flat bottom (see Fig. 21b). In the case of *ice*, librations are more restricted. Indeed, as seen in Fig. 21a, the potential bottom is rather narrow.

It appears that in the structure, corresponding to the hat potential, hydrogen bonds are strongly bent or broken. Correspondingly, the law of motion of a dipole, governed by such a potential, substantially differs from the law determined by a harmonic-like potential.

The lifetime τ_{or} can be identified with the time, during which a *neutral* rigid polar molecule participates in a *cooperative* motion. This cooperation is provided mainly due to a fast reorientation of protons inside a rather stable tetrahedral-like set of oxygen atoms. It is appropriate to recall the assertion by Robinson et al. [13] that "a key to the understanding of dynamical processes in liquid water may be the stability, on characteristic time scales, of the five-membered quasitetrahedral arrangement of the oxygen atoms. While the oxygen atoms form a more or less static background, water molecules, through mainly hydrogen motions, carry out librations and hindered rotations on much faster time scales. Such motions of the hydrogen atoms are strongly dependent *in a cooperative way* on the behavior of other water molecules in the general vicinity, that is to say, whether these molecules are rotating or librating. It is this last feature that lowers the librational frequency[13] and activation barrier with increasing temperature (or pressure)." When a "favorable" configuration arises, a proton jump with attached water molecule to another one may occur as an activationless quantum process (a qualitative description of such a process is given in Appendix I).

Returning to a rather free libration of a dipole in the hat well, we remark that a curved part of the well's bottom is characterized by the form factor f and by the *spread* $= (1 - f) r_{H_2O} \beta$, where $r_{H_2O} \approx 1.5$ Å is the radius of a water molecule. The case of ice, in which $f \approx 0.15$ and *spread* ≈ 0.52 Å, substantially differs from the case of water, in which $f \approx 0.8$ and *spread* ≈ 0.12 Å. Hence, short-range interactions of H_2O molecules are revealed in ice at longer distances than in water.

Second, let us consider in more detail the vibration mechanisms of H-bonded molecules governed by elastic force constants. In view of our rough two-fractional description of water spectra, the T- and V-bands appear mainly due to mechanisms b and c and partly due to mechanism d—namely:

- Mechanism b, responsible for the T-peak depicted in Figs. 20a and 20b, concerns elastic oscillation of two charged water molecules along the H-bond; this oscillation is governed by the longitudinal force constant k.

- Mechanism c, responsible for the V-peak shown in the same Figures, concerns elastic reorientation of a rigid dipole about the H-bond; this motion is governed by the dimensionless angular force constant c_{rot}.

---

[13]Such a decrease of the librational frequency is demonstrated in Section IV in a wide temperature interval 20–60°C (except the break-of-continuity point near 27°C).

- Mechanism d concerns nonelastic vibration of a nonrigid dipole in a direction perpendicular to a nondisturbed H-bond, with such vibration being governed by the transverse force constant $k_\perp$.

In Figs. 23a and 23b we show contributions of various mechanisms, respectively, to the absorption and loss of ice in the frequency region 50–300 cm^{-1}, where V- and T-bands are placed.

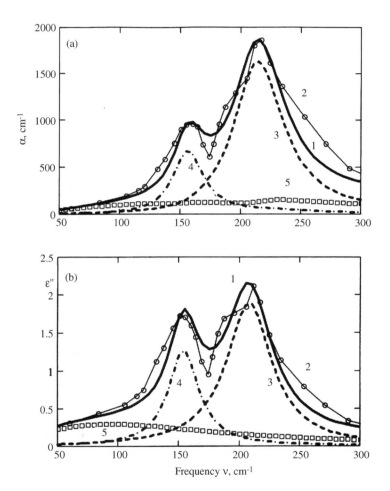

Figure 23 Absorption (a) and loss (b) frequency dependences in T- and V-bands calculated for ice at the temperature −7°C. Solid lines (1) refer to the total absorption or loss, open circles (2) refer to the experimental data by Warren [49], and lines 3–5 refer to contributions to spectra due to elastic translation along the H-bond (3), elastic reorientation about the H-bond (4), and transverse vibration (5).

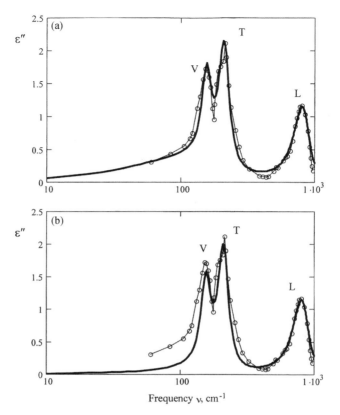

Figure 24 Comparison of the loss spectra calculated for ice with (a) and without (b) account of transverse vibration. Temperature −7°C.

To understand the role of mechanism d, we compare in Figs. 24a and 24b the loss spectra of ice calculated, respectively with and without account of mechanism d. We see that the latter substantially modifies the *left wing* of the V-band. As will be demonstrated in Section D, the transverse vibration *dominates* at still lower frequencies (in the nonresonance region).

For ice, unlike for water, we use slightly *different* lifetimes τ_q and τ_μ, which pertain to elastic vibrations and are ascribed, respectively, to the mechanisms (b) and (c), namely, $\tau_q = 0.23$ ps and $\tau_\mu = 0.34$ ps. If $\tau_q = \tau_\mu$, the calculated spectrum shows worse agreement with experiment. The proximity of the lifetimes τ_q and τ_μ indicates that *elastic* translation and reorientation of H-bonded molecules are determined by the same (or by a very similar) state of molecular structure, characterized by elastic constants k and c_{rot}.

On the other hand, in the case of ice a great difference between the lifetimes τ_q and τ_μ and the fitted hat-well lifetime τ_{or} indicates a cardinal difference between

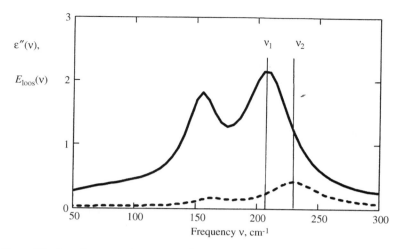

Figure 25 Loss frequency dependences $\varepsilon''(v)$ (solid curve) and energy loss function $E_{loss}(v)$ (dashed curve) calculated for ice at $-7°C$.

the elastic-vibration mechanisms (b, c) and the nonelastic reorientation mechanism (a). Similarly, mechanisms b,c and d substantially differ due to the large distinction of times τ_q and τ_μ from the transverse-vibration time $\tau_\perp$.

It appears that without involving both mechanisms b and c, pertinent to elastic vibration, it would be difficult to calculate and interpret the double 200-cm^{-1} band in ice. Note that recent studies (Amo and Tominaga [26], Klug et al. [50]; Nielsen et al. [27]) with use of Raman spectroscopy and incoherent inelastic neutron scattering reveal, unlike FIR spectroscopy, existence also *in water* of different bands in the region 50–300 cm^{-1}.

Now let us estimate (see Fig 25) the so-called transverse optic–longitudinal optic splitting [8,50] characteristic for ice at $v \approx 230$ cm^{-1}. Namely, the loss curve $\varepsilon''(v)$ is shifted on the frequency scale with respect to the "energy loss function"

$$E_{loss}(v) \equiv \text{Im}[-\varepsilon(v)^{-1}] \tag{71}$$

It follows from Fig. 25 that the right solid and right dashed peaks are located, respectively, at $v_1 \approx 207$ and $v_2 = 230$ cm^{-1}, so the above-mentioned shift $v_2 - v_1$ comprises about 23 cm^{-1}. Comparison with Fig. 19, calculated by Marchi [8] for the very low temperature (100 K), shows a similar picture of this shift but in terms of a rather 'noise-like' pattern.

C. Nonresonance Spectrum

If the radiation frequency v comprises several tens of cm^{-1} or less, the ice spectrum loses its resonance character. In Fig. 26a we represent, by solid line 1,

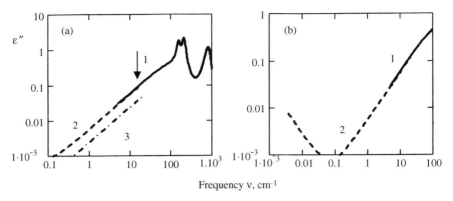

Figure 26 The loss frequency dependence, calculated for ice from the composite molecular model (1) and from the empirical formula (72) with the fitting coefficients $c_{fit} = 2.35$ (line 2) and with $c_{fit} = 1$ (line 3). The temperature $- 7°C$. The arrow depicts the point placed at $v = 20\,cm^{-1}$, where the lines 1 and 2 have equal loss values.

the loss spectrum of ice at $-7°C$ calculated in the resonant and nonresonant regions. The dashed line 2 depicts the same $\varepsilon''(v)$ dependence, provided by the empirical formula by Hufford [20]

$$\varepsilon = c_{fit}\left[\alpha_F(T_C)F^{-1} + \beta_F(T_C)F\right] \tag{72}$$

in terms of frequency F expressed in GHz. Here the parameters α_F and β_F (also expressed in GHz) are given by

$$\alpha_F = (50.4 + 62\Theta) \times 10^{-4}\exp(-22.1\Theta), \quad \beta_F = (0.445 + 0.00211T_C) \times 10^{-4}$$

$$+\frac{0.585 \times 10^{-4}}{(1 - T_C/29.1)^2} \tag{73}$$

T_C is the temperature in $°C$, Θ is the dimensionless inverse temperature,

$$\Theta = \left[300(273.15 + T_C)^{-1} - 1\right] \tag{74}$$

and c_{fit} is the fitting coefficient of the order of unity.

We introduce this coefficient with the purpose of equality at some frequency, say at $v = 20\,cm^{-1}$, we identify the loss ε'', found for our composite model (see solid line 1), to the loss estimated from Eq. (72). This procedure gives us the dashed line 2 found for the coefficient $c_{fit} = 2.35$. Note in the original cited work by Hufford the coefficient $c_{fit} = 1$. In Fig. 26a the dash–dotted line 3 corresponds to the latter c_{fit} value.

We see from Fig. 26a (solid line 1) that the loss spectrum, calculated for our model with the same parameters, as chosen above (Table IX), exhibits *resonance lines* at the frequencies $v < 50\ \mathrm{cm}^{-1}$. At $v < 20\ \mathrm{cm}^{-1}$ the calculated solid loss curve 1, becoming nonresonant, coincides with the nonresonant dashed curve 2 calculated from Eqs. (72)–(74) with $c_{fit} = 2.35$. Both loss ε'' curves 1 and 2 decrease linearly with v (in the log–log plot) in the interval from 50 to $\approx 0.1\ \mathrm{cm}^{-1}$. For further decrease of frequency the empirical dependence (72) exhibits a minimum at v about $0.1\ \mathrm{cm}^{-1}$ (viz, in the millimeter wavelength region). Near this minimum and at lower frequencies, our molecular model should not be applied.

In the low-frequency range the Debye formula

$$\varepsilon' + i\varepsilon'' = (\varepsilon_0 - \varepsilon_1)(1 - iF_D g_D^{-1})^{-1} + \varepsilon_1 \tag{75}$$

for the complex permittivity could be used instead of (72)–(74), where the frequency F_D and the parameter g_D are expressed in kHz. According to Hufford [20], the temperature dependence of the inverse reduced relaxation time g_D and of the static permittivity ε_0 are given by the following empirical relationships:

$$g_D = 64.1\exp(-22.1), \qquad \varepsilon_0 = 81.8 + 96, \varepsilon_1 = 3.15 \tag{76}$$

where Θ is given by (74). So, the Debye relaxation time τ_D may be estimated as

$$\tau_D \approx (10^3 \times 2\pi g_D)^{-1} \tag{77}$$

In Figs. 27a and 27b we show the low-frequency dependences of the dielectric constant ε' and the loss ε'', calculated from Eqs. (75) and (80). Here the solid lines

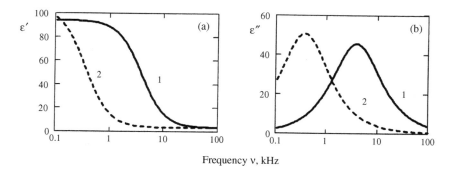

Figure 27 Low-frequency (Debye) spectra $\varepsilon'(v)$ (a) and $\varepsilon''(v)$ (b) of ice calculated from the formulas (75) and (76) for the temperatures $-7°\mathrm{C}$ (1) and $-30°\mathrm{C}$ (2).

refer to the temperature $-7°C$ and the dashed lines to $-30°C$. The loss maxima in Fig. 27b, referring to

$$\tau_D(-7°C) = 4.1 \times 10^{-5} \text{ s}, \qquad \tau_D(-30°C) = 4.4 \times 10^{-4} \text{ s} \qquad (78)$$

exceed the Debye relaxation times of water, respectively, by 8 and 7 orders of magnitude.

D. Spectrum at Low Temperature

1. Experimental Data for $T = 100$ K

For this calculation we use as the starting point the well known experimental data by Bertie et al. [51]. Since the dielectric constant ε' given in this work is negative in a few frequency points, we correct the above data by adding 2.5 to the value ε' represented in Table III of this work. In the frequency range of interest the so-corrected data by Bertie et al. [51] are presented in Table XI. They are also shown in Fig. 28 by solid lines. The original (uncorrected by us) $\varepsilon'(v)$ dependence is depicted in Fig. 28a by the dashed line.

2. Parameterization of the Model

Comparing Figs. 28 and 20, we see that at low temperature (100 K) the T-band maximum loss ε''_m is about twice that, measured at 266 K, while the maximum absorption coefficients α_m, observed in the L- and T-bands, are

TABLE XI
Experimental Dielectric Parameters of Ice I_h at the Temperature 100 K Used for Comparison with the Results of Our Calculations

v, cm^{-1}	α, cm^{-1}	ε'	ε''	v, cm^{-1}	α, cm^{-1}	ε'	ε''
1000	399	3.91	0.0756	225	3210	3.92	3.74
950	2053	3.71	0.383	220	2380	4.8	3.02
900	3615	3.88	0.778	210	1688	5.37	2.32
850	4052	4.17	1.02	200	1248	5.36	1.75
800	3300	4.71	1	190	1123	5.41	1.66
750	1535	4.93	0.51	180	723	5.37	1.10
700	950	4.78	0.327	170	965	5.04	1.50
600	471	4.66	0.184	160	1051	5.32	1.84
500	112	4.53	0.0509	140	620	6.07	1.35
400	112	4.28	0.0595	120	260	6.14	0.661
300	525	3.74	0.313	100	153	5.87	0.447
250	1414	3.49	0.983	80	108	5.76	0.388
240	1679	3.20	1.12	60	62.9	5.81	0.304
230	3610	2.14	2.73	40	18	5.8	0.130
229	3772	2.28	3.21	30	9	5.7	0.0855
							5

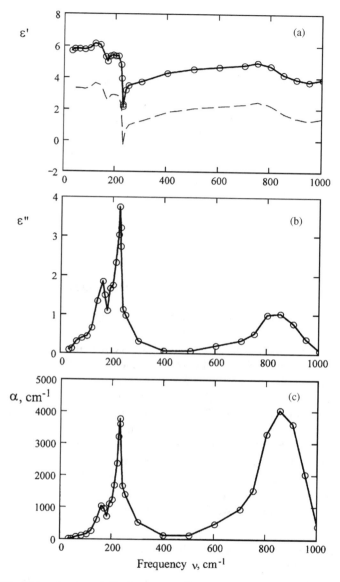

Figure 28   Dielectric spectra of ice I_h measured at the temperature 100 K by Bertie et al. [51] with correction of dielectric constant ε' explained in the text. The original uncorrected $\varepsilon'(v)$ dependence is shown in part a by the dashed line.

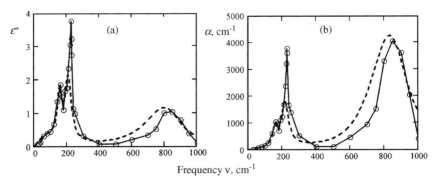

Figure 29 Comparison of the loss (a) and absorption (b) ice spectra measured at the temperature 100 K (solid lines) and at 266 K (dashed lines).

approximately equal:

$$\varepsilon_m''(100\,\text{K}) \approx 2\varepsilon_m''(266\,\text{K})\ in\ the\ T\text{-band} \tag{79a}$$

$$\alpha_m''(100\,\text{K}) \approx \alpha_m(266\,\text{K})\ in\ the\ L\text{-band} \tag{79b}$$

In Fig. 29 we compare the experimental ice spectra at these two temperatures: The solid lines refer to 100 K and dashed lines to 266 K. The relationship (79a) could in principle be satisfied if (i) we increase the proportion r_{vib} of vibrating particles as compared with $r_{\text{vib}}(266\,\text{K})$ given in Table X. Next, to avoid the decrease of the absorption $\alpha_m(100\,\text{K})$, occurring due to decrease of the proportion $r_{\text{or}} \equiv 1 - r_{\text{vib}}$ of the particles, reorienting in the hat well, (ii) we may increase their lifetime τ_{or}. Then the libration band will become narrower and its absorption maximum α_m will increase. This may restore the condition (79b). Narrowing of the L-band also corresponds to the observed distinction of the dashed curves from the solid, seen in the right parts of Figs. 29a and 29b.

To achieve these results, we do the following:

(1) We set $r_{\text{vib}}(100\,\text{K}) = 45\%$, while $r_{\text{vib}}(266\,\text{K}) \approx 30\%$. Hence, *the concentration of almost freely rotating molecules (librators) falls as i freezes out due to a decrease in temperature.*

(2) We make the libration lifetime about two times longer: $\tau_{\text{or}}(100\,\text{K}) \approx 0.12$ ps, while $\tau_{\text{or}}(266\,\text{K}) \approx 0.055$ ps. It is interesting that even after a very intense cooling the rotation lifetime still remains rather short (an order of magnitude less than 1 p).

To parameterize in this low-T case the reduced well depth u, we assume that the hat-well depth U_{or} is almost independent of T. Since $U_{\text{or}} \equiv uT$, we roughly may estimate u as

$$u(\text{low } T) \approx \frac{u(\text{higher } T) \, T(\text{higher})}{T(\text{low})} \tag{80a}$$

so that

$$u(100 \text{ K}) \approx \frac{u(266\text{K}) \, 266}{100} \tag{80b}$$

Since $u(266 \text{ K}) \approx 8.5$ (see Table X), (80b) gives $u(100 \text{ K}) \approx 22.6$. However, a better agreement with experiment is obtained for $u(100 \text{ K}) \approx 20$. Therefore, our criterion that $U_{\text{or}} \approx \text{const}(T)$ holds.

To find the transverse-vibration (TV) intensity factor $g_\perp$, we should have the loss factor $\varepsilon'' \approx 0.1$ near the high-frequency boundary ν_{bond} of the nonresonance region ($\nu_{\text{bond}} \approx 30 \text{ cm}^{-1}$). The parameterization yields $g_\perp(100 \text{ K}) \approx 2$. Note that this value is much less than $g_\perp$ at higher $T = 266 \text{ K}$, since $g_\perp(266 \text{ K}) \approx 8$ (see Table X). Thus, *a lowering of the ice temperature leads to a decrease of the association of polar H_2O molecules*.

The parameters of the model fitted for $T = 100 \text{ K}$ are presented in Table XII (compare the latter with Table X). We also see from Table XII that for our parameterization we have the following characteristics:

(i) considerable (by a factor of about three) reduction of the longitudinal/transverse force constants and

(ii) still more reduction (by a factor of about five) of the TV frequency parameter $p_\perp$.

Item (i) means that *the hydrogen-bond elasticity strongly decreases with cooling*, and the item is relevant to the high-frequency shift of the transverse HB-vibration and to the increase of its intensity (see Fig. 30).

TABLE XII
Parameters of the Molecular Model Fitted/Estimated for Ice at 100 K

$\varepsilon_s = 94$	$\rho = 0.92 \text{ g cm}^{-3}$	$\delta_\infty = 4.3$	$k_\mu = 1.1$
$\mu_0 = 1.45 \text{ D}$	$\beta = 20°$	$f = 0.12$	$\tau_{\text{or}} = 0.12 \text{ ps}$
$u = 20$	$r_{\text{or}} = 55\%$	$\mu_{\text{or}} = 1.6 \text{ D}$	$\theta_\tau = 13.2°$
$k = 22,480 \text{ dyn cm}^{-1}$	$\tau_q = \tau_\mu = 0.37 \text{ ps}$	$\nu_q = 220 \text{ cm}^{-1}$	$\mu_q = 5.3 \text{ D}$
$\mu_\mu = 3.3 \text{ D}$	$c_{\text{rot}} = 0.16$	$k_\perp = 6740 \text{ dyn cm}^{-1}$	$\tau_\perp = 0.013 \text{ ps}$
$p_\perp = 0.08$	$g_\perp = 2$		

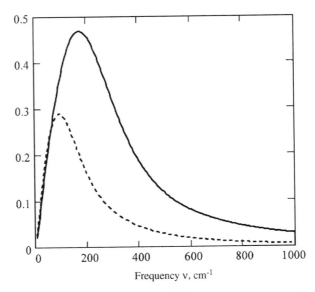

Frequency v, cm^{-1}

Figure 30 Frequency dependence of transverse vibration loss pertinent to the HB molecules. Calculation for ice at 100 K (solid curve) and 266 K (dashed curve).

3. The Calculated Spectra

The computed absorption and loss factor are shown, respectively, in Figs. 31a and 31b; Fig. 31c represents the low-frequency wing of the $\varepsilon''(v)$ curve characteristic for the non-resonance loss spectrum. Fig. 32 represents the frequency dependence of the dielectric constant $\varepsilon'(v)$ and the Cole–Cole plot $\varepsilon''[\varepsilon'(v)]$. The latter two graphs also agree with experiment. We see that our theory agrees reasonably well with the spectra observed by Bertie et al. [51]. Note the empirical formulas (72)–(74) by Hufford [20] could be applied for describing of the far IR ice loss spectrum (viz., at $v < 100$ cm^{-1}), if the constant c_{fit} in Eq. (72) is fitted properly (see Fig. 33). For $T = 100$ K $c_{\text{fit}} \approx 17$.

In Fig. 34 we depict the energy loss function $E_{\text{loss}}(v)$ defined by Eq. (71). This function is calculated for the temperature 100 K with the parameters of the model presented in Table X. The splitting $\Delta v_E = v_2 - v_1$, comprising about 13 cm^{-1}, is close to the experimental value [8]. This splitting is substantially less than at the temperature 266 K (cf. Figs. 34 and 25).

E.   Prediction of Ice Spectrum at $-30°$C

Let us introduce the dimensionless inverse temperature $\theta(T) \equiv 1000\, T^{-1}$ and denote two particular values of θ as $\theta_1 \equiv 1000\, T_1^{-1}$ and $\theta_2 \equiv 1000\, T_2^{-1}$, where $T_1 = 100$ K and $T_2 = 266$ K are the temperatures considered above. We seek the

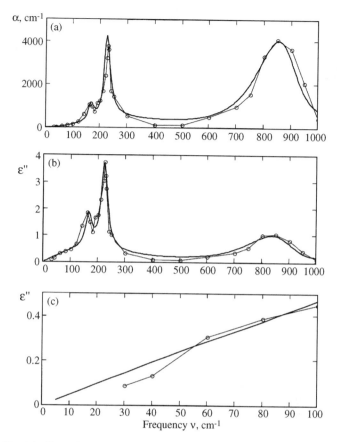

Figure 31 (a) Absorption and (b) loss. (c) Ice spectra calculated (solid lines) for the temperature 100 K. Open circles refer to the experimental data by Bertie et al. [51].

temperature dependence $X(\theta)$ of any chosen parameter X of our model, with X_1 and X_2 being denoted as the values of X at T_1 and T_2, which are known (fitted in the previous study). To predict the dielectric spectra in terms of our theory, we should find the $X(\theta)$ values for a set of our parameters, with θ being any chosen θ. In this section we take $T = 273.16 - 30 \approx 243$ K and $\theta = 1000/T \approx 4.1$. For simplicity we determine $X(\theta)$ using the proportionality relationship, namely, the equation

$$\frac{X(\theta) - X_1(\theta)}{X_2(\theta) - X_1(\theta)} = \frac{\theta - \theta_1}{\theta_2 - \theta_1} \tag{81}$$

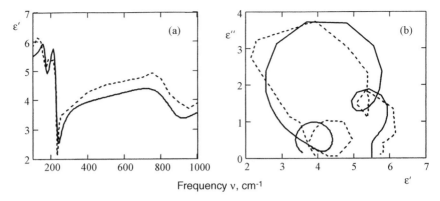

Figure 32   Frequency dependence of the dielectric constant ε' and the Cole–Cole plot $\varepsilon''[\varepsilon'(v)]$ of ice at $T = 100$ K. Designations as in Fig. 30.

The first step. We find (see Table XIII) from Eq. (81) the set $X(\theta)$ of the parameters, which determine the dimensionless arguments of the spectral functions involved in our theory. These are:

For mechanism a, the parameters: u, β, f, d, y_{or}.

For mechanism b, c, the parameters: k, c_{rot}, v_q, y_q, y_μ .

For mechanism d, the parameters: $k_\perp$, u_{lim} ,w, s_{lim}, $y_\perp$. $\eta_\perp$, $p_\perp$, $g_\perp$.

The second step. For mechanisms a–c we calculate all other parameters, which determine the dielectric spectra, using, in particular, the results of the first step and the formulas, given in Appendix IV. The predicted spectra, shown in Fig. 35,

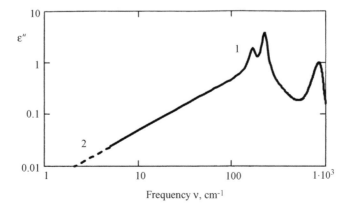

Figure 33   Wideband loss frequency dependence calculated for ice, $T = 100$ K (curve 1). Curve 2 depicts the empirical relationship (72), in which the fitted parameter $c_{fit} = 17$.

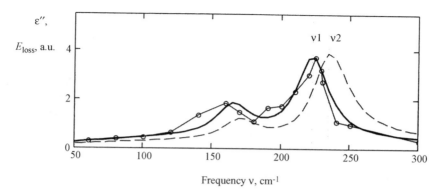

Figure 34 The same as in Fig. 24 but for the temperature 100 K.

resemble the spectra calculated above for 266 K (Figs. 20a,b and 22a,b) and for 100 K (Figs. 31 and 32).

F. Loss Gap in Ice Spectrum

The empirical formula (72)–(73) predicts a deep "loss gap" comprising two orders of magnitude, where the loss factor ε'' varies from 0.1 to 0.01. However, the term 'gap' looks to be conditional, since the change of the frequency v comprises five orders (!) in this gap.

The experimental data by Warren confirm this behavior. As seen from Fig. 36, when the temperature decreases from -1 to $-60°C$, then:

(i) the minimum loss value ε''_{min}, attained in the bottom of the gap, strongly decreases and

(ii) the point of this minimum shifts to lower frequencies.

In Table XIV we partially reproduce the Warren's data for the temperatures $-1°C$, $-5°C$, $-20°C$, and $-60°C$. As is seen from this table, the dielectric constant ε' exerts extremely low change for the wavenumber v varying from 0.001

TABLE XIII
Parameters Pertinent to the Temperature $-30°C$ (Obtained from Parameterization of the Model at the Temperatures 100 and 266 K)

Mechanism a Librational Band							Mechanism b T-Band	
β (deg)	u	f	y_{or}	r_{or}	k_μ	δ_∞	y_q	v_q (cm^{-1})
23.3	9.16	0.15	0.805	0.69	1.07	2	0.3	206

Mechanism c V-Band			Mechanism d THz-Band		
Y_μ		c_{rot}	$k_\perp$	p	$g_\perp$
0.206		0.16	6504	0.287	7.66

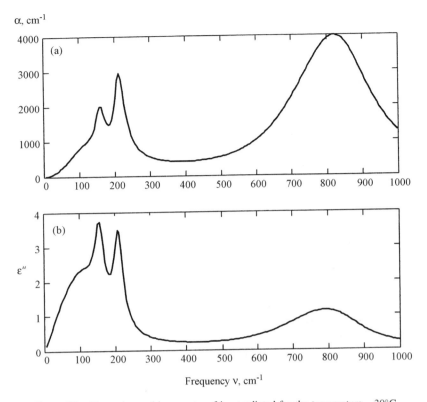

Figure 35 Absorption and ice spectra of ice predicted for the temperature −30°C.

to 50. However, in accord with the Kramers–Cronig rule, this change produces a strong frequency dependence of the loss factor in the wavelength range extending from 10 m to 0.1 mm. If the factor c_{fit} is chosen properly, the empirical formula (72)–(74) qualitatively describes the profile of the $\varepsilon''(v)$ gap (c_{fit} decreases with the fall of the temperature). However, as is seen in Fig. 36b, the difference between the solid line, representing the experiment, and the dashed line, representing empirical estimation, is noticeable.

Although the loss gap presents a phenomenon specific for ice, a certain analogy still exists between ice and water with respect to a slow variation of the dielectric constant ε' and loss ε'' in a wide frequency region. Thus, Figs. 37a and 37b demonstrate how slow the variation of water parameters ε'' and ε' occurs with change of v in the THz range from 10 to 100 cm^{-1}.

In Fig. 37c we mark by curves 1–4 the contributions to this variation of ε'' due to

(i) libration of a polar molecule in an intermolecular hat well (1),

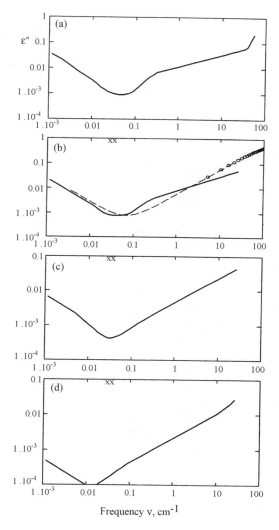

Figure 36 The experimental loss spectrum of ice, represented from Warren [53] for the temperatures $-1°C$, $-5°C$, $-20°C$, $-60°C$ (solid lines, from top to bottom). In part b the dashed line depicts estimation of ε'' from (72) and (73) with $c_{fit} = 2.35$ for the temperature $-7°C$ and open circles depict calculation of ε'' from the composite model for the same temperature.

(ii) vibration of a nonrigid dipole along the H-bond (2),

(iii) reorientation of a permanent HB dipole about this bond (3),

(iv) vibration of a nonrigid dipole in a direction perpendicular to the H-bond (4).

TABLE XIV
Real (n) and Imaginary (κ) Components of Refractive Index of Ice

Frequency	Temperature $-1°C$		Temperature $-5°C$		Temperature $-20°C$		Temperature $-60°C$	
v (cm^{-1})	n	κ	n	κ	n	κ	n	κ
1.163E-3	1.7880	9.70E-3	1.7872	9.70E-3	1.7845	9.70E-3	1.7780	9.70E-3
5.000E-3	1.7872	1.95E-3	1.7865	1.95E-3	1.7840	1.95E-3	1.7772	1.95E-3
1.111E-2	1.7872	8.20E-4	1.7865	8.20E-4	1.7837	8.20E-4	1.7772	8.20E-4
1.250E-2	1.7872	6.92E-4	1.7865	6.92E-4	1.7837	6.92E-4	1.7772	6.92E-4
1.389E-2	1.7872	5.95E-4	1.7865	5.95E-4	1.7837	5.95E-4	1.7772	5.95E-4
1.562E-2	1.7872	5.00E-4	1.7865	5.00E-4	1.7837	5.00E-4	1.7772	5.00E-4
2.000E-2	1.7872	3.82E-4	1.7865	3.82E-4	1.7837	3.82E-4	1.7772	3.82E-4
2.500E-2	1.7872	3.15E-4	1.7865	3.15E-4	1.7837	3.15E-4	1.7772	3.15E-4
2.857E-2	1.7872	2.89E-4	1.7865	2.89E-4	1.7837	2.89E-4	1.7772	2.89E-4
3.448E-2	1.7872	2.64E-4	1.7865	2.64E-4	1.7837	2.64E-4	1.7772	2.64E-4
5.000E-2	1.7872	2.49E-4	1.7865	2.49E-4	1.7837	2.49E-4	1.7772	2.49E-4
5.882E-2	1.7872	2.58E-4	1.7865	2.58E-4	1.7837	2.58E-4	1.7772	2.58E-4
6.667E-2	1.7872	2.70E-4	1.7865	2.70E-4	1.7837	2.70E-4	1.7772	2.70E-4
7.692E-2	1.7872	2.94E-4	1.7865	2.94E-4	1.7837	2.94E-4	1.7772	2.94E-4
9.009E-2	1.7872	3.40E-4	1.7865	3.40E-4	1.7837	3.40E-4	1.7772	3.40E-4
1.429E-1	1.7871	6.95E-4	1.7865	6.95E-4	1.7838	6.95E-4	1.7772	6.95E-4
2.000E-1	1.7870	1.16E-3	1.7865	1.16E-3	1.7840	1.16E-3	1.7772	1.16E-3
2.500E-1	1.7869	1.50E-3	1.7865	1.50E-3	1.7840	1.50E-3	1.7772	1.50E-3
3.125E-1	1.7866	1.88E-3	1.7866	1.88E-3	1.7840	1.88E-3	1.7772	1.88E-3
1.000	1.7852	3.24E-3	1.7852	3.24E-3	1.7838	3.24E-3	1.7773	3.24E-3
4.000	1.7830	6.21E-3	1.7830	6.21E-3	1.7822	6.21E-3	1.7789	6.21E-3
1.000E+01	1.7816	9.54E-3	1.7816	9.54E-3	1.7814	9.54E-3	1.7807	9.54E-3
1.585E+01	1.7820	1.18E-2	1.7820	1.18E-2	1.7820	1.18E-2	1.7820	1.18E-2
1.995E+01	1.7832	1.32E-2	1.7832	1.32E-2	1.7832	1.32E-2	1.7832	1.32E-2
2.512E+01	1.7860	1.47E-2	1.7860	1.47E-2	1.7860	1.47E-2	1.7860	1.47E-2
3.163E+01	1.7921	1.62E-2	1.7921	1.62E-2	1.7921	1.62E-2	1.7921	1.62E-2
3.549E+01	1.7983	1.76E-2	1.7983	1.76E-2	1.7983	1.76E-2	1.7983	1.76E-2
3,981E+01	1.8070	2.24E-2	1.8070	2.24E-2	1.8070	2.24E-2	1.8070	2.24E-2
4.466E+01	1.8172	3.14E-2	1.8172	3.14E-2	1.8172	3.14E-2	1.8172	3.14E-2
5.01E+01	1.8275	4.56E-2	1.8275	4.56E-2	1.8275	4.56E-2	1.8275	4.56E-2

Source: Reproduced from Warren [49].

We should remark that no one of the mechanisms previously elaborated for water enables analytical description of the above fall of $\varepsilon''(v)$ for the case of ice. This phenomenon probably characterizes the transition from *collective motions* in a fluid typical for Debye relaxation, characterized by the relaxation time τ_D, to *vibration of individual molecules*, characterized by much shorter lifetimes τ_{or}, τ_q, τ_μ. The latter times have the same order of magnitude in water and ice, but τ_D drastically differs in these fluids. That is why the theory of far-IR spectra is rather similar in ice and water. On the contrary, it appears that the molecular theory pertaining to the low-frequency ice spectra (which still is not elaborated)

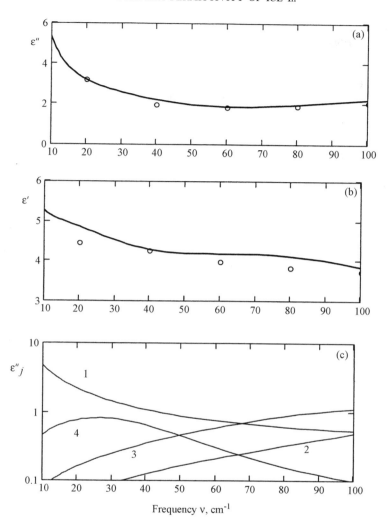

Figure 37 Frequency dependences of the loss factor (a) and of the dielectric constant (b) calculated (solid curve) and measured (open circles) in water. (c) Contributions to loss due to: libration of a permanent dipole in the hat well (1), vibration of a nonrigid dipole along the H bond (2), reorientation of polar molecules about this bond (3), and transverse vibration of a nonrigid dipole with respect to the H bond (4). Temperature 300 K.

should strongly differ from the Debye theory of molecular diffusion in liquids. Hence,

  (i) extremely small variation of ε' with ν, demonstrated by Table XIV, cannot now be described on physical grounds and

(ii) extremely slow variation of ε'' with ν in the loss gap is only roughly described by empirical formulas (72)–(74).

APPENDIX IV. COMPLEX PERMITTIVITY OF ICE: LIST OF FORMULAS

A. Spectral Functions

The *librational* SF L_{or} (z_{or}), where for simplicity we write $z_{or} \equiv z$, is found as an integral over the dimensionless energy h of rigid dipoles-librators, with h being less than the reduced hat-well depth u:

$$L_{or}(z) = \left(\frac{2}{\pi}\right)^5 \frac{4u\,(\beta f)^4}{D} \int_0^u S(h,z)\,[1 + \sigma(h)]^7 \exp(-h)\,h\,dh \qquad (A14)$$

$$S(h,z) = \sum_{n=1}^{\infty} \frac{\sin^2\left[\dfrac{2n-1}{2}\dfrac{\pi}{1+\sigma(h)}\right]}{(2n-1)^2\left[(2n-1)^2\sigma^2 - (1+\sigma)^2\right]^2\left[(2n-1)^2h - (2z\beta f/\pi)^2(1+\sigma)^2\right]} \tag{A15}$$

$$\sigma \equiv \sigma(h) = \frac{\pi}{2}\frac{1-f}{f}\sqrt{\frac{h}{u}}, \quad D = \frac{4\beta^2 fud}{\pi}\int_0^u\left[f + \frac{\pi}{2}(1-f)\sqrt{\frac{h}{u}}\right]\exp(-h)\,dh \quad (A16)$$

Here β is the angular half-width of the hat well, f is its form factor, and d is the fitted constant ($d = 0.7$).

The dimensionless complex frequency is given by

$$z = z_{or} = x_{or} + iy_{or}, \qquad x_{or} = \omega\eta_{or}, \qquad \omega = 2\pi\nu, \qquad y_{or} = \eta_{or}/\tau_{or} \quad (A17)$$

where x_{or} and ω are the dimensionless and angular radiation frequency, τ_{or} is the lifetime of the hat well, and y_{or} is the frequency of strong collisions.

The timescale, connecting the reduced and physical quantities, is

$$\eta_{or} = \sqrt{I_{or}(2k_BT)^{-1}} \tag{A18}$$

where $I_{or}(H_2O) = 1.483 \times 10^{-40}$ is the moment of inertia of a free rotating water molecule represented as a symmetric top. The permanent moment μ_{or} of a librating dipole is expressed through that (μ_0) of such a molecule, the fitted dipole-moment factor k_μ (close to unity), and the optical permittivity n_∞^2 by

$$\mu_{or} = k_\mu \mu_0 (n_\infty^2 + 2)/3 \qquad (A19)$$

The simplified *translational and orientational spectral functions* L_q and L_μ, pertinent to elastic vibration of H-bonded molecules, are represented each in the form of a Lorentz line:

$$L_q(z_q) = \frac{E}{B} \frac{f_q^2}{\left(f_q^2 - z_q^2 \right)}, \qquad L_\mu(z) = \frac{E}{A} \frac{p_q^2}{\left(p_q^2 - z_\mu^2 \right)} \qquad (A20)$$

Here the constants are expressed in terms of the temperature T, the dimer's dimensions (a, l), and the translational (k) and dimensionless angular (c_{rot}) force constants:

$$p_q^2 = \frac{18A(1-a)^2}{71(1+a^2) - 2a}, \quad f_q^2 = \frac{16Ba^2(1+a^2)}{71(1+a^2) - 2a}, \quad A = \frac{2kc_{rot}l^2a^2}{k_B T(1-a)^2}, \quad B = \frac{kl^2}{2k_B T} \qquad (A21)$$

where $a = r_{OH}/l$ is the ratio of the covalent and equilibrium hydrogen bonds.

The E-factor is determined by the H-bond energy U_{HB} as

$$E = 1 - \frac{u_{lim}e^{-u_{lim}}}{1 - e^{-u_{lim}}}, \qquad u_{lim} \equiv U_{HB}(RT)^{-1} \qquad (A22)$$

where $R \approx 0.002 \text{ kcal mol}^{-1} \text{ deg}^{-1}$ is the gas constant and $U_{HB} \approx 2.6 \text{ kcal mole}^{-1}$. The corresponding complex frequencies are defined as

$$z_q = \eta_{vib}(\omega + i\tau_q^{-1}) = 2\pi c v \eta_{vib} + i y_q \text{ for elastically translating charges} \qquad (A23)$$

$$z_\mu = \eta_{vib}(\omega + i\tau_\mu^{-1}) = 2\pi c v \eta_{vib} + i y \text{ for elastically reorienting rigid dipoles} \qquad (A24)$$

$$\eta_{or} = \sqrt{I_{vib}(2k_B T)^{-1}} \quad I_{vib}(H_2O) = 2m_H r_{OH}^2 \qquad (A25)$$

where m_H is the mass of the hydrogen atom.

The SF of the *transverse vibrators* $L_\perp(z_\perp)$ is determined by the integral

$$L_\perp(z_\perp) = 3w \int_0^{s_{lim}} \frac{(18Dws^2 - z_\perp^2)s^6 \exp(-ws^4)\, ds}{12w^2s^4 + z_\perp^4} \Bigg/ \int_0^{s_{lim}} s^2 \exp(-ws^4)\, ds \qquad (A26)$$

where $s_{lim} \approx 0.2$, $D \equiv [2\mathbf{E}(1/\sqrt{2})/\mathbf{K}(1/\sqrt{2})] - 1 \approx 0.627$, and $\mathbf{K}(\cdot)$ and $\mathbf{E}(\cdot)$ are the full elliptic integrals of the, respectively, first and second kind.

The transverse complex frequency $z_\perp$ is defined as

$$z_\perp = x_\perp + iy_\perp, \quad \text{where } x_\perp = \omega\eta_\perp, \quad y_\perp = \eta_\perp/\tau_\perp, \quad \eta_\perp \equiv p_\perp(l+r)\sqrt{\frac{m}{4k_BT}} \quad (A27)$$

where $m \equiv m(H_2O)$ is the mass of the water molecule [not to be confused with m_H in (A25)], and $p_\perp$ is the fitting constant of the order of unity, which shifts the frequency $v_\perp$, that is, the position of the transverse-loss maximum.

B. Formulas for the Complex Permittivity of Ice

$$\varepsilon'(v) = \text{Re}[\Delta\varepsilon_{or}(v) + \Delta\varepsilon_q(v) + \Delta\varepsilon_\mu(v) + \Delta\varepsilon_\perp(v)] + \delta_\infty \quad (A28)$$

$$\varepsilon''(v) = \text{Im}[\Delta\varepsilon''_{or}(v) + \varepsilon''_q(v) + \varepsilon''_\mu(v) + \varepsilon''_\perp(v)] \quad (A28)$$

$$\Delta\varepsilon_{or}(v) = 6\pi GL_{or}(z_{or}), \qquad \varepsilon''_{or}(v) = 6\pi G \ \text{Im}[L_{or}(z_{or})],$$
$$G = \frac{\mu^2_{or}(1-r_{vib})N}{3k_BT}, \qquad N = \frac{N_A\rho}{M} \quad (A29)$$

$$\varepsilon_q(v) + \varepsilon_\mu(v) = 6\pi\left[G_qL_q[z_q(v)] + G_\mu L_\mu[z(v)]\right] \quad (A30)$$

$$\Delta\varepsilon_\perp(v) = g_\perp G_\perp L_\perp(z_\perp) \quad (A31)$$

where

$$G_\perp = \pi(k_BT)^{-1}\mu^2_\perp r_{vib}N \quad (A32)$$

and where N_A is the Avogadro number; ρ is the density; M is the molecular mass; the constant δ_∞ is found from the condition that the total dielectric constant ε' at its minimum (near the end of the far-IR region) should slightly exceed unity; and $g_\perp$ is the fitting constant, which relates squares of longitudinal and transverse elastic dipole moments as $\mu^2_\perp = g_\perp\mu^2_q$ and determines the association factor as

$$\zeta = \sqrt{g_\perp} \quad (A33)$$

The total absorption coefficient $\alpha(v)$ and partial coefficients $\alpha_j(v)$ pertinent to various specific mechanisms are given by

$$\alpha(v) = \frac{2\pi v}{n(v)}\varepsilon''(v) \quad \text{and} \quad \alpha_j(v) = \frac{2\pi v}{n(v)}\varepsilon''_j(v), \quad j \equiv \text{or, q, } \mu, \perp \quad (A34)$$

where the refractive index is given by

$$n(v) = \text{Re}\{[\varepsilon'(v) + i\varepsilon''(v)]^{1/2} \quad (A35)$$

APPENDIX V. CLASSICALLY ESTIMATED MINIMAL ACTION AND PHASE TRANSITIONS IN ICE AND WATER

Here we continue the estimates given in Appendix III.

In view of Table XII, at a very low temperature (100 K) the lifetime $\tau_q \approx 0.37$ ps. Hence, the decrease of temperature from $-7°C$ to 100 K leads to the increase of the parameter A_{break} in (A7) from $6.23h$ to $10h$. Consequently, application of classical theory for calculation of the far IR ice spectra has at least the same permissibility as in the case of water at room temperature.

On the other hand, if we set in (A11) the minimum possible value A_{break}, equal to h, then we obtain the following estimate for the minimum bandwidth:

$$\Delta v_{min} \approx \Delta h c \bar{v}^2 (U_{HB})^{-1} \tag{A36}$$

We take for the T-band the mean frequency $\bar{v}$ equal to 220 cm^{-1}; then Eq. (A36) yields

$$\Delta v_{min} \approx 3 \times 10^{10} \times 220^2 \ 6.7 \times 10^{-27} \ (2.6 \times 6.95 \times 10^{-14})^2 \approx 54 \, cm^{-1} \tag{A37}$$

In view of Fig. 28b or 28c, we find that the limiting linewidth (A37) is rather close to the bandwidth $\Delta v_{T\text{-band}}$ observed (or estimated) at the level ½ at $T = 100$ K.

From our data, obtained above for the temperatures $-7°C$, $-30°C$, and 100 K, we show in Fig. 38a that the bandwidth $\Delta v_{T\text{-band}}$ decreases with cooling of ice. However, at very low temperature this parameter practically does not change with T. Our estimates show that at a given pressure the *criterion (A36) roughly corresponds to the phase transition of ice I_h to other ice modification.*

Let us turn now to water. We shall compare the classically estimated in (A8) action A_{break} with its limiting value h. We redraw Fig. 13 in terms of the inverse temperature. Then we see from Fig. 38b that the ratio A_{break}/h gradually decreases

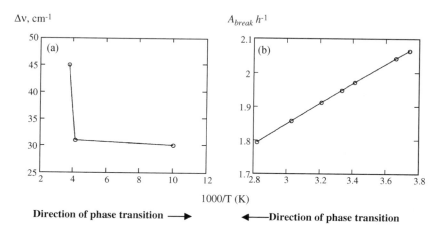

Figure 38 Temperature dependence of the bandwidth $\Delta v_{T\text{-band}}$ calculated for ice (a) and the ratio A_{break}/h calculated for water (b).

with the rise of temperature. *We assume that the theoretical limit $A_{break}/h = 1$ could probably be connected with the liquid–vapor phase transition.*

Since our estimates are based on fitting the model parameters using the theory, which is far from rigorous, we would like to outline here only a tendency of molecular events to be provoked by the change of temperature. Concluding this discussion, we suggest the following:

The vapor–water phase transition occurs when classically determined action (A8) approaches to the minimal vallue h due to too short lifetime τ_q of longitudinally vibrating dipoles. On the other hand, the transition of ice I_h to other ice modification occurs when the translational bandwith $\Delta v_{-t-band}$ approaches to classically estimated limit (A36)   (A38).

VI. ELASTIC LONGITUDINAL VIBRATION OF A NONRIGID AND ELASTIC REORIENTATION OF RIGID DIPOLES[14]

A. Main Assumptions of Two-Fraction Model. Scheme of a Dimer

First we consider in more detail our two-fraction model of water. Absorption of electromagnetic radiation arises due to interaction of an external electric field **E** with the medium in which radiation propagates. In our model the medium is composed (i) of polar molecules represented as reorienting permanent (rigid) dipoles and (ii) of oppositely charged molecules represented as vibrating nonrigid dipoles (dimers), (i) and (ii) represent, respectively, the fractions LIB and VIB. Water, as compared with other polar fluids, is characterized by a large permanent dipole moment μ of an individual water molecule and by a still greater time-varying moment of a nonrigid dipole (the latter property follows from our estimations). Electromagnetic radiation is absorbed in water mainly due to two processes.

The first process prevails at relatively low frequencies. The electric component **E** of radiation orients dipole moments μ along the field direction, while chaotic molecular motions hinder this orientation; μ and **E** are the vectors, and the field **E** is assumed to vary harmonically with time t. Due to inertia of reorienting molecules the time dependence of the polarization *lags behind* the time dependence **E**(t), so that heating of the medium occurs (the heating effect is not considered in this work). The dielectric spectrum obeys the Debye relaxation, for which the absorption monotonically increases with frequency.

The second process concerns the *resonance-like* absorption of radiation, when the ac frequency is close to the mean frequency of dipole oscillations. In liquids such a resonance usually rather degrades due to strong collisions of a dipole with the surrounding medium. These collisions influence the full energy of a particle. We suggest that the particle's velocities have Boltzmann distribution. In our theory

---

[14]The material of this Section VI is based on the work by Tseitlin et al. [52]

we involve also collisions with the "walls" of an intermolecuiar potential well. These collisions are assumed to change only the direction of a particle velocity.

A distinction of water from other polar fluids is determined mainly by a large dipole moment μ of a water molecule and by H-bonds connecting them. These bonds constantly break and restore. One isolated water molecule may be termed *monomer*. Two water molecules, connected by the HB, form a *dimer*. For the mean H-bond lifetime τ_{HB}, namely for the dimer H-bond lifetime, our estimation gives the value about 0.1 ps. Note that in the literature, for the HB lifetime a greater value (near 1 ps) is often given.

We deal with the two-fraction model, whose dielectric properties are described by four molecular mechanisms (a–d), which are illustrated in Fig. 39a.

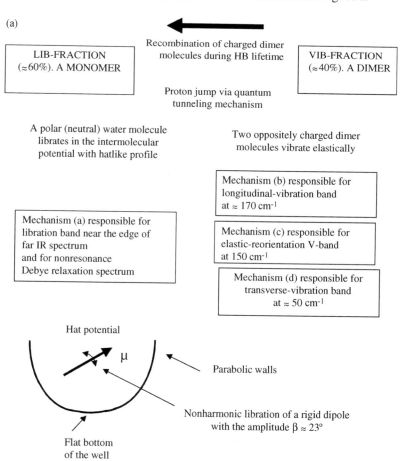

Figure 39 Scheme of a mixed (two-fraction) model applied to water (a); dimer scheme (b); orientation with respect to the high-frequency field direction (c); and diagram of the linear velocities of the hydrogen atoms (d).

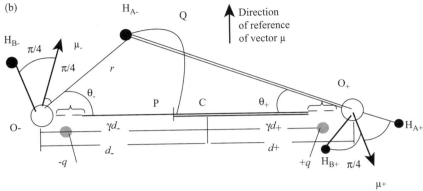

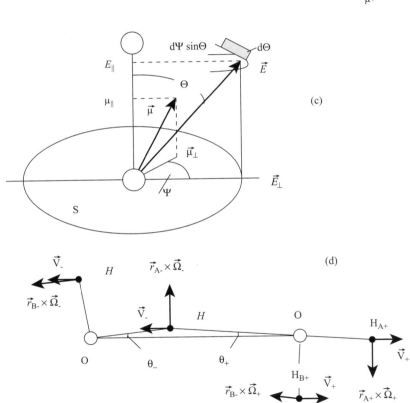

Figure 39 (*Continued*).

The LIB fraction (the fraction of monomers) containing about 55–70% of all water molecules comprises a permanent dipole, namely, a neutral molecule librating in a tight surrounding of other molecules in a condensed medium. This motion, governed by mechanism a, is responsible for the librational band, placed near the boundary of the IR region, and for the low-frequency Debye relaxation band, placed at microwaves. The Debye spectrum is not considered in this work.

The VIB fraction, containing $\approx$30–45% of water molecules, comprises a dimer. Time variation (vibration) of HB lengths and directions is of interest in this work, since it determines the dielectric and Raman spectra of water in the vicinity of the T-band. The dimer vibration and its response are ruled by mechanisms b–d.

On the straight line between two oxygen atoms, depicted in Fig. 39b, a hydrogen atom is placed. It forms the covalent bond with its own (left) O atom and the hydrogen bond with the foreign (right) O atom. The left and right water molecules, to which both oxygen atoms belong, are assumed to be *oppositely charged* (negatively left and positively right) due to the *proton jump* occurring via the LIB$\rightarrow$VIB transition. This jump is assumed to happen via quantum tunneling, when the hydrogen atoms moving chaotically acquire dispositions favorable. Our estimate shows that for such a jump in water the "waiting" time comprises for such a jump a few parts of a picosecond. In the course of the reverse VIB$\rightarrow$LIB transition, occurring during as $\approx$0.1 ps, the recombination of water ions takes place.

This picture of molecular events implies that a noticeable concentration (about 40%) of short-lived charged dimer molecules is commensurable with concentration of all water molecules. Hence, microscopically small *charged* regions of the fluid are assumed to be constantly substituted by the regions composed *of neutral* water molecules, and vice versa.

Response of the longitudinal vibration, directed along the equilibrium HB, is described by mechanism b; mechanism c describes the response arising due to a turn of polar HB molecules about this bond. Mechanisms b and c lead to dielectric loss, respectively, in the T- and V-bands (the V-band is placed near the T-band at a little lower frequency). These mechanisms also determine the intense 170-cm^{-1} band in the Raman spectrum (RS). For simplicity, both these bands often are termed the "T-band" in spite of their somewhat different origin.

Vibration of charged molecules, transverse to the equilibrium HB direction, which leads to the HB bending, is governed by mechanism d. The latter affects the low-frequency wing of the V-band. We regard this mechanism as that responsible for the 50-cm^{-1} peak of the Raman spectrum.

Hence, we may summarize the above brief introduction as follows:

1. The librational (LIB) fraction is attributed to an *individual* neutral polar H2O or D$_2$O molecule *(monomer)*, reorienting nonharmonically in tight surrounding of water medium. The molecule is represented as a rigid dipole

having form of a symmetric top. The spectrum of the LIB fraction is described by the *mechanism (a)*.

2. The vibration (VIB) fraction is represented as a *dimer* of oppositely charged molecules. The spectrum of the VIB fraction is described by *mechanisms b, c, and d*. The first two are attributed to *harmonic* elastic nonrigid-dipole translation of charges coupled with rigid-dipole reorientation. In a good approximation, employed in this work for calculation of water spectra, these types of motion are regarded as *uncoupled*. Mechanism d is attributed to elastic *nonharmonic* vibration of a nonrigid dipole in direction, perpendicular to the H-bond.

3. After a short period, when a neutral monomer exists, practically instantaneous LIB→VIB transition occurs due to a proton jump. This period results by appearance of two charged particles, forming a dimer. Inverse VIB→LIB transition reveals as a recombination of two charges, which takes place roughly during a short period, when the hydrogen bond exists. Thus, in each point of aqueous medium, a charged form of matter is constantly replaced by neutral form, as illustrated by Fig. 39a.

In Section VI we study in detail two fast short-lived vibration mechanisms b and c, which concern item 2. The dielectric response to the elastic rotational vibrations of hydrogen-bonded (HB) polar molecules and to translational vibrations of charges, formed on these molecules, is revealed in terms of two interrelated Lorentz lines. A proper force constant corresponds to each line. The effect of these constants on the spectra of the complex susceptibility is considered. The dielectric response of the H-bonded molecules to elastic vibrations is shown to arise in the far IR region. Namely, the translational band (T-band) at the frequency v about 200 cm^{-1} is caused by vibration of charges, while the neighboring V-band at v about 150 cm^{-1} arises due to elastic rigid-dipole reorientations. In the case of water these bands overlap, and in the case of ice they are resolved due to longer vibration lifetime.

Mechanism d, also referring to the VIB fraction, is studied in Section VII.

As far as it concerns the mechanisms b and c, the equilibrium and strained states of a dimer are shown in Fig. 39b. The second state differs from the first one, because:

(i) the centers of the mass of molecules vibrate along the horizontal line O_-O_+, on which negatively and positively charged oxygen atoms are located, and

(ii) the O_-H_{A-} covalent bond and O_+H_{A-} hydrogen bond rotate about this line, respectively, through angles θ_- and θ_+.

The left molecule and quantities, relevant to it, are denoted by index A, and the right molecule and quantities, relevant to it, are denoted by index B. Thus, the hydrogen bond is strained due to elastic forces, accompanying its stretching-compression and rotation. The left molecule is negatively and the right molecule positively charged[15].

We assume the following:

1. The centers of mass of the water molecules and oxygen atoms coincide. The distances between the centers of mass and charge centers vary with time. These distances are denoted, respectively, by d and γd. Note that γ is set to be constant ($\gamma < 1$).

2. The molecules, constituting the dimer, are in the same plane and rotate in it.

3. The centers of mass of the molecules, as well as the charges, move only in the equilibrium direction of the hydrogen bond, so that the O_-O_+ line does not change its direction.

4. The angles, formed by the covalent bonds, are right; for the equilibrium direction of the hydrogen bond, the O_+H_{A+} and O_-O_{A-} bonds are on the same line with it.

5. The masses of both molecules are the same; we neglect the difference between these masses due to different particles, "attached" to the left and right oxygen atoms.

The reference point for radius vectors is placed at the point C (see Fig. 39b), where the center of mass of the dimer is located (i.e., according to assumption 5) in the middle of the equilibrium distance $r + l$ between the oxygen atoms. Here r is the length of the covalent bond O—H and l the length of the hydrogen bond *at equilibrium*. The centers of mass of the molecules vibrate along the symmetry axis coinciding with the equilibrium H-bond direction. This vibration is marked by the dashed lines and by curly brackets. The radius vectors, pertaining to these molecules, are denoted as $\vec{d}_\pm$.

We assume that the dimer represents a closed system, so that its momentum is conserved. It follows from the foregoing and assumption 5 that the linear velocities and displacements of the left and right molecules are equal in absolute value and opposite in direction. The equilibrium position of the hydrogen atom H_{A-} is denoted by point P on the O_-O_+ axis. The O_-H_{A-} covalent bond deflects through the angle θ_- with respect to the symmetry axis. The H-bond is elongated by Δl and rotates through the angle θ_+ about the symmetry axis (see Fig. 39b). It follows from assumption 4 that the O_+H_{A+} covalent bond rotates through the angle $(-\theta_+)$ so that at each time this bond takes a position on the same line with

---

[15] It should be mentioned that appearance of charges on the ends of a nonrigid dipole has probably essentially quantum origin, as generally noticed by Sharma et al. [53]. The relevant covalent-bond deformations are not shown in Fig. 39b; these bonds are assumed to be the same as those of isolated molecules.

the hydrogen bond for any direction of the latter. The line with the arrows, connecting point P with point Q and connecting point Q with the resulting position of the atom H_{A-}, refers to the rotation and translation displacements of this atom.

B. The Autocorrelator Transformation

A spectroscopically active isotropic medium is treated as a medium consisting of rigid dipoles and of *pairs* of hydrogen-bonded molecules. In *each* pair the latter are oppositely charged. The point is that it is difficult to represent the resulting spectral function $L(z)$ as the *sum of separate* spectral functions (SF), one of which corresponds to positively and the other to negatively charged water molecules. It is more correct to seek the SF directly *for the pair*. This feature reflects collective interactions arising in our molecular system.

The number of such pairs per unit volume of the medium is $N_{vib}/2$, where N_{vib} is the concentration of molecules, experiencing elastic vibrations (the index 'vib' denotes the quantities relevant to such molecules).

We introduce the normalized concentration

$$G_{vib} = \frac{1}{2} \frac{\mu^2 N_{vib}}{3k_B T} \tag{82}$$

where μ is the permanent dipole moment of an HB water molecule (which generally differs from the rigid moment μ_{or} of a molecule librating in the hat well), k_B is the Boltzmann constant, and T is the temperature. The 'vibration' (i.e., pertaining to vibrations) complex susceptibility of the medium $\chi = \chi' + i\chi''$ will be found *in the high-frequency approximation*. According to Eq. (3.5) in GT1, in this approximation the following relation holds:

$$\chi = G_{vib} L(z) \tag{83}$$

$$z = x + iy, \qquad x = \omega \eta_{vib}, \qquad y = \eta_{vib}/, \tau_{vib}, \qquad \eta_{vib} = \sqrt{I_{vib}(2k_B T)^{-1}} \tag{84}$$

where z is dimensionless complex frequency, I_{vib} is the moment of inertia, τ_{vib} is the H-bond lifetime, and ω is the angular radiation frequency.

In our variant of the response theory the "spectral function" (SF) $L(z)$ of the model is linearly related to the *spectrum* of the dipole autocorrelation function (ACF).

As the authors of GT1 show, the susceptibility $\chi* = \chi' + i\chi''$ in the high-frequency approximation reduces to $\chi_B^* \equiv 3\langle q^2 \rangle GL(z)$. In our case of an isotropic medium, we have $3\langle q^2 \rangle = 1$. Index B means that the distributions of particle velocities and coordinates at the instants immediately after strong collisions are of Boltzmann type.[16] The calculation of the vibration SF presents the main task solved in Section VI.

In our calculation scheme the quantities χ and $L(z)$ are proportional. We shall show that the frequency dependence (83) represents the sum of *two* Lorentz lines.

The resulting dipole moment is the sum of the moments $q\gamma\vec{d}$ and $\vec{\mu}$. The first is the moment of the *elastic* nonrigid dipole, since this moment depends on the distance d between charges. The second is the moment of rigid dipoles, whose orientation varies in time.

As applied to the problem under study, we shall employ expression (27b) in GT2 for the spectral function, namely,

$$L(z) = 3iz\left\langle Q(0) \int\limits_{0}^{\infty} [Q(\varphi) - Q(0)]e^{iz\varphi}\,d\varphi \right\rangle \tag{85}$$

where $\varphi = t/\eta_{\mathrm{vib}}$ denotes the dimensionless time, angular brackets denote the ensemble average, and instead of q in GT2 we write Q. This quantity is proportional to the projection (denoted by index E) of the resulting dipole moment onto time-varying field $\vec{E}$:

$$Q = (\mu_E + q\gamma d_E)\mu^{-1} \tag{86}$$

Here q is the absolute value of the charge, and the vectors $\vec{d} \equiv \vec{d}_+ - \vec{d}_-$ and $\gamma\vec{d}$ connect, respectively, the centers of mass of the molecules and their charges.

We now average the product $Q(0)Q(\varphi)$ over the orientations of the symmetry axis O_-O_+ (the dimer orientation with respect to the radiation field is shown in Fig. 39b). For this, we express the projections μ_E and d_E in terms of the angles Θ and Ψ, where Θ is the angle between the direction of radiation field $\vec{E}$ and the symmetry axis and Ψ is the angle formed by the projections $\mu_\perp$ and $\vec{E}_\perp$ of the vectors $\vec{\mu}$ and $\vec{E}$ on the plane S, transverse to this axis Using the relation

$$\mu_E = \big[\text{(the projection of } \vec{\mu} \text{ on symmetry axis)}_E$$
$$+ \text{(the projection of } \vec{\mu} \text{ on the plane S)}_{E_\perp}\big]_E$$

and taking assumption 3 into account, we have

$$\mu_E = \mu_\| \cos\Theta + \mu_\perp \sin\Theta \cos\Psi \tag{87}$$

---

[16]For other collision models (e.g., for the Gross model, see GT1) the susceptibility χ is related to the spectral function by a *rational relation* different from (83). This leads to an essential change of the low-frequency spectrum.

We note that

$$d_E = d \cos \Theta \tag{88}$$

The expression for the average over the indicated angles is written, noting that the element of the unit-sphere surface is equal to $\sin \Theta d\Theta d\Psi$ (see Fig. 39b):

$$3\mu^2 \langle Q(0)Q(\varphi) \rangle_{\Theta,\Psi}$$

$$= \langle (\mu_{\parallel} \cos \Theta + \mu_{\perp} \sin \Theta \cos \Psi + q\gamma \, d \cos \Theta)(\mu_{\parallel} \cos \Theta + \mu_{\perp} \sin \Theta \cos \Psi$$

$$+ q\gamma d \cos \Theta)_0 \rangle_{\Theta,\Psi} \tag{89}$$

Since

$$\Theta \in [0, \pi/2], \quad \langle \cos^2 \Theta \rangle = \int_0^{\pi/2} \cos^2 \Theta \sin \Theta \, d\Theta = \frac{1}{3}, \quad \Psi \in [0, 2\pi]$$

$$\langle \sin^3 \Theta \cos^2 \Psi \rangle_{\Theta,\Psi} = \int_0^{\pi/2} \sin^3 \Theta \, d\Theta \int_0^{2\pi} \cos^2 \Psi \frac{d\Psi}{2\pi} = \frac{1}{3}, \quad \langle \cos \Theta \cos \Psi \rangle_{\Theta,\Psi} = 0$$

Eq. (89) reduces to

$$3\mu^2 \langle Q(0)Q(\varphi) \rangle_{\Theta,\Psi} = \langle [\mu_{\parallel}(0) + q\gamma \, d(0)][\mu_{\parallel}(\varphi) + q\gamma \, d(\varphi)] \rangle + \langle \mu_{\perp}(0)\mu_{\perp}(\varphi) \rangle$$

$$= \langle \vec{\mu}(0) \bullet \vec{\mu}(\varphi) \rangle + \langle q\gamma[d(0)\mu_{\parallel}(\varphi) + d(\varphi)\mu_{\parallel}(0)] \rangle + (q\gamma/\mu)^2$$

$$\langle d(0)d(\varphi) \rangle \tag{90}$$

We have to find the difference

$$\mu^2 \langle Q(0)Q(\varphi) - Q(0)^2 \rangle_{\Theta,\Psi} \tag{91}$$

Expression (90) can be transformed by representing the dipole moment as the sum $\vec{\mu} = \vec{\mu}_+ + \vec{\mu}_-$ According to Fig. 39a and assumption 4, we have

$$\mu_{\pm \parallel}/\mu = \cos(\theta_{\pm} + \pi/4), \quad \mu_{\pm \perp}/\mu = \mp \sin(\theta_{\pm} + \pi/4) \tag{92}$$

Summing the quantities with the same projection indices, we find the projections of the rigid dipole moment. Introducing

$$\theta \equiv (1/2)(\theta_- - \theta_+) \tag{93}$$

we have from (92)

$$
\frac{1}{\mu}\left\{ \begin{matrix} \mu_\| \\ \mu_\perp \end{matrix} \right\} = \left\{ \begin{matrix} \cos \\ \sin \end{matrix} \right\} (\theta_- + \pi/4)
$$

$$
+ \left\{ \begin{matrix} \cos \\ -\sin \end{matrix} \right\} (\theta_+ + \pi/4) = 2\cos\left(\frac{\theta_- + \theta_+}{2} + \pi/4 \right) \left\{ \begin{matrix} \cos \\ \sin \end{matrix} \right\} \theta
$$

$$
= \sqrt{2}\left(\cos\frac{\theta_+ + \theta_-}{2} - \sin\frac{\theta_+ + \theta_-}{2} \right) \left\{ \begin{matrix} \cos \\ \sin \end{matrix} \right\} \theta \tag{94}
$$

We now introduce the dimensionless displacement of the oxygen atoms

$$
s \equiv (d - r)/l - 1 \tag{95}
$$

At equilibrium (i.e., for $d = r + l$), this quantity vanishes, as it must.

Assuming that longitudinal and linear displacements are small, we approximate the normalized projection Q of the total dipole moment by the linear function of θ and s. Accordingly, $\cos\frac{1}{2}(\theta_+ + \theta_-)$ is equated to unity, and $\sin\frac{1}{2}(\theta_+ + \theta_-)$ is equated to zero. Thus we have

$$
\mu_0/\mu \approx \sqrt{2}\cos\theta, \qquad \mu_0/\mu \approx \sqrt{2}\sin\theta \tag{96}
$$

The first term in (90) contributes to the autocorrelator equal to the averaged difference $\frac{1}{2}\left[\frac{\bar{\mu}(0)\bar{\mu}(\varphi)}{\mu^2} - 1 \right]$. We obtain the following expression for this quantity:

$$
2\cos\theta(0)[\cos\theta(\varphi) - \cos\theta(0)] + 2\sin\theta(0)[\sin\theta(\varphi) - \sin\theta(0)]
$$
$$
\to 2\theta(0)[\theta(\varphi) - \theta(0)] \tag{97}
$$

where the arrow denotes reduction due to averaging. Indeed, $\langle\cos\theta(\varphi)\rangle \equiv \langle\cos\theta(0)\rangle$. Consequently, the first term vanishes after averaging, while the sine function in the remaining term is replaced by its argument.

Taking (91), into account, we show that after averaging, the second term in (90) vanishes. We denote $\theta_0 \equiv \theta(\varphi)$ and $s_0 \equiv s(\varphi)$. According to (95), we have

$$
d = l(1 + a + s), \qquad a \equiv r/l \tag{98}
$$

Taking (101), into account, we write

$$
\frac{q\gamma}{\mu}[d(0)\mu_\|(\varphi) + d(\varphi)\mu_\|(0) - 2d(0)\mu_\|(0)]
$$

$$
\propto \{(1 + a + s_0)\cos\theta + (1 + a + s)\cos\theta_0 - 2(1 + a + s_0)\cos\theta_0\}
$$

$$
\propto (1 + a)[\cos\theta + \cos\theta_0 - 2\cos\theta_0]_1 + [s_0\cos\theta + s\cos\theta_0 - 2s_0\cos\theta_0]_2
$$

The averaging over s and θ can be performed independently. Having averaged over θ, we find that $[\cdot]_1$ vanishes. After averaging over s, $\cos\theta$ and $\cos\theta_0$ are replaced by unity. Thus, $[\cdot]_2$ vanishes. Our statement is proved.

We note that the same result can also be obtained by taking into account the multiplier $\sin[(\theta_+ + \theta_-)/2]$ in (94).

After ensemble averaging, the autocorrelator, except (97) contains also the term quadratic in s, which is obtained as the result of averaging of the last term in (90):

$$\left(\frac{q\gamma}{\mu}\right)^2 d(0)\langle[d(\varphi) - d(0)]\rangle \rightarrow \left(\frac{q\gamma l}{\mu}\right)^2 \langle[1 + a + s(0)][s(\varphi) - s(0)]\rangle$$

$$\approx \left(\frac{q\gamma l}{\mu}\right)^2 \langle s(0)[s(\varphi) - s(0)]\rangle \qquad (99)$$

because the product $(1 + a)\,[s(\varphi) - s(0)]$ vanishes after averaging.

Thus according to (97) and (99) in our approximation the spectral function is reduced to the sum of the rigid-dipole and nonrigid dipole (charge) components.

Now we modify the obtained formulas, taking into account the fact that the integral absorption is as many times less than the correct one as the number (defined by the model) of the spectroscopically active degrees of freedom of the particle is less than the total number of degrees of freedom. Since we have *two rotational and three translational* degrees of freedom, we double expression (97) and treble expression (99):

$$L(z) \approx 2L_\mu(z) + (3/2)(q\gamma l/\mu)^2 L_q(z) \qquad (100a)$$

$$L_\mu(z) \equiv 2iz\left\langle \theta(0) \int_0^\infty [\theta(\varphi) - \theta(0)]e^{iz\varphi}\mathrm{d}\varphi \right\rangle \qquad (100b)$$

$$L_q(z) \equiv 2iz\left\langle s(0) \int_0^\infty [s(\varphi) - s(0)]e^{iz\varphi}\mathrm{d}\varphi \right\rangle \qquad (100c)$$

C. Strained-State Energy of the Hydrogen Bond

This quantity is equal to the dimer potential energy, which is the sum of two quantities. One is quadratic in the hydrogen-bond elongation Δl and another quadratic in the turn through the angle θ_+ about the HB equilibrium direction.

We first express these quantities in terms of angular deflection θ_- of the atom H_{A-} and of variable s defined by (95). Introducing the normalized covalent-bond length $a = r/l$ and using the cosine theorem in the approximation of small linear and angular displacements, we find in view of Fig. 39a, that

$$\left(\frac{l+\Delta l}{l}\right)^2 = \left(\frac{r}{l}\right)^2 + \left(\frac{d}{l}\right)^2 - 2\frac{rd\cos\theta_-}{l^2} = a^2 + (1+a+s)^2$$

$$- 2a(1+a+s)\cos\theta_-$$

$$\approx (1+s)^2 + a(1+a+s)\theta_-^2 \qquad (101)$$

According to the sine theorem, in the same approximation we obtain

$$\frac{r}{\sin\theta_+} = \frac{l+\Delta l}{\sin\theta_-}; \quad \sin\theta_+ \approx \theta_+ = \frac{a\sin\theta_-}{\sqrt{(1+s)^2 + a(1+a+s)\theta_-^2}} \approx \frac{a\theta_-}{1+s}$$

The comparison with (93) yields

$$2\theta = \theta_- - \frac{a\theta_-}{1+s} = \theta_-\frac{1-a+s}{1+s}$$

$$\theta_- \approx 2\theta\frac{1+s}{1-a+s} \approx \frac{2\theta}{1-a}, \qquad \theta_+ \simeq \frac{2a\theta}{1-a+s} \approx \frac{2a\theta}{1-a} \qquad (102)$$

From (101) we find the relative elongation squared:

$$\left(\frac{\Delta l}{l}\right)^2 \equiv \left(\frac{l+\Delta l}{l}\right)^2 + 1 - 2\frac{l+\Delta l}{l} = (1+s)^2 + a(1+a)\theta_-^2$$

$$+ 1 - 2\sqrt{(1+s)^2 + a(1+a)\theta_-^2}$$

$$\approx 1 + 2s + s^2 + a(1+a)\theta_-^2 + 1 - 2\left[1+s+\frac{a}{2}(1+a)\theta_-^2\right] = s^2 \qquad (103)$$

Thus, the hydrogen-bond elongation Δl and rotation angle θ_+ are respectively proportional to the small quantities s and θ_+.

We denote the longitudinal force constant by k and the dimensionless rotary constant by c. In terms of these constants the strained-state potential energy is expressed as

$$U = k(\Delta l)^2/2 + kcl^2\theta_+^2/2$$

Inserting $(\Delta l)^2$ (103) and θ_+^2 (102) into this expression and dividing the result by $k_B T$, we obtain the expression for the dimensionless dimer potential energy:

$$u(\theta,s) \equiv \frac{U}{k_B T} \simeq A\theta^2 + Bs^2, \qquad A \equiv \frac{2kcl^2a^2}{k_B T(1-a)^2}, \qquad B \equiv \frac{kl^2}{2k_B T} \qquad (104)$$

D. Kinetic Energy of the Dimer

We now introduce the translational velocities $\vec{v}_{\pm} \equiv d\vec{d}/dt$ and the angular velocities $\vec{\Omega}_{\pm} \equiv \mp\vec{d}\theta_{\pm}/dt$. The minus sign is here involved, because the right hydrogen atoms deflect through[17] the angle $(-\theta)$. It follows from the equality $\vec{d}_{+} + \vec{d}_{-} = 0$ and the definition $\vec{d} \equiv \vec{d}_{+} - \vec{d}_{-}$ that

$$\vec{d}_{\pm} = \pm\vec{d}/2, \qquad \vec{v}_{\pm} = \pm\frac{1}{2}\frac{d\vec{d}}{dt} \equiv \pm\vec{v}/2 \tag{105}$$

Next, we introduce the vectors $\vec{r}_{A\pm}$ and $\vec{r}_{B\pm}$ directed toward the hydrogen atoms in the same way as the corresponding covalent bonds. The absolute values of these vectors are equal to the covalent-bond lengths. The linear velocities, referring to rotations, are equal to $\vec{r}_{A\pm} \times \vec{\Omega}_{\pm}$ and $\vec{r}_{B\pm} \times \vec{\Omega}_{\pm}$.

The diagram for velocities, given in Fig. 39d, is drawn with account of two above-mentioned conditions, namely, (i) mutual orthogonality of the $O_{+}H_{A\pm}$ and $O_{+}H_{B\pm}$ bonds and (ii) smallness of the angular displacements. Consequently, for the $H_{\pm}$ atoms, the rotational components of linear velocities, directed transversely to the covalent bonds, are almost orthogonal and for the $H_{B\pm}$ atoms almost parallel to the translational components $\vec{v}_{\pm}$ directed along the symmetry axis. Accounting for this property and relation (105), we write

$$v_{A\pm}^2 \simeq v_{\pm}^2 + r^2\Omega_{\pm}^2 = (v/2)^2 + r^2\Omega_{\pm}^2, \qquad v_{B\pm}^2 \simeq (v_{\pm} + r\Omega_{\pm})^2 = (v/2 \pm r\Omega_{\pm})^2$$

Next, we multiply the sum of these quantities by $m/2$ and add the result to the kinetic energy of two oxygen atoms, equal to $(m/2)2 \cdot 16v_{\pm}^2 = (m/2)8v^2$, where m is the hydrogen-atom mass.

So, we find the following expression for the dimer kinetic energy:

$$K_{H_2O} = \frac{m}{2}\left[\frac{v^2}{4} + r^2\Omega_{+}^2 + \frac{v^2}{4} + r^2\Omega_{-}^2 + \left(\frac{v}{2} + r\Omega_{+}\right)^2 + \left(\frac{v}{2} - r\Omega_{-}\right)^2 + 8v^2\right] \tag{106}$$

$$= \frac{m}{2}[9v^2 + 2r^2(\Omega_{+}^2 + \Omega_{-}^2) + rv(\Omega_{+} - \Omega_{-})]$$

$$= \frac{I_{vib}^{H_2O}}{2}\left[\Omega_{+}^2 + \Omega_{-}^2 + \frac{9}{2r^2}v^2 + \frac{v}{2r}(\Omega_{+} - \Omega_{-})\right] \tag{107}$$

[17]The notation $\vec{\Omega}_{\pm} \equiv \mp\vec{d}\theta_{\pm}/dt$ is the kinematic relation for the angular-velocity vector expressed as the derivative of the angular displacement. In this section the velocity v (Roman letter) should not be confused with Greek letter ν, which denote wavenumber or, in other words, frequency.

Here I_{vib} is the moment of inertia:

$$I_{vib} = I_{vib}^{H_2O} = 2mr^2 \qquad (108)$$

For the case of ordinary water, according to assumption 1 and Fig. 39c, this moment is introduced assuming that *only the hydrogen atoms* contribute to I_{vib}.

In the case of heavy water, we double $I_{vib}^{H_2O}$ and all terms in [·] entering (106), except the last term. Then instead of (107) we have the following expression for D_2O:

$$K_{D_2O} = \frac{m}{2}[10v^2 + 4r^2(\Omega_+^2 + \Omega_-^2) + 2rv(\Omega_+ - \Omega_-)]$$

$$= \frac{I_{vib}^{D_2O}}{2}\left[\Omega_+^2 + \Omega_-^2 + \frac{5}{2r^2}v^2 + \frac{v}{2r}(\Omega_+ - \Omega_-)\right] \qquad (109)$$

where now

$$I_{vib}^{D_2O} = 4mr^2 \qquad (110)$$

We denote the derivative with respect to the normalized time φ, introduced in (85), as $\varphi = t/\eta_{vib}$ by putting dot above the corresponding letter. Then $\Omega_\pm = \mp d\theta_\pm/dt = \mp\eta_{vib}^{-1}\dot\theta_\pm$. The expression for the dimensionless kinetic energy of H_2O and D_2O is found by dividing (107) or (109) by k_BT and by expressing the angular and linear velocities $\Omega_\pm$ and v in terms of the derivatives $\dot\theta$ and $\dot s$ according to definition (95) and relations (102):

$$\Omega_\pm = 2\begin{Bmatrix} -a \\ 1 \end{Bmatrix}\frac{\dot\theta}{(1-a)\eta_{vib}}; \qquad v = \frac{d}{d\varphi}d(\varphi) = \frac{l\dot s}{\eta_{vib}};$$

$$\begin{Bmatrix} \eta_{vib}(H_2O) \\ \eta_{vib}(D_2O) \end{Bmatrix} = \frac{1}{\sqrt{2k_BT}}\begin{Bmatrix} I_{vib}(H_2O) \\ I_{vib}(D_2O) \end{Bmatrix} \qquad (111)$$

Then we have

$$\frac{K_{H_2O}}{k_BT} = \frac{4(1+a^2)}{(1-a)^2}\dot\theta^2 + \frac{9\dot s^2}{2a^2} - \frac{1+a}{a(1-a)}\dot\theta\dot s,$$

$$\frac{K_{D_2O}}{k_BT} = \frac{4(1+a^2)}{(1-a)^2}\dot\theta^2 + \frac{5\dot s^2}{2a^2} - \frac{1+a}{a(1-a)}\dot\theta\dot s \qquad (112)$$

Adding this expression to the potential energy (104), we find the normalized total energy (Hamiltonian)

$$h(\theta, \dot{\theta}, s, \dot{s}) = \frac{4(1+a^2)}{(1-a)^2}\dot{\theta}^2 + \left\{\begin{matrix} 9 \\ 5 \end{matrix}\right\}\frac{\dot{s}^2}{2a^2} - \frac{1+a}{a(1-a)}\dot{\theta}\dot{s} + A\theta^2 + Bs^2 \quad \text{for } \left\{\begin{matrix} H_2O \\ D_2O \end{matrix}\right\}$$

(113)

where A and B are given by (104).

Below we write all formulas bearing in mind *light water* (for heavy water the relevant modification is given in Section I).

E. Dielectric Response to Translational Vibrations

We now eliminate the rotational degree of freedom—that is, omit the following terms entering (113) and containing θ and $\dot{\theta}$:

$$h(\theta, \dot{\theta}, s, \dot{s}) \rightarrow h(s, \dot{s}) = \frac{9\dot{s}^2}{2a^2} + Bs^2$$

(114)

After differentiation, we obtain the second-order equation in the variable s:

$$\ddot{s} + f_q^2 s = 0, \qquad f_q^2 \equiv (2/9)a^2B$$

(115)

In this approximation we should omit the first term in (100a). According to (100b), we now have

$$L(z) = 3(q\gamma l/\mu)^2\langle s_0\hat{s}(z)\rangle$$

(116)

$$\hat{s}(z) \equiv iz\int_0^\infty [s(\varphi) - s_0)]e^{-iz\varphi}\,\mathrm{d}\varphi = s_0 + iz\int_0^\infty s(\varphi)e^{-iz\varphi}\,\mathrm{d}\varphi$$

(117)

We transform (117) by integrating by parts:

$$\hat{s}(z) = -\int_0^\infty \dot{s}(\varphi)e^{-iz\varphi}\,\mathrm{d}\varphi = -\frac{i}{z}\dot{s}(0) - \frac{i}{z}\int_0^\infty \ddot{s}(\varphi)e^{-iz\varphi}\,\mathrm{d}\varphi$$

(118)

The first term is omitted, because it, together with the multiplier s_0, gives no contribution to the average $\langle s(0)\hat{s}(z)\rangle$. Next, we transform (118) by taking equation of motion (115) into account. Comparing the result with (117), we have

$$\hat{s}(z) = \frac{if_q^2}{z}\int_0^\infty s(\varphi)e^{-iz\varphi}\,\mathrm{d}\varphi = s(0) + iz\int_0^\infty s(\varphi)e^{-iz\varphi}\,\mathrm{d}\varphi$$

(119)

Representing the integral in (119) as $\int_0^\infty s(\varphi)e^{iz\varphi}\,d\varphi = (iz)^{-1}(\hat{s} - s_0)$, we obtain from (118) and (116):

$$\hat{s}(z) = \frac{f_q^2}{f_q^2 - z^2}s_0$$

and

$$L(z) = 3\left(\frac{q\gamma l}{\mu}\right)^2 \frac{f_q^2}{f_q^2 - z^2}\langle s^2 \rangle \tag{120}$$

Thus, the spectral function presents a Lorentz line and its intensity is proportional to the average squared relative elongation s defined by (103).

According to definition of the average

$$\langle s^2 \rangle = \int_0^{s_{\mathrm{lim}}} s^2 \exp(-Bs^2)\,ds \bigg/ \int_0^{s_{\mathrm{lim}}} \exp(-Bs^2)\,ds - \int_0^{u_{\mathrm{lim}}} e^{-u}\sqrt{u}\,du \bigg/ B\int_0^{u_{\mathrm{lim}}} e^{-u}\frac{du}{\sqrt{u}} \tag{121a}$$

We have expressed s in terms of the potential energy using the relation $u(s) \equiv u(0,s) = Bs^2$, to which (104) is reduced because of elimination of the rotational degree of freedom.

The limiting value u_{lim} corresponds to a *broken* H-bond, since the strained-state energy cannot exceed the H-bond energy U_{HB}. Thus, we have

$$u(s) \leq u_{\mathrm{lim}} \equiv U_{\mathrm{HB}}(RT)^{-1} \tag{121b}$$

where U_{HB} is approximately equal to 2.6 kcal/mol and $R \approx 0.002\,\mathrm{kcal}$ $\mathrm{mol}^{-1}\cdot\mathrm{deg}^{-1}$ is the gas constant.

Taking into account (121a), we obtain now the formula for the average squared relative elongation:

$$\langle s \rangle^2 = S/(2B)$$

$$S \equiv 2\int_0^{u_{\mathrm{lim}}} e^{-u}\sqrt{u}\,du \bigg/ \int_0^{u_{\mathrm{lim}}} e^{-u}\frac{du}{\sqrt{u}} = -2\sqrt{u_{\mathrm{lim}}}e^{-u_{\mathrm{lim}}} \bigg/ \int_0^{u_{\mathrm{lim}}} e^{-u}\frac{du}{\sqrt{u}} + 1$$

$$= 1 - \frac{2\sqrt{u_{\mathrm{lim}}}e^{-u_{\mathrm{lim}}}}{\sqrt{\pi}\,\mathrm{erf}\left(\sqrt{u_{\mathrm{lim}}}\right)} \tag{121c}$$

so that

$$\frac{\langle \Delta l^2 \rangle}{l^2} = \langle s^2 \rangle = \frac{S}{2B} = \frac{S k_B T}{k l^2} \tag{122}$$

Accounting for (120), we finally derive the expression for the translational spectral function:

$$L(z) \simeq 3 \left(\frac{q\gamma l}{\mu} \right)^2 \frac{S}{2B} \frac{f_q^2}{f_q^2 - z^2} \tag{123}$$

Inserting it into (83) and taking (82) into account, we obtain the formula for the high-frequency susceptibility:

$$\chi \simeq G_{\text{vib}} L(z) = \frac{N_{\text{vib}}(q\gamma l)^2}{4 k_B T} \frac{S}{B} \frac{\Omega^2}{\Omega^2 - \hat{\omega}^2} = \frac{N_{\text{vib}}(q\gamma)^2}{2\bar{m}} \frac{S}{\Omega^2 - \hat{\omega}^2} \tag{124}$$

Here the reduced mass $\bar{m} = m_{\text{H}_2\text{O}}/2 = m/9$ and the complex frequency $\hat{\omega} = \omega + i\tau^{-1}$. Noting (115), we find that the angular frequency $\Omega = f_q/\eta$, pertaining to translations, is expressed by a standard fashion through k and $\bar{m}$:

$$\Omega^2 = \frac{f_q^2}{\eta_{\text{vib}}^2} = \frac{2a^2}{9} \frac{2k_B T}{I_{\text{vib}}} \frac{k l^2}{2 k_B T} = \frac{2 k l^2}{9 a^2 \cdot 2 m r^2} = \frac{k}{\bar{m}} \tag{125}$$

Note that this formula allows us to relate the force constant k with the T-band resonance frequency ν_q by putting $\Omega = 2\pi c_{\text{light}} \nu_q$, where $\nu_q \approx 200 \, \text{cm}^{-1}$ and c_{light} is the speed of light (in other sections of this work c_{light} is denoted, as usually, c, while the rotary force constant is denoted c_{rot}).

The result (124) will reduce to that described by Eq. (350) in GT2, if:

(a) We take one-third of Eq. (125), since the derivation in GT2 was drawn for a one-dimensional ensemble.

(b) The upper limits of the integrals in (121a) and (121c) tend to infinity (then $S \to 1$).

(c) We put $\gamma = 1$, so that the positions of charges on molecules should coincide with the centers of mass of these molecules.

Represented here the derivation of the formulas (124) and (125) allows to justify application (though with essential changes) of the harmonic oscillator model, given in GT2, pp. 250–268, for calculation of the water/ice band, located at ≈ 200 cm^{-1}.

F. Equations of Motion and Expression for Spectral Function

1. Transformation of Hamilton Equations

We put expression (113) for the dimensionless total energy into the Hamilton equations

$$\frac{1}{2}\frac{\mathrm{d}}{\mathrm{d}\varphi}\frac{\partial h}{\partial \dot\theta} = -\frac{\partial h}{\partial \theta}, \qquad \frac{\mathrm{d}}{\mathrm{d}\varphi}\frac{\partial h}{\partial \dot s} = -\frac{\partial h}{\partial s}$$

expressed in terms of orientational ($\dot\theta$) and translational ($\dot s$) generalized velocities. So, we obtain the system of equation

$$\frac{4(1+a^2)}{(1-a)^2}\ddot\theta - \frac{1+a}{2a(1-a)}\ddot s = -A\theta, \qquad -\frac{1+a}{2a(1-a)}\ddot\theta + \frac{9}{2a^2}\ddot s = -Bs \quad (126)$$

Its physical meaning becomes clear if the term with $\ddot s$ is rearranged to the right-hand side of the first equation and the term with $\ddot\theta$ to that of the second one. The first equation then means that the angular acceleration $\ddot\theta$ is due to the elastic force $-A\theta$ and inertial force. The latter appears, because rotation occurs in a non-inertial reference system translating with the acceleration $\ddot s$. According to the second equation, the translational acceleration $\ddot s$ is due to a similar inertial force and the elastic force ($-Bs$). This inertial force, proportional to the angular acceleration $\ddot\theta$, appears because translations occur in a non inertial (non uniformly rotating) reference system.

Rewriting system (131) in a general form

$$a_{11}\ddot\theta + a_{12}\ddot s = b_1, \qquad a_{21}\ddot\theta + a_{22}\ddot s = b_2 \quad (127)$$

we solve it for the derivatives:

$$\ddot\theta = \begin{vmatrix} b_1 & a_{12} \\ b_2 & a_{22} \end{vmatrix} \Big/ \begin{vmatrix} a_{11} & a_{12} \\ a_{21} & a_{22} \end{vmatrix}, \qquad \ddot s = \begin{vmatrix} a_{11} & b_1 \\ a_{21} & b_2 \end{vmatrix} \Big/ \begin{vmatrix} a_{11} & a_{12} \\ a_{21} & a_{22} \end{vmatrix} \quad (128)$$

Substituting the coefficients from (126) into (127), we obtain the system of two second-order linear differential equations

$$\ddot\theta + p^2\theta = -ws, \qquad \ddot s + f^2 s = -g\theta \quad (129)$$

with the following constant coefficients:

$$p^2 \equiv \frac{9}{2a^2} A \frac{4a^2(1-a)^2}{72(1+a^2)-(1+a)^2} = \frac{18A(1-a)^2}{71(1+a^2)-2a} \tag{130a}$$

$$f^2 \equiv 4 \frac{(1+a^2)}{(1-a)^2} B \frac{4a^2(1-a)^2}{72(1+a^2)-(1+a)^2} = \frac{16Ba^2(1+a^2)}{71(1+a^2)-2a} \tag{130b}$$

$$g \equiv \frac{1+a}{2a(1-a)} A \frac{4a^2(1-a)^2}{72(1+a^2)-(1+a)^2} = \frac{2Aa(1-a^2)}{71(1+a^2)-2a} \tag{130c}$$

and

$$w \equiv \frac{1+a}{2a(1-a)} B \frac{4a^2(1-a)^2}{72(1+a^2)-(1+a)^2} = \frac{2Ba(1-a^2)}{71(1+a^2)-2a} \tag{130d}$$

2. Analytical Formula for Spectral Function

To obtain the expression for the spectral function without solving the system (129), we introduce the function

$$\hat{\theta}(z) \equiv iz \int_0^\infty (\theta - \theta_0)e^{-iz\varphi}\, d\varphi = \theta_0 + iz \int_0^\infty \theta e^{-iz\varphi}\, d\varphi \tag{131}$$

similar to the complex $\hat{s}(z)$ (117) (index 0 denotes the values for $\varphi = 0$). Using equations of motion (129) and performing transformations, similar to those in Section E, we have

$$\hat{\theta}(z) = -\int_0^\infty \dot{\theta}e^{-iz\varphi}\, d\varphi = -\frac{i}{z}\dot{\theta}_0 - \frac{i}{z}\int_0^\infty \ddot{\theta}e^{-iz\varphi}\, d\varphi$$

$$\to \frac{i}{z}\int_0^\infty (p^2\theta + ws)e^{-iz\varphi} d\varphi = \frac{p^2}{z^2}\left(\hat{\theta} - \theta_0\right) + \frac{w}{z^2}(\hat{s} - s_0)$$

and

$$\hat{s} = -\int_0^\infty \dot{s}e^{-iz\varphi}\, d\varphi = -\frac{i}{z}\dot{s}_0 - \frac{i}{z}\int_0^\infty \ddot{s}e^{-iz\varphi}\, d\varphi$$

$$\to \frac{i}{z}\int_0^\infty (f^2 s + gs)e^{-iz\varphi}\, d\varphi = \frac{f^2}{z}(\hat{s} - s_0) + \frac{g}{z^2}(\hat{\theta} - \theta_0)$$

Thus we have derived the system

$$\left(1 - \frac{p^2}{z^2}\right)\hat{\theta} - \frac{w}{z^2}\hat{s} = -\frac{1}{z^2}(p^2\theta_0 + ws_0), \quad -\frac{g}{z^2}\hat{\theta} + \left(1 - \frac{f^2}{z^2}\right)\hat{s} = -\frac{1}{z^2}(f^2s_0 + g\theta_0)$$

$$(132)$$

of linear algebraic equations for $\hat{\theta}$ and $\hat{s}$ of the form (127). Substituting the coefficients entering (132) into the relationships of the type (128), we find

$$\hat{\theta} = \frac{(f^2 - z^2)(p^2\theta_0 + ws_0) - w(f^2s_0 + g\theta_0)}{(p^2 - z^2)(f^2 - z^2) - gw} \rightarrow \frac{p^2(f^2 - z^2) - gw}{(p^2 - z^2)(f^2 - z^2) - gw}\theta_0$$

$$(133a)$$

and

$$\hat{s} = \frac{(p^2 - z^2)(f^2s_0 + g\theta_0) - g(p^2\theta_0 + ws_0)}{(p^2 - z^2)(f^2 - z^2) - gw} \rightarrow \frac{f^2(p^2 - z^2) - gw}{(f^2 - z^2)(p^2 - z^2) - gw}s_0$$

$$(133b)$$

It is easy to see that the "partial" spectral functions L_μ (100b) and L_q (100c) are expressed in terms of the complex functions $\hat{\theta}(z)$ (131) and $\hat{s}(z)$ (117) as

$$L_\mu(z) = 2\langle\hat{\theta}(z)\theta_0\rangle = 2\left\langle\frac{\hat{\theta}(z)}{\theta_0}\theta_0^2\right\rangle, \quad L_q(z) = 2\left\langle\frac{\hat{s}(z)}{s_0}s_0^2\right\rangle \quad (134a)$$

According to (133), the ratio $\hat{\theta}/\theta_0$ does not depend on θ_0, and the ratio $\hat{s}/s_0$ on s_0. From (134a), we then obtain

$$L_\mu(z) = 2\frac{\hat{\theta}(z)}{\theta_0}\langle\theta^2\rangle, \quad L_q(z) = 2\frac{\hat{s}(z)}{s_0}\langle s^2\rangle \quad (134b)$$

Thus, the partial spectral functions are proportional to the averaged squared translational and orientational displacements θ and s. The range $[\theta, s]$ of averaging over these phase variables is determined according to the criterion, similar to that used for specifying the limiting potential energy $u_{\lim}$ in (121b). As applied for two phase variables, the generalization of this criterion, according to (104), yields the inequality

$$u(\theta, s) = A\theta^2 + Bs^2 \le u_{\lim}, \quad u_{\lim} \equiv U_{HB}(RT)^{-1} \quad (135a)$$

Substituting its left-hand side into the Boltzmann exponent, similar to that (121a), we write

$$
\left\{ \begin{array}{c} \langle\theta\rangle^2 \\ \langle s\rangle^2 \end{array} \right\} = \left\{ \begin{array}{c} A^{-1} \\ B^{-1} \end{array} \right\} \int_{[\theta,s]} \left\{ \begin{array}{c} \theta^2 \\ s^2 \end{array} \right\} \exp\left(-A\theta^2 - Bs^2\right) d\theta ds \Bigg/ \int_{[\theta,s]} \exp\left(-A\theta^2 - Bs^2\right) d\theta ds
$$

(135b)

We see from (135a) that the boundary of the interval $[\theta, s]$ in the coordinates $\theta\sqrt{A}$ and $s\sqrt{B}$ presents a circle of radius $\sqrt{u_{\lim}}$. Therefore it is convenient to pass in (135b) to the polar coordinates ρ and ψ, related to the former ones by $\sqrt{A}\theta = \rho\cos\psi$ and $\sqrt{B}s = \rho\sin\psi$. Then $u = \rho^2$, $d\theta ds \propto \rho d\rho d\psi$, and relation (135b) can be reduced to the form

$$
\left\{ \begin{array}{c} \langle\theta\rangle^2 \\ \langle s\rangle^2 \end{array} \right\} = \left\{ \begin{array}{c} A^{-1} \\ B^{-1} \end{array} \right\} \int_0^{\sqrt{u_{\lim}}} e^{-\rho^2}\rho^3 d\rho \int_0^{2\pi} \left\{ \begin{array}{c} \cos^2\psi \\ \sin^2\psi \end{array} \right\} d\psi \Bigg/ \int_0^{\sqrt{u_{\lim}}} e^{-\rho^2}\rho\, d\rho \int_0^{2\pi} d\psi
$$

$$
= \left\{ \begin{array}{c} A^{-1} \\ B^{-1} \end{array} \right\} E/2
$$

(135c)

where

$$
E \equiv \langle u\rangle = \frac{\int_0^{\sqrt{u_{\lim}}} e^{-\rho^2}\rho^3\, d\rho}{\int_0^{\sqrt{u_{\lim}}} e^{-\rho^2}\rho\, d\rho} = \frac{-1/2\rho^2 e^{-\rho^2}\big|_0^{\sqrt{u_{\lim}}}}{\int_0^{\sqrt{u_{\lim}}} e^{-\rho^2}\rho\, d\rho} + 1 = 1 - \frac{u_{\lim}e^{-u_{\lim}}}{e^{-\rho^2}\big|_0^{\sqrt{u_{\lim}}}} = 1 - \frac{u_{\lim}e^{-u_{\lim}}}{1 - e^{-u_{\lim}}}
$$

(136)

The coefficient E is similar to S in (121c). For small $\langle\theta^2\rangle$ and $\langle s^2\rangle$, these quantities can be estimated by assuming $E = 1$. Then

$$
\langle\theta^2\rangle \simeq (2A)^{-1}, \qquad \langle s^2\rangle \simeq (2B)^{-1}
$$

(137)

Substituting (135c) into (134b), we obtain the formula for the partial spectral functions

$$
L_\mu(z) = \frac{E}{A}\frac{p^2(f^2 - z^2) - gw}{(p^2 - z^2)(f^2 - z^2) - gw}
$$

(138a)

and

$$L_q(z) = \frac{E}{B} \frac{f^2(p^2 - z^2) - gw}{(f^2 - z^2)(p^2 - z^2) - gw}$$ (138b)

According to (83) and (100), the dielectric susceptibility is then given by

$$\chi \equiv \chi_{\mathrm{vib}}(z) = G_{\mathrm{vib}}L(z) = G_\mu L_\mu(z) + G_q L_q(z)$$ (139a)

where

$$G_\mu \equiv \frac{r_{\mathrm{vib}}\mu^2 N}{3k_B T}, \qquad G_q \equiv \frac{r_{\mathrm{vib}}(q\gamma l)^2 N}{4k_B T}$$ (139b)

In the limit of "pure" translations ($G_\mu \to 0$), this leads to formulas (123) and (124).

Expressions (139) are used in this chapter for calculation of the far IR water/ice spectra.

3. Reduction of Spectral Function to Sum of Lorentz Lines

We now transform the denominator of expressions (138) in order to reduce them to a form more convenient for interpretation:

$$\left(\frac{p^2 + f^2}{2} - z^2\right)^2 + p^2 f^2 - \left(\frac{p^2 + f^2}{2}\right)^2 - gw = (b^2 - z^2)^2 - D^2$$ (140a)

where

$$b^2 \equiv \frac{p^2 + f^2}{2}, \qquad D^2 \equiv gw + \left(\frac{p^2 - f^2}{2}\right)^2$$ (140b)

From this expression and formula (138a) we find

$$\frac{A}{E}L_\mu(z) = \frac{p^2(f^2 - z^2) - gw}{2D}\left(\frac{1}{b^2 - D - z^2} - \frac{1}{b^2 + D - z^2}\right)$$

$$\frac{2DAL_\mu(z)}{E} = \frac{p^2 f^2 - gw + p^2(b^2 - D - z^2 - b^2 + D)}{b^2 - D - z^2}$$

$$- \frac{p^2 f^2 - gw + p^2(b^2 + D - z^2 - b^2 - D)}{b^2 + D - z^2}$$

$$L_\mu(z) = \frac{E}{2AD}\left[\frac{p^2(f^2 - b^2 + D) - gw}{b^2 - D - z^2} + \frac{p^2(-f^2 + b^2 + D) + gw}{b^2 + D - z^2}\right]$$

(141a)

Interchanging p and f and replacing A by B and index μ by q, we obtain the formula

$$L_q(z) = \frac{E}{2BD} \left[\frac{f^2(p^2 - b^2 + D) - gw}{b^2 - D - z^2} + \frac{f^2(-p^2 + b^2 + D) + gw}{b^2 + D - z^2} \right] \qquad (141b)$$

similar to (141a). We note that $b^2 + D > 0$ and, according to (140), the condition $b^2 - D \geq 0$ is equivalent to the inequality $\left(\frac{p^2 + f^2}{2} \right)^2 - gw - \left(\frac{p^2 - f^2}{2} \right)^2 \geq 0$, that is, to the condition $(pf)^2 \geq gw$. The validity of the latter can easily be proved by using (130).

Thus, $b^2 + D$ and $b^2 - D$ are both positive. Hence, *each* partial spectral function can be represented as the sum of two Lorentz lines with the centers at the frequencies

$$v_m = x_m (2\pi c v_{\text{vib}})^{-1} \qquad (142a)$$

where

$$x_m \approx \sqrt{b^2 \pm D} = \sqrt{(p^2 + f^2)/2 \pm \sqrt{(p^2 - f^2)^2/4 + gw}} \qquad (142b)$$

We recall that the law of motion itself was *not* used in deriving the formulas (141). From general reasoning, it is clear that the *time dependence* of orientational and translational coordinates, corresponding to *each* spectral function, (141a) or (141b), can be represented as the sum of *two* harmonics with frequencies (142). However, since the resonance denominators coincide in formulas (141a) and (141b) *pairwise*, the resulting susceptibility (139a) represents the sum of only *two* Lorentz lines (not of four lines, as might seem at the first sight). The intensity of *each* such line is determined by two spectral functions: one contributed by rigid dipoles and the other contributed by a nonrigid dipole. This fact (it is discussed below in more detail with an example) reflects an interrelation between translations and reorientations suffered by the H-bonded molecules.

G. Integral Absorption

Using (100a) and (141), we derive the formula for the normalized integral absorption

$$\Xi \equiv \int_{-\infty}^{\infty} x \chi''(x)\, dx = \Xi_q + \Xi_\mu, \quad \Xi_q \equiv \frac{3}{2} \left(\frac{q \gamma l}{\mu} \right)^2 \int_{-\infty}^{\infty} x \operatorname{Im} L_q(z)\, dx,$$

$$\Xi_\mu \equiv \int_{-\infty}^{\infty} x \operatorname{Im}[2L_\mu(z)]\, dx \qquad (143)$$

Noticing that for any real d the integral $\int_{-\infty}^{\infty} x \operatorname{Im}\left[\frac{1}{d^2-(x+iy)^2}\right] dx = \pi$ and accounting for (130), (141), and (143), we obtain the following expressions for the partial dimensionless integral absorptions:

$$\Xi_q = \frac{3}{2}\left(\frac{q\gamma l}{\mu}\right)^2 \frac{\pi f^2 E}{B} = 24\pi E\left(\frac{q\gamma l}{\mu}\right)^2 \frac{a^2(1+a^2)}{71(1+a^2)-2a},$$

$$\Xi_\mu = 2\pi E \frac{p^2}{A} = 36\pi E \frac{(1-a)^2}{71(1+a^2)-2a} \qquad (144)$$

The ratio of these quantities is proportional to the squared ratio of dipole moments, pertaining to rigid and nonrigid H-bonded dipoles.

$$\Xi_\mu/\Xi_q = \frac{3}{2}\left(\frac{\mu}{q\gamma l}\right)^2 \frac{(1-a)^2}{1+a^2} \qquad (145)$$

Using (83), (108), and (144), we find the absolute values of the partial integral absorptions:

$$\Pi_q \equiv \frac{G_{\text{vib}}}{\eta^2}\Xi_q = 8\pi N_{\text{vib}}(q\gamma l)^2 I^{-1}E \frac{a^2(1+a^2)}{71(1+a^2)-2a}$$

$$= 4\pi E N_{\text{vib}}(q\gamma)^2 m^{-1}\frac{1+a^2}{71(1+a^2)-2a} \qquad (146)$$

and

$$\Pi_\mu \equiv \frac{G_{\text{vib}}}{\eta^2}\Xi_\mu = 12\pi E N_{\text{vib}}\mu^2 I^{-1}\frac{1+a^2}{71(1+a^2)-2a} \qquad (147)$$

Thus, the total integral absorption Π is determined by the mass, charge, dipole moment, and structure parameter $a = r/l$. We emphasize that Π does *not* depend on the force constants and potential-well lifetime, as it must.

H. Spectral Functions in Particular Cases

1. The Case of Independency Between Translations and Reorientations of H-Bonded Molecules

In the system (129) we omit[18] ws in the first equation and $g\theta$ in the second one. Then we find that each partial spectral function is described by *only one* Lorentz

---

[18]Strictly speaking, the used approximation contradicts initial assumptions of the model. Nevertheless, as will be seen, the resulting frequency dependences (148) agree satisfactorily with those obtained from more rigorous formulas (138).

line, namely,

$$L_q(z) = \frac{E}{B} \frac{f^2}{f^2 - z^2}, \qquad L_\mu(z) = \frac{E}{A} \frac{p^2}{p^2 - z^2} \tag{148}$$

The center x_q of the line, born by translation of charges, and the center x_μ of the other line, born by elastic reorientations of rigid dipoles, are determined by dimer parameters as

$$x_q \approx f \quad \text{and} \quad x_\mu \approx p \tag{148a}$$

With decreasing of the rotary force constant c the center of the "reorientation" line L_μ shifts toward lower frequencies. The ratio of line intensities could be *roughly* estimated as

$$\frac{\Im_q}{\Im_\mu} \approx \frac{G_q A \; \text{Im}\left\{ f^2 \left[(f - x_q - iy)(f + x_q) \right]^{-1} \right\}}{G_\mu B \; \text{Im}\left\{ p^2 \left[(p - x_\mu - iy)(p + x_\mu) \right]^{-1} \right\}}$$

$$\approx \frac{3Af}{4Bp} \left(\frac{q\gamma l}{\mu} \right)^2 \approx \frac{a^2 \sqrt{2c(1 + a^2)}}{(1 - a)^2} \left(\frac{q\gamma l}{\mu} \right)^2 \tag{149}$$

For a numerical estimation we take (like in above sections) the distance between neighboring oxygen atoms $r_{OO} = 2.85 \, \text{Å}$ and the covalent-bond length $r = 1.02 \, \text{Å}$. Then we have the H-bond length $l = 2.85 - 1.02 = 1.83 \, \text{Å}$ and the parameter $a = r/l = 0.557$. It follows from (149) that the intensities of both lines have the same order of magnitude (the line L_q, stipulated by translation of charges is generally more intense). We shall show that approximately the same result holds also for a more rigorous solution, accounting for interrelation between elastic translation and reorientation.

2. Translations for "Frozen" Rotation

If the rotary force constant c is very large, so that the coefficient A in (104) tends to infinity, then an infinitely large restoring moment inhibits rotation. In this case, according to (130a) and (130c), $p \to \infty$ and $g \to \infty$. Spectral function (138a) turns to zero, and the relationship (138b) reduces to

$$L_q(z) = \frac{E}{B} \frac{f^2 - gw/p^2}{f^2 - gw/p^2 - z^2} \tag{150}$$

From (130), we have

$$(pf)^2 - gw = AB\frac{288a^2(1+a^2)(1-a)^2 - 4a^2(1-a^2)^2}{[71(1+a^2) - 2a]^2}$$

$$= 4ABa^2(1-a)^2\frac{72(1+a^2) - (1+a)^2}{[71(1+a^2) - 2a]^2} = \frac{4ABa^2(1-a)^2}{71(1+a^2) - 2a}$$

$$(151)$$

Dividing this expression by p^2, given by (130a), we find that

$$f^2 - gw/p^2 = f_q^2 \tag{152}$$

We see that the left-hand side of Eq. (152) coincides with the quantity f_q^2 in (115). Therefore, transition to infinitely large rotary constant is equivalent to excluding the rotational degree of freedom from Hamiltonian (113). Exactly this assumption was employed in Section E.

3. Librations for "Frozen" Translations

In order to pass on to "pure" elastic *reorientations* [we have in mind that the latter occur in the parabolic (and *not* in the hat) potential], we let f and w tend to infinity. Using (138a), (130b), and (130d), we find analogously to (150) and (151) the corresponding spectral function L_μ:

$$L_\mu(z) = \frac{E}{A}\frac{p_\mu^2}{p_\mu^2 - z^2}, \qquad p_\mu^2 \equiv p^2 - gw/f^2 \tag{153}$$

Comparison with (153) gives

$$p_\mu^2 = (p/f)^2(f^2 - gw/p^2) = (2/9)Ba^2(p/f)^2 \tag{154a}$$

Noting (130a) and (130b), we then have

$$p_\mu^2 = \frac{9A(1-a)^2}{8Ba^2(1+a^2)}\frac{2}{9}Ba^2 = A\frac{(1-a)^2}{4(1+a^2)} \tag{154b}$$

and

$$L_\mu(z) = \frac{(1-a)^2}{4(1+a^2)}\frac{E}{p_\mu^2 - z^2} \tag{154c}$$

Let us relate the dimensional reorientation frequency $\Omega_{lib} = p_\mu/\eta_{vib}$ to the translational frequency Ω in (127). Using expression (110) for the moment of inertia of H_2O, namely, $I_{vib} = 2mr^2 = \frac{2}{18}\bar{m}r^2 = \frac{1}{9}\bar{m}a^2l^2$, we find

$$\Omega_{lib}^2 = \frac{2kcl^2a^2}{k_BT(1-a)^2} \frac{2k_BT}{I_{vib}} \frac{(1-a)^2}{4(1+a^2)} = \frac{9\Omega^2c}{2(1+a^2)} = c_{crit}\Omega^2 \qquad (154d)$$

where

$$c_{crit} = \frac{9c}{2(1+a^2)} \qquad (154e)$$

is a certain "critical" rotary force constant. We shall return to this point in Section VI.J.2.

I. Relations for Heavy Water

The ratio between deuterium and hydrogen masses, as well as between the moments I_{vib} of inertia (110) and (108) of heavy and light water, is 2. On the other hand, the ratio between the kinetic energies (112) of these molecules and correspondingly between the Hamiltonians (113) is *less*. Namely, instead of coefficient $9\,\dot{s}^2$ for H_2O we have for D_2O $5\,\dot{s}^2$. Taking this into account, we replace the system (128), whose coefficients are given in equations (126), by the following system:

$$\ddot{\theta} = \left[\frac{5A\theta}{2a^2} - \frac{Bs(1+a)}{2a(1-a)}\right]\Delta^{-1}, \qquad \ddot{s} = -\left[\frac{4Bs(1+a^2)}{(1-a)^2} + \frac{A\theta(1+a)}{a(1-a)}\right]\Delta^{-1} \qquad (155)$$

where

$$\Delta = \begin{vmatrix} \dfrac{4(1+a^2)}{(1-a)^2} & -\dfrac{(1+a)}{2a(1-a)} \\[3mm] -\dfrac{(1+a)}{2a(1-a)} & \dfrac{5}{2a^2} \end{vmatrix}$$

Instead of (130), we then obtain

$$p^2 = \frac{10A(1-a)^2}{39(1+a^2) - 2a}, \qquad f^2 = \frac{16Ba^2(1+a^2)}{39(1+a^2) - 2a} \qquad (156a)$$

and

$$g = \frac{2Aa(1-a^2)}{39(1+a^2) - 2a}, \qquad w = \frac{2Ba(1-a^2)}{39(1+a^2) - 2a} \qquad (156b)$$

Other changes in formulas, pertaining to D_2O, are presented in Table XV.

TABLE XV
Relations for Light and Heavy Water

Formula	H$_2$O	D$_2$O
	M is the hydrogen-atom mass	$2m$ is the deuterium-atom mass
	$18m$ is light-water molecule mass	$20m$ is the mass of heavy-water molecule
	$\bar{m} = 9m$ is the reduced H$_2$O mass	$\bar{m}$ is the reduced D$_2$O mass
(108)	$I_{\text{vib}} = 2mr^2 = \frac{2}{9}\bar{m}r^2$	$I_{\text{vib}} = 4mr^2 = \frac{2}{5}\bar{m}r^2$
(115)	$f_q^2 \equiv \dfrac{2}{9a^2}B$	$f_q^2 \equiv \dfrac{2}{5a^2}B$
(125)	$\Omega^2 = \dfrac{k}{\bar{m}(\text{H}_2\text{O})}$	$\Omega^2 = \dfrac{k}{\bar{m}(\text{D}_2\text{O})}$
(144)	$\Xi_q = 24\pi\left(\dfrac{q\gamma l}{\mu}\right)^2 \dfrac{Ea^2(1+a^2)}{71(1+a^2)-2a}$	$\Xi_q = 24\pi\left(\dfrac{q\gamma l}{\mu}\right)^2 \dfrac{Ea^2(1+a^2)}{39(1+a^2)-2a}$
(144)	$\Xi_\mu = 36\pi\dfrac{E(1-a)^2}{71(1+a^2)-2a}$	$\Xi_\mu = 20\pi\dfrac{Ea^2(1+a^2)}{39(1+a^2)-2a}$
(145)	$\Xi_\mu/\Xi_q = \dfrac{3}{2}\left(\dfrac{\mu}{q\gamma r}\right)^2\dfrac{(1-a)^2}{1+a^2}$	$\Xi_\mu/\Xi_q = \dfrac{5}{6}\left(\dfrac{\mu}{q\gamma r}\right)^2\dfrac{(1-a)^2}{1+a^2}$
(147)	$\Pi_q = \dfrac{8\pi N_{\text{vib}}(q\gamma l)^2}{I_{\text{vib}}}\dfrac{Ea^2(1+a^2)}{71(1+a^2)-2a}$	$\Pi_q = \dfrac{20\pi N_{\text{vib}}(q\gamma l)^2}{3I_{\text{vib}}}\dfrac{Ea^2(1+a^2)}{39(1+a^2)-2a}$
(147)	$\Pi_\mu = \dfrac{12\pi E}{I_{\text{vib}}}N_{\text{vib}}\mu^2\dfrac{1+a^2}{71(1+a^2)-2a}$	$\Pi_\mu = \dfrac{20\pi E}{3I_{\text{vib}}}N_{\text{vib}}\mu^2\dfrac{1+a^2}{39(1+a^2)-2a}$
(154b)	$p_\mu^2 == \dfrac{2Ba^2}{9}\left[\dfrac{p(\text{H}_2\text{O})}{f(\text{H}_2\text{O})}\right]^2 = A\dfrac{(1-a)^2}{4(1+a^2)}$	$p_\mu^2 == \dfrac{2Ba^2}{5}\left[\dfrac{p(\text{D}_2\text{O})}{f(\text{D}_2\text{O})}\right]^2 = A\dfrac{(1-a)^2}{4(1+a^2)}$
(154d)	$\Omega_{\text{lib}}^2 = \dfrac{9\Omega^2 c}{2(1+a^2)}$	$\Omega_{\text{lib}}^2 = \dfrac{5\Omega^2 c}{2(1+a^2)}$

J. Effects of Rotary Elasticity

1. Dimensionless Absorption

The resonance ice spectra, shown in Section V.B, are calculated for one chosen value of the dimensionless force constant ($c = 0.16$). We now consider the connection between this parameter c and the features of theoretical absorption spectra. In the example, given below, the other model parameters are taken the same as in Section V.B.

We calculate the dimensionless frequency dependences of the partial absorption coefficients

$$A_q(x) = x\text{Im}[G_q L_q(x)], \qquad A_\mu(x) = x\text{Im}[G_\mu L_\mu(x)] \qquad (157)$$

The first dependence, $A_q(x)$, refers to elastic translation of charges (of a nonrigid dipole) and the second, $A_\mu(x)$, to elastic reorientation of H-bonded rigid dipoles. For calculation of dependences (157) the *rigorous* formulas (138) and (139) are used.[19] We take the following values of the rotary force constant c: 0.4, 0.2, 0.15, and 0.05. For $c = 0.2$ the dependences (157) are calculated also from the simplified formulas (148).

Below we consider the absorption ice spectrum found for the temperature $-7°C$ (this spectrum is located near frequency 200 cm^{-1}). In this example, reflected in Fig. 40a–40d, we employ for calculation the following parameters: $A/c = 650$, $B = 102.4$, and $E = 0.96$.

In Fig. 40b, which pertains to $c = 0.2$, curve 1 refers to translation and curve 2 to reorientation, which concern, respectively, the T- and V-bands. Both (solid) curves calculated from the rigorous formulas (138) and (139) are *double-humped*. On the contrary, each of the dashed curves, 3 and 4, which are found from the simplified Eqs. (148), has *one peak*. This distinction arises, since in the second case (for curves 3 and 4) the interrelation between the elastic translations and reorientations is disregarded.

If the force constant c is sufficiently *small*, in the "translational" component the smaller peak is located to the left of the larger one. In the "orientational" component the smaller peak is located to the right of the larger one. This is seen in Fig. 40b for $c = 0.2$ and in Fig. 40c for $c = 0.15$.

For a sufficiently *large* constant c, as in Fig. 40a, where $c = 0.4$, the situation changes in two points: (i) The "translational" peaks shift to the left of the "orientational" one and (ii) the smaller and larger peaks change places. Since the case (i) does *not* correspond to experiment, we draw the conclusion that for obtaining reasonable ice/water spectra the rotary force constant c should *not* be too large.

In Section VI.J.2 we shall show that on the other hand the constant c should not be too small. The restriction for a small c-value also becomes evident in Fig. 40d, where the chosen rotary constant c is 0.05. Now, as seen from the dashed line, the response to elastic reorientations vanishes, while the response to elastic translation weakens. This is depicted by the solid line: The intensity and frequency of the translational peak substantially decreases.

Figures 40a–c also demonstrate that if the constant c varies in a rather narrow interval of values, the positions and heights of the absorption maxima in the T- and V-bands change only slightly.

Comparison of curves 3,4 with curves 1,2 in Fig. 40b demonstrates that the simplified description of elastic-vibration spectra, given by Eq. (148), yields a rather good agreement with more accurately calculated spectra. described by the

---

[19]Formulas (141), equivalent to the latter and representing the spectral function as a sum of Lorentz lines, gives the same result, as it must.

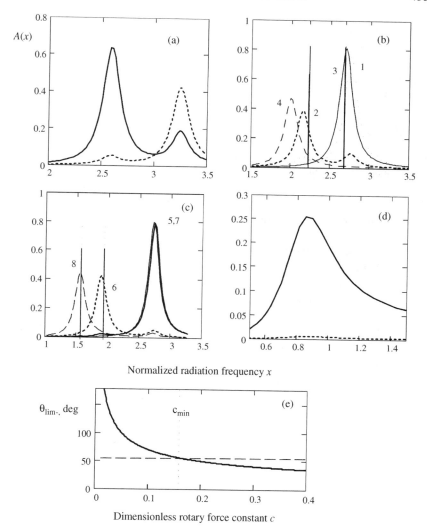

Figure 40 (a–d) Frequency dependences of the dimensionless absorption $A(x)$. Solid lines denote absorption due elastic translations of charges and dashed lines due to elastic reorientations. Calculation according to strict theory: (a) $c = 0.4$; (b) $c = 0.2$ for curves 1 and 2; (c) $c = 0.15$ for curves 5 and 6 and $c = 0.1$ for curves 7 and 8; (d) $c = 0.05$. Approximate calculation: (b) for $c = 0.2$ (curves 3 and 4). Vertical lines refer to the Lorentz line centers estimated as $x_q = f$, $x_\mu = p$. (e) Amplitude of angular vibration versus rotary force constant, horizontal line depicts the quantity (158).

formulas (138). The positions $x_q \approx f$ and $x_\mu \approx p$ of the respective translational and orientational loss maxima, estimated from (147), are denoted in Fig. 40b by the vertical lines (the vertical lines in Fig. 40c denote the frequencies x_μ).

The frequencies x_q and x_μ, pertaining, respectively, to translation and elastic reorientation, approximately coincide in Fig. 40b and 40c with the frequencies of the *main maxima*, precisely obtained from Eqs. (148) and (149).

Finally, as seen in Fig. 40b, the smaller maximum of curve 2 is located at the same frequency as the main maximum of curve 1. Analogously, the main maximum of curve 2 coincides with the smaller maximum of curve 1. This property arises due to a pairwise equality of the resonance denominators in the formulas (141a) and (141b). For this reason the dielectric-loss frequency dependence is described by only *two* Lorentz lines. The difference of the *strict* theory from the approximate is revealed in that the intensity of each line is determined by vibrations of *both* types (longitudinal and rotational). As for the *simplified* representation (148), the intensity of each line is determined by *only one* vibration type.

Let us now touch the *interrelation* between the elastic translations and reorientations of the HB molecules. Mathematically, this phenomenon can be ascribed to the "crossed" members in the right-hand sides of differential equations (129). Due to these members, the time dependence of the orientational (θ) and translational (s) displacements could be described by *two* harmonics, rather than by only one harmonic in the case of approximate theory. Due to indicated interrelation, the curves 1 and 2 do not coincide in Fig. 40b with the curves 3 and 4. Analogously, the curve 1 does not coincide with curve 2 in Fig. 41.

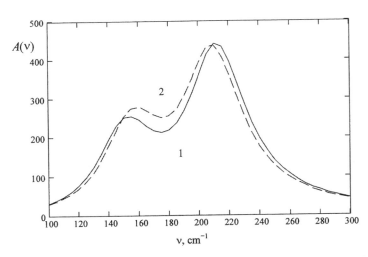

Figure 41   Frequency dependence of the normalized vibration absorption, obtained for ice at $-7°C$. Solid line is obtained for the strict theory and dashed line for the simplified theory, in which the interrelation between elastic translations and reorientations is disregarded.

2. Dependence of Angular Vibration on Rotary Force Constant

We now estimate the minimum value, c_{min}, of the rotary force constant c. For $c < c_{min}$ the above-used approximation of small oscillation amplitudes becomes inapplicable The point is in the following. The root-mean-square angular deflection θ_-, pertaining to the covalent bond of the *lefthand molecule*[20] (see Fig. 39a), according to (102) and (137), increases with decreasing of c. As follows from the indicated formulas, the limiting deflection is

$$\theta_{\lim -}(c) = \frac{2}{1-a}\sqrt{\frac{u_{\lim}}{A(c)}} \tag{158}$$

We assume that for the minimal value c_{min} the amplitude (158) becomes equal to the *half* of the tetrahedral angle Θ_{cr} (note, $\Theta_{cr} \approx 109°$). Then

$$\theta_{\lim -}(c_{crit}) = (1/2)\Theta_{cr} \tag{159}$$

The dependence (158) is shown in Fig. 40e. Equation (159) is satisfied at $c_{crit} = 0.16$. Precisely this value, depicted in Fig. 40e by the dashed line, was chosen in Section V.B for calculation of the resonance ice spectra at the temperature $-7°$C.

Thus, although the mean-square amplitude $\sqrt{\langle \Delta l^2 \rangle}$ of elastic longitudinal vibrations, performed by charged molecules along the H-bond, is small,[21] our estimate (159) shows *that the angular-vibration amplitude is large*. This example illustrates the mobility of molecules in the water medium.

3. On Approximate Description of Vibration Spectrum

In Section VI.J we have already pointed out the small difference between the spectra, calculated according to the rigorous formulas (138) and to the simpler formulas (148) (we remind that the latter were found disregarding the interrelation between elastic reorientation and translation). For an estimate we use the relation of constants A and B to other model parameters given by (130a) and (130b). The results of this estimation are as follows.

As applied to the far IR *ice* spectrum, the agreement of the approximate calculation with the precise one turns out to be sufficiently good, though their difference is *principally* noticeable (compare the curves 1 and 2 in Fig. 41). As

---

[20]According to (102), this deflection is greater by the factor of a^{-1} than the covalent bond deflection pertaining to the *righthand molecule*.

[21]Using Eq. (122) for estimate, we set the following parameters of our model, corresponding to ice at $-7°$C: $B \approx 100$ and $S \approx 1$. Then the relative vibration amplitude turns out to be less than 10% of the H-bond length l. Namely, $\sqrt{\langle \Delta l^2 \rangle}/l \approx 0.07$.

applied to *water*, the calculation using rigorous and approximate formulas coincides with the graphical accuracy! Consequently, in practical calculation this interrelation can be neglected. So, one may use simpler formulas (148) for "partial" spectral functions, and not to employ more rigorous formulas (138)–(139). This is one of the main results of our work.

4. Restriction for H-bond Length

Turning to a particular case of "frozen" translations, in which the spectral function is given by (153) and the effective reorientation frequency by (154d), one can see that, if the rotary constant c exceeds a certain critical value

$$c_{\text{crit}} = (2/9)(1 + a^2) = \frac{2}{9}\frac{r^2 + l^2}{l^2} \qquad (160)$$

then the translational minimum is located to the *left* of the vibration one. On the contrary, for $c < c_{\text{crit}}$ this maximum is located to the right. The rigorous calculation, shown by the solid lines in Fig. 40a–c, confirms this pattern. Indeed, for the example, corresponding to this figure, the formula (160) gives $c_{\text{crit}} \approx 0.29$. Comparing Fig. 40a and 40b, where the respective values $c = 0.4 > c_{\text{crit}}$ and $c = 0.2 < c_{\text{crit}}$ are taken, we see that just as the simplified theory predicts, the *main* orientational maximum is located to the left of the main translational one in the first figure (for $c = 0.4$) and to the right in the second figure (for $c = 0.2$).

Since experimentally observed spectra give the patterns similar to those represented in Figs. 40b or 40c (and *not* as in Fig. 40a), we conclude that *the effective rotary force constant should not noticeably exceed the value (160)*. This means that in a reasonable molecular model the average H-bond length l should not be too short.

K. Conclusions

The analytical formulas (138) and (139) or equivalent formulas (141) describe the harmonic-vibration contribution χ to the total complex susceptibility. A simpler variant of this model, in which the partial spectral functions are represented by the formulas (148), gives for *water* a graphical coincidence with the results of the above-mentioned rigorous theory. For *ice*, both approaches agree well (cf. the solid and dashed curves in Fig. 41).

In Sections III, IV, and V, these results allowed us calculation of ice/water elastic-vibration spectra, which are located in the translational band and in the nearby placed V-band. Each band is represented in terms of one Lorentz line. For this calculation the simple formulas (148) were used.

VII. NONHARMONIC TRANSVERSE VIBRATION OF A NONRIGID DIPOLE[22]

The classical model is described for a nonrigid dipole (dimer), forming the hydrogen bond (HB) and performing nonharmonic vibration in a direction, transverse to this bond. A nonlinear equation of motion is derived for this dimer, comprising two charged water molecules. The THz dielectric response to dimer vibration is found. An explicit expression is derived for the autocorrelator that governs the spectrum generated by transverse vibration (TV) of dimer. This expression is obtained by analytical solution of the truncated set of recurrence equations. The analytical formula for the complex susceptibility $\chi_\perp$ is obtained (index $\perp$ refers to quantities related to transverse vibration).

In the first part of this work (Sections II through V) we have combined the formula for $\chi_\perp$ given there without derivation, with the formulas for χ_q, χ_μ, and χ_{or}, accounting for dielectric response, arising, respectively, from elastic harmonic vibration of charged molecules along the H-bond (HB), from elastic reorientation of HB permanent dipoles about this bond, and from a rather free libration of a permanent dipole in a "defect" of water/ice structure modeled by the hat well. The set of four frequency dependences, namely of $\chi_{or}(v)$, $\chi_q(v)$, $\chi_\mu(v)$, and $\chi_\perp(v)$, allows us to describe the water/ice wideband spectra. For these dependences and those similar to them—namely $\varepsilon_{or}(v)$, $\Delta\varepsilon_q(v)$, $\Delta\varepsilon_\mu(v)$, and $\Delta\varepsilon_\perp(v)$ for the partial[23] complex permittivity—we refer to mechanisms a, b, c, and d.

The TV-dielectric relaxation mechanism allows us to (i) remove the THz "deficit" of loss ε'' inherent in previous (see GT2) theoretical studies, (ii) explain the THz loss and absorption spectra in supercooled (SC) water, (iii) describe, in agreement with the experiment, the low- and high-frequency tails of the two bands of ice H_2O located in the range 10–300 cm^{-1}, and (iv) describe the nonresonance loss spectrum in ice in the submillimetric region of wavelengths. Specific THz dielectric properties of SC water are ascribed to association of water molecules, revealed in our study by transverse vibration of the HB charged molecules.

Later in Section VII, we, starting from a simplified water structure, relevant to the hydrogen-bonded molecules, shall describe the frequency dependence of the susceptibility $\chi_\perp(v)$ typical of the THz region, bordering with the translational band and on the nearby placed V-band, centered at $v \approx 150$ cm^{-1}. This mechanism, pertaining to the $\chi_\perp(v)$ spectrum, is used in our calculation scheme

---

[22]Section VII is based on the work by Gaiduk and Crothers [23].

[23]We recall the difference in notations: The symbol Δ denotes that the corresponding quantity is found in the high-frequency approximation, so that its real part vanishes at $v \to \infty$, while $\varepsilon_{or}(\infty)$ differs from zero. For water we have $\varepsilon_{or}(\infty) = n_\infty^2 \approx 1.7$; for ice this quantity is greater.

for describing the water/ice spectra in terms of a two fraction (mixed) model, based on consideration in each fluid of the librational (LIB) and vibrational (VIB) states.

A. Transverse Susceptibility as Autocorrelator

We consider the dimer comprising two oppositely charged water molecules (see Fig. 42a). The left-hand molecule (with the oxygen atom O_-) is charged negatively, and the right-hand one (with the atom O_+) is charged positively. This

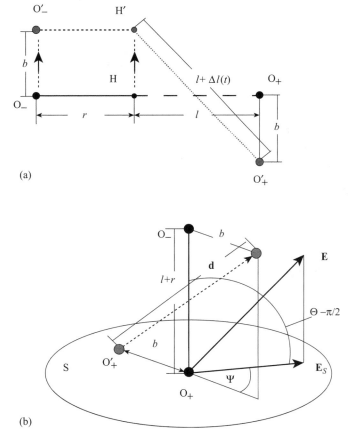

Figure 42 (a) Scheme of transverse vibrations of the HB molecules pertaining to the dimer O_-O_+. The broken line, whose length is l, depicts the H-bond at equilibrium. In a strained state the covalent bond O–H takes the position $O'-H'$ due to transverse shift on the distance $b(t)$, with the lines O–H and $O'-H'$ being parallel. The right oxygen atom O+ shifts on the same distance b and takes the position O'_+. The disturbed H-bond takes position $H'O_+'$, its length $l + \Delta l(t)$ undergoes periodic vibration. (b) The scheme pertaining to derivation of Eq. (167).

system is in some respect kindred to that shown in Fig. 39a. The difference consists in the following. We do not regard reorientation (turn) of the covalent bond O_-H_{A-} (shown in the latter figure), which leads to appearance of the time-varying angles θ_- and θ_+.

In Section VII we consider vibrational motions of oxygen atoms, transverse to the equilibrium HB direction O_-O_+. In the course of such motions the H-bond turns and its length changes in time as $l + \Delta l(t)$, where l is the H-bond at equilibrium (see Fig. 42a).

The law of motion is determined by the elastic force, which is assumed to be proportional to the transverse force constant $k_\perp$. Thus, the charged oxygen molecules constitute a nonrigid dipole. Regarding the dimer as a closed system, we note that if b is the displacement of the left H-atom, then that of the right one is $(-b)$. Let $d(t)$ be the time-dependent distance between the oxygen atoms, let $\gamma d(t)$ be the distance between the charges (the proportionality factor γ is assumed to be constant), and let $\mathbf{d}$ be the vector connecting these atoms (outgoing from the position of the oxygen atom $O_+{}'$). We *neglect coupling*[24] of the longitudinal (considered in Section VI) and transverse vibrations of the HB charged molecules.

The spectroscopic-active medium is regarded as that comprising *pairs* of the HB molecules. Hence, we consider a model of their *collective motions* and the relevant dielectric relaxation. The point is that for the dimer scheme, shown in Fig. 42a, it *would be incorrect* to represent the spectral function (SF) $L_\perp$ as a *sum* of functions corresponding, respectively, to the positively and negatively charged molecules suffering vibrations (a similar reasoning was used in Section VI).

Let a unit volume of an isotropic medium comprise $N_{vib}/2$ of such pairs (nonrigid dipoles). We shall calculate the generated complex susceptibility $\chi_\perp$ by using the *high-frequency approximation* for which it is assumed that at the instant just after a strong collision the velocities and position coordinates are given by the Boltzmann distribution (marked by the subscript B). Then, in view of Eq. (3.5) in GT1, the complex susceptibility χ is proportional to the spectral function L:

$$\chi^* = \chi' + i\chi'' \equiv \chi_B^* = GL, G = \mu^2 N/(3k_B T) \qquad (161)$$

where the constant moment μ pertains to the reorienting permanent dipole.

As applied to our problem, pertaining to a nonrigid "elastic" dipole, we rewrite Eq. (163), bearing in mind the susceptibility $\chi_\perp$ generated by transverse vibration and contributed by both positive and negative *charges* of each nonrigid dipole. In (163), replacing μ by the moment $\bar{\mu}$ of elastic dipole at equilibrium and omitting the complex-conjugation symbol and the subscript B, we have

$$\chi_\perp = G_{vib}L(\hat{\omega}), \qquad G_{vib} = (1/3)(\bar{\mu}^2 N_{vib})/(k_B T) \qquad (162)$$

---

[24]A permissibility of similar approximation (formulated as uncoupling of longitudinal vibration and elastic reorientation) was proved in Section VI.

$$\bar{\mu} = q\gamma r_{OO} = q\gamma(r+l) \tag{163}$$

where $r_{OO} \equiv r + l$ is the distance between two neighboring oxygen atoms, $\hat{\omega}$ denotes the complex frequency

$$\hat{\omega} = \omega + i\tau_{\perp}^{-1} \tag{164}$$

$\tau_{\perp}$ is the TV lifetime, and G_{vib} is the normalized concentration of vibrating particles.

In accord with the general definition [GT2, Eq. (22b)] the dipolar autocorrelator (spectral function) can be written in our notation as

$$L(z) = 3i\hat{\omega}\left\langle Q_0 \int\limits_0^\infty (Q - Q_0)\, \exp(i\hat{\omega}t)\, dt \right\rangle \tag{165}$$

where angular brackets denote an ensemble average and

$$Q \equiv Q(t) = \frac{\mu_E(t)}{\bar{\mu}} = \frac{q\gamma d_E(t)}{\bar{\mu}} = \frac{d_E(t)}{l+r}, \quad Q_0 \equiv Q(0) \tag{166}$$

We emphasize the change of notation as compared with GT2 and Gaiduk et al. [56]: Instead of $q(t)$ and z/η, the variable $Q(t)$ and the complex frequency $\hat{\omega}$ are employed, where $Q(t)$ is the quantity proportional to the projection (denoted by the subscript E) of the total dipole moment onto the time-varying external field $\mathbf{E}$.

Now we shall express the projection d_E through the angles Θ and Ψ pertinent to orientation of the symmetry axis O_-O_+. Here Θ is the angle between field $\mathbf{E}$ and the O_-O_+ axis and Ψ is the angle between the projections d_S and E_S of vectors $\mathbf{d}$ and $\mathbf{E}$ on the plane S normal to O_-O_+ axis (see Fig. 42b). In view of this figure, projections of $\mathbf{d}$ onto the symmetry axis and plane S are, respectively, $l + r$ and $2b$. Using the relationship

$$d_E = (\text{projection of } \mathbf{d} \text{ onto the symmetry axis})_E$$
$$+ [(\text{projection of } \mathbf{d} \text{ onto plane S})_{E_\perp}]_E$$

we have

$$d_E = (l+r)\cos\Theta + 2b\sin\Theta\cos\Psi = (l+r)(\cos\Theta + s\sin\Theta\cos\Psi) \tag{167}$$

Here we introduce a small dimensionless transverse displacement s of a water molecule from the equilibrium position, occurring at a rather small time interval:

$$s \equiv 2b(l+r)^{-1} \tag{168}$$

To obtain the average in (165) over the angles Θ and Ψ, we take into account that:

(i)   the element of a unit sphere surface is equal to $\sin\Theta d\Theta d\Psi$ (see Fig. 42b);

(ii) $\Theta \in [0, \pi/2]$; $\langle \cos^2 \Theta \rangle = \int_0^{\pi/2} \cos^2 \Theta \sin \Theta d\Theta = \frac{1}{3}$; $\Psi \in [0, 2\pi]$, $\langle \cos^2 \Psi \rangle = \frac{1}{2}$; and

(iii) the angles Θ and Ψ do not change with time.

Hence,

$$3\langle Q_0 Q(\varphi)\rangle_{\Theta,\Psi} = \langle(\cos\Theta + s_0\sin\Theta\cos\Psi)[\cos\Theta - \cos\Theta + (s - s_0)\sin\Theta\cos\Psi]\rangle_{\Theta,\Psi}$$
$$= 3s_0(s - s_0)\langle\sin^2\Theta\cos^2\Psi\rangle_{\Theta,\Psi} = s_0(s - s_0)$$

$$(169)$$

Then Eq. (165) reduces to

$$L_\perp(\omega) = i\hat{\omega}\left\langle\int_0^\infty s_0(s - s_0)\exp(i\hat{\omega}t)\,dt\right\rangle \qquad (170)$$

Thus we have transformed the autocorrelator (165). Now we shall calculate the averages over coordinates and velocities of the particles with account of their law of motion, governed by elastic force. It is convenient to employ the following representation of Eq. (170):

$$L_\perp(\omega) = \langle s_0 \hat{S}_1(\omega)\rangle \qquad (171)$$

where

$$\hat{S}_n(\omega) = i\hat{\omega}\int_0^\infty (s^n - s_0^n)\exp(i\hat{\omega}t)\,dt, \qquad n = 1, 2, 3, \dots \qquad (172)$$

B.   Nonlinear Equation of Motion

To calculate the average (171) we first express the strained-state potential energy U through the variable s (168). The energy U is related to the HB lengthening Δl as

$$U = (1/2)k_\perp(\Delta l)^2 \qquad (173)$$

Noting (see Fig. 42a) that $l + \Delta l = \sqrt{l^2 + 4b^2} \approx l + 2b^2 l^{-1}$, we have in view of (168):

$$\Delta l \approx \frac{2b^2}{l} = \frac{2(l+r)^2}{l} \frac{s^2}{4} = l(1+a)^2 \frac{s^2}{2}$$

Substitution into (173) gives the expression

$$u = \frac{U}{k_B T} = w s^4, \qquad w \equiv \frac{k_\perp l^2 (1+a)^4}{8 k_B T} \tag{174}$$

for the dimensionless potential energy u. Noticing that the kinetic energy H_{kin} of two vibrating particles is $2(1/2)\, m(db/dt)^2$, we derive the formula for the full dimensionless energy $h_\perp$:

$$h_\perp = \frac{H}{k_B T} = \frac{H_{\text{kin}} + U}{k_B T} = \frac{m}{4 k_B T}(l+r)^2 \left(\frac{ds}{dt}\right)^2 + w s^4 \tag{175}$$

Now we introduce the dimensionless time φ and complex frequency $z_\perp$:

$$\varphi = \frac{t}{\eta_\perp}, \qquad z_\perp \equiv \hat{\omega}\eta_\perp = x_\perp + i y_\perp, \qquad x_\perp = \omega \eta_\perp, \qquad y_\perp = \eta_\perp / \tau_\perp \tag{176a}$$

where

$$\eta_\perp \equiv p_\perp (l+r) \sqrt{\frac{m}{4 k_B T}} = \frac{1}{2} p_\perp l (1+a) \sqrt{\frac{m}{k_B T}}, \qquad \dot{s} \equiv ds/d\varphi \tag{176b}$$

Here we have additionally introduced the *frequency factor* $p_\perp$. Varying the latter allows us to control the position $\nu_\perp$ of the maximum of the loss curve $\chi_\perp''(\nu)$.

Noticing that the full (constant in time) energy $h_\perp$ is expressed as

$$h = \dot{s}^2 + w s^4 \tag{177}$$

(we further omit the symbol $\perp$), we derive the nonlinear equation of motion

$$\ddot{s} = -2 w s^3 \tag{178}$$

which describes a nonharmonic transverse vibration of the HB molecules.

C.   Transverse Spectral Function

1.   *Transformation of Eq. (172) to Infinite System of Recurrence Equations*

Using the set of linear recurrence equations, we calculate approximately the integral (172) with account of the expression (177) for energy h and of equation of motion (178).

First, we rewrite Eq. (172), replacing $\hat{\omega}$ by $z_\perp/\eta_\perp$ and $\hat{S}_n(\hat{\omega})$ by $\hat{S}_n(z_\perp)$. Below, to simplify notation, we set $z_\perp \to z$ and $\hat{s}_n(z_\perp) \to \hat{s}_n$. Then we integrate the so modified Eq. (172) by parts, noting that the surface term vanishes at both limits (at $x \to \infty$ due to the existence of a positive imaginary part y of the complex frequency $z = x + \mathrm{i}y$).

Next, we integrate by parts once more:

$$\hat{s}_n = iz \int_0^\infty (s^n - s_0^n) \exp(iz\varphi)\, d\varphi = -n \int_0^\infty s^{n-1}\dot{s}\, \exp(iz\varphi)\, d\varphi$$

$$= \frac{n}{iz} \left\{ -s^{n-1}\dot{s}\, \exp(iz\varphi)\Big|_0^\infty + \int_0^\infty \left[s^{n-1}\ddot{s} + (n-1)\dot{s}^2 s^{n-2} \right] \exp(iz\varphi)\, d\varphi \right\}$$

$$(179)$$

The surface term vanishes again (due to its oddness) after ensemble averaging over velocity $\dot{s}$ in the infinite interval $[-\infty, \infty]$. Substituting $\ddot{s}$ from (178) and $\dot{s}$ from (177), we have

$$\hat{s}_n = -\frac{in}{z} \int_0^\infty \left[(n-1)s^{n-2}(h - ws^4) - 2ws^3 s^{n-1} \right] \exp(iz\varphi)\, d\varphi$$

Here we rearrange terms as follows:

$$\hat{s}_n = \frac{in}{z} \int_0^\infty \left[(n+1)w\left(s^{n+2} - s_0^{n+2} + s_0^{n+2}\right)\frac{iz}{iz} - (n-1)h\left(s^{n-2} - s_0^{n+2} + s_0^{n+2}\right)\frac{z}{z} \right] \exp(iz\varphi)\, d\varphi$$

Accounting for the definition (179), we may rewrite this relationship as

$$\hat{s}_n = \frac{n}{z^2}(n+1)w(\hat{s}_{n+2} - s_0^{n+2}) - n(n-1)\frac{h}{z^2}(\hat{s}_{n-2} - s_0^{n+2}) \qquad (180)$$

Thus, we have obtained an infinite system of linear recurrence equations for the complex variables $\hat{s}_n$, where the deflection s_0 is small (the subscript 0 is further omitted).

For calculation of the average (171), we restrict our consideration by equations with $n = 1$ and $n = 3$, omitting terms of the order of s^4 (and smaller).

Then we have

$$\hat{s}_1 = 2\frac{w}{z^2}(\hat{s}_3 - s^3) \quad \text{for } n = 1; \quad \hat{s}_3 = -6\frac{h}{z^2}(\hat{s}_1 - s) \quad \text{for } n = 3$$

whence we get the system of two linear equations for $\hat{s}_1$ and $\hat{s}_3$.

$$\hat{s}_1 - 2\frac{w}{z^2}\hat{s}_3 = -2\frac{w}{z^2}s^3, \qquad 6\frac{h}{z^2}\hat{s}_1 + \hat{s}_3 = 6\frac{h}{z^2}s \qquad (181)$$

Its solution for $\hat{s}_1$ is given by

$$\hat{s}_1 = \left(12\frac{hws}{z^4} - 2\frac{ws^3}{z^2}\right)\left(1 + 12\frac{hw}{z^4}\right)^{-1} = \frac{2w(6hs - z^2s^3)}{z^4 + 12hw} \qquad (182)$$

Now we combine this expression with Eq. (171), where we replace s_0 by s, replace designation $L_\perp(\hat{\omega})$ by $L_\perp(z)$ and restore designation $z_\perp$, viz. $z \to z_\perp$. Hence, we obtain the expression for the transverse spectral function $L_\perp(z_\perp)$ as a *correlator*:

$$L_\perp(z_\perp) = 2w\left\langle\frac{6hs^2 - z_\perp^2 s^4}{12hw + z_\perp^4}\right\rangle \qquad (183)$$

2.   The Norm C

To write down Eq. (183) in an analytical form, we first take the phase-volume element $d\Gamma$ as the product of two arbitrary constants of integration, h and φ_0. They constitute a pair of canonically conjugated phase variables:

$$d\Gamma = dh\, d\varphi_0 \qquad (184)$$

where φ_0 is an "initial instant" of time. The law of motion of a conservative system (178) generally can be represented as $s(\varphi, \varphi_0, h)$. Since φ_0 comes additionally with the current time φ, one may replace this as $s(\varphi + \varphi_0, h)$ and replace the element (184) by

$$d\Gamma = dh\, d\varphi \qquad (185)$$

In view of Eq. (177) the velocity $\dot{s}$ and coordinate s are related by

$$\dot{s} = \sqrt{h - ws^4} \qquad (186)$$

so the maximum transverse displacement $s = s_m$ will be reached at zero velocity, when

$$s = s_m(h) = (h/w)^{1/4} \tag{187}$$

The solution $s(\varphi)$ of the equation of motion (178) is generally a periodic function. Let $\Phi(h)$ be its period. Integration in Eq. (183) over the phase volume Γ should be taken over all representative initial instants (phases) φ_0. Therefore we choose the time interval $\varphi \in [0, \Phi/4]$, during which the coordinate s varies from 0 to the amplitude value $s_m(h)$. Let the energy h be distributed in the interval $[0, u_{\lim}]$. In accord with the definition of the Boltzmann distribution, the following equality holds:

$$\int W \, d\Gamma = C \int_0^{u_{\lim}} \exp(-h) dh \int_0^{\Phi(h)/4} d\varphi = \frac{1}{4} C \int_0^{u_{\lim}} \exp(-h) \Phi(h) dh = 1 \tag{188}$$

Here C is the norm to be found. Solving Eq. (186), we find the period $\Phi(h)$:

$$\Phi(h) = 4 \int_0^{(h/w)^{1/4}} \frac{ds}{\sqrt{h - ws^4}} = \frac{4}{(wh)^{1/4}} \int_0^1 \frac{dt}{\sqrt{1 - t^4}} = 2\sqrt{2}\,(wh)^{-1/4} \mathbf{K}(1/\sqrt{2}) \tag{189}$$

where $\mathbf{K}$ is the full elliptic integral of the first kind, $\mathbf{K}(1/\sqrt{2}) \approx 1.35$.
Hence, the norm C, found from Eqs. (188) and (194), is given by

$$C = 4 \left[\int_0^{u_{\lim}} \Phi(h) \exp(-h) \, dh \right]^{-1} = \left[2\sqrt{2w}\, \mathbf{K}(1/\sqrt{2}) \int_0^{s_{\lim}} s^2 \exp(-ws^4) \, ds \right]^{-1} \tag{190}$$

We have formally introduced here the limiting displacement $s_{\lim}$, related to the energy $u_{\lim}$ analogously with (187):

$$s_{\lim} = (u_{\lim}/w)^{1/4} \tag{191a}$$

so that conversely

$$u_{\lim} = ws_{\lim}^4 \tag{191b}$$

We regard $s_{\lim}$ as a free parameter of the model. Note our theory holds, if $s_{\lim} \ll 1$.

Now we write down the formula for the dimensional transverse-vibration angular frequency $\Omega_\perp(h)$. Accounting for Eq. (189), we have

$$\Omega_\perp(h) = \frac{2\pi}{\eta_\perp \Phi(h)} = \frac{\pi (wh)^{1/4}}{\eta_\perp \sqrt{2} \, \mathbf{K}(1/\sqrt{2}) \, \Phi(h)} \tag{192a}$$

This frequency depends on the energy h as $h^{1/4}$ and, in view of Eq. (174), on the force constant $k_\perp$ as $(k_\perp)^{3/4}$. To estimate the mean $\Omega_\perp$-value and the corresponding mean frequency $\langle v_\perp \rangle = \langle \Omega_\perp \rangle/(2\pi c_{\text{lt}})$, expressed in cm^{-1}, we find the average Eq. (189) over h:

$$\langle \Phi(h) \rangle \approx \int_0^{h_{\lim}} \Phi(h) \, \exp(-h) \, dh \bigg/ \int_0^{h_{\lim}} \exp(-h) \, dh = \frac{4}{C[1 - \exp(-u_{\lim})]}$$

or, equivalently,

$$\langle \Phi(h) \rangle = \frac{\langle T_\perp \rangle}{\eta_\perp} \approx \frac{2\pi}{\langle \Omega_\perp \rangle \eta_\perp} = \frac{c}{\langle v_\perp \rangle \eta_\perp} \tag{192b}$$

Comparing with Eq. (190), we derive the relationship[25]

$$\langle v_\perp \rangle \approx [1 - \exp(u_{\lim})] \left[8\sqrt{2} \, \mathbf{K}\!\left(1/\sqrt{2}\right) c \eta_\perp \int_0^{s_{\lim}} s^2 \, \exp(-ws^4) \right]^{-1} \tag{193}$$

which could be used for parameterization of the model.

3. Analytical Formula for Transverse Complex Susceptibility

To find the average in (183), we choose another pair of canonically conjugated phase variables—the coordinate s and the velocity $\dot{s}$. Hence, the element of the phase volume is represented now as:

$$d\Gamma = ds \, d\dot{s} \tag{194}$$

For clarity, for the example of a harmonic oscillator let us demonstrate the equivalence of the representations (185) and (194). The latter obeys the

---

[25]The expression (193) is applicable for rather rare collisions, when the collision frequency $y_\perp$ introduced in Eq. (179b) is not very large.

equation of motion $\ddot{s} + p^2 s = 0$ with $p = \text{const}$ and $u_{\text{lim}} \to \infty$. Now the Hamiltonian is $h(s, \dot{s}) = (1/2)\dot{s}^2 + (1/2)p^2 s^2$. For the representation (185) the period $\Phi = 2\pi/p$, so that Eq. (188) gives $C = 4/\Phi = 2p/\pi$. On the other hand, we obtain *the same result* $C = 2p/\pi$ using (194) and taking into account that

$$C \int_0^\infty \int_0^\infty \exp\left(-\frac{\dot{s}^2}{2} - \frac{p^2 s^2}{2}\right) d\dot{s}\, ds = C\frac{2}{p}\left(\int_0^\infty \exp(-t^2)\, dt\right)^2 = 1$$

Equivalence of the forms (185) and (194) is proved.

Now we shall calculate the average (183). It follows from Eq. (186) that $d\dot{s} = (1/2)(h - ws^4)^{-1/2} dh$. Then Eq. (194) gives

$$d\Gamma = (1/2)(h - ws^4)^{-1/2} dh\, ds \tag{195}$$

We write down an explicit formula for the ensemble average of Eq. (183):

$$L_\perp(z) = 2wC \int_0^{u_{\text{lim}}} \int_0^{s_m(h)} \frac{6hs^2 - z_\perp^2 s^4}{12hw + z_\perp^4} \frac{\exp(-h)\, dh\, ds}{2\sqrt{h - ws^4}}$$

$$= wC \int_0^{u_{\text{lim}}} \frac{dh\, \exp(-h)}{12hw + z_\perp^4} \left[6h \int_0^{s_m(h)} \frac{s^2 ds}{\sqrt{h - ws^4}} - z_\perp^2 \int_0^{s_m(h)} \frac{s^4 ds}{\sqrt{h - ws^4}} \right]$$

$$\tag{196}$$

where the norm C is given by Eq. (190). The integrals over s could be taken analytically. The first integral in the square brackets in Eq. (196) is

$$\int_0^{s_m} \frac{s^2 ds}{\sqrt{h - ws^4}} = \frac{1}{\sqrt{h}} \int_0^{s_m} \frac{s_m^3 \left(s/s_m\right)^2 d^s/s_m}{\sqrt{1 - \left(s/s_m\right)^4}} = \frac{1}{\sqrt{w}} \int_0^1 \frac{t^2 dt}{\sqrt{1 - t^4}} = \frac{s_m D}{\sqrt{2w}} \mathbf{K}\left(1/\sqrt{2}\right)$$

$$\tag{197}$$

where $E(\cdot)$ is the full elliptic integral of the second kind, $D \equiv 2\dfrac{\mathbf{E}\left(1/\sqrt{2}\right)}{\mathbf{K}\left(1/\sqrt{2}\right)} - 1$ $= 0.627$.

To find the second integral in square brackets in Eq. (196), we write

$$S \equiv \int_0^{s_m} \frac{s^4 ds}{\sqrt{h - ws^4}} = -\int_0^{s_m} \frac{s}{2w} \, d\sqrt{h - ws^4}$$

Integrating by parts and omitting the vanishing "surface" term, we have for S the linear equation

$$S = \frac{1}{2w} \int_0^{s_m} \frac{(h - ws^4) ds}{\sqrt{h - ws^4}} = -\frac{S}{2} + \frac{h}{2w} \int_0^{s_m} \frac{ds}{\sqrt{h - ws^4}}$$

The integral on the right-hand side is found with account of (197):

$$S = (3\sqrt{2w})^{-1} \mathbf{K}\left(1/\sqrt{2}\right) s_m^3$$

Substituting (190), (197) and the expression for S into (196) and tripling the result for generalization to three dimensions, we express the spectral function $L_\perp$ in analytic form as an ordinary integral over the variable s_m. Omitting the subscript m, we finally have

$$L_\perp(z_\perp) = 3w \int_0^{s_{\lim}} \frac{(18Dws^2 - z_\perp^2) \, s^6 \, \exp(-ws^4) \, ds}{12w^2 s^4 + z_\perp^4} \Bigg/ \int_0^{s_{\lim}} s^2 \, \exp(-ws^4) \, ds$$

$$(198)$$

In view of Eqs. (162) and (163), the transverse vibration complex susceptibility $\chi_\perp$ is

$$\chi_\perp = (6\pi)^{-1} g_\perp G_\perp L_\perp(z_\perp) \tag{199a}$$

where

$$G_\perp = \pi (k_B T)^{-1} \mu_\perp^2 N_{\mathrm{vib}} \tag{199b}$$

In Eq. (199a) we have introduced the *intensity parameter* $g_\perp$ connecting the transverse ($\mu_\perp$) and longitudinal $\bar\mu$ dipole moments of vibrating HB particles as

$$\mu_\perp^2 = g_\perp \bar\mu^2 \equiv g_\perp [q\gamma(r + l)]^2 \tag{199c}$$

Note that, as follows from the parameterization of the model, the moments $\bar\mu$ and $\mu_\perp$ may substantially differ. We interpret this property as a manifestation of molecules association in aqueous medium.

For practical calculations of the transverse susceptibility $\chi_\perp$ and of the relevant contribution $\Delta\varepsilon_\perp$ of transverse vibrations to the total permittivity, given in the high-frequency approximation by

$$\varepsilon_\perp(v) \approx 6\pi\chi_\perp(v) \tag{200}$$

we should specify the dimer dimensions $r_{OO} = r + l$, $a = r/l$, the mass m, and the parameters of the current TV model: moment $\mu_\perp$, limiting displacement s_{lim}, frequency parameter $p_\perp$ involved in Eq. (176), force constant $k_\perp$ and normalized strong-collision frequency $y_\perp$, as well as the molecular constants, involved in the relationship

$$N = r_{vib}N_A M^{-1} \tag{201}$$

Here N_A is the Avogadro number, ρ the density, M the molecular mass, and r_{vib} the proportion of the vibrating particles.

We take into account the mean vibration period $<T_\perp> \approx <\Phi> /\eta_\perp$. During the lifetime $\tau_\perp \approx \eta_\perp/y_\perp$, when the molecular configuration under consideration exists, approximately $m_\perp = \tau_\perp / <T_\perp> \approx (y_\perp <\Phi>)^{-1}$ vibration cycles occur. It follows from Eqs. (190) and (192b) that

$$m_\perp \approx C[1 - \exp(-u_{lim})] (4y_\perp)^{-1} = [1 - \exp(-u_{lim})]$$

$$\times \left[8\sqrt{2w}\, \mathbf{K}\left(1/\sqrt{2}\right) y_\perp \int_0^{s_{lim}} s^2 \exp(-ws^4)\, ds \right]^{-1} \tag{201}$$

Just as for Eq. (193), this formula is applicable provided the parameter $y_\perp$ is not too large.

D.   Influence of Frequency Factor on Transverse-Vibration Loss Spectrum

Such illustration, calculated from Eqs. (199) and (200), is given in Fig. 43 for water at the temperature 27°C. The employed model parameters are close to those presented in Table I for H_2O at this temperature. In Fig. 43a we depict the loss-contribution $\varepsilon_\perp''(v)$ to the total complex permittivity ε. We take two values of the frequency factor $p_\perp$, introduced in Eq. (176b). The solid curve 1 refers to $p_\perp = 0.65$. The dashed curve 2 in Fig. 4a refers to $p_\perp = 1$ (without change of the rest of the parameters).

For greater $p_\perp$, the frequency $v_{\perp m}$ of the loss maximum is lower. The ratio $v_{\perp m}(p_2)/v_{\perp m}(p_1)$ is equal to the ratio p_1/p_2. The factor w, estimated from Eq. (174), is equal to ≈ 36. The limiting displacement $s_{lim} \approx 0.2$, introduced in Eq. (191a), determines the total (viz., in both directions relative to the symmetry axis) transverse shift with the amplitude $2b_m \approx 0.57$ Å. This value comprises

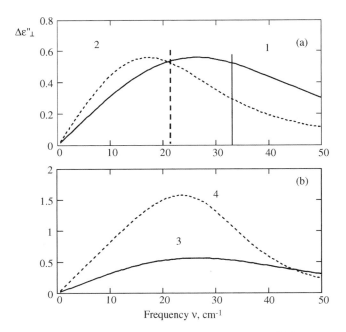

Figure 43 Frequency dependence of the transverse-vibration loss $\varepsilon''_\perp(v)$ calculated for water H_2O at the temperature 27°C. $p_\perp = 0.65$ (for curves 1, 3, 4) and $p_\perp = 1$ (for curve 2); $y_\perp = 2.8$ (for curves 1, 2, 3) and 2 (for curve 4). The vertical lines mark the TV-band center frequency, estimated from Eq. (193).

about half of the covalent-bond length O–H. The reduced energy $u_{\lim}$, evaluated from Eq. (191b), *is very small* (≈ 0.042). Such a smallness of $s_{\lim}$ allows application of the theory given above for transverse vibration. The normalized strong-collision frequency $y_\perp$ is rather large[26] ($y_\perp = 2.8$); in accord with Eq. (176b) the lifetime $\tau_\perp$ is then about 0.1 ps.

The dielectric response shown in Fig. 43a resembles a highly damped resonance curve with the peak frequency $v_{\perp m}$ located in the THz region. The $\varepsilon''_\perp(v)$ spectra in Fig. 43b are compared for $y_\perp = 2.8$ (curve 1) and $y_\perp = 2$ (curve 2), for which the lifetimes $\tau_\perp$ are respectively ≈ 0.089 and 0.14 ps. For greater $y_\perp$ (i.e., for shorter lifetime $\tau_\perp$) the absorption curves become shallower. The frequency $v_{\perp m}$ of the loss-peak intensity, marked in Fig. 43b by the vertical lines, agrees with the estimate (193) only approximately. It is worthwhile to emphasize the following:

(i) The transverse force constant is less than the longitudinal one: $k_\perp/k \approx 0.3$.

---

[26]For such great $y_\perp$ Eqs. (193) and (201) still give rather correct estimates.

(ii) Closeness of the lifetimes τ_q and $\tau_\perp$, pertinent to translational and transverse vibrations of the HB molecules (0.076 and 0.089 ps, respectively).

(iii) A very small energy interval (up to $0.04\, k_B T$), in which the transversely vibrating particles are distributed.

The smallness of s_{lim} could be interpreted as follows. It is perhaps a steric effect due to existence in a condensed medium of neighboring molecules that brings to a stop slow molecules moving transversely. Result (ii) possibly allows us to employ to a good approximation the same lifetime τ_{vib} for both vibration motions.[27] Item (iii) suggests the idea that a certain hierarchy of energies could be a characteristic feature of elastic or quasi-elastic vibrations: The smallest energy interval (from 0 to u_{lim}) corresponds to vibrations transverse to the H-bond, while the longitudinal vibrations are connected with a much larger energy interval extending up to the energy u_{HB} that is related to the HB break.

E. Manifestation of Transverse Vibrations in Water/Ice Spectra

As seen from Fig. 43a, varying the frequency factor $p_\perp$ we, indeed, may change the TV band center frequency $v_\perp$. Correspondingly, variation of the intensity factor $g_\perp$ allows changing the TV loss maximum $\varepsilon''_\perp$.

Now we shall briefly list the effects produced by transverse vibration (TV effects) regarding the water/ice spectra (more details were given in the first part of this work—in Sections II–V).

1. For *water* at 27°C, one may compare the position of the TV band (Fig. 3e) with positions of the translational (Fig. 3a) and V-band (Fig. 3c). We see that the TV band is located to the left of the other two. For *ice* at −7°C a similar comparison could be made from Figs. 3f, 3b, and 3d. Evidently, the TV-band is two orders of magnitude wider in ice than in water and is characterized by a several times smaller amplitude $\varepsilon''_{\perp m}$.

   Comparison of the curve 4 in Figs. 4 and 5 pertaining to water at 20.2°C shows that the TV band, unlike the librational band, remains approximately the same in ordinary and heavy water.

2. The change of temperature affects *weakly* the TV-loss curve $\varepsilon''_\perp$ (cf. dash–dotted lines in Figs. 8b and 8d). On the contrary, in supercooled (SC) water the above curve strongly widens without a noticeable change of the maximum intensity (see Fig. 8f). It appears that in this respect, SC water and ice are revealed as kindred fluids.

3. The *resonance* ice peaks to a good approximation could be described in the far-IR region, as quasi-resonance peaks in water, without account of

---

[27]This assertion does not hold for the case of ice.

transverse vibration. The latter plays a leading role in the formation *in ice* of the V-band left wing (at frequencies lower than ≈ 100 cm^{-1}). It becomes clear if one compares Figs. 24a and 24b, calculated for ice at $-7°C$. This is especially true with respect to the *nonresonance* ice spectrum (see curves 1 in Figs. 26). At very low temperature (Fig. 30) the TV-loss curve becomes still more wide and intense.

4. Finally, perhaps the most noteworthy feature of transverse vibration concerns a very small kinetic energy pertaining to TV vibrating molecules. This energy is much lower than the H-bond energy U_{HB}. Small-energy phenomena generally play a very important role in aqueous fluids—for examples, regarding water clusters. One may suggest that noticeable just mentioned transverse-vibration effects originate, in spite of their low-energy level, since they are concerned with an *ensemble* of interacting molecules. To a certain extent, this feature is revealed in the fact that for the proper describing of the THz spectra it becomes necessary to involve in our theory the "effective" frequency and intensity factors ($p_\perp$ and $g_\perp$) instead of "rigorous" coefficients $p_\perp = 1$ and $g_\perp = 1$, relevant to only two transversally vibrating molecules.

VIII. SPATIAL CONFIGURATION PERTAINING TO HAT WELL. SUMMARY PERMITTIVITY

A. About Evolution of Hat Potential

In our "nomenclature" of physical mechanisms, introduced in Section II, *mechanism (a)* plays a key role in dielectric-relaxation theory. We remark that a potential, in which a particle performs a *nonharmonic* motion, is relatively seldom met in molecular spectroscopy.

The following three circumstances are characteristic for the hat-potential model:

(i) The motion equations are formulated for a *conservative system* pertaining to free (i.e., *without friction*) spatial libration of a rigid dipole in a certain hat-like potential well. Its walls represent a good model of the surrounding medium. The mathematical treatment of the problem is *not* rigorous, but the approximation used satisfactorily agrees with experiment.

(ii) The curvature of the potential-well bottom could be varied. As our calculations show, a properly fitted potential profile strongly depends (unlike its depth) on temperature.

(iii) We employ an analytically calculated ACF spectrum, specified as the "spectral function" (SF) $L(z)$, where $z \equiv x + iy$ is the complex frequency. Its imaginary part y determines the *lifetime τ of the potential well*. Varying this τ allows us to *directly* control the width of the loss curve $\varepsilon''(v)$ *without* involving the friction coefficient in the equations of motion. Note that such a possibility is lacking in the molecular-dynamics simulation method, where the loss bandwidth is determined only by the NVT data (see, e.g., Section X.A).

The hat-like potential model evolved from consideration of rotation of a permanent dipole between two perfectly reflecting plates inclined one with respect to the other through the angle 2β. The history of the hat potential is described in detail in the context of dielectric relaxation in liquids in GT2.

The hat potential has a flat bottom followed by steeply rising parabolic walls. The corresponding spectral function $L_{or}(z)$ was first given (without derivation) in [12]; a more sophisticated theroy is presented in Gaiduk et al. [56] and in GT2. In this work, substantially the simplified formulas (8)–(10) for this SF are used.

1. Planar Libration—Regular Precession Approximation

First, we mention briefly a "sophisticated theory" from GT2.

Let θ be the angular deflection of a dipole from the symmetry axis of the potential $U(\theta)$. The latter is characterized by a small ($\ll \pi/2$) angular halfwidth and by the welldepth U_0. Its reduced value $u = U_0/(k_B T)$ is assumed to be $\gg 1$. For brevity, we consider now a quarter-arc of the circle, where half of the well is localized. The bottom of the potential well is *flat* at $0 \leq \theta \leq f\beta$, where f is the form factor, defined as a ratio of this flat-part width to the whole width of the well. In the remainder of the well we take the *parabolic* dependence U on θ. Thus an assumed hat-potential profile (as illustrated by Fig. 44) is taken as

$$\frac{U(\theta)}{k_B T} = \left\{ \begin{array}{l} 0 \\ u\ (\theta - \beta f)^2 (\beta - \beta f)^{-2} \\ u \end{array} \right\} \quad \text{at} \quad \left\{ \begin{array}{l} 0 \leq \theta \leq \beta f \\ \beta f \leq \theta \leq \beta \\ \beta \leq \theta \leq \pi/2 \end{array} \right\} \tag{202}$$

In this figure the flat part $2\beta f$ and the total angular width 2β are shown as typical for water. The rectangular profile for which dielectric properties of fluids previously were rather successfully treated (see VIG and GT2) corresponds to $f = 1$.

In order to determine specific phase regions for a librator (a polar linear molecule moving in a well), we express the full energy H of this molecule as a function of the polar coordinate θ and of two angular velocities:

$$H(\theta, \frac{d\theta}{dt}, \frac{d\phi}{dt}) = \frac{I}{2}\left(\frac{d\theta}{dt}\right)^2 + \frac{I}{2}\left(\frac{d\phi}{dt}\right)^2 \sin^2\theta + U(\theta) \tag{203}$$

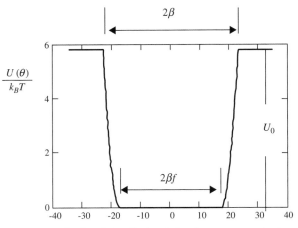

Deflection θ of a dipole relative the symmetry axis, deg

Figure 44   The form of the hat curved potential well. The chosen parameters are typical for water: the half-width of the well, $\beta = 23°$; the reduced well depth, $u = 5.8$; and the form factor, $f = 0.75$.

Polar angle θ denotes the angle between the dipole moment $\boldsymbol{\mu}$ and the symmetry axis of the potential. Then the first term in (203) refers to the change of the polar angle. The second term in (203) describes *precession* of a dipole about the symmetry axis, when the uthal coordinate ϕ varies, P_ϕ being a constant momentum. Indeed, since H does not depend on ϕ, we have $dP_\phi/dt = \partial H/\partial\phi = 0$, so that

$$P_\phi = \frac{\partial H}{\partial(d\phi/dt)} = I\frac{d\phi}{dt}\sin^2\theta = \text{const}, \qquad \frac{I}{2}\left(\frac{d\phi}{dt}\right)^2\sin^2\theta \equiv \frac{P_\phi^2}{2I\sin^2\theta} \quad (204)$$

We rewrite Eq. (204) for the energy H by taking into account Eqs. (202) and (203), dividing both sides of (204) by k_BT, replacing *inside the well* (where $\theta \leq \beta \ll 1$)$\sin^2\theta$ by θ^2. Introducing the reduced full energy h and other quantities as follows:

$$h = \frac{H}{k_BT}, \quad \varphi = \frac{t}{\eta}, \quad \dot\theta \equiv \frac{d\theta}{d\varphi}, \quad \dot\phi \equiv \frac{d\phi}{d\varphi} = \frac{l}{\sin^2\theta}, \quad l = \frac{P_\phi}{\sqrt{2Ik_BT}}, \quad \eta = \sqrt{\frac{I}{2k_BT}} \quad (205)$$

we have instead of (203) the equation of motion in the form

$$V(\theta) = h - \dot\theta^2 = \begin{cases} l^2/\theta^2 \\ l^2/\theta^2 + \dfrac{u(\theta - \beta f)^2}{(\beta - \beta f)^2} \\ l^2/\sin^2\theta + u \end{cases} \quad \text{at} \quad \begin{cases} 0 \leq \theta \leq \beta f \\ \beta f \leq \theta \leq \beta \\ \beta \leq \theta \leq \pi/2 \end{cases} \quad (206)$$

We have introduced here the *effective potential* $V(\theta)$. Equation (206) formally coincides with the similar equation for a one-dimensional rotation of a dipole in such the potential $V(\theta)$. The potential (206) attains its minimum at some angle $\bar{\theta}$, which generally is less than β. We remark that in the case of rectangular well V is minimal at $\theta = \beta$.

The differential equation (206) determines the trajectory of any given dipole. Since $\dot{\theta} = \sqrt{h - V(\theta)}$ and $V(\theta) > 0$, it follows from (206) that the solution of its equation exists *only if* $h \leq v(\theta)$. Therefore the energy h of a particle, *localized in the well*, cannot exceed $V(\beta)$. The laws of motion $\{\theta(t), \phi(t)\}$ are such, that the orientation θ of a dipole relative the symmetry axis *oscillates* between two values $\theta_{\min}$ and $\theta_{\max}$, at which $\dot{\theta} = 0$, namely $h = V(\theta)$.

An analysis given in GT2 shows that three phase regions $\mathcal{L}, \mathcal{P}, \mathcal{R}$ exist. The quantities related to them are marked, respectively, by the superscripts $\cup$, $\sim$, and $\circ$.

The regions $\mathcal{L}$ and $\mathcal{P}$, to which the spectral functions $\check{L}(z)$ and $\tilde{L}(z)$ correspond, are occupied by the particles localized *in the well*. Any $\mathcal{L}$-particle penetrates into the flat part of the well. We may conditionally call it *"librator."* The $\mathcal{P}$-particle, conditionally called *"precessor,"* move only in the parabolic part of the well and has energy h close to that, for which the effective potential $V(\theta)$ undergoes its minimum, in which

$$V(\bar{\theta}) = [V(\theta)]_{\min} \tag{207}$$

Motion of a $\mathcal{P}$-particle resembles regular precession of a dipole around the symmetry axis.

In region $\mathcal{R}$ the trajectory of an $\mathcal{R}$-particle (we conditionally call it *"rotator"*) is *not* limited by the width of the well. Such a particle performs *complete rotation*, its spectral function is denoted L.

Estimates show that the potential minimum is low only for *small l values* (such tha $l^2 < 0.1$). Therefore, the particles with small axial momenta yield the main contribution to the spectrum stipulated by reorienting dipoles. At such small l-values the angle θ usually falls into the range 15–18 degrees; that is, θ is turned out to be rather close to the libration amplitude β equal to $\approx 23°$.

To *simplify* the analytical expressions for the SF $L(z)$, we may distribute all the particles, moving *in* the well, among two groups, specified by the following values of the reduced momentum l.

1. The first group, with $l = 0$, to which the SF $\check{L}(z)$ corresponds, comprises all the $\mathcal{L}$-particles. They move in the *planes* including the symmetry axis (viz. in the cross section of the well), so that the minimum θ of the effective potential is placed in the middle of the well. Then Eqs. (206) reduce to

$$V(\theta) = h - \dot{\theta}^2 = \begin{Bmatrix} 0 \\ (\theta - \beta f)^2 s^2 \end{Bmatrix} \quad \text{at} \quad \begin{Bmatrix} 0 \leq \theta \leq \beta f \\ \beta f \leq \theta \leq \beta \end{Bmatrix} \tag{208}$$

where the dimensionless steepness S of the hat well is defined as

$$S \equiv (\beta - \beta f)^{-1} \sqrt{u} \qquad (209)$$

2. The second group, with $\dot{\theta} \equiv 0$, and $\theta \equiv \bar{\theta}$ for which the spectral function is denoted $\tilde{L}(z)$, comprises all the $\mathcal{P}$-*particles*. They *precess with the constant azimuthal velocity* $\dot{\Phi}$ at the 'frozen' polar angle $\theta \equiv \bar{\theta}$, in which the effective potential is minimal. Particles of this group, unlike the librators, *do not* occupy the flat part of the hat curved potential well.

Assumptions 1 and 2 constitute the *planar libration–regular precession* approximation. In Gaiduk et al. [56] and in GT2 the corresponding spectral functions $\check{L}(z)$ and $\tilde{L}(z)$ are found in analytical form, as well as the SF $\mathring{L}(z)$ for the rotators. These SFs are expressed as simple integrals from elementary functions over the full energy of a dipole. The total spectral function is thus represented as

$$L(z) = \check{L}(z) + \tilde{L}(z) + \mathring{L}(z) \qquad (210)$$

2. Planar Libration Approximation

In recent versions of our theory, described after publication of GT2, it was shown that in water about 2/3 of permanent dipoles are librators and therefore librate rather freely in the hat-like potential, formed by strongly bent or torn hydrogen bonds. Such a motion is termed *reorientation* (or *libration*) in order to distinguish it from *angular turn* pertaining to rotation of a permanent dipole *about the H-bond*. The latter motion is far from being free: It is governed by the dimensionless rotary force constant c_{rot}.

Application of the relationship (210) for calculation of the libration band in water and in ice made evident the barest necessity to *simplify* the hat model, specified in Section VIII.A.1.

The features of this model, which initially were formulated due to a rather *abstract* reasoning, hardly could be reconciled with the structures of these fluids. It was supposed that a more realistic intermolecular potential should be used, in which one should disregard contributions to spectra (i) of the *"precessors,"* performing *rotation* in a parabolic part of the well around its symmetry axis and (ii) of the *rotators* moving with constant angular velocity *over* a potential barrier through a complete circle.

In our calculations, made in Sections II through V, the neglect of item (i) allows to avoid *worsening* of the theoretical absorption spectrum. One may suppose that the trajectories, ascribed to these precessors, cannot simulate "real" trajectories. It appears that the latter are mostly related to the change of polar coordinate θ in a *diametric* cross section of the well.

Item (ii) also appears to be excessively abstract. In the configuration, pertaining to water/ice, one hardly can find space, in which a water molecule, regarded as a rigid dipole, might rotate freely over a complete circle. That is why we have left in expression (210) only the first term. Hence, we have used for calculations the formulas (8)–(10) represented in Section II.B. The derivation of expression (8) is given in Gaiduk et al. [56] and in GT2, p. 166.

Hence, for the transition from the "full" expression (210) to the "simplified" (8)–(10) (a) in Eq. (175), containing the normalizing coefficient C, the contributions of precessors and rotators were disregarded. Moreover, we removed a misprint committed in GT2 and inserted additionally in the formula for D the factor d equal to ≈ 0.7. One may interpret this factor as that, which accounts for influence of surrounding molecules on the spectral function (8).

We remark additionally to the formulas given in Section I.B that the mean number m_{or} of the reorientation cycles, performed by a dipole in the hat well during its lifetime τ_{or}, is usually less than 5 and can be estimated as

$$
m_{\text{or}} \approx \frac{\tau_{\text{or}}\sqrt{3/2}}{4\beta\eta_{\text{or}}}\left[f + \frac{\pi}{2}(1-f)\sqrt{\frac{3}{2u}}\right]^{-1}
\tag{211}
$$

B. Variants of Spatial Configuration for Planar Libration

We consider here possible alternative structural schemes of librational motion of rigid dipoles. These schemes should, in principle, be in line with possible configurations of molecules in water and ice. For simplicity, we regard the case of *one dimensional motion*.

Except consideration of purely geometrical (steric) factors, we shall derive mathematical expressions describing the response of a water molecule, librating between other molecules. Hence, we consider a one-dimensional hat well fitted to strongly simplified to water/ice structure. We emphasize that such an attempt is made for the first time.[28]

We suggest two variants pertaining to interpretation of hat model.

1. Confined Precession of a Dipole around Hydrogen Bond (Fig. 45a)

Let us assume that a hydrogen atom H_a of a water molecule $H_aO_AH_b$ constitutes the H-bond (it is shown by a dashed line) with the other molecule, assumed to be immobile (the oxygen atom O_B of the latter is shown in Fig. 45a). A hydrogen

---

[28]The Sections VIII.B and VIII.C are written in collaboration with B. M. Tseitlin.

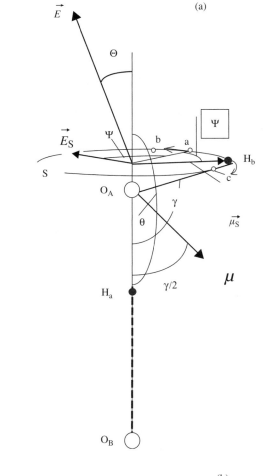

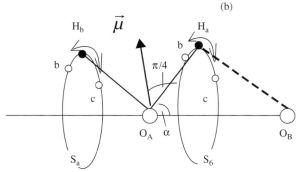

Figure 45 Structural schemes for the hat potential. The restrictions for the libration angle 2β are marked by symbols a and c. (a) The scheme illustrating libration of the hydrogen atom H_b about an immobile hydrogen bond H_a–O_B. (b) the scheme illustrating libration of hydrogen atoms H_a and H_b and a strained hydrogen bond H_a–O_B about the line O_A–O_B.

atom H_b performs rotary-recurrent motion around the H-bond, that is to say, around the line O_AO_B in the potential well $U(\psi)$. Here ψ is a current angular deflection of the covalent bond $O-H_b$ from its equilibrium direction, that is, from the middle of the well. Its width is marked by the symbols b and c. One may assume that such a width is determined by steric barriers constituted by neighboring molecules.

Regarding our problem, $U(\psi)$ represents the hat potential. The line O_B-H_a pertaining to the H-bond does not change its direction when the dipole moment $\vec{\mu}$ changes its orientation. (just as the line O_A-H_a of the covalent bond does not change its position). The atom H_b moves in the plane S, transversal to the cone axis, along the arc in the cone base. The hydrogen bond connecting the atom H_b with the third molecule (not shown in Fig. 45a) is assumed to be torn. We assume that deflection of the atom H_b from the middle of the well $U(\psi)$ could be noticeable. However, this deflection cannot exceed (again from steric restrictions) a quarter of circle. In view of Fig. 45a, the moment of inertia of a rotating H_b atom is

$$I = mr_{OH}^2 \sin^2 \theta \approx \frac{1}{2} mr_{OH}^2 \qquad (212)$$

where r_{OH} is the length of the covalent-bond O–H and m is the hydrogen-atom mass. Hence, the loss factor will substantially change if the hydrogen atom is replaced by the heavy-hydrogen one, which indeed is typical for the librational band.

2. Confined Precession of a Hydrogen Bond around $O \cdots O$ Direction (Figure. 45b)

Now we suggest that the atom H_a of the molecule $H_aO_AH_b$ precesses about the symmetry axis, namely about the straight line, on which the oxygen atoms O_A and O_B are located (see Fig. 45b). We assume that the precession angle is limited by steric restrictions, which are again marked by symbols b and c. We suppose that the like precession is characteristic also for the hydrogen atom H_A. Similar to the Fig. 45a, the hydrogen bond, connecting the atom H_A with the third molecule, is assumed to be torn. Both covalent bonds, H_aO_A and H_bO_A, move on conical surfaces, just as the hydrogen bond H_aO_B. Therefore the atoms H_a and H_b move in vertical planes along the same arcs of circles S_a and S_b.

Hence, unlike the scheme in Fig. 45a, now two (not one) hydrogen atoms turn and the H-bond direction does not coincide with the symmetry axis O_AO_B. We make the reservation that the H-bond length is suggested to be invariant, thus disregarding the elastic vibrations of molecules due to stretching-compression of this bond.

It follows from Fig. 45b that the moment of inertia, equal to

$$I = mr_{OH}^2 \left[\sin^2 \alpha + \sin^2 (\pi - \alpha - \pi/2) \right] = mr_{OH}^2 \qquad (213)$$

is about twice than that given by (212).

C. Spectral Function of Planar Hat Model

1. General Expression

In the first part of our work (Sections II through V) we used formulas (8)–(10) for the SF $L_{or}(z)$ of the hat potential describing the librational band in water and ice. Now we shall show that the spatial librations, which lead to these formulas (8)–(10), are "compatible" with planar motion of rigid dipoles, if we use the schemes, discussed in Section VIII.B.

Let us assume that all librators are localized in the well. At the bottom the potential

$$U(\theta) = 0 \quad \text{for} \quad -\beta f \le \theta \le \beta f$$

where θ is the deflection of a dipole from the symmetry axis in the plane of libration. At $|\theta| > \beta f$ the hat potential is assumed to be parabolic:

$$\frac{U(\theta)}{k_B T} = \frac{u(\theta - \beta f)^2}{\beta^2 (1 - f)^2} \qquad \text{for} \quad \beta f \le \theta \le \beta \qquad (214)$$

It is suggested that for $\theta \in [(-\beta, -\beta f]$ the dependence $U(\theta)$ is analogous. The smaller the f, the narrower the calculated absorption loss spectrum. Substituting Eq. (214) into the Hamiltonian for planar rotation, we have[29]

$$h(\theta, \dot{\theta}) = \dot{\theta}^2 + U(\theta)(k_B T)^{-1} \qquad (215)$$

where $\dot{\theta}$ means the derivative of θ over the normalized time $\varphi \equiv t/\eta$ and $\eta \equiv I(2k_B T)^{-1}$, I being the moment of inertia of a molecule. Differentiating (215) over time φ with use of (205), we come from (215) to the equation of motion:

$$\ddot{\theta} = 0 \quad \text{for} \quad 0 \le \theta \le \beta f \qquad \text{and} \qquad \ddot{\theta} = -\frac{u(\theta - \beta f)^2}{\beta^2 (1 - f)^2} \quad \text{for} \quad \beta f \le \theta \le \beta$$

$$(216)$$

---

[29]Equations (213) coincides with (202) if we exclude from consideration in the last formula the dipoles (termed 'rotators') moving outside the well and having energy greater than u. Setting also to zero the projection l of the momentum onto the plane of rotation, we find that the formula (160) from GT2 reduces to (215).

Since Eqs. (214)–(216), derived for one-dimensional motion, are applicable also for planar libration approximation, considered in GT2, we state that the obtained there expression for the libration period Φ and for Fourier amplitudes b_{2n-1}, are also valid. Thus we may use the formulas

$$\Phi = \frac{2\beta f}{\sqrt{h}}(1 + \sigma), \quad \sigma \equiv \frac{\pi}{2}\frac{1-f}{2f}\sqrt{h}/u \tag{217}$$

and

$$b_{2n-1} = -\frac{8\beta f(1+\sigma)^3 \sin\left(\dfrac{\pi}{2}\dfrac{2n-1}{1+\sigma}\right)}{\pi^2(2n-1)^2[(2n-1)^2\sigma^2 - (1+\sigma)^2]} \tag{218}$$

derived in GT2.

Now we find the norm C of the Boltzmann distribution $W = C \exp[-h(\Gamma)]$, where $h(\Gamma)$ represents dependence of the normalized energy $h \equiv H(k_B T)^{-1}$ on the phase variables Γ. We may choose for these variables the energy h and its canonically conjugated quantity—"initial" time φ_0. The latter is involved in the law of motion additively with the time variable φ. According to definition, C is inversely to the statistic integral st:

$$C \simeq \text{st}^{-1} \tag{219}$$

where

$$\text{st} = \iint \exp(-h)\, d\varphi_0 dh \equiv \int_0^u \Phi \exp(-h)\, dh \tag{220}$$

The dimensionless libration period Φ is determine by the formula in (217). Substitution of Φ in (A2) gives

$$\text{st} = 2\beta \int_0^u \left(\frac{f}{\sqrt{h}} + \pi\frac{1-f}{2\sqrt{u}}\right) e^{-h} dh$$

$$= 2\beta\sqrt{\pi}\left[f\,\text{erf}\left(\sqrt{u}\right) + \sqrt{\pi}\frac{1-f}{2\sqrt{u}}(1 - e^{-u})\right] \tag{221}$$

Then, combining this with (A.1) we have

$$C = \left(2\sqrt{\pi}D_{\text{flat}}\right)^{-1} \tag{222}$$

where D_{flat} is given further by Eq. (224b). Doubling expression (3.36) in GT1 for transversal spectral function, we write $\Phi = 2\pi/\breve{p}$ instead of $\breve{p}$, taking into account that in our case $K_{\parallel} \approx 0$ and $K_{\perp} \approx L$. Then the spectral function we seek is

$$L(z) = \pi^2 C \int_0^u e^{-h}\frac{dh}{\Phi}\sum_{n=1}^{\infty}\frac{(2n-1)^2 b_{2n-1}^2}{(2n-1)^2(\pi/\Phi)^2-z^2} \tag{223}$$

where Fourier amplitudes b_{2n-1} are found from (218). Using (217) and (218), we write

$$\frac{\pi^2}{\Phi}\frac{(2n-1)^2 b_{2n-1}^2}{(2n-1)^2\left(\pi/\breve{\Phi}\right)^2-z^2}$$

$$= \frac{\pi^2\sqrt{h}(2n-1)^2 64\beta^2 f^2(1+\sigma)^6\sin^2\left(\dfrac{\pi}{2}\dfrac{2n-1}{1+\sigma}\right)}{\pi^4 2\beta f(1+\sigma)(2n-1)^4\left[(2n-1)^2\sigma^2-(1+\sigma)^2\right]^2\left[\dfrac{(2n-1)^2\pi^2 h}{4\beta^2 f^2(1+\sigma)^2}-z^2\right]}$$

$$= \frac{128\beta^3 f^3(1+\sigma)^7\sqrt{h}\sin^2\left(\dfrac{\pi}{2}\dfrac{2n-1}{1+\sigma}\right)}{\pi^4(2n-1)^2\left[(2n-1)^2\sigma^2-(1+\sigma)^2\right]^2\left[(2n-1)^2 h-(1+\sigma)^2\left(\dfrac{2\beta f}{\pi}z\right)^2\right]}$$

Substituting this result and (222) in (223), we finally find the planar spectral function:

$$L(z) = \frac{64\beta^3 f^3}{\pi^{9/2}D_{\text{flat}}}\int_0^u S(h)[1+\sigma(h)]^7\exp(-h)\sqrt{h}\,dh \tag{224a}$$

where

$$D_{\text{flat}} \equiv \beta\left[f\,\text{erf}\left(\sqrt{u}\right)+\frac{1-f}{2}\sqrt{\frac{\pi}{u}}(1-e^{-u})\right] \tag{224b}$$

$$\sigma(h) \equiv \frac{\pi}{2}\frac{1-f}{f}\sqrt{\frac{h}{u}} \tag{224c}$$

and

$$S(h,z)=\sum_{n=1}^{\infty}\frac{\sin^2\left[\dfrac{2n-1}{2}\dfrac{\pi}{1+\sigma(h)}\right]}{(2n-1)^2\left[(2n-1)^2\sigma^2(h)-(1+\sigma(h))^2\right]^2\left[(2n-1)^2 h-(2z\beta f/\pi)^2\left((1+\sigma(h))^2\right)\right]} \tag{224d}$$

The expressions (224), being similar to Eqs. (8)–(10), differ from them by less power of h before the differential and by less of the factor βf, as well as by other coefficients of the type D_{flat} before the integral.

Now we shall modify the spectral function (224a) by taking into account a one-dimensional libration scheme proposed in Section VIII.B.

2. Corrected SF for Libration about H-Bond

Starting from the general expression (5.35) in VIG for the spectral function, we replace χ^*_B by LG and q by μ_E/μ, where μ_E is the projection of the dipole moment $\boldsymbol{\mu}$ on the external field $\mathbf{E}$. Then we write

$$
L(z) = \frac{3iz}{\mu^2} \left\langle \mu_E(0) \int_0^\infty [\mu_E(\varphi) - \mu_E(0)]e^{iz\varphi}d\varphi \right\rangle
\tag{225}
$$

Vector $\mathbf{E}$ in Fig. 45a, which we take equal to the amplitude $\mathbf{E}_m$ of the time varying field, is inclined at the angle Θ to the dimer axis $O_A O_B$, while the projection $\mathbf{E}_s$ of this vector on the plane S is shifted by the angle Ψ with respect to the equilibrium projection of $\boldsymbol{\mu}$ onto S. In this figure, θ and γ denote, respectively, inclination of the vector $\boldsymbol{\mu}$ to the axis $O_A O_B$ and the valency angle $H_a O_A H_b$. Then

$$
\theta = \pi - \gamma/2 \approx (3/4)\pi
\tag{226}
$$

and

$$
\left|\boldsymbol{\mu}_\parallel\right|_{E_m} = \mu \cos\frac{\gamma}{2}\cos\Theta; \quad \left|\boldsymbol{\mu}_\perp\right|_{E_m} = \mu \,\sin\frac{\gamma}{2}\,\sin\Theta\,\cos(\Psi + \psi)
\tag{227}
$$

where indices $\parallel$ and $\perp$ denote, respectively, projection of the dipole moment vector onto the symmetry axis of the potential and onto direction of the field component in the plane, transverse to this axis. Projection μ_E is found from (227) as the following sum:

$$
\frac{\mu_E}{\mu} = \cos\Theta \cos\frac{\theta}{2} + \sin\Theta \sin\frac{\theta}{2}\cos(\Psi + \psi)
\tag{228}
$$

In view of our assumptions, the first term is constant and therefore does not contribute to the (225). This could be proved easily by averaging over Θ and Ψ. Taking into account the contribution to the autocorrelator of the remaining term, we have

$$
L(z) = 3iz \left\langle \int_0^\infty \sin^2\Theta \sin^2\frac{\theta}{2}[\cos(\Psi + \psi)\cos(\Psi + \psi_0) + \cos^2(\Psi + \psi_0)]e^{iz}d\varphi \right\rangle
$$

We take account that the direction of the field $\mathbf{E}_m$ is arbitrary, so

$$\langle \sin^2 \Theta \rangle = 2/3, \quad \langle \cos(\Psi + \psi) \cos(\Psi + \psi_0) \rangle = \frac{1}{2} \langle \cos(\psi - \psi_0) \rangle$$

since after averaging over Ψ the term $\cos(2\Psi + \psi + \psi_0)$ vanishes. Next,

$$\cos^2(\Psi + \psi_0) = \frac{1}{2}$$

As a result, we have

$$L(z) = iz \sin^2 \frac{\theta}{2} \left\langle \int_0^\infty e^{iz\varphi} [\cos(\psi - \psi_0) - 1] \, d\varphi \right\rangle$$

The expression in square brackets we represent as follows:

$$\cos \psi_0 (\cos \psi - \cos \psi_0) + \sin \psi_0 (\sin \psi - \sin \psi_0)$$

In our small-amplitude approximation with respect to the variable ψ, the average of the first term could be omitted, since it is on the order of ψ^4.

In the second term we replace the sine by its argument. Doubling the result in order to obtain the correct (corresponding to two-dimensional rotation) integral absorption, just as it was made for the derivation of the formulas (219), we have

$$L(z) = iz(1 - \cos \theta) \left\langle \int_0^\infty \psi_0 [\psi(\varphi) - \psi_0] e^{iz\varphi} \, d\varphi \right\rangle \tag{229}$$

This result reduces to the expression (5.51b) in VIG, if the latter is rewritten in the small-amplitude approximation and replaced by 2 the constant multiplier $(1 - \cos \theta)$. Thus,

$$L_{\text{or}}(z) = \frac{1 - \cos \theta}{2} L(z)|_{\text{from (224a)}} \tag{230a}$$

Note that

$$\frac{1 - \cos \theta}{2} \approx \frac{1}{2} \left(1 + \frac{1}{\sqrt{2}} \right) \tag{230b}$$

For one-dimensional rotation the Hamiltonian is given by

$$H = \frac{I}{2} \left(\frac{d\psi}{dt} \right)^2 + U(\psi) \tag{231}$$

where

$$I = mr_{OH}^2 \sin^2\theta \approx \frac{1}{2}mr_{OH}^2 \tag{232}$$

Note that the equation for the reduced Hamiltonian $h = H(k_BT)^{-1}$, equivalent to (231), coincides with (215), if to replace ψ for θ.

Thus, we can prove the validity of the structurally-dynamic scheme, represented by Fig. 45a, if we, (i) introduce the multiplier $(1/2)(1-\cos\theta)$ into Eq. (224a) for the spectral function, (ii) substitute $U(\psi)$ from (214) into (231) and (iii) calculate the moment of inertia I from Eq. (232).

3. Corrected SF for Librating H-Bond

Let the covalent bond O_AH_a be inclined to the symmetry axis at the angle α. We again denote the angle between the covalent bonds by γ (further we set $\gamma = \pi/2$). Dipole moment $\boldsymbol{\mu}$ is directed along the bisector between the lines of these bonds. Therefore, this moment is inclined to the symmetry axis at the angle $\alpha + \pi/5$. Contrasting the scheme shown in Fig. 45a with that in Fig. 45b, we conclude that the angle θ in (230) now should be replaced by $\alpha + \pi/4$:

$$L_{or}(z) = \frac{1}{2}\left[1 - \cos\left(\alpha + \frac{\pi}{4}\right)\right]L(z)\Big|_{\text{from (229a)}} \tag{233}$$

We note that the moment of inertia for the scheme in Fig. 45b is

$$I_{or} = mr_{OH}^2\left[\sin^2\alpha + \sin^2(\pi - \alpha - \pi/2)\right] = mr_{OH}^2 \tag{234}$$

4.   Results of Modeling

In the second structural scheme (Fig. 45b) the moment of inertia is greater than in the scheme of Fig. 45a. An analysis shows that the former scheme, unlike the latter one, gives a *qualitative* agreement with the experimental [49] ice spectrum.

In Fig. 45a we show the *dimensionless* absorption calculated for the librational band for the parameters, typical for liquid water. Here distinction of the planar model (solid lines) from the spatial one (dashed line) turns out to be noticeable. For ice (Fig. 45b) the indicated distinction becomes less.

In Fig. 46 the absorption coefficient, calculated for ice in this band, is expressed in cm^{-1}. The frequency dependences, found for spatial (a) and planar (b) schemes, are compared. In the planar variant the maximum absorption is less than in the spatial variant. This defect of the theory could be reduced by increasing the dipole moment of the reorienting molecules.

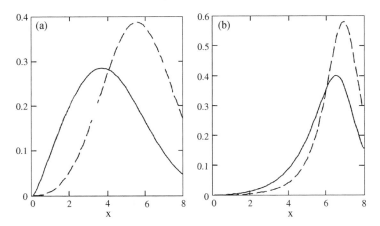

Figure 46 Dimensionless absorption frequency dependences, calculated for the planar (solid lines) and spatial (dashed lines) hat models. (a) Calculation for liquid water at 27°C with the parameters $\beta = 23°$, $u = 8$, $f = 0.8$, $y = 0.3$. (b) Calculation for ice at –7°C with the parameters $\beta = 23.5°$, $u = 8.5$, $f = 0.15$, $y = 0.8$

Thus, we have elaborated the *planar* hat model by considering the water/ice structure in the "defect-sites," where, unlike the "HB-sites," *a polar molecule reorients rather freely,* Since for such a simplified model we have obtained a *qualitative* agreement between the theory and experiment, we thereby proved that the previously proposed "hat model" indeed mirrors in "real" fluids (in water and ice) the dielectric response of dipole orientational motion.

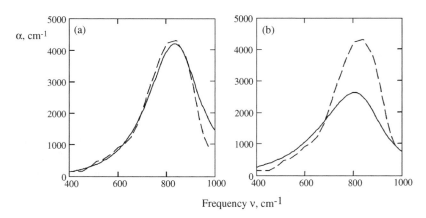

Figure 47 Absorption frequency dependence in the libration band calculated for ice at the temperature –7°C for the spatial (a) and planar (b) hat models with the parameters $u = 8.5$, $f = 0.15$, $y = 0.8$; $\beta = 23.5°$ (spatial model) and 21.2° (planar model)

Therefore, we supported the result obtained earlier that the hat model can be used for calculation of the librational spectrum in water and ice.

We have also shown that some originally proposed (e.g., in GT2) properties of the hat potential turn out to be excessively abstract, so that we disregarded them. We mean here (i) absence of free rotors, performing circular motion, since in the water/ice structure there is no place for such a motion, and (ii) nonexistence of "precessors," since, as it appears, the libration of dipoles in a diametric cross section of the hat-like potential well dominates. On the other hand, the well's curvature is important for modeling the spectra, since, unlike the well depth, this curvature strongly depends on temperature and fluid phase state.

D. Summary Permittivity of Water

We shall make here several elucidations to Sections II.B and II.E.

In view of Eq. (28), the key aspect of our two-fraction model is reveals division of the total complex permittivity ε into the part $\varepsilon_{or}(v)$, including contribution of librations performed in the hat well, and the part $\Delta\varepsilon(v)$, including contribution of vibrating H-bonded molecules. These parts are arranged in such a way that when v tends to infinity, $\varepsilon_{or}(v)$ tends to optic permittivity n_∞^2 and $\Delta\varepsilon(v)$ vanishes. In another limit, at zero frequency, $\varepsilon_{or}(0)$ almost coincides with the static permittivity ε_s, while $\Delta\varepsilon(0)$ is much less:

$$\Delta\varepsilon(0) \ll \varepsilon_{or}(0) \qquad (235)$$

Since at *low* frequencies—at $\omega < \omega_D$ or, equivalently, at $v < v_D$, where $\omega_D = \tau_D^{-1}$ and $v_D = \omega_D(2\pi c)^{-1}$—the contribution to ε of the part $\Delta\varepsilon(v)$ is small, we take into account the *internal field* of the medium *only* the first term $\varepsilon_{or}(v)$ of the right-hand part of (28). Then Eq. (4) holds, which connects the complex permittivity $\varepsilon_{or}(v)$ of librators with their complex susceptibility $\chi_{or}(v)$ found from Eq. (5).

Generally, to relate the susceptibility χ provided by motions of rigid dipoles, one may suggest [see GT1, p.140; GT2, p. 200] that a polar medium under study is influenced by an external macroscopic time-varying electric field $\mathbf{E}_e(t) = \mathrm{Re}\left[\hat{\mathbf{E}}_m \exp(i\omega t)\right]$, where $\hat{\mathbf{E}}_m$ is the complex amplitude. This field induces some local field $\mathbf{E}_{ind}(t) = \mathrm{Re}\left[\hat{\mathbf{E}}_i \exp(i\omega t)\right]$ in a cavity surrounding each polar molecule. A given molecule directly experiences the latter field.

If we neglect the difference between two complex amplitudes, $\hat{\mathbf{E}}_m$ and $\hat{\mathbf{E}}_i$, then ε and χ would be related as (we omit complex conjugation symbols)

$$\chi(\omega) = \frac{\varepsilon - n_\infty^2}{4\pi} \qquad (236)$$

A more rigorous theory, accounting for an internal field correction, yields the following ratio of two field complex amplitudes:

$$\frac{\hat{E}_i}{\hat{E}_m} = \frac{3\varepsilon}{2\varepsilon + n_\infty^2}$$

Then Eq. (236) transforms to

$$\chi(\varepsilon) = \frac{\varepsilon - n_\infty^2}{4\pi} \frac{2\varepsilon + n_\infty^2}{3\varepsilon} \tag{237}$$

Obviously, at zero frequency we have the following relation between the static values ε_s and χ_s:

$$\chi_s = \frac{\varepsilon_s - n_\infty^2}{4\pi} \frac{2\varepsilon_s + n_\infty^2}{3\varepsilon_s} \tag{238}$$

On the other hand, the dependence, inverse to (237), gives the quadratic equation for $\varepsilon(\chi)$, having the solution (4). Now we replaced χ by χ_{or} and ε by ε_{or}, since in our reasoning we do *not* touch all the complex permittivity but only the part that is related to the hat model. Analogously, in the relationship (238) we should replace ε_s and χ_s by the quantities $\varepsilon_{s,or} = \varepsilon_s - \Delta\varepsilon_s$ and $\chi_{s,or}$, related to the hat model, where

$$\Delta\varepsilon_s \equiv \Delta\varepsilon(v)|_{v=0} \tag{239}$$

is the static permittivity generated by HB vibrating molecules. Then we get instead of (243):

$$\chi_{or,s} = \frac{\varepsilon_s - \Delta\varepsilon_s - n_\infty^2}{4\pi} \frac{2(\varepsilon_s - \Delta\varepsilon_s) + n_\infty^2}{3(\varepsilon_s - \Delta\varepsilon_s)} \tag{240}$$

where ε_s is the total static permittivity, which we consider to be known from the experimental data.

Hence, Eq. (5) gives the frequency dependence of the hat susceptibility $\chi_{or}(v)$, while Eq. (4) yields such a dependence for the permittivity $\varepsilon_{or}(v)$.

However, Eq. (240) shows that these quantities cannot be found until we find the vibration static permittivity (239). Thus, the "vibration" and "libration" parts of the spectra, being formally independent one from the other, are actually connected via total static permittivity ε_s.

To release these part we

(1) involve the proportion r_{vib} of vibrating molecules as a fitting parameter (see Eq. (6)),

(2) find the frequency dependence $\Delta\varepsilon(v)$ and its static part $\Delta\varepsilon_s$ from Eqs (30), (32), and (33),

(3) estimate the static librational permittivity from (Eq. (31),

(4) find the librational static susceptibility from Eq. (240), and

(5) calculate the librational permittivity from Eq. (4).

Varying the parameters of the model with account of the relationship (11), which guarantees the coincidence of the calculated low-frequency and Debye spectra (we regard the latter as correct), we repeat this procedure several times (usually only a few times) until the results of calculations become satisfactory.

Note, finally, that setting frequency x to zero in Eq. (5), we may relate the static librational susceptibility $\chi_{or,\,s}$ to the Kirkwood correlation factor g_{or} as

$$\chi_{or,s} = g_{or}G_{or} \qquad (241)$$

Combining it with (240), we obtain Eq. (7) for g_{or}. If the parameters of the model are fitted reasonably, the value of g_{or} falls usually between the values 1 and 3.

The fitting procedure, described for ice in accordance with Eqs. (7b) and (7c), is simpler, since we do *not* calculate here the static susceptibility generated by librating molecules.

IX.   PRINCIPLES OF ANALYTICAL DIELECTRIC RELAXATION THEORY[30]

A.   General Qualitative Features

The method described in this chapter appears on the border common to physical electronics, molecular, and chemical physics. This method was founded in joint researches of one of the authors (V.I.G.) with Yuri Kalmykov and Boris Tseitlin. During a long-standing improvement of the theory, after considering a multitude of specific models, we at last succeeded in calculation of water spectra and, quite recently, calculation of ice spectra regarded in "water-like" manner. The models, which yield these two main results, are described in detail in the preceding sections of this work. In Section IX a fundament of our method will be briefly reviewed. Its latest version belongs mainly to B.M. Tseitlin and V.I.G. The final general expressions obtained for the dipolar complex susceptibility agree with modern descriptions, based on application of dipolar autocorrelation function

---

[30]In this section we follow in general outline the review article GT2.

(ACF). However, the results of our theory, based completely on application of classical physics, are found originally.

As distinct from "pure" electrodynamics and "pure" mechanics, the complex permittivity $\varepsilon(v)$ of a polar fluid is determined by self-consistent interrelation of Maxwell and Newton equations. Indeed, trajectories of the particles depends on applied electromagnetic field $E(t)$. Since the latter disturbs these trajectories, they are not known a priori.

To find observable and rather simple analytical solutions of the problem we assume the following:

(i) Electromagnetic field $\mathbf{E}$ and current density $\mathbf{J}$ in a fluid are represented in the form of a transverse homogeneous plane wave:

$$\{\mathbf{E}(t), \mathbf{J}(t)\} = \frac{1}{2}\left\{\widehat{\mathbf{E}}, \widehat{\mathbf{J}}\right\} \exp[i(\omega t - \mathbf{k}\mathbf{r})] + \frac{1}{2}\left\{\widehat{\mathbf{E}}^*, \widehat{\mathbf{J}}^*\right\} \exp[-i(\omega t - \mathbf{k} * \mathbf{r})]$$

$$(242)$$

with the wavevector $\mathbf{k}$ normal to the electric field $(\mathbf{k} \perp \mathbf{E})$; $\widehat{\mathbf{E}}$ and $\widehat{\mathbf{J}}$ being the complex amplitudes; $i = \sqrt{-1}$; ω is angular radiation frequency. Note, in Section IX we retain the complex-conjugation symbol (it is omitted in other sections of this chapter).

(ii) The space dispersion (that is, the variation of the field amplitude $|\widehat{\mathbf{E}}|$ on lengths of the order 1 Å) is neglected. All nonlinear phenomena (e.g., excitation of higher field harmonics) are disregarded.

(iii) Usually we do *not* calculate the trajectories of individual particles. We study the combined a.c. field action on polar medium in terms of the averaged consumed *complex power*. Its real part is related to the loss factor $\chi''(\omega)$ of the susceptibility, and the imaginary part is related to the dielectric constant $\chi'(\omega)$.

Let us consider, for instance, in the medium irradiated by the field (242), an *infinite* sequence of charges q, appearing in this medium due to collisions at the instants

$$\ldots, \tau, \ldots, \tau - 2\pi/\omega, \ldots, \tau - 4\pi/\omega, \ldots \qquad (243)$$

As is shown in GT2, Eq. (5c), the complex power, consumed due to interaction of this sequence with radiation, is equal to

$$j(\tau) \equiv qv_E(\tau) \exp(i\omega\tau) \qquad (244)$$

where v_E denotes the projection of the particle velocity on the field $\mathbf{E}$ and the instant τ is chosen arbitrary $(0 < \omega\tau < 2\pi)$.

(iv) A particle is assumed to move in *vacuum*. Its motion is determined by the *static potential*, which we consider to be known. In different models such potentials are involved intuitively and from experience.

(v) The lifetime t_v of each particle, during which it interacts with the field (242), is *limited*, since any particle is subjected to *strong collisions* with other particles of the medium. After such a collision a particle randomly acquires a new velocity corresponding to Boltzmann distribution. Equally, we may regard that t_v is the lifetime of the potential, in which a particle moves. For simplicity, we take the *exponential*[31] distribution $\Psi(t_v)$ of particles over time t_v between strong collisions. Namely, we suppose that in the interval $[0, t_v]$ a chosen particle moves without collisions with other particles, but experiences it during time dt_v in the interval $[t_v, t_v + dt_v]$. Evidently,

$$\Psi(t_v) = \frac{1}{t_v} \exp\left(-\frac{t_v}{\tau}\right), \quad \text{so that} \quad \int_0^\infty \Psi(t_v)\, dt_v = 1 \qquad (245)$$

The parameter τ is called the "mean time between strong collisions" (or simply "mean time"); τ plays a role of the *friction coefficient*. However, in our theory, friction is *not* introduced explicitly.

We should remark that periodic collisions of a particle with *walls* of the chosen potential well, in which a particle oscillates, are *not* regarded as strong collisions, since energy of a particle conserves after such collisions.

(vi) The Boltzmann distribution over the particle velocities v_0 established at the instants t_0 "just after strong collisions" is taken into account. As for the distribution $F(\Gamma)$ over orientations of dipoles at such instants t_0, it may slightly differ from the Boltzmann distribution. Here Γ denotes the set of phase variables, pertaining to a particle in the medium.

(vii) The complex susceptibility $\chi(\omega)$ is related to the propagation constant k as

$$k^2 = \left(\omega^2/c^2\right)\left(4\pi\chi + n_\infty^2\right)$$

where c is the speed of light and n_∞^2 is the optical permittivity. On the other hand, $\chi(\omega)$ is determined by the mean *complex power*, consumed by a propagating wave,

(viii) The equations of motion of individual particles are used but *not* solved in this work.

---

[31]Index "v" in the variable t_v is involved as a brief notation of the word "vive."

(ix) Summations of the contributions (244) of various τ-sequences via integration over phase $\omega\tau$ in the interval $[0, 2\pi]$ gives the total complex power and therefore the complex susceptibility $\chi(\omega)$. However, in this calculation scheme the variables in the above-mentioned integral depend on the external field amplitude $\hat{E}$ and calculations are very complicated.

(x) Therefore it is reasonable to radically change this scheme aimed to perform calculations were with *nondisturbed* variables. For this purpose the underlying expression for $\chi(\omega)$ is transformed to the correlator $L(z)$ termed *spectral function* (SF). Here z is the complex frequency, $z = x + iy$. Its real part is proportional to the frequency ω, and its imaginary part is proportaional to τ^{-1}. This SF $L(z)$ is connected, by a *rational* relationship, with the *ACF spectrum* $\tilde{C}(\omega)$. However, the time ACF $C(t)$ is *not* used in this work, since the spectra of fluids are found directly by the SF.

Hence, this first step allows calculating the Boltzmann susceptibility $\chi_B(\omega)$, which is proportional the spectral function $L(z)$ and is found via integration of undamped variables, so that the integrand comprises now only the variables, determined by undamped motion of dipoles. Particularly, this integrand is proportional to the Boltzmann orientational distribution F_B.

It should be noted that the SF $L(z)$ is also widely used in this work for describing the complex permittivity in the far-IR region.

(xi) The next step in dielectric relaxation theory concerns transformation of the relevant integral for the complex susceptibility $\chi(\omega)$ allowing us to employ the orientational distribution F differing from F_B with again nondamped variables in the integrand. Existence of $F \neq F_B$ is taken into account in the so-called "collision models." In this work except Boitzmann (with $F = F_B$) we employ the *Gross collision model* with $F = F_G$.

The obtained relationship for such $\chi(\omega)$ turned out to be a *rational function* of the SF $L(z)$. The obtained formula allows calculation of $\chi(\omega)$ (for any specific molecular model) in all the frequency region, including the low-frequency part of the spectrum. In its high-frequency part such $\chi(\omega)$ coincides with the Boltzmann susceptibility $\chi_B(\omega)$.

In this work the mentioned generalization of the theory is applied to the hat model.

Hence, transformations of Maxwell equations, the change of orientational distribution functions to the form, close to the Boltzmann distribution with account of classical equations of motions yield the complex susceptibility $\chi(\omega)$ determined by *unperturbed collision-free motion* of an individual particle in a given static potential well. In our approach, the complex permittivity $\varepsilon(\omega)$ is found as a simple rational function of this susceptibility $\chi(\omega)$.

Below a brief list of main formulas will be given, illustrating the above-mentioned qualitative features of our method used in this work for calculation of the susceptibility $\chi(\omega)$.

B. Complex Power Consumed on Perturbed Trajectories

We start from the first pair of Maxwell equations connected with electric charges moving in vacuum. Let vector $\mathbf{J}$ be the electric current density produced by these charges. In the case of a polar medium, positive (q) and negative $(-q)$ charges of each dipole are bound one to the other. Therefore we may represent the complex power (244) as

$$j(t) = q\frac{d}{dt}\left[r_E^+(t) - r_E^-(t)\right]\exp(i\omega t) = \frac{d\mu_E}{dt}\exp(i\omega t) \qquad (246)$$

$\mathbf{r}^+$ and $\mathbf{r}^-$ are the coordinates of positively and negatively charged particle and μ_E is the projection of a dipole moment $\boldsymbol{\mu} = q(\mathbf{r}^+ - \mathbf{r}^-)$ of a pair of coupled charges onto the a.c. field direction. Furthermore, we consider dielectric response of a *rigid* dipole in such frequency interval, in which one may regard that a polar particle reorients *as a whole*.

After transformations, one may obtain the following relationship for the complex susceptibility (see GT2, Eq. (14c)):

$$\chi^*(\omega) = \frac{2iN}{\omega\tau\hat{E}^*}\int_\Gamma W(\Gamma)\,d\Gamma\int_0^{2\pi}F(\gamma)\frac{d\gamma}{2\pi}\int_0^\infty e^{-t_v/\tau}\frac{dt_v}{\tau}\int_{t_0}^{t_0+t_v}e^{i\omega t}\frac{d\mu_E(t)}{dt}dt \qquad (247)$$

Here $\gamma \equiv \omega t_0$ is an "initial phase," $0 \le \omega t_0 \le 2\pi$. The first (from the left) integral presents averaging over phase variables; the second one presents averaging over instants t_0 after strong collisions, and $F(\gamma)$ is the orientational distribution function (DF); the third one presents averaging over lifetimes t_v of the potential well; and the last (fourth) integral appears, since according to the so-called "t_o theorem" (see GT2, p. 86) the complex susceptibility is proportional to the *average over t_v value of the complex power.* Its value generated by one τ-sequence is represented by Eq. (244). Changing the time variable $(t_v + t_0 \to t_v)$ and denoting t_v simply as t, we may get rid of one integral in Eq. (247) by partial integration over t:

$$\chi^*(\omega) = \frac{2iN}{\omega\tau\widehat{E}*}\int_\Gamma W(h)\,d\Gamma\int_0^{2\pi}F(\gamma)e^{i\gamma}\frac{d\gamma}{2\pi}\int_0^\infty \frac{d\mu_E(t)}{dt}e^{i\hat{\omega}t}dt \qquad (248)$$

We have introduced here the dimensionless energy h and the complex frequency $\hat{\omega}$, where $h = H/(k_B T)$, H is the full energy of a particle, and $\hat{\omega} \equiv \omega + i/t$.

The relationship (248) indicates that the complex susceptibility $\chi^*(\omega)$ is determined by the averaged complex power, consumed in the unit volume of the medium by electromagnetic field. The averaging is performed over time, over "initial" (after strong collisions) instants and over phase variables of the particle. We emphasize that just such an averaging produces *smooth* frequency dependences of the real (χ') and imaginary (χ'') parts of $\chi(\omega)$. Note that the real part of the complex power, absorbed by an individual particle, chosen as *a point* in the phase space Γ, may be positive or negative. However, at *equilibrium*, due to Boltzmann distribution, this provides only a positive *resulting* imaginary part χ'' of the susceptibility (the loss factor) *at any frequency* ω, as we saw in this work at for *all* examples given above.

A distinguishing feature of the expression (248) is that its *integrand depends on the radiation field* $|\hat{E}|$, since the particle trajectories are disturbed by it. This fact embarrasses calculations and strongly restricts the practical effectiveness of this calculation scheme. However, we know that if the field amplitude tends to zero, the quantity $|\hat{E}|$ involved in the variable $\mu_E(t)$ (248), will be canceled by $|\hat{E}|$ in the denominator of (248). Then the right-hand part of (248) will *not* depend on the field. In the next step we shall certain of that.

C.   Spectral Function and Its Relation to ACF Spectrum

Spectral function (SF) plays a key role in this work. Let us introduce the dimensionless parameters: complex frequency z, related to $\hat{\omega}$ by the microscopic timescale η; concentration G; and the ratio Q of the dipole-moment projection μ_E^0 to the value $|\mu|$ of this moment:

$$z = x + iy \equiv \eta \left(\omega + \frac{i}{\tau} \right) = \eta \hat{\omega}, \quad \varphi = \frac{1}{\eta}, \quad \eta = \sqrt{\frac{I}{2k_B T}}, \quad G \equiv \frac{\mu^2 N}{3k_B T}, \quad Q = \frac{\mu_E^0}{\mu}$$

$$(250)$$

where the superscript 0 in μ_E^0 denotes that the relevant variable refers to a *steady-state trajectory*, which, in distinction from (247) and (248), is *not* disturbed by the external $E(t)$ field. In (250), I is the moment of inertia and $\hat{\omega}$ is defined in (249).

Starting from the *Boltzmann collision model*, we write down the formula for the Boltzmann orientational distribution F_B pertaining to weak electromagnetic field:

$$F_B = \exp \left(-\frac{\mu_E(0)E(0)}{k_B T} \right) \approx 1 - \frac{\mu(0)E(0)}{k_B T}$$

$$(251a)$$

It is assumed that this distribution is established in an isotropic medium immediately after the instant $t = 0$ of a strong collision. Equation (251a) may be represented in the form

$$F_{\mathrm{B}} \approx 1 - \frac{\mu(0)}{k_{\mathrm{B}}T} \left(\widehat{E} e^{i\gamma} + \widehat{E}^* e^{-i\gamma} \right) \tag{251b}$$

Here $\gamma = \omega t_0$ is an arbitrary "initial phase," $0 \leq \gamma \leq 2\pi$.

Let us denote χ_{B} the "Boltzmann" complex susceptibility, corresponding to the orientational distribution F_{B}. It may be shown (see, e.g., GT2) that this susceptibility can be represented as the *correlator* depending on the nondisturbed dimensionless dipole-moment projection Q, introduced in (251):

$$\chi_{\mathrm{B}}^* = 3Giz \left\langle Q(0) \int_0^\infty [Q(\varphi) - Q(0)] \exp(iz\varphi) \, d\varphi \right\rangle \tag{252}$$

or, equivalently,

$$\chi_B^*(z) = GL(z) \tag{253}$$

where the correlator $L(z)$ is termed *spectral function* (SF):

$$L(z) = 3i \frac{z}{\eta} \int_\Gamma Q(0) W(h) \, d\Gamma \int_0^\infty [Q(t) - Q(0)] \exp\left(\frac{izt}{\eta}\right) dt \tag{254a}$$

In a more compact representation

$$L(z) \equiv 3iz \left\langle Q_0 \int_0^\infty (Q - Q_0) e^{iz\varphi} d\varphi \right\rangle \tag{254b}$$

In (252) and here, angled brackets denote ensemble averaging, $Q_0 \equiv Q(0)$.

Hence, the Boltzmann susceptibility χ_{B} and the SF $L(z)$ are proportional to a certain integral over time and over phase volume Γ. Expression (253) is also equivalent to

$$\chi_B(z) = 3G\langle Q^2 \rangle L(z) \tag{255}$$

since due to spatial isotropy

$$\langle Q^2 \rangle = \int\limits_0^\pi \cos \Psi \, \sin^2 \Psi \, d\Psi = \frac{1}{3} \qquad (255a)$$

where Ψ is the angle between any chosen direction in *space* and direction of the external-field amplitude $\mathbf{E}_m$. Note that $\mathbf{E}_m$ is a real quantity and that the amplitude $\hat{E}$ in (251b) is generally a complex one.

In our work the SF has a twofold application. *First*, the formula (254) is used for description of the *far-IR spectrum in the high-frequency approximation*. Such a spectrum accounts for resonance or, more often, quasi-resonance interaction of dipoles with radiation. An absorption peak arises when the radiation frequency ω is near the *mean* thermal frequency determined by the Boltzmann energy distributions.

Second, the SF (254) is used in the case of the *Gross collision model* as a constitutive element of the formula for the complex susceptibility χ_G. In this case the orientational distribution F_G, differing from F_B, changes radically the calculated *low-frequency spectrum*, while the far-IR spectrum is very close to that given by the Boltzmann susceptibility χ_B (252). We shall return to this point in Section IX.D.

We remark that Eq. (252) is derived in terms of classical mechanics, provided that the condition $\left\langle \frac{d}{dt} [Q(t)Q(t')] \right\rangle = 0$ holds. Therefore the formula (252) is valid *irrespectively* of the type of the projection $\mu_E^0 = |\mathbf{\mu}|Q$. Namely, this projection may correspond to reorientation of a rigid dipole or to vibration of bound charges of a nonrigid dtpole.

Now we introduce the *time-varying* normalized ACF $C(t)$, defined as

$$C(t) = \frac{\langle Q(0)Q(t) \rangle}{\langle Q(0)^2 \rangle} = \left\langle \frac{\mu_E(0)\mu_E(t)}{[\mu_E(0)]^2} \right\rangle \qquad (256)$$

Evidently, the ACF *spectrum* $\tilde{C}(\omega)$ is related to the time ACF $C(t)$ as

$$\tilde{C}(\omega) = \int\limits_0^\infty C(t) \, \exp(i\omega t) \, dt \qquad (257)$$

Replacing ω by the complex frequency $\hat{\omega}$, given by Eq. (249), we have

$$\tilde{C}\left(\omega + \frac{i}{\tau}\right) = \int\limits_0^\infty C(t) \, \exp\left[i\left(\omega + \frac{i}{\tau}\right)t\right] dt \qquad (258)$$

Noting definition of φ and η, given in (250), we now represent Eq. (254) as a sum:

$$L(z) = 3i\left(\omega + \frac{i}{\tau}\right)\langle Q(0)Q(t)\rangle \int_0^\infty \exp\left[\left(\omega + \frac{i}{\tau}\right)t\right]dt$$

$$- 3\left\langle\int_0^\infty [Q(0)]^2 \exp(iz\varphi)diz\varphi\right\rangle \qquad (259)$$

In view of (255), $\langle Q(0)Q(t)\rangle = C(t)\langle Q(0)^2\rangle$. Then, according to (256), the first term in (259) reduces to $3i\left(\omega + \frac{i}{\tau}\right)\tilde{C}\left(\omega + \frac{i}{\tau}\right)\langle[Q(0)]^2\rangle$, while the second term gives $\langle[Q(0)]^2\rangle$. Thus,

$$L(z) = L\left[\eta\left(\omega + \frac{i}{\tau}\right)\right] = 3i\left(\omega + \frac{i}{\tau}\right)\tilde{C}\left(\omega + \frac{i}{\tau}\right)\left\langle[Q(0)]^2\right\rangle dt + \left\langle[Q(0)]^2\right\rangle$$

Comparing this with (260) and noting that $\langle[Q(0)]^2\rangle \equiv \langle Q^2\rangle$, we find that the SF $L(z)$ is *linearly* related to the *spectrum* $\tilde{C}(\omega)$ of the normalized ACF; also, the SF is linearly related to the integral over time taken from the ACF $C(\varphi)$:

$$L(z) = 1 + i\left(\omega + i\tau^{-1}\right)\tilde{C}\left(\omega + i\tau^{-1}\right) \qquad (260)$$

or equivalently

$$L(z) = 1 + iz\int_0^\infty C(\varphi)e^{iz\varphi}d\varphi \qquad (261)$$

D. Complex Susceptibility for Gross Collision Model

1. Transition to Undisturbed Variables

In Section IX.C the dynamic quantity $Q(t)$ refers to undisturbed trajectory; that is, Q depends on particles moving at *equilibrium*. That is why the Boltzmann susceptibility $\chi_B(z)$, as well as the spectral function $L(z)$, do not depend on the external field amplitude $\hat{E}$. Now we turn to calculation of the susceptibility in a more general case. We shall transform Eq. (248) in order to exclude such a dependence of χ on $\hat{E}$.

Namely, the drawback of Eq. (248) is that its integrand comprises the dynamic quantity $\mu_E(t)$ and the induced distribution $F(\gamma)$. Both are *perturbed by radiation field and therefore depend on $\hat{E}$*. As a result, further calculations become cumbersome. It is possible to overcome this drawback on the basis of a linear-response approximation. The field-induced difference $\delta\mu_E$ of the law of motion $\mu_E(t)$ from the steady-state law $\mu_E^0(t)$ is proportional to the field amplitude $\hat{E}^*$. The same is supposed regarding the difference $F(\gamma) - 1$ of induced and homogeneous distributions (for the latter $F \equiv 1$). A steady-state dipole trajectory does not depend on phase γ and therefore does not contribute to the integral (248) at $F = 1$. Then, in a linear approximation, we may represent the average of $\mu_E(t)F(\gamma)$ over γ as a sum:

$$\langle \mu_E(t)\,F(\gamma) \rangle_\gamma = \mu_E^0(t)\,\langle [F(\gamma) - 1] \rangle_\gamma + \langle \delta\mu_E(t) \rangle_\gamma \qquad (262)$$

Here this average is denoted by subscript γ. Correspondingly, the right-hand side of Eq. (248) can be represented as the sum:

$$\chi^*(\omega) = \chi_{std}^*(\omega) + \chi_{dyn}^*(\omega) \qquad (263)$$

where subscripts "std" and "dyn" mean "steady" and "dynamic".

The expression (248) is reduced to the first term of (263), if one replaces $\mu_E(t)$ by the function $\mu_E^0(t)$ determined for the steady-state law of motion but for a *rigorous* induced distribution F (differing from 1). On the other hand, $\chi^*(\omega)$ is reduced to the second term if one sets in (248) the *perturbed* $\mu_E(t)$ dependence but assumes the distribution F to be homogeneous ($F = 1$). Thus, the first and second terms of (262) and (263) are stipulated, respectively, by effect of radiation on *statistic distributions* and on *dynamics of art individual particle*. That is why we use the subscripts "std" and "dyn" in Eq. (263).

To find the component $\chi_{dyn}^*(\omega)$, we apply the *average perturbation (AP) theorem*, formulated and proved in GT, pp. 374–376 (see also VIG, pp. 82–87, and GT2, pp. 90–93). This theorem allows us to express the ensemble average of the quantity $\delta\mu_E$ induced by a.c. electric field, through an integral including *unperturbed* time dependence $\mu_E^0(t)$. The simplified formulation of the theorem is

$$\int_\Gamma \delta Q(t)W(h)\,d\Gamma = (k_B T)^{-1} \int_\Gamma Q(t)\,A(t)\,W(h)\,d\Gamma \qquad (264)$$

Here $A(t)$ is the work produced in the interval $[0, t]$ by some time varying FORCE(t), which is proportional to the a.c. electric field $E(t)$. In our case,

$$Q = \mu_E^0/\mu \quad \text{and} \quad A(t) = \int_0^t E(t')\frac{d}{dt'}\mu_E^0(t')\,dt' \qquad (265)$$

Generally, the AP theorem is applicable to *any steady-state dynamic quantity* $Q(t)$ [of the sort discussed above, viz., like $\mu_E^0(t)$]. The theorem permits to express through undisturbed variables the increment δQ induced *by any external force*, if the following conditions hold:

(a) The value $|A(t)/(k_B T)|$ is small at any time t, so that the increment $\delta Q(t)$ is proportional to the FORCE(t), which performs the *work* $A(t)$.

(b) The canonically conjugated quantities S_i and p_i, which generally may differ from the integrals of motion, are used as the phase variables of a relevant ensemble of particles.

(c) The one-particle Boltzmann distribution function determined by these variables is used, so that $W(h) \propto \exp[-h(s_i, p_i)]$.

Thus we set the variable $Q = \mu_E^0/\mu$ in (264) and express through this quantity the work $A(t)$ given by Eq. (265), where we represent $E(t')$ in view of Eq. (242). We take into account that (i) $\mathbf{kr} = 0$ and (ii) the variable t in (247) is replaced by $t - t_0$. Then the expression (242), used in GT2 (p. 85) for transformation of Maxwell equations of motion, reduces to

$$\mathbf{E}(t) = \frac{1}{2}\left\{\widehat{E}\exp[i(\omega t + \gamma)] + \widehat{E}*\exp[-i(\omega t + \gamma)]\right\} \qquad (266)$$

The first term in $\{\cdot\}$ does not contribute to the average over γ *in* (248). Substituting (265) in (264), we have

$$\int_\Gamma \delta q(t)W(h)\,\mathrm{d}\Gamma = \frac{1}{2k_B T}\int_\Gamma e^{i\widehat{\omega}t}\mu_E^0(t)W(h)\,\mathrm{d}\Gamma \int_0^t e^{-(i\omega t'+\gamma)}\frac{\mathrm{d}}{\mathrm{d}t'}\mu_E^0(t')\,\mathrm{d}t' \qquad (267)$$

To extract the susceptibility-component $\chi_{\mathrm{dyn}}^*(\omega)$ from (248) (1.14c), we use with account of (267) the following replacement:

$$F\int_\Gamma \ldots\frac{\mathrm{d}\mu_E}{\mathrm{d}t}W\,\mathrm{d}\Gamma \rightarrow \int_\Gamma \ldots\frac{\mathrm{d}}{\mathrm{d}t}\delta\mu_{\widehat{E}}W\,\mathrm{d}\Gamma$$

$$= \frac{1}{2k_B T}\int_\Gamma \ldots e^{i\widehat{\omega}t}\frac{\mathrm{d}\mu_E^0}{\mathrm{d}t}W\,\mathrm{d}\Gamma \int_0^t e^{-(i\omega t'+\gamma)}\frac{\mathrm{d}}{\mathrm{d}t'}\mu_E^0(t')\mathrm{d}t' \qquad (268)$$

It is convenient to represent the result by involving the *spectral function* (SF) $L(z)$ as follows:

$$\chi_{\mathrm{dyn}}(\omega) = \frac{\widehat{\omega}}{\omega}GL(z) \qquad (269)$$

where

$$L(z) \equiv \frac{3i}{\widehat{\omega}\tau} \int_{\Gamma} D(z)W(h)\,d\Gamma \qquad (270)$$

$$D(z) \equiv \int_{0}^{\infty} \frac{dq(t)}{dt} \exp\left(i\frac{zt}{\eta}\right) \int_{0}^{t} \frac{dq(t')}{dt'} \exp\left(-i\frac{xt'}{\eta}\right) d\,t' \qquad (271)$$

Further transformation described in GT2 (pp. 92–93) yields just expression (254a) given above without derivation.

2. *Expression of Gross Susceptibility Trough Spectral Function*

Now we consider an important self-consistent *Gross* collision model. In this model just as in the Boltzmann one (with $F = F_B$) the velocity distribution function is assumed to be unperturbed by a.c. field. We represent the Gross induced distribution with $F = F_G$ as

$$F_G(\gamma) - 1 \simeq \frac{\mu_E(0)}{2k_B T}\left[\widehat{A}^*(z)\,\widehat{E}\,e^{i\gamma} + \widehat{A}(z)\,\widehat{E}^*\,e^{-i\gamma}\right] \qquad (272)$$

To find out an unknown complex factor $\widehat{A}(z)$, we write down the standard connection

$$\widehat{P} = \chi\widehat{E} \qquad (273)$$

between the *complex polarization* $\widehat{P}$ of an isotropic polar medium and the complex field amplitude, with the factor of proportionality (in (273)) being equal to the complex susceptibility.

On the other hand, by definition, the polarization at an instant $t = 0$ after a strong collision is determined by the total dipole moment of a unit volume, comprising N polar molecules:

$$P(0) = \mathrm{Re}\left(\widehat{P}e^{i\gamma}\right) \equiv Ng \int_{\Gamma} \mu_E(0)\, W F(\gamma)\,d\Gamma \qquad (274)$$

Up to now we neglected *interaction between the particles of a medium*. To take roughly this interaction into account, we have introduced in Eq. (274) the Kirkwood correlation factor g. The latter, in general, differs from 1, since the total dipole moment of a unit volume can be larger (if $g > 1$) or less (if $g < 1$), than the dipole moment found as sum of electric moments of non-interacting particles. The coefficient g approximately accounts for distinction of the statistical distribution from a *one-particle* distribution function *WF*, which is employed in this work.

Substituting (272) into (247) with account of (254a) we have the following relationships between two components of the complex susceptibility:

$$\chi_{std}^* = -\frac{iy}{z}\widehat{A}(z)\chi_{dyn}^* = -\frac{iy}{x}G\widehat{A}(z)L(z) \tag{275}$$

whence the resulting complex susceptibility is connected with the complex factor $\widehat{A}$ by

$$\chi^* = \frac{G}{x}[z - iy(z)L(z)] \tag{276}$$

Finally, we obtain the formulas for the susceptibilities, corresponding to the Boitzmann and Gross orientational distributions, F_B and F_G.

The first one, given by (251), is found from (276), if we set $\widehat{A} = 1$. Then we get

$$\chi^* = GL(z) \tag{277}$$

Consequently, we have confirmed the result (253).

To find $\widehat{A}$ at $F = F_G$, we take into account the definitions of Q in (265) and of G in (250), along with the equality $\langle Q^2 \rangle = 1/3$ pertinent to an isotropic medium (see Eq. (255a)). Thus from Eqs. (272) and (274) we have

$$P(0) = \frac{G}{2}\left[\widehat{A}^*(z)\widehat{E}e^{i\gamma} + \widehat{A}(z)\widehat{E}^*e^{-i\gamma}\right] \tag{278}$$

We express the left-hand part of this formula through the complex amplitudes and then equate the expressions multiplied by the same exponential factors $\exp(\pm i\gamma)$. In view of (273) we find

$$\widehat{A}(z) = \chi^*(Gg)^{-1} \tag{279}$$

Substituting this relation into Eq. (276), we finally express the Gross susceptibility χ_G as the following ratio of expressions comprising the spectral function $L(z)$;

$$\chi^*(\omega) = \frac{gGzL(z)}{gx + iyL(z)} \tag{280}$$

Just as the SF $L(z)$, the Gross susceptibility is determined by undamped trajectories of dipoles, not depending of the field amplitude $\widehat{E}$.

It should be noted that in the original work by Gross [57], where mathematical representation of the model is quite different from that given above, only the case $g = 1$ was originally considered. Here the term "Gross collision model" is ascribed to an arbitrary g-value, which is involved in Eq. (280). Note that previously (e.g., in GTI, VIG) the orientational distribution F_G due to the factor g employed was termed the "correlation-orientation" or "orientation-correlation" distribution.

We should also remark that Kalmykov and Limonova, using a similar collision model applied to an isotropic dipolar medium, have obtained [58, 59] the following expression for the complex susceptibility:

$$\chi^* = \frac{GzL(z)}{gx + (z - gx)L(z)} \tag{280a}$$

It coincides with (280) only for $g = 1$. Calculation (see GT1, p. 192) made for nonassociated liquids showed coinciding spectra, found from (280) and (280a).

The relation (280) is intensively employed in this chapter for calculation of wideband water spectra, obtained with use of the hat model.

It follows from (280) that at high frequencies, such that $x \gg y$, the spectral function $L(z)$, which is proportional to the dimensionless absorption coefficient, equal to $x \, \text{Im}[\chi^*(x)]$.

In other limits at low frequencies, one may neglect the frequency dependence of the spectral function $L(z)$ by setting $L(z) \to L(iy)$. In this approximation, Eq. (280) yields the Debye-relaxation formula:

$$\chi \cong \chi_\infty + \frac{\chi_s - \chi_\infty}{1 - ix/x_D} \tag{281}$$

where $x_D = \eta/\tau_D$, τ_D is the Debye relaxation time and χ_∞ and χ_s are, respectively, the susceptibilities at the high-frequency edge of the Debye relaxation band and at $x = 0$.

Let us now denote as χ_D'' the maximum loss value χ_{max}'', attained at $x = x_D$. It is shown in VIG, Section 6, that for the Gross collision model our approximation $L(z) \approx L(iy)$ gives

$$\tau_D = \frac{g}{y \, L(iy)} \tag{282}$$

and

$$\chi_D'' = \frac{G}{2}[g - L(iy)] \tag{283}$$

From (282) we find the dimensionless frequency x_D of the loss maximum:

$$x_D = yL(iy)/g \tag{284}$$

Thus, at low frequencies the Gross susceptibility (280) reduces to the Debye relaxation spectrum.

If y (imaginary part of z) is much less than x (real part of z), then in the opposite limit (in the far IR range), Eq. (280) reduces to the quasi-resonance frequency dependence

$$\chi^*(\omega) = \frac{G\left(1 + \frac{iy}{x}\right)L(z)}{1 + \frac{iy}{gx}L(z)} \approx GL(z) \tag{285}$$

which is described by the Boltzmann susceptibility.

X. FINAL DISCUSSION

A. Comparison of Analytical Theory with Molecular Dynamics Simulation

This project is suggested by one of the authors (V.I.G.) of this work. The MD calculations were performed by A.Yu. Zasetsky, who kindly agreed to elucidate here briefly the results of his study. Section XI.A includes an extract from the article by Zasetsky and Gaiduk [58] entitled "Study of Temperature Effect on Far Infrared Spectra of Liquid H_2O and D_2O by Analytical Theory and Molecular Dynamic Simulations."

1. Calculation Details

The extended simple point charge (SPC/E) model [59] is used. This model is known to give reasonably accurate values of static dielectric permittivity of liquid water at ambient conditions [60]. The MD simulations were performed for both H_2O and D_2O with the system size of 1024 particles at 220 K, 240 K, 267 K, 273 K, 300 K, and 355 K. The parallel molecular dynamics code for arbitrary molecular mixtures (DynaMix) is implemented by Lyubartsev and Laaksonen [61]. The simulations have been carried out on a Linux cluster built on the Tyan/Opteron 64 platform, which enables calculations of relatively long trajectories for a system of 1024 water molecules. The simulation run lengths depend on temperature and are in the range between 1 ns and 4 ns for the warmest and coldest simulation, respectively. As the initial condition was a cubic lattice, the equilibration time was chosen to be temperature dependent in the range from 200 ps at 355 Kto1 ns at 200K.

The motion equations have been solved by the Verlet Leap-frog algorithm subject to periodic boundary conditions in a cubic simulation cell and a time step of 2 fs. The simulations have been performed in the NVT ensemble with the Nose–Hoover thermostat [62]. The SHAKE constraints scheme [65] was used. The spherical cutoff radius comprises 1.2 nm. The Ewald sum method was used to treat long-range electrostatic interactions.

For periodically replicated nonconducting molecular systems, the complex dielectric permittivity $\varepsilon(\omega)$ can be computed from the time autocorrelation function (ACF) of the total dipole moment [64] during the course of a simulation run as follows:

$$\frac{\varepsilon(\omega) - n_\infty^2}{4\pi} = \frac{\langle \mathbf{M}^2 \rangle}{3VkT} \left[1 + \int_0^\infty dt e^{iat} \frac{\partial C(t)}{\partial(t)} \right] \tag{286}$$

with the complex refractive index $n^* = n + ik$ given by

$$n^*(\omega) = \frac{2\pi}{\lambda} \operatorname{Im}\left(\sqrt{\varepsilon(\omega)} \right) \tag{287}$$

Here $\varepsilon(\omega) \equiv \varepsilon'(\omega) + i\varepsilon''(\omega)$ is the permittivity with the complex-conjugation symbol omitted, n_∞ is the refractive index at infinite frequency, V is the volume of the simulation cell, k_B is the Boltzmann constant, λ is the wavelength, and T is temperature. M is the total dipole moment of the simulation cell with the angular brackets indicating the ensemble average. The dipole moment is computed as

$$\mathbf{M} = \sum_{i=1}^N \sum_{\alpha=1}^3 \mathbf{r}_{i\alpha} q_\alpha \tag{288}$$

where the summation runs over all the molecules (index i) and three atom sites (index α) carrying the charge value of $q_\alpha \cdot C(t)$ is the normalized ACF of the total dipole moment, $C(t) = \mathbf{M}(t)(\mathbf{M})(0)/\mathbf{M}^2(0)$. Note that we compute the function $C(t)$ and the value of the total dipole moment $\mathbf{M}$ independently, because sampling of the value $\mathbf{M}$ from trajectories generated by the MD method can be done much more often, which provides better statistics and thus more accurate value of the static dielectric constant ε_s that is computed through the following relation:

$$\frac{\varepsilon_s - n^2}{4\pi} = \frac{1}{3VkT} \left(\langle \mathbf{M}^2 \rangle - \langle \mathbf{M} \rangle \langle \mathbf{M} \rangle \right) \tag{289}$$

Velocity time ACFs $\langle \mathbf{v}(t)\mathbf{v}(0) \rangle$ for the center-of-mass (COM) of a water molecule have been computed in the local (molecular) coordinate frame, making it possible to look at the COM motion alone each local (Cartesian) coordinate X, Y, and Z independently [65]. All three atoms O, H1, and H2 are placed on the X0Z (molecular) plane and the $H1 - O - H2$ median coincides with the Z axis.

2. Influence of Temperature on Loss Spectra

We will compare the results of the MD simulations with those obtained by our analytical *two-fraction* composite model and applied to water H_2O in Section IV.

In the left-hand column of Fig. 48 the solid lines show the loss spectra of water calculated for the temperatures 273, 300, 330, and 355 K. The circles depict the experimental data. As is seen, the results *quantitatively* agree with experiment (the set of model parameters and references for the experimental data are given in Section IV).

The MD calculations, as shown in the right-hand column of Fig. 48, yield a rather qualitative agreement with the experimental data. Note that the fluctuations pertinent to the MD data are smoothed out with increasing temperature (cf. upper and lower spectra in the right column of Fig. 46) in a wide frequency (from 0 to 1000 cm^{-1}) and temperature (from 220 to 355 K) ranges. *However, the employed MD calculation scheme, briefly observed in Section X.I, is incapable of describing the water spectra in the 200-cm^{-1} band.* This band, marked by arrow in the left-hand column, does not appear in the right-hand column.

Comparison with a number of other works shows that this drawback is generally *typical* for the currently used MD method.

In view of Sharma et al., [29] the classical simulations address the FIR activity of the translational modes by means of force fields, where *each* water molecule is polarized according to a local field so that the induced polarization is *intramolecular*. On the contrary, in the above-cited work it was shown in terms of the first-principle quantum approach that for a tetrahedral coordination of oxygen atoms the liquid is regarded as an assembly of electrons and nuclei and that quantum mechanical evidence is that the FIR spectra depend on concerted tetrahedral fluctuations of the H bonds. In this assembly "the polarization effects depend on the environment in a strong *intermolecular* way."

In our classical approach, based on intuitively introduced molecular potentials, depending on an appropriate force constant (regarded as fitted model parameters), the translational band is ascribed to interaction of adjacent charged molecules, hence, generally to the same origin is suggested as by Sharma et al. [29]. However, we calculate the spectra generated by *two* elastically vibrating molecules instead of considering the assembly of molecules. Perhaps the main advantage of our approach consists in application of the *two-fraction* model, for which two features are characteristic: (1) The main drop with frequency v of the permittivity constant $\varepsilon'(v)$ from the static value to $\varepsilon_\infty \approx 3$ is determined by motion of a polar water molecule, regarded as an *isolated* rigid dipole, noninteracting with other molecules and moving rather freely in a certain (hat) intermolecular potential, and (2) the translational band, which presents (in this context) a peak of our interest, is determined by intermolecular vibrations governed by the longitudinal force constant, whose value is fitted to obtain the 200-cm^{-1} band.

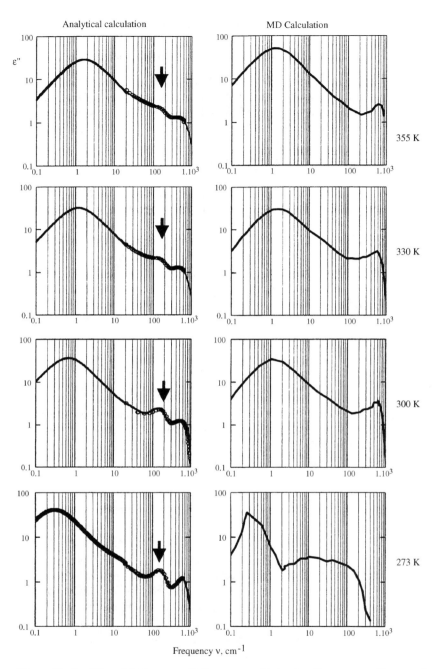

Figure 48 Evolution of the loss water H_2O spectrum with the rise of temperature. The arrow in the left-hand column denotes the 200-cm^{-1} band. This band does not appear in the right-hand column calculated by MD simulations [58].

508

One may suggest that the item 2 is *not* properly regarded in the existing MD approach. We emphasize, however, inestimable advantage of the MD simulation method regading its ability to *predict* the spectra (as well as many other important properties of investigated fluids) by using *only* NVT data. The analytical theory employed in this work is generally capable of parameterizing the model only *ad hoc* by using to a greater or lesser extent the experimentally obtained spectra. Therefore it is reasonable to employ both methods for calculation of the water/ice spectra.

B. Comparison of Water and Ice Spectra

Following the works by Gaiduk and Crothers [23, 31] and by Gaiduk and Kutuza [65], an attempt was undertaken in Section V to apply for ice I_h the model, originally elaborated for calculation of water spectra. At first sight, this idea is seemingly strange due to difference of phase states of these fluids and because in the known molecular dynamics simulations, quite different approaches are used for water and ice. However, two facts can serve as foundation for such an attempt: evident similarity of structures of these fluids and of their far-IR spectra.

Our calculations described in this work show distinctly that the similarity of the models applied to water and ice either diminishes or disappears when the frequency is reduced from far-IR to microwave ranges and even farther to very low frequencies.

1. Resonance Far-IR Ice Spectrum

Reorientation of rigid dipoles in a rather empty space, which we term the "structural defect" (see specific molecular configurations proposed in Section VIII), is responsible for the librational band of ice and water. This band is generated by the LIB fraction of the hat model in the far-IR region (from 400 to $1000 \, cm^{-1}$), adjoining with the IR range. Our analytical description of the response in the former region agrees satisfactorily with the experimental data.

In terms of the model used for ice, the difference between the liquid and solid states is exhibited mainly in the following:

(a) The hat potential, modeling the dipole reorientations in a "structural defect," in the case of ice is close to the parabolic potential, while in the case of water its bottom is much more flat than in ice.

(b) Dipole reorientations in the hat potential in ice are characterized by smaller lifetimes than in water and by substantially longer lifetimes of the hydrogen-bonded molecules.

The harmonic-vibration model of the HB molecules, described in Section VI, was first successfully applied in Sections III and IV for calculation of the permittivity spectra in water in the range $50–300 \, cm^{-1}$. Then it turned out that this model satisfactorily agrees with experiment also for ice. Moreover, in ice the translational band is observed independently from the so-called V-band. This property of the ice spectrum gave grounds for involving "mechanism c" in our theory related to reorientation of the polar molecules about the hydrogen bond. Note that in water the T- and V-bands overlap.

Another important property of the vibration model concerns involvement of mechanism d, which relates to the transverse vibration of the H-bonded molecules. This allows obtaining for *ice* a much better description of the low-frequency wing of the far-IR band. This is evident from Fig. 24a, where the loss spectrum is calculated for ice with account of transverse vibration, while in Fig. 24b this vibration is disregarded.

2.   Nonresonance Fa-IR Ice Spectrum

The model described in this chapter can be applied to the calculation of the permittivity spectra *in water* in the broad frequency range $0-1000 \, cm^{-1}$ and to calculation of the *ice* far-IR spectra in the resonance region $50-1000 \, cm^{-1}$. As seen in Fig. 26 (curves I), in a nonresonance ice spectrum *only* the transverse-vibration mechanism (d) works. Indeed, we see from Fig. 24b that at $v < 50 \, cm^{-1}$, namely in the submillimeter wavelength region and at lower frequencies, mechanisms a–c practically vanish.

3.   Transition to Low-Frequency Ice Spectrum

In the microwave region and at lower frequencies the high frequency approximation, used in this work for ice, becomes inapplicable. In particular, this approach does not describe a deep loss minimum $\varepsilon''(v)$, denoted in Fig. 26b by the dashed line. Such a behavior is typical for ice in a wide temperature range (Fig. 36), but has no analogy in the case of water, where the loss minimum of the $\varepsilon''(v)$ curve is not emphasized. On the contrary, curve $\varepsilon''(v)$ exhibits a shallow minimum in a much narrower frequency range (Fig. 37a). Moreover, in ice, only transverse vibration constitutes the loss minimum, while as depicted in Fig. 37c, in water all mechanisms, especially mechanisms a and d, contribute to the loss minimum.

The hat model also describes the low-frequency (Debye) spectrum of water placed at the frequency v about $10 \, cm^{-1}$. The *ice* spectrum is characterized by quite other behavior. In *this* frequency range we see *no signs* of mechanism a pertaining to libration of a permanent dipole in a certain intermolecular well.

4.   On Possibility of New Paradigm for Calculation of Water and Ice Spectra

Hence, the low-frequency spectra of water and ice drastically differ. We did not solve this problem in this work, but we propose here the following alternative for future investigation.

(i) *The First Variant.* The microscopic models underlying the water and ice spectra are similar in the far-IR region and *drastically differ* in the low-frequency range, in which *for ice* a very deep loss minimum and very wide relaxation region are typical (the latter is depicted in Fig. 27).

As we saw in this work, the existing model describes the water spectra very well in all regions under interest (from 0 to $1000 \, cm^{-1}$). Therefore, we need to elaborate a new model expressly for ice. This model may possibly be based on a mechanism(s) other than that employed above, since the new model should be capable of describing the ice spectra in the vast range—for example, from $10 \, cm^{-1}$ to, say, the kHz region.

(ii) *The Second Variant.* The new model should be *universal, applicable for ice and water.* Since it is hardly possible to apply for ice the *existing* model in 1-khz to 100-cm^{-1} range, it is needed to elaborate the new model for both ice and water, notwithstanding the fact that the existing model of water spectra is very good.

The latter variant (ii) appears to be more preferable. If we choose this variant, it means that *a new paradigm* should be employed for calculation of water/ice spectra.

It would be reasonable to

(a) *assume* that the water and ice the spectra arise due to existence of polar molecules with large and rather close dipole moments, since the values of the static permittivity ε_s in these fluids are near;

(b) *propose* that the *high-frequency approximation* should hold not only for ice, as it is now, but also for water; this allows a very good description of the ice and water spectra in the far-IR region, say, for $v > 10 \, cm^{-1}$ and

(c) consequently, *set a task* that one has to describe the spectra in the region, spreading from $\approx 1 \, cm^{-1}$ to very low frequencies, comprising several (for water) or many (for ice) orders of magnitude.

The key aspect of the new model concerns the change of the very notion of the mixed model, describing the water/ice spectra.

Now, the separating principle for two fractions consists in *the type of motion* of dipoles. Indeed, in the LIB fraction, almost freely librating polar molecules are considered and in the VIB fraction the vibrating hydrogen-bonded molecules are considered.

In the future model the separating principle for two fractions comprises the *timesales of molecular motions*, that is, the scales for used *specific (or relaxation) times*.

The first fraction may be constituted by the molecules with *long relaxation time*, and the second fraction may be constituted by the *fast vibrating/librating molecules*, characterized by much shorter times.

A specific property of *ice* is that the distinction between the characteristic times of molecules, comprising the first and second fractions, is tremendous. In the case of *water*, such a distinction may comprise only one-two orders of magnitude.

In the new calculation scheme the hat model may be involved twice. *First*, this model may be applied for describing the relaxation frequency band, characterized by the Debye time τ_D. *Second*, this model may be used for describing the libration band, characterized by the lifetime τ_{or}. Therefore, the parameters of the two used hat models (or similar to them) should be quite different. We emphasize that in the present calculation scheme, applied to water, we employ the hat model only once—for describing both Debye and libration bands, for which the same set of the model parameters is used.

It would be interesting to demonstrate an applicability of the here-proposed new scheme (or similar to it) in the near future. For completeness we include without discussion, references [66–76].

Finally, we remark that in the present work the molecular model, used for calculation of the ice spectra, is constructed analogously to such a model employed for water spectra. Since the main aspect of the proposed new approach (ii) concerns describing the low-frequency ice spectra and (in the case of success) describing the water spectra, the new idea may be termed the *ice-like model of water spectra*.

Acknowledgements

We would like to acknowledge Science Foundation Ireland Research Frontiers Programme for financial support for this work. V.I.G. expresses sincere gratitude to Professor W. T Coffey for his hospitality in Dublin, appreciates the long-termed collaboration with Boris Tzeitlin, whose materials were, in particular, included in this work, and is grateful to A. Yu. Zasetsky for the rendered possibility to discuss in this work the joint paper by Zasetsky and Gaiduk [58].

References

1. D. Eisenberg and W. Kauzmann, *The Structure and Properties of Water*, Clarendon Press, Oxford, 1969.
2. G. E. Walrafen, *Water, A Comprehensive Treatise*, Vol. 1, F. Franks, ed., Plenum, New York, 1972, p. 151.
3. B. Guillot, *J. Chem. Phys* **95**, 1543 (1991).
4. P. L. Silvestrelli, M. Bernasconi, and M. Parrinello, *Chem. Phys. Lett.* **277**, 478 (1997).
5. B. Bursulaya and H. J. Kim, *J. Chem. Phys.* **109**, 4911 (1998).
6. E. Whalley, in *Chemistry and Physics of Ice*, E. Whalley, S. J. Jones, and L. W. Gold, eds., Royal Society of Canada, Ottawa, 1973, pp. 73–81.
7. G. Nielson, R. M. Townsend, and S. A. Rice, *J. Chem. Phys.* **81**, 5288 (1984).
8. M. Marchi, *J. Chem. Phys.* **85**, 2414 (1986).
9. V. I. Gaiduk and B. M. Tseitlin, *Adv. Chem. Phys.* **87**, 125–378 (1994) (cited throughout as GT1).
10. V. I. Gaiduk and B. M. Tseitlin, *Adv. Chem. Phys.* **127**, 65–331 (2003) (cited throughout as GT2).
11. V. I. Gaiduk, *Dielectric Relaxation and Dynamics of Polar Molecules* World Scientific, Singapore, 1999 (cited throughout as VIG).

COMPLEX PERMITTIVITY OF ICE Ih

12. V. I. Gaiduk and J. K. Vij, *Phys. Chem. Chem. Phys.* **4**, 5289 (2002).

13. G. W. Robinson, J. Lee, K. G. Casey, and D. Statman, *Chem. Phys. Lett.* **123**, 483 (1986).

14. V. I. Gaiduk, B. M. Tseitlin, and D. S. F. Crothers, *J. Mol. Struct.* **738**, 117 (2005).

15. J. Teixeira, M.-C. Bellissent-Funel, S. H. Chen, and A. J. Dianoux, *Phys. Rev.* A, **31**, 1913 (1985).

16. J. Teixeira, Private communication, 2005.

17. M. Maeno, *Science About Ice*, Translation from Japanese, MIR, Moscow (1988) (in Russian).

18. N. Agmon, *J. Phys. Chem.* **100**, 1072 (1996).

19. H. J. Liebe, G. A. Hufford, and T. Manabe, *Int. J. Infrared Millimeter Waves* **12**, 659 (1991).

20. G. Hufford, *Int. J. Infrared Millimeter Waves* **12**, 677 (1991).

21. H. R. Zelsmann, *J. Mol. Struct.* **350**, 95 (1995).

22. H. D. Downing and D. William, *J. Geophys. Res.* **80**, 1656 (1975).

23. V. I. Gaiduk and D. S. F. Crothers, *J. Mol. Liquids*, **128**, 145–160 (2006).

24. V. I. Gaiduk, O.F. Nielson and D. S. F. Crothers, *J. Mol. Liquids*, **137**, 92–103 (2008).

25. V. I. Gaiduk, unpublished (2007).

26. T. Amo and Y. Tominaga, *Physics A* **276**, 401 (2000).

27. O. F. Nielsen, C. Johansson, K. L. Jakobsen, D. H. Christensen, M. R. Wiegell, T. Pedersen, M. Gniadecka, H. C. Wulf, and P. Westh, *Proc. SPIE.* **4098**, 160 (2000).

28. C. Rønne, P.-O. Åstrand, and S. R. Keiding, *Phys. Rev. Lett.* **82**, 2888 (1999).

29. M. Sharma, R. Resta, and R. Car, *Phys. Rev. Lett.* **95**, 187401 (2005).

30. V. I. Gaiduk and D. S. F. Crothers, *J. Mol. Struct.* **798**, 75–88 (2006).

31. V. I. Gaiduk and D. S. F. Crothers, *J. Phys. Chem.* A **110**, 9361–9369 (2006).

32. J. K. Vij, D. R. J. Simpson, and O. E. Panarina, *J. Mol. Liq.* **112**, 125 (2004).

33. A. Yu. Zasetsky, A. F. Khalizov, and J. J. Sloan, *J. Chem. Phys.* **121**, 6941 (2004).

34. V. P. Voloshin, E. A. Zheligovskaya, G. G. Malenkov, Yu. I. Naberukhin, and D. L. Tytik, *Russian Chemical Journal (Journal of the Russian Mendeleev Chemical Society* **45**(3), 31 (2001) (in Russian).

35. Yu. M. Kessler and V. E. Petrenro, in *Water: Structure, State, Solvation. Achievements of the Last Years*, A. M. Kutepov, ed., Nauka, Moscow, 2003, p. 77 (in Russian).

36. G. Ruocco and F. Sette, *J. Phys. Condens. Matter* **11**, R259 ((1999).

37. M. Sampoli, G. Ruocco, and F. Sette, *Phys. Lett.* **79**, 1678 (1997).

38. V. Krasnoholovets, *CEJP.* **2**(4), 678 (2004).

39. G. E. Walrafen, M. R. Fisher, M. S. Hokmabadi, and W.-H. Yang, *J. Chem. Phys.*, **85**, 6970 (1986).

40. M.-C. Bellissent-Funel, *Il Nuovo Chimento* **20 D**, 2107 (1998).

41. V. I. Gaiduk and B. G. Emetz, *Vestn. Khar'k. Univ., Ser. Radiofiz. Elektron.* **7**, 95 (1978).

42. A. A. Vashman, and I. S. Pronin, *Nuclear Magnetic Relaxation Spectroscopy*, Energoatomizdat, Moscow, p. 114 1986, (in Russian).

43. N. Dass and N. C. Varshoeya, *J. Phys. Soc. Japan* **26**(3), 823 (1969) (cited from Ref. 42).

44. V. I. Gaiduk and S. A. Nikitov, *Opt. Spectrosk.* **98**, 919 (2005).

45. G. E. Walrafen, *J. Chem. Phys.* **48**, 244 (1968).

46. M. G. Mikhailov, V. A. Solov'ev, and Yu. P. Syrnikov, *Fundamentals of Molecular Acoustics*, Nauka, Moscow (1964), (in Russian).

47. Yu. V. Gurikov, in *Structure and Role of Water in a Living Organism*, Leningrad University, 1966, p. 3.

514   VLADIMIR I. GAIDUK AND DERRICK S. F. CROTHERS

48. Yu. V. Gurikov, *J. Strukt. Khimii*, **12**(2), 208 (1971) (in Russian).

49. S. G. Warren, *Applied Optics* **23**, 1206 (1984).

50. D. D. Klug, E. Whalley, E. C. Svensson, J. H. Root and V. F. Sears *Phys Rev E*, **44**, 841 (1991).

51. J. E. Bertie, H. J. Labbé, and E. Whalley, *J. Chem. Phys.* **50**, 4501–4520 (1969).

52. B. M. Tseitlin, V. I. Gaiduk, and S. A. Nikitov, *Radiotekh. Electron*, **50**, 1085 (2005). [*J. Commun. Technol. Electron.* **50**, 1002 (2005)]. (cited throughout as TGN).

53. R. R. Sharma *Phys Rev Lett*, **54**, 1964 (1985).

54. V. I. Gaiduk, B. M. Tseitlin, and D. S. F. Crothers, *J. Mol. Liquids* **114**, 63 (2004).

55. E. P. Gross, *Phys. Rev.* **97**, 395 (1955).

56. Yu. P. Kalmykov and S. V. Limonova, *J. Mol. Liq.* **43**, 71 (1989).

57. Yu. P. Kalmykov, *Khim. Fiz.* **7**, 1341 (1988). [*Sov. J. Chem. Phys.* 1991. **7**, 2415 (1991)].

58. A. Yu. Zasetsky and V. I. Gaiduk, *Journal of Physical Chemistry A* III, 5599 (2007).

59. H. J. C. Berendsen, J. R. Grigera, and T. P. Straatsma, *J. Phys. Chem.* **91**, 6269–6271 (1987).

60. M. R. Reddy and M. Berkowitz, *Chem. Phys. Lett.* **155**, 173 (1989).

61. A. P. Lyubartsev, and A. Laaksonen, *Comput. Phys. Commun.* **128**, 565–589 (2000).

62. J. P. Ryckaert, G. Ciccotte and H. S. C. Berendsen, *J. Comput. Phys.* **23**, 327–341 (1977).

63. J. M. Caillol, D. Levesque, and J. J. Weiss, *J. Chem. Phys.* **85**, 6645 (1986).

64. P. A. Madden and R. W. Impey, *Chem. Phys. Lett.* **123**, 502–506 (1986).

65. V. I. Gaiduk and B. G. Kutuza, *Opt. and Spectrosc.* **101**(5), 696–707 (2006). [Opt. Spektrosk. **101** (5), 744–754 (2006)].

66. M.-C. Bellissent Funel, and L. Bosio, *J. Chem. Phys* **102**, 3727 (1995).

67. D. Bertolini, M. Cassettari, M. Ferrario, P. Grigolini, and G. Salvetti, *Adv. Chem. Phys.* **62**, 277 (1985).

68. S. Bratos, M. Diraison and G. Tarjus, J.-Cl. Leieknam, *Phys. Rev. A.* **45**, 5556 (1992).

69. Ch. M. Briskina, V. I. Gaiduk, and B. M. Tseitlin, *Radiotekh. Electron.* **49**, 790 (in Russian) (2004).

70. V. I. Gaiduk, B. M. Tseitlin, Ch. M. Briskina, and D. S. F. Crother, *J. Mol. Struct*, **606**, 9 (2002).

71. V. I. Gaiduk, *Opt. Spektrosk.* **94**, 228 (2003) (in Russian).

72. V. I. Gaiduk, D. S. F. Crothers, Ch. M. Briskina, and B. M. Tseitlin, *J. Mol. Struct.* **689**, 11 (2004).

73. V. A. Golunov, V. A. Korotkov, and E. V. Sukhonin, *Overall Results of Science and Technique*, Radio Engineering Series, VINITI, Moscow, **41**, 68 (1990) (in Russian).

74. B. Guillot, *J. Chem. Phys.* **95**, 1543 (1991).

75. S. Nose, *Mol. Phys.* **52**, 255–268 (1984).

76. A. Yu. Zasetsky, A. F. Khalizov, M. E. Earle, and J. J. Sloan, *J. Phys. Chem.* **109**, 2760 (2005).

AUTHOR INDEX

Numbers in parentheses are reference numbers and indicate that the author's work is referred to although his name is not mentioned in the text. Numbers in *italic* show the page on which the complete references are listed.

Advances in Chemical Physics, Volume 141, edited by Stuart A. Rice
Copyright © 2009 John Wiley & Sons, Inc.

SUBJECT INDEX

Advances in Chemical Physics, Volume 141, edited by Stuart A. Rice
Copyright © 2009 John Wiley & Sons, Inc.